Recent Development in

Theories & Numerics

International Conference on Inverse Problems

Recent Development in
Theories & Numerics

International Conference on Inverse Problems

Hong Kong, China 9 — 12 January 2002

Editors

Yiu-Chung Hon
City University of Hong Kong, China

Masahiro Yamamoto
University of Tokyo, Japan

Jin Cheng
Fudan University, China

June-Yub Lee
Ewha Woman's University, Korea

World Scientific
New Jersey • London • Singapore • Hong Kong

Published by

World Scientific Publishing Co. Pte. Ltd.
5 Toh Tuck Link, Singapore 596224
USA office: Suite 202, 1060 Main Street, River Edge, NJ 07661
UK office: 57 Shelton Street, Covent Garden, London WC2H 9HE

British Library Cataloguing-in-Publication Data
A catalogue record for this book is available from the British Library.

INTERNATIONAL CONFERENCE ON INVERSE PROBLEM — RECENT DEVELOPMENT IN THEORIES AND NUMERICS

ISBN 981-238-366-2

Printed in Singapore by Mainland Press

Foreword

The first International Conference on Inverse Problems – Recent Theoretical Development and Numerical Approaches was held at the City University of Hong Kong from January 9 – 12, 2002.

The conference was jointly organized by Dr. Yiu-Chung Hon (City University of Hong Kong, HKSAR), Professor Masahiro Yamamoto (University of Tokyo, Japan), Professor Jin Cheng (Fudan University, China), and Professor June-Yub Lee (Ewha Womans University, Korea). It was agreed to hold the conference alternatively among the above four places once per two years. The next conference has been scheduled to be held at the Fudan University, Shanghai in April 2004.

The purpose of this conference was to establish a first and strong collaboration link among the universities of Hong Kong and worldwide leading researchers in inverse problems. The conference addressed both theoretical (mathematics), applied (engineering) and development aspects on inverse problems. It was intended to nurture Asian-American-European collaborations in this evolving interdisciplinary area and envisioned that the conference will lead to a long-term commitment and collaboration among the participated countries and researchers.

There were a total of more than 120 participants. A call for the submission of paper was sent after the conference. A total of 45 papers were submitted for publication at the proceedings. All papers were formally sent for external referring and finally we obtained this volume of proceedings, though it took more than a year after the conference.

The papers included in the proceedings cover wide scopes, which reflect the current flourishing theoretical and numerical researches for inverse problems. For the sake of convenience of the readers, we classify the papers into the following five sections:

1. Surveys (8 papers): overview of specific inverse problems and methods.
2. Theoretical aspects (11 papers): mathematical analyses such as the uniqueness, the stability and numerical schemes on the basis of the theoretical results.
3. Numerical methods (10 papers): focuses on new applications or proposals of new methods for inverse problems.
4. Solutions to applied inverse problems (11 papers): concrete numerical solutions to applied inverse problems related to image reconstruction, meteorology, medical diagnostics, clustering, and finance problems.
5. Related topics (2 papers).

Finally, as the organizers of the Inverse Problems Conference 2002 we wish to express our cordial thanks to all the plenary and invited speakers, members of the International Scientific Committee, and the Advisory Board. We would also like to thank the Croucher Foundation, the Hong Kong Mathematical Society, and the City University of Hong Kong for their financial supports. Particularly, we thank Ms. Jane Hui and Ms. Siu-Ki Wong for their excellent administrative and clerical works in making all the arrangements for the conference.

January 2003

Yiu-Chung Hon, City University of Hong Kong, HKSAR
Masahiro Yamamoto, University of Tokyo, Japan
Jin Cheng, Fudan University, China
June-Yub Lee, Ewha Womans University, Korea

List of Plenary/Invited Speakers

Invited Plenary Speakers on keynote talks:

- G. Bao (Michigan State University, USA)
- J. Cheng (Fudan University, China)
- V. Isakov (Wichita State University, KS, USA)
- P.C. Sabatier (Université des Sciences et Techniques du Languedoc, France)

Invited Speakers on introductory talks:

- D. Anikonev (Institute of Applied Mathematics, Vladivostok, Russia)
- H. T. Banks (North Carolina State University, Raleigh, USA)
- J. Frankel (Tennessee University, USA)
- H. Fujiwara (Kyoto University, Japan)
- P. Maass (Universität Bremen, Germany)
- A. Neubauer (Johannes Kepler Universitat, Linz, Austria)
- J. S. Pang (The Johns Hopkins University, USA)
- J. K. Seo (Yonsei University, Korea)
- K. Tanuma (Osaka Kyoiku University, Japan)
- D. D. Trong (HoChiMinh City University, Vietnam)
- G. Uhlmann (University of Washington, USA)
- J. Z. Zhang (City University of Hong Kong, HKSAR)

- G. Nakamura (Hokaido University, Japan)
- M. K. Ng (Hong Kong University, HKSAR)
- J, Onishi (Ibaraki University, Japan)
- Y. J. Tan (Fudan University, China)
- M. Tanaka (Shinshu University, Japan)
- C. S. Tong (Hong Kong Baptist University, HKSAR)
- J. M. Yong (Fudan University, China)
- G. Q. Zhang (Beijing Academy of Science, China)
- J. Zou (Chinese University of Hong Kong, HKSAR)

Organizing Committee

- Y.C. Hon (City University of Hong Kong , HKSAR)-Co-Chair Organizer
- M. Yamamoto (University of Tokyo, Japan)- Co-Chair Organizer
- J. Cheng (Fudan University, PRC)
- J.Y. Lee (Ewha Womans University, Korea)

CONTENTS

Section II: Theoretical Aspects

Section III: Numerical Methods

Section I

Surveys

RECOVERY OF SMALL ELECTROMAGNETIC INHOMOGENEITIES

H. AMMARI

Centre de Mathématiques Appliquées, CNRS UMR 7641 & Ecole Polytechnique, 91128 Palaiseau Cedex, France

E-mail: ammari@cmapx.polytechnique.fr

We survey recent results on the inverse problem of identifying locations and certain properties of the shapes of small dielectric inhomogeneities in a homogeneous background medium from boundary measurements.

1 The Helmholtz Equation

1.1 Problem Formulation

Let Ω be a bounded C^2-domain in $\mathbf{R}^d$, $d \geq 2$ and ν be the outward unit normal to $\partial\Omega$. Assume that Ω contains a finite number of inhomogeneities, each of the form $z_j + \alpha B_j$, where $B_j \subset \mathbf{R}^d$ is a bounded, smooth domain containing the origin. The total collection of inhomogeneities is $\mathcal{B}_\alpha = \cup_{j=1}^m (z_j + \alpha B_j)$. The points $z_j \in \Omega, j = 1, \ldots, m$, which determine the location of the inhomogeneities, are assumed to satisfy the following assumptions:

$$|z_j - z_l| \geq c_0 > 0, \forall\, j \neq l \quad \text{and} \quad \text{dist}(z_j, \partial\Omega) \geq c_0 > 0, \forall\, j. \tag{1}$$

Assume that $\alpha > 0$, the common order of magnitude of the diameters of the inhomogeneities, is sufficiently small, that these inhomogeneities are disjoint, and that their distance to $\mathbf{R}^d \setminus \overline{\Omega}$ is larger than $c_0/2$. Let μ_0 and ε_0 denote the permeability and the permittivity of the background medium, and assume that $\mu_0 > 0$ *and* $\varepsilon_0 > 0$ *are positive constants.* Let $\mu_j > 0$ and $\varepsilon_j > 0$ denote the permeability and the permittivity of the j-th inhomogeneity, $z_j + \alpha B_j$, these are also assumed to be positive constants. Introduce the piecewise-constant magnetic permeability

$$\mu_\alpha(x) = \begin{cases} \mu_0, & x \in \Omega \setminus \bar{\mathcal{B}}_\alpha, \\ \mu_j, & x \in z_j + \alpha B_j,\ j = 1 \ldots m. \end{cases}$$

If we allow the degenerate case $\alpha = 0$, then the function $\mu_0(x)$ equals the constant μ_0. The piecewise constant electric permittivity, $\varepsilon_\alpha(x)$ is defined analogously.

Let E_α be the electric field in the presence of the inhomogeneities. It solves the Helmholtz equation

$$\nabla \cdot (\frac{1}{\mu_\alpha} \nabla E_\alpha) + \omega^2 \varepsilon_\alpha E_\alpha = 0 \ \text{ in } \Omega, \tag{2}$$

with the boundary condition $E_\alpha = f$ on $\partial\Omega$, where $\omega > 0$ is a given frequency. The electric field, E_0, in the absence of any inhomogeneities, satisfies the following equation:

$$\Delta E_0 + k^2 E_0 = 0 \ \text{ in } \Omega, \tag{3}$$

where $k^2 = \omega^2 \mu_0 \varepsilon_0$, with $E_0 = f$ on $\partial\Omega$. In order to insure well-posedness (also for the α-dependent case for α sufficiently small [18]) we shall assume that k^2 is not an eigenvalue for the operator $-\Delta$ in $L^2(\Omega)$ with the Dirichlet boundary conditions. It has been shown in Vogelius and Volkov[18] that the following asymptotic formula holds uniformly on $\partial\Omega$:

$$\begin{aligned} &\frac{\partial E_\alpha}{\partial \nu}(x) - \frac{\partial E_0}{\partial \nu}(x) - 2\int_{\partial\Omega} (\frac{\partial E_\alpha}{\partial \nu} - \frac{\partial E_0}{\partial \nu})(y) \frac{\partial G(x,y)}{\partial \nu} \, ds(y) \\ &= 2\alpha^d \sum_{j=1}^m (\frac{\mu_j}{\mu_0} - 1) \nabla_y \frac{\partial G(x, z_j)}{\partial \nu(x)} \cdot M_j(\frac{\mu_j}{\mu_0}) \nabla E_0(z_j) \\ &-2\alpha^d k^2 \sum_{j=1}^m (1 - \frac{\varepsilon_j}{\varepsilon_0}) \frac{\partial G(x, z_j)}{\partial \nu(x)} |B_j| E_0(z_j) + o(\alpha^d), \end{aligned} \tag{4}$$

where the remainder $o(\alpha^d)$ is independent of the set of points $\{z_j\}_{j=1}^m$ provided that (1) holds, $G(x, y)$ is a free space Green's function for $\Delta + k^2$, and each $M_j(\frac{\mu_j}{\mu_0})$ is a $d \times d$, symmetric, positive definite matrix associated with the j-th inhomogeneity [14,12]. $M_j(\frac{\mu_j}{\mu_0})$ is the polarization tensor, that is given by

$$(M_j)_{ll'}(\frac{\mu_j}{\mu_0}) = |B_j| \delta_{ll'} + \left(\frac{\mu_j}{\mu_0} - 1\right) \int_{\partial B_j} y_l \frac{\partial \phi_{l'}^+}{\partial \nu_j} \, ds(y),$$

where, for $1 \le l' \le d$, $\phi_{l'}(y)$ is the unique function which satisfies

$$\begin{cases} \Delta \phi_{l'} = 0 \quad \text{in} \quad B_j \ \text{and} \ \mathbf{R}^d \setminus \overline{B_j}, \\ \dfrac{\mu_j}{\mu_0} \dfrac{\partial \phi_{l'}^-}{\partial \nu_j} - \dfrac{\partial \phi_{l'}^+}{\partial \nu_j} = -\nu_j \cdot e_{l'} \quad \text{on} \quad \partial B_j, \end{cases}$$

with $\phi_{l'}$ continuous across ∂B_j and $\lim_{|y| \to \infty} \phi_{l'}(y) = 0$. Here $\{e_{l'}\}_{l'=1}^d$ is an orthonormal basis of $\mathbf{R}^d$, ν_j denotes the outward unit normal to ∂B_j, superscripts $-$ and $+$ indicate the limiting values as the point approaches ∂B_j from outside B_j, and from inside B_j, respectively.

The inverse problem considered in this section is to identify the locations $\{z_j\}_{j=1}^m$ and the polarization tensors $\{M_j(\frac{\mu_j}{\mu_0})\}_{j=1}^m$ of the small inhomogeneities $\mathcal{B}_\alpha$ from the boundary measurements of $\frac{\partial E_\alpha}{\partial \nu}$ on a given part $\Gamma \subseteq \partial\Omega$. We want to reduce this reconstruction problem to the calculation of an inverse Fourier transform.

1.2 Identification Reconstruction Procedure from Complete Boundary Measurements

Assume $\Gamma = \partial\Omega$. Before describing our identification procedure, let us introduce the sets $N(\Omega) = \{v : v \in H^1(\overline{\Omega}) \cap H^2(\Omega), \Delta v + k^2 v = 0 \text{ in } \Omega\}$. The general approach we use to recover the locations and the polarization tensors of the small inhomogeneities is to integrate the solution E_α against special test functions in the set $N(\Omega)$ [5]. This approach is similar to the ideas used by Calderón [11] in his proof of uniqueness of the linearized conductivity problem and later, by Sylvester and Uhlmann in their important work [17] on uniqueness of the three-dimensional inverse conductivity problem. See also Isaacson and Isaacson [15] for exact calculation of the Calderón's approximation for the case of homogeneous concentric disks in $\mathbf{R}^2$.

Let v be any function in $N(\Omega)$. According to Ammari *et al* [5] the following estimate can be derived from (4):

$$\int_{\partial\Omega} \frac{\partial E_\alpha}{\partial \nu} v \, ds - \int_{\partial\Omega} \frac{\partial v}{\partial \nu} E_\alpha \, ds = \alpha^d \sum_{j=1}^m (1 - \frac{\mu_j}{\mu_0}) \nabla E_0(z_j) \cdot M_j(\frac{\mu_j}{\mu_0}) \nabla v(z_j)$$
$$+ \alpha^d k^2 \sum_{j=1}^m (1 - \frac{\varepsilon_j}{\varepsilon_0}) |B_j| E_0(z_j) v(z_j) + o(\alpha^d), \tag{5}$$

where $|B_j|$ stands for the volume of the set B_j. We want to make suitable choices for the test functions v in $N(\Omega)$ and the boundary condition $E_\alpha|_{\partial\Omega}$ in order to get simple equations for the unknown parameters, namely, for the points $\{z_j\}_{j=1}^m$ and the polarization tensors $\{M_j(\frac{\mu_j}{\mu_0})\}_{j=1}^m$.

Let us describe our inversion method. Take η to be a vector in $\mathbf{R}^d$, $\eta^\perp$ a unit vector in $\mathbf{R}^d$ which is orthogonal to η, and γ a complex number. Then $e^{i(\eta + \gamma\eta^\perp)\cdot x}$ is a solution to the Helmholtz equation in $\mathbf{R}^d$ if and only if $\gamma^2 = k^2 - |\eta|^2$ and in this case $e^{i(\eta - \gamma\eta^\perp)\cdot x}$ is also a solution to the Helmholtz equation in $\mathbf{R}^d$. For simplicity, let us consider the case where all the B_j are balls. In this case all the matrices $M_j(\frac{\mu_j}{\mu_0})$ are multiples of the identity matrix which makes our analysis simpler.

If $\frac{\partial E_\alpha}{\partial \nu}$ is known on the whole boundary $\partial\Omega$ then taking $E_\alpha = e^{i(\eta+\gamma\eta^\perp)\cdot x}$ on $\partial\Omega$ and $v = e^{i(\eta-\gamma\eta^\perp)\cdot x}$ in Ω we have from Ammari *et al* [5] the following asymptotic result.

Theorem 1 *The following asymptotic expansion holds:*

$$\begin{aligned}\Lambda_\alpha(\eta) &= \int_{\partial\Omega} \frac{\partial E_\alpha}{\partial \nu}(y) e^{i(\eta-\gamma\eta^\perp)\cdot y} \, ds(y) - \int_{\partial\Omega} \frac{\partial}{\partial \nu}(e^{i(\eta-\gamma\eta^\perp)\cdot y}) E_\alpha(y) \, ds(y) \\ &= \alpha^d \sum_{j=1}^{m} e^{2i\eta\cdot z_j} \Big[(1 - \frac{\mu_j}{\mu_0}) M_j(\frac{\mu_j}{\mu_0})(2|\eta|^2 - k^2) + k^2(1 - \frac{\varepsilon_j}{\varepsilon_0})|B_j|\Big] + o(\alpha^d).\end{aligned} \tag{6}$$

Recall that $e^{2i\eta\cdot z_j}$ (up to a multiplicative constant) is the Fourier transform of the Dirac function δ_{-2z_j} (a point mass located at $-2z_j$). Multiplication by powers of η in the Fourier space corresponds to differentiation of the Dirac function. Therefore, the function $\Lambda_\alpha(\eta)$ is the inverse Fourier transform of a distribution supported at the points z_j. A numerical Fourier inversion of a sample of $\Lambda_\alpha(\eta)$ will yield the points z_j with a small error as $\alpha \to 0$. It is natural to use a fast Fourier transform for this inversion. One can estimate the number of the sampling points needed for an accurate discrete Fourier inversion using the Shannon's sampling theorem [13]. This number is of order $(\frac{h}{r})^d$. One needs this amount of sampled values of η to reconstruct, with resolution r, a collection of inhomogeneities that lie inside a square of side h. Once the points $\{z_j\}_{j=1}^m$ are found, one may find $\{M_j(\frac{\mu_j}{\mu_0})\}_{j=1}^m$ by solving appropriate linear system arising from the asymptotic formula (6). If B_j are general domains, our calculations become more complicated, and eventually we have to deal with pseudo-differential operators (independent of the space variable x) applied to the same Dirac functions. Associated numerical experiments to this inversion procedure have been successfully conducted in the case of the conductivity problem [5,19]. In view of the asymptotic results derived in Ammari *et al* [8] a similar approach may be applied to the full Maxwell equations in the presence of small dielectric inhomogeneities. The reader is referred to the recent papers [1,4] for rigorous derivations of asymptotic formulas for the displacement vector in an elastic medium consisting of finitely many imperfections of small diameter, embedded in a homogeneous reference medium and the resolution of the associated inverse problem.

1.3 Incomplete Boundary Measurements

Assume that $\Gamma \subset\subset \partial\Omega$. Let $\tilde{N}(\Omega) = \{v : v \in H^1(\overline{\Omega}) \cap H^2(\Omega), \Delta v + k^2 v = 0 \text{ in } \Omega, v = 0 \text{ on } \Gamma_c\}$, where $\Gamma_c = \partial\Omega \setminus \overline{\Gamma}$.

The following asymptotic formula can be derived from (4):

$$\int_\Gamma \frac{\partial E_\alpha}{\partial \nu} v \, ds - \int_{\partial\Omega} \frac{\partial v}{\partial \nu} E_\alpha \, ds = \alpha^d \sum_{j=1}^m (1 - \frac{\mu_j}{\mu_0}) \nabla E_0(z_j) \cdot M_j(\frac{\mu_j}{\mu_0}) \nabla v(z_j) + \alpha^d k^2 \sum_{j=1}^m (1 - \frac{\varepsilon_j}{\varepsilon_0}) |B_j| E_0(z_j) v(z_j) + o(\alpha^d). \tag{7}$$

The main difficulty in generalizing our approach to the case when $\frac{\partial E_\alpha}{\partial \nu}$ is known only on a part $\Gamma \subset\subset \partial\Omega$ is to construct a function $w_\alpha(x)$ in $\tilde{N}(\Omega)$, that is asymptotically $e^{i(\eta - \gamma\eta^\perp)\cdot x}$ as α approaches 0. The following lemma holds [6].

Lemma 1 *Let $\Omega' \subset\subset \Omega$ be a C^2-domain. Let $\eta \in \mathbf{R}^d$ and $\eta^\perp$ be a unit vector in $\mathbf{R}^d$ that is orthogonal to η. There exists $w_\alpha \in \tilde{N}(\Omega)$ such that*

$$w_\alpha(x) = e^{i(\eta - \gamma\eta^\perp)\cdot x} + o(\alpha^d) \text{ and } \nabla w_\alpha(x) = i(\eta - \gamma\eta^\perp) e^{i(\eta - \gamma\eta^\perp)\cdot x} + o(\alpha^d),$$

uniformly in Ω'.

This lemma is an immediate corollary of the following density result [6].

Proposition 1 *The set $\tilde{N}(\Omega)$ is dense, in the $L^2(\Omega')$ norm, in the set $N(\Omega)$.*

Now, if we choose $E_\alpha = e^{i(\eta + \gamma\eta^\perp)\cdot x}$ on $\partial\Omega$ and $v = w_\alpha$ in Ω then, since the points $\{z_j\}_{j=1}^m$ are away from the boundary $\partial\Omega$, it follows from (7) and Lemma 1 that the following asymptotic expansion holds.

Theorem 2 *We have:*

$$\begin{aligned} \Lambda_\alpha(\eta) &= \int_\Gamma \frac{\partial E_\alpha}{\partial \nu} w_\alpha \, ds - \int_{\partial\Omega} \frac{\partial w_\alpha}{\partial \nu} E_\alpha \, ds \\ &= \alpha^d \sum_{j=1}^m e^{2i\eta\cdot z_j} \left[(1 - \frac{\mu_j}{\mu_0}) M_j(\frac{\mu_j}{\mu_0})(2|\eta|^2 - k^2) + k^2 (1 - \frac{\varepsilon_j}{\varepsilon_0}) |B_j| \right] + o(\alpha^d), \end{aligned}$$

where the function w_α is chosen according to Lemma 1.

2 The Wave Equation

Let $\tau = \sqrt{\varepsilon_0\mu_0}$. For an arbitrary (non null) $\eta \in \mathbf{R}^d$, the function $E_0(x,t) = e^{i(\eta\cdot x+\frac{|\eta|}{\tau}t)}$ satisfies the following wave equation:

$$\tau^2\partial_t^2 E_0 - \Delta E_0 = 0 \quad \text{in } \Omega\times\mathbf{R}.$$

Consider now the initial boundary value problem for the wave equation in the presence of the small inhomogeneities

$$\begin{cases} \varepsilon_\alpha\partial_t^2 E_\alpha - \nabla\cdot(\dfrac{1}{\mu_\alpha}\nabla E_\alpha) = 0 & \text{in } \Omega\times(0,T), \\ E_\alpha|_{t=0} = E_0|_{t=0}, \partial_t E_\alpha|_{t=0} = \partial_t E_0|_{t=0} & \text{in } \Omega, \\ E_\alpha|_{\partial\Omega\times(0,T)} = E_0|_{\partial\Omega\times(0,T)}. \end{cases} \tag{8}$$

Here $0 < T < +\infty$ is a final observation time. It can be shown that there exists a unique solution $E_\alpha \in L^\infty(0,T;H^1(\Omega)) \cap W^{1,\infty}(0,T;L^2(\Omega))$ to the initial boundary value problem (8), and this solution satisfies the following estimate as $\alpha \to 0$:

$$\begin{aligned} &||E_\alpha - E_0||_{L^\infty(0,T;H_0^1(\Omega))} + ||\partial_t(E_\alpha - E_0)||_{L^\infty(0,T;L^2(\Omega))} \\ &\quad + ||\varepsilon_\alpha\partial_t^2(E_\alpha - E_0)||_{L^\infty(0,T;H^{-1}(\Omega))} \le C\alpha, \end{aligned} \tag{9}$$

where the constant C is independent of α and the set of points $\{z_j\}_{j=1}^m$ provided that assumption (1) holds. Also, $\frac{\partial E_\alpha}{\partial\nu}|_{\partial\Omega\times(0,T)}$ belongs to $L^2(0,T;L^2(\partial\Omega))$.

Our goal in this section is to identify the locations $\{z_j\}_{j=1}^m$ and the polarization tensors $\{M_j(\frac{\mu_j}{\mu_0})\}_{j=1}^m$ from the dynamic boundary measurements of $\frac{\partial E_\alpha}{\partial\nu}$ on $\Gamma\times(0,T)$. As for the Helmholtz equation, we want to reduce this reconstruction problem to the calculation of an inverse Fourier transform. To do so, let us now rewrite in $\Omega\times(0,T)$ the first equation in (8) as follows

$$\tau^2\partial_t^2 E_\alpha - \Delta E_\alpha = \sum_{j=1}^m(\tau^2-\tau_j^2)\partial_t^2 E_\alpha\chi(z_j+\alpha B_j) - [\frac{\partial E_\alpha}{\partial\nu}]_{\partial(z_j+\alpha B_j)}\delta_{\partial(z_j+\alpha B_j)}, \tag{10}$$

where $\tau_j = \sqrt{\varepsilon_j\mu_j}$, $\chi(z_j+\alpha B_j)$ is the characteristic function of the domain $z_j+\alpha B_j$, $[\frac{\partial E_\alpha}{\partial\nu}]_{\partial(z_j+\alpha B_j)}$ denotes the jump of $\frac{\partial E_\alpha}{\partial\nu}$ across $\partial(z_j+\alpha B_j)$, and $\delta_{\partial(z_j+\alpha B_j)}$ is the surface Dirac measure at $\partial(z_j+\alpha B_j)$. We now reduce our inverse problem to an inverse source one by expanding the right-hand side of (10) in terms of the small parameter α. The leading order term in this asymptotic expansion contains information on the locations $\{z_j\}_{j=1}^m$ and the domains $\{B_j\}_{j=1}^m$ and thus, our inverse problem becomes to obtain this information from knowledge of $\frac{\partial E_\alpha}{\partial \nu}|_{\Gamma\times(0,T)}$. The following asymptotic expansions can be derived [2].

Lemma 2 *We have the following asymptotic expansions*

$$\begin{aligned}
\partial_t^2 E_\alpha \chi(z_j+\alpha B_j) &= -\tfrac{|\eta|^2}{\tau^2} e^{i(\eta\cdot x+\frac{|\eta|}{\tau}t)}\chi(z_j+\alpha B_j)+o(1),\\
[\frac{\partial E_\alpha}{\partial \nu}]_{\partial(z_j+\alpha B_j)} &= i(1-\tfrac{\mu_0}{\mu_j})\Big[\nu_j\cdot\eta+(\tfrac{\mu_j}{\mu_0}-1)\tfrac{\partial \Phi_j}{\partial \nu_j}\Big|_+ +(\tfrac{x-z_j}{\alpha})\cdot\eta\Big]e^{i(\eta\cdot z_j+\frac{|\eta|}{\tau}t)}\\
&\quad +o(1),
\end{aligned} \tag{11}$$

for all $x\in\Omega, t\in(0,T)$, where the remainder $o(1)$ is independent of the set of points $\{z_j\}_{j=1}^m$ provided that (1) holds.

2.1 Complete Dynamic Boundary Measurements

Suppose that $\Gamma=\partial\Omega$. Suppose that $T>\tau$ diam (Ω). Let $\omega\in\mathbf{R}^d$ be a unit vector. We introduce $\varphi\in C_0^\infty(\mathbf{R})$ satisfying $0\le\varphi\le 1, \int_{\mathbf{R}}\varphi\, dt=1, \varphi=0$ for $|t|\ge T-\tau$ diam (Ω) to obtain the approximation of the plane wave $\delta(\frac{t}{\tau}-x\cdot\omega)$ of direction ω:

$$\frac{1}{\alpha}\varphi(\frac{\frac{t}{\tau}-x\cdot\omega}{\alpha})\to\delta(\frac{t}{\tau}-x\cdot\omega) \text{ as } \alpha\to 0. \tag{12}$$

Note that $\delta(\frac{T}{\tau}-x\cdot\omega)=0$ for all $x\in\Omega$.

Multiplying (10) by $\dfrac{1}{\alpha}\varphi(\dfrac{\frac{t}{\tau}-x\cdot\omega}{\alpha})$, integrating by parts over $\Omega\times(0,T)$ and using (11) and (12), one gets:

$$\begin{aligned}
&-\int_0^T\int_{\partial\Omega}\frac{\partial}{\partial\nu}(E_\alpha-E_0)(y)(\frac{1}{\alpha}\varphi(\frac{\frac{t}{\tau}-y\cdot\omega}{\alpha}))\,ds(y)\,dt\\
&=\alpha^d\sum_{j=1}^m\Big[(\frac{\tau_j^2}{\tau^2}-1)|\eta|^2\\
&\quad +(1-\frac{\mu_0}{\mu_j})\frac{|\eta|}{\tau}\eta\cdot\int_{\partial B_j}(\nu_j+(\frac{\mu_j}{\mu_0}-1)\frac{\partial\Phi_j}{\partial\nu_j}\Big|_+(y))y\,ds_j(y))\cdot\omega\Big]e^{i(\eta+\frac{|\eta|}{\tau}\omega)\cdot z_j}\\
&\quad +o(\alpha^d).
\end{aligned} \tag{13}$$

Recall that $\frac{\partial E_\alpha}{\partial\nu}|_{\partial\Omega\times(0,T)}$ belongs to $L^2(0,T;L^2(\partial\Omega))$. Thus, if we choose $\omega=\frac{\eta}{|\eta|}$, it follows from (13) and (12) that the following theorem holds:

Theorem 3 *Suppose that $T > \tau$ diam (Ω). The following asymptotic expansion holds:*

$$\begin{aligned}\Lambda_\alpha(\eta) &= \int_0^T \int_{y\in\partial\Omega, y\cdot\frac{\eta}{|\eta|}=\frac{t}{\tau}} \frac{\partial}{\partial\nu}(E_\alpha - E_0)(y)\, ds(y)\, dt \\ &= \alpha^d \sum_{j=1}^m \Big[(\frac{\tau_j^2}{\tau^2} - 1)|\eta|^2 - \frac{1}{\tau}(1 - \frac{\mu_0}{\mu_j})(M_j(\frac{\mu_j}{\mu_0})\eta)\cdot\eta \Big] e^{i(1+\frac{1}{\tau})\eta\cdot z_j} + o(\alpha^d),\end{aligned} \tag{14}$$

where the remainder $o(\alpha^d)$ is independent of the set of points $\{z_j\}_{j=1}^m$.

2.2 Incomplete Dynamic Boundary Measurements

Assume that $\Gamma \subset\subset \partial\Omega$. Suppose that T and Γ are such that they geometrically control Ω which roughly means that every geometrical optic ray, starting at any point $x \in \Omega$ at time $t = 0$ hits Γ before time T at a non diffractive point [9]. Using as weights particular background solutions constructed by a geometrical control method, as it has been done in the original work [20,21], we have developed in Ammari[2] a similar asymptotic method based on appropriate averaging of the partial dynamic boundary measurements for finding $\{z_j\}_{j=1}^m$ and $\{M_j(\frac{\mu_j}{\mu_0})\}_{j=1}^m$. The following holds from Ammari[2].

Theorem 4 *Let $\eta \in \mathbf{R}^d$. Let E_α be the unique solution in $C^0(0,T;H^1(\Omega)) \cap C^1(0,T;L^2(\Omega))$ to the wave equation (8). Suppose that Γ and T geometrically control Ω then we have*

$$\begin{aligned}&\int_0^T \int_\Gamma \Big[\theta_\eta(\frac{\partial E_\alpha}{\partial\nu} - \frac{\partial E_0}{\partial\nu}) + \partial_t\theta_\eta \partial_t(\frac{\partial E_\alpha}{\partial\nu} - \frac{\partial E_0}{\partial\nu})\Big] \\ &= \alpha^d \sum_{j=1}^m \Big[(\frac{\tau_j^2}{\tau^2} - 1)|\eta|^2 |B_j| - (1 - \frac{\mu_0}{\mu_j})(M_j(\frac{\mu_j}{\mu_0})\eta)\cdot\eta \Big] e^{2i\eta\cdot z_j} + o(\alpha^d),\end{aligned} \tag{15}$$

where θ_η is the unique solution to the ODE

$$\begin{cases} \partial_t^2\theta_\eta - \theta_\eta = e^{i\frac{|\eta|}{\tau}t}\partial_t(e^{-i\frac{|\eta|}{\tau}t} g_\eta) & \text{for } x\in\Gamma, t\in(0,T), \\ \theta_\eta(x,0) = 0, \partial_t\theta_\eta(x,T) = 0 & \text{for } x \in \Gamma. \end{cases}$$

with the boundary control $g_\eta \in H_0^1(0,T;L^2(\Gamma))$ is defined in such a way that the unique weak solution w_η in $C^0(0,T;L^2(\Omega)) \cap C^1(0,T;H^{-1}(\Omega))$ of the wave equation

$$\begin{cases} (\tau^2\partial_t^2 - \Delta)w_\eta = 0 \quad \text{in } \Omega\times(0,T), \quad w_\eta|_{t=0} = \beta(x)e^{i\eta\cdot x} \in H_0^1(\Omega), \\ \partial_t w_\eta|_{t=0} = 0 \quad \text{in } \Omega, \quad w_\eta|_{\Gamma\times(0,T)} = g_\eta, \quad w_\eta|_{\partial\Omega\setminus\overline{\Gamma}\times(0,T)} = 0, \end{cases}$$

satisfies $w_\eta(T) = \partial_t w_\eta(T) = 0$. *Here* β *is a cut-off function vanishing in a neighborhood of* $\partial\Omega$.

3 Discussion

We have provided a new algorithm for reconstructing small dielectric inhomogeneities. The reader is referred to Friedman and Vogelius[14], Cedio-Fengya *et al* [12], Brühl *et al* [10], and Kwon *et al* [16] for other algorithms. Numerical experiments in Ammari *et al* [5] for the two-dimensional (time-independent) inverse conductivity problem seem to suggest that our algorithm is very effective and also quite stable with respect to noise in measurements and errors in the different approximations. Our approach could be carried out to recuperate the locations of the inhomogeneities with a very high resolution and capture further properties of the geometries of the small domains B_j (namely, the generalized polarization tensors defined in Ammari and Kang[3]) by use of higher-order terms in the asymptotic expansion of the boundary measurements with respect to the small parameter α. We refer the reader to the recent papers [7,4] for such approach.

References

1. C. Alves and H. Ammari, *Boundary integral formulae for the reconstruction of imperfections of small diameter in an elastic medium,* SIAM J. Appl. Math. **62** ,94–106(2002).
2. H. Ammari, *An inverse initial boundary value problem for the wave equation in the presence of imperfections of small volume,* to appear in SIAM J. Control Optim.
3. H. Ammari and H. Kang, *High-order terms in the asymptotic expansions of the steady-state voltage potentials in the presence of conductivity inhomogeneities of small diameter,* submitted.
4. H. Ammari, H. Kang, G. Nakamura, and K. Tanuma *Complete asymptotic expansions of solutions of the system of elastostatics in the presence of inhomogeneities of small diameter*, submitted.
5. H. Ammari, S. Moskow, and M. Vogelius, *Boundary integral formulas for the reconstruction of electromagnetic imperfections of small diameter,* to appear in ESAIM: Cont. Opt. Calc. Var.
6. H. Ammari and A. Ramm, *Recovery of small electromagnetic inhomogeneities from partial boundary measurements,* to appear in C. R. Acad. Sci. II.

7. H. Ammari and J. K. Seo, *An accurate formula for the reconstruction of conductivity inhomogeneities,* submitted.
8. H. Ammari, M. Vogelius, and D. Volkov, *Asymptotic formulas for perturbations in the electromagnetic fields due to the presence of imperfections of small diameter II.The full Maxwell equations*, J. Math. Pures Appl. **80**, 769–814(2001).
9. C. Bardos, G. Lebeau, and J. Rauch, *Sharp sufficient conditions for the observation, control and stabilization of waves from the boundary,* SIAM J. Control Opt. **30** , 1024–1065(1992).
10. M. Brühl, M. Hanke, and M. Vogelius, *A direct impedance tomography algorithm for locating small inhomogeneities,* submitted.
11. A. P. Calderón, *On an inverse boundary value problem,* Seminar on Numerical Analysis and its Applications to Continuum Physics, Soc. Brasileira de Matemática, Rio de Janeiro,65–73 (1980).
12. D. J. Cedio-Fengya, S. Moskow, and M. Vogelius, *Identification of conductivity imperfections of small diameter by boundary measurements. Continuous dependence and computional reconstruction,* Inverse Problems **14** , 553–595 (1998).
13. I. Daubechies, *Ten Lectures on Wavelets,* SIAM, Philadelphia, (1992).
14. A. Friedman and M. Vogelius, *Identification of small inhomogeneities of extreme conductivity by boundary measurements: a theorem on continuous dependence*, Arch. Rat. Mech. Anal. **105** , 299–326(1989).
15. D. Isaacson and E. L. Isaacson, *Comments on Calderón's paper: "On an inverse boundary value problem"*, Math. Compt. **52** , 553–559 (1989).
16. O. Kwon, J. K. Seo, and J. R. Yoon, *A real-time algorithm for the location search of discontinuous conductivities with one measurement,* Commun. Pure Appl. Math. **55**, 1–29 (2002).
17. J. Sylvester and G. Uhlmann, *A global uniqueness theorem for an inverse boundary value problem,* Ann. Math.**125** , 153–169(1987).
18. M. Vogelius and D. Volkov, *Asymptotic formulas for perturbations in the electromagnetic fields due to the presence of inhomogeneities*, Math. Model. Numer. Anal. **34**, 723–748 (2000).
19. D. Volkov, *An inverse problem for the time harmonic Maxwell equations,* Ph.D. Thesis, Rutgers University, New Brunswick, NJ, (2001).
20. M. Yamamoto, *Well-posedness of some inverse hyperbolic problems by the Hilbert uniqueness method,* J. Inverse Ill-posed Problems **2** , 349–368(1994).
21. M. Yamamoto, *Stability, reconstruction formula and regularization for an inverse source hyperbolic problem by a control method,* Inverse Problems **11** , 481–496(1995).

INVESTIGATION OF INVERSE AND OTHER NONCLASSICAL PROBLEMS IN THE RUSSIAN FAR EAST

D.S. ANIKONOV, A.E. KOVTANUYK, D.S. KONOVALOVA,
V.G. NAZARENE AND I.V. PROTHORAX

Institute of Applied Mathematics, FEB RAS, Vaginoscope, Russia,
E-mail: anik@iam-mail.febras.ru

The report contains the short review of author s book Transport Equation and Tomography published in Russian and English. Also some latest author s results are represented. Mainly, the book is devoted to problems of X-ray tomography, which are meant as problems of determination of the internal structure of an unknown medium by use the results of passing radiation through the medium. From mathematical point of view, these problems are treated as nonclassical problems for the steady-state transport equations (linear Boltzman equations). Research of classical problems for the same equations is considered as a preliminary necessary stage of investigation.

1 Introduction

At present, many specialists from various countries use the term "inverse problems" in a generalized sense. We try to show certain common properties of similar problems:

a) as a rule, they include nonlinear dependence between known data and information which is sought for;
b) mostly, they are incorrect problems from classical point of view;
c) they directly answer to questions of a given applied problem, instead of many corresponding forward problems.

Historically, this term (inverse problems) came into the common use in the 60-th years of the 20-th century, mainly due to the ideas of A.N. Tikhonov and M.M. Lavrentjev. At that period research of inverse problems was really considered as investigation in an inverse direction with respect to classical boundary value problems for known differential equations of mathematical physics. Just at the same moment, the mentioned classical problems obtained the new title: "forward problems". Such approach seemed to be the shortest way for solving certain actual applied problems. In fact, an investigator could use an authorized mathematical model and corresponding knowledge obtained earlier for the forward problems. Note, that a part of these applied problems was considered as a pressing matter in view of their defense significance.

As a rule, inverse problems turn out incorrect in a classical sense, i.e. they belong to the set of ill-posed problems. This circumstance was a base for opponents to inverse problems. In particular, a famous Russian mathematician S.L. Sobolev once told that the incorrect problems just appeared due to wrong settings of the actual questions. Therefore he offered to set similar problems somehow differently.

Even at present, it is difficult to be a judge in that interesting discussion. We only note, that the opponents to inverse problems hardly take into account necessity of fast solving some actual problems. However that may be, the inverse problems have formed a wide scientific field, which is being investigated in many countries of the world. Moreover, the term "inverse problems" is applied rather generally including the cases, where the respective forward problems are absent. By the way, we do not take a risk to call all our problems by this term.

Perhaps, the idea expressed by S.L. Sobolev is embodied in direct mathematical modeling that means investigation of a mathematical model created specially for a given question. This approach does without any forward or inverse problems and allows one to decrease mathematical difficulties on this way. Simultaneously, it requires from a researcher a high level of knowledge in neighboring sciences, besides mathematics, needed for creation of an appropriate mathematical description of a natural phenomena. In view of this reason, the authors of this paper prefer to use the approach connected with forward and inverse problems in spite of multiple mathematical difficulties. In other words, we do not dare to create a new mathematical model of a high quality.

It is interesting to note that research of inverse and other nonclassical problems for complex cases may prove to be easier than for simple ones. At least, we know a few similar examples. Simultaneously, good adequacy between a complex mathematical model and a corresponding natural process is more probable.

2 Short Review of Various Investigations

A significant part of investigation of nonclassical problems is connected with tomography problems understood in a wide sense. Generally, we mean tomography as determining the characteristics of an unknown medium by the given results of passage a physical signal through the medium. In the Russian part of the Far East, tomography investigations are implemented in institutes of the Far Eastern Branch of the Russian Academy of Sciences (FEB RAS) and in Universities. Thus, the Pacific Oceanological Institute carries out experimental investigations in acoustic tomography. This research is led by the director of this institute professor V.A. Akulichev. In particular, such experiments were implemented in coastal areas of the Japan Sea to determine water temperature and dynamics[1,2,3]. An important part of this work was creation a special equipment complex for sounding and measurements of acoustic signals[1]. This activity is supported by certain American and Chinese organizations.

In the Far Eastern Technical University tomography problems of experimental reconstruction of magnetic and electrical fields are studied. These problems are solved by use of fiber optic measuring lines[4]. The mathematical technique consists in integration by parts (the formulae of Gauss – Ostrogradsky and Stokes) and the inverse Radon transformation.

By our opinion the experimental investigations mentioned above can hardly be called a pure tomography. Although the media are sounded by signals and methods used are typical for tomography, the quantities which are sought for, coincide with ones in appropriate forward problems. Remark that tomography problems often can be formulated as inverse problems and hardy as any forward ones. Moreover, knowledge about water dynamics and temperature and the distribution of magnetic and electrical fields usually cannot give the final answer to any actual question of practice. Similar knowledge seems us to be useful and important preliminary information. Such property also corresponds to forward problems. Thus, we offer to call such experimental (as well as similar theoretical) investigations by the term "half–tomography" or "tomography by methods".

As far as theoretical investigations are concerned, the most part of them are implemented in the Institute of Applied Mathematics and in the Far Eastern State University.

The scientific group led by professor G.V. Alekseev researched inverse problems for the equation of Helmholtz treated as problems of underwater acoustics[5,6,7]. They determine the sources of sound in water and substantiate an active control of acoustic fields. The according mathematical methods use analytical properties of a generalized acoustic potential and theory of extremal problems. The final results are represented as the appropriate algorithms which have been successfully tested in many numerical experiments.

An associated professor A.Yu. Chebotarev (Far Eastern State University) studies inverse problems of hydrodynamics for an equation of the Navier-Stokes type. The stationary and evolutionary cases are considered. Mainly, the according mathematical technique consists in use of variational inequalities. The results consist in theorems of correctness and algorithms. From physical point of view, the researched problems can be treated as determining external conditions providing the required properties of the water flow[8,9].

3 Radiation Tomography

Probably, this scientific direction is the most advanced one among similar investigations of nonclassical problems in the Russian Far East.

Radiation tomography uses certain kinds of radiation for sounding of unknown objects. Probably, it gives the best distinguish ability among various tomography methods, as well as certain disadvantages, for example, a danger for an organism or a short depth of possible sounding in some substances.

Firstly, we mention a significant scientific direction devoted to a liquid radioactive waste management which has been implemented in the Institute of Chemistry FEB RAS[10,11,12]. This activity is led by professor V.I.Sergienko, who is now the president of the FEB RAS. One fragment of this investigation was a tomography problem of determination of the total attenuation coefficient for a substance. Unfortunately, this problem was solved by an experimental way only. By our opinion, use of the modern tomography methods would have made the investigation shorter and cheaper. This case seems to be a typical example of a lack of cooperation among specialists of various fields of science.

The Institute for Automation and Control Processes FEB RAS, headed by professor V.P. Myasnikov, carries out research connected with measurements of infrared radiation by special earth satellites [13,14]. Partially, it is a joint Russian-Japanese investigation. In particular, the optical depth of clouds and sea surface temperature have been determined. The measurement were implemented with respect to different angles of observation and the corresponding differences allowed one to apply a gradient method. Like in the previous example, the experimental part of the investigation strongly prevails over the theoretical part.

In radiation tomography, the most part of the theoretical results belongs to the scientific group headed by professor D.S. Anikonov. The authors of this article represent the main body of the group. They are science researchers from the Institute of Applied Mathematics of FEB RAS and also teachers of the Far Eastern State University. All of them have reached the scientific rank a doctor of sciences.

Many their results (excluding the latest) are collected in the Russian book[15] and its English version[16]. Below, we give a detailed review of these books and a short review of the current investigations.

We use a known transport equation (linear Boltzmann equation) as a main part of an appropriate mathematical model of a natural process. Some boundary conditions supplement the model. We study forward and inverse problems for the transport equation. The first of them have auxiliary character for the further investigations which often prove to be certain inverse and other nonclassical problems.

Our investigations consist in the theoretical foundation of new algorithms and notions in X-ray and γ-ray tomography. Also, we examined our algorithms on simulated and real tests. In principle, the results may be used much more widely, because the transport equation describes not only photon migration but many other kinds of radiation.

Let us introduce the following designations: a vector r belongs to a convex bounded domain $G \subset R^3$ (a medium), $\Omega = \{\omega : \omega \in R^3,\ |\omega| = 1\}$, a numerical variable E (energy) belongs to an interval $[E_1, E_2]$; a function $f(r, \omega, E)$ means a density of a particle flux at a point r, moving in a direction ω and having energy E. Consider the following steady-state transport equation [15,16]

$$\omega \cdot \nabla_r f(r, \omega, E) + \mu(r, E) f(r, \omega, E)$$
$$= \int_\Omega \int_{E_1}^{E_2} k(r, \omega \cdot \omega', E, E') f(r, \omega', E') dE' d\omega' + J(r, \omega, E), \tag{1}$$

where the function $\mu(r, E)$ is treated as the coefficient of total attenuation; $k(r, \omega \cdot \omega', E, E')$ is the indicatrix of scattering and $J(r, \omega, E)$ is the density of the internal sources of radiation.

We use a partition of the domain G:

$$\overline{G} = \bigcup_{i=1}^{p} \overline{G}_i, \quad G_i \bigcap G_j = \emptyset,\ i \neq j; \ G_0 = \bigcup_{i=1}^{p} G_i, \quad \partial G_0 = \bigcup_{i=1}^{p} \partial G_i,$$

where the domains G_i are called inclusions, zones or heterogeneities in G. They can be interpreted as the internal bodies in the medium G, formed by different substances. The coefficients of equation (1) are smooth enough with respect to $r \in G_0$ and they can have nonzero jumps for $r \in \partial G_0$. Let z be a common point of two boundaries ∂G_i and ∂G_j, $i < j$. We use the following designation for a jump:

$$[\![\mu(z, E)]\!] = \lim_{r' \to z} \mu(r', E) - \lim_{r'' \to z} \mu(r'', E), \ r' \in G_i, \ r'' \in G_j.$$

Analogously the jumps for other functions are defined.

Define a function $d(r, \omega)$, $r \in \overline{G}$, $\omega \in \Omega$, as a distance from point r to the boundary of G in direction ω and consider the sets:

$$\Gamma^{\pm} = \{r : r = r_0 \pm d(r_0, \pm\omega)\omega, \ r_0 \in G_0, \ \omega \in \Omega\}.$$

It is clear, that $\Gamma^+ \subset \partial G \times \Omega$ and $\Gamma^- \subset \partial G \times \Omega$.

Now we can supplement the equation (1) by the following boundary conditions:

$$f(\xi,\omega,E) = h(\xi,\omega,E), (\xi,\omega) \in \Gamma^-, \; E \in [E_1,E_2], \tag{2}$$

$$f(\eta,\omega,E) = H(\eta,\omega,E), (\eta,\omega) \in \Gamma^+, \; E \in [E_1,E_2], \tag{3}$$

where the function $h(\xi,\omega,E)$ describes the density of the input flux on the boundary of the medium G and $H(\eta,\omega,E)$ is the output one.

The forward problem consists in determination of the function $f(r,\omega,E)$, $r \in G$, $\omega \in \Omega$, $E \in [E_1,E_2]$ from the equation (1) and the boundary condition (2) when the functions μ, k, J, h are known.

It is well-known, that under certain general restrictions the forward problem has a unique solution, if the following inequality takes place:

$$\mu(r,E) \geq \mu_s^*(r,E), \; \mu_s^*(r,E) = \int_\Omega \int_{E_1}^{E_2} k(r,\omega\cdot\omega',E,E')dE'd\omega'.$$

The function μ_s^* can be called the duel coefficient of scattering.

Tomography Problem 1 *Determine the coefficient $\mu(r,E)$, $r \in G$, $E \in [E_1,E_2]$ from the transport equation (1) and the boundary conditions (2) and (3), when only the functions $h(\xi,\omega,E)$ and $H(\eta,\omega,E)$ are known.*

Note that other coefficients k and J are unknown here and, at the same time, are not sought for. From the physical point of view, this problem is aimed to determination of the internal structure of the unknown medium G by the given measurements of the radiation density on the boundary of G.

At least, two methods have solved the problem 1. One of them belongs to I.V. Prothorax. It is called the method of the discontinuous input sources. Another method of multiple irradiation of G belongs to A.E. Kovtanyuk. The authors proved the theorems of uniqueness and stability. The corresponding algorithms were created and tested on many simulated examples. Moreover, the Moscow Department of the International Union of Engineering successfully tested the algorithms on its real data.

It is important to remark that both these methods operate with special external sources that cannot be realized in certain simple mathematical models. In particular, the well-known plane-parallel case in transport theory proves to be both the simplest mathematical description and the most difficult case for investigation of nonclassical problems. At least, all our methods proved for tomography cannot be formulated by the poor set of quantities admitted in the plane-parallel case. This circumstance confirms the conception presented in the introduction.

Tomography Problem 2 *Determine the surface ∂G_0 from the transport equation (1) and the boundary conditions (2), (3) where only the function $H(\eta, \omega, E)$ is known.*

This problem differs from the problem 1 because of using less data and less information is to be determined. From the physical point of view the contact interfaces between various materials in the medium G are to be found by the measurements of output radiation on the boundary of G.

While this problem has been solved in the monoenergetic case, when all quantities do not depend on energy E. In other words, we consider here the case when the loss of energy in interactions is sufficiently small. Such description is realistic for soft X-rays, for example.

Accordingly, we omit the dependence of E here and introduce the following designations. Let $z \in \partial G_0$, i.e. the point z belongs to the boundaries ∂G_i and ∂G_j. Consider the function

$$m(z, \omega) = -[\![\mu(z)]\!] f(z, \omega) + [\![P(z, \omega)]\!] \tag{4}$$

where P is the right-hand side of the transport equation in the monoenergetic case.

D.S. Anikonov introduced the function called him by the indicator of heterogeneity[17]:

$$\mathrm{Ind}(r) = |\mathrm{grad} \int_{\Omega} H(r + d(r, \omega)\omega, \omega) d\omega|, \; r \in G_0, \tag{5}$$

and proved that $\mathrm{Ind}(r)$ is continuous in G_0 and $\mathrm{Ind}(r) \to \infty$, when $r \to z$, if the following inequality holds:

$$m(z, \omega) \neq 0, \;\; z \in \partial G_0, \;\; \omega \in \Omega. \tag{6}$$

In other words, the equality $\mathrm{Ind}(z) = \infty$ may take place only on ∂G_0 which is sought for in the problem 2.

This property easy implies the theorem of uniqueness and the corresponding algorithm. The algorithm was successfully tested on simulated examples and on real data by the International Union mentioned above.

Now consider the equality which is opposite to (6)

$$m(z, \omega) = 0, \;\; z \in \partial G_i, \;\; 1 \leq i \leq p, \;\; \omega \in \Omega, \tag{7}$$

where i is a fixed number.

D.S. Anikonov and I.V. Prothorax proved that, if the equality (7) takes place, then the problem 2 cannot have the unique solution[15,16].

Therefore the condition (7) is called the condition of invisibility and the corresponding medium G is called invisible under its radiography. We can easy present mathematical examples with such property.

However, the condition (7) seems too complex for its practical application. Consider the following much more simple condition

$$[\![\mu_a(z)]\!] = 0, \ z \in \partial G_i, \ 1 \leq i \leq p. \tag{8}$$

where μ_a is the coefficient of absorption. It was shown [15,16] that the simple equality (8) approximates the complex equality (7). Thus, a medium satisfying the property (8) is called a poorly visible medium.

V.G. Nazarene continued these investigations for the case of soft X-ray [18]. He found many real poorly visible media and established a method of artificial creation of similar media. In this article we represent a small part of his results in table 1 and table 2. The needed values of the coefficient of absorption for real substances are taken from the well-known tables of J.H. Hubbell and S.M. Seltzer.

The table 1 contains a list of some pairs of materials whose contact boundary is poorly visible for some X-ray energy E' within interval [1 Kev – 6 Kev]. $\mu_1(E')$ and $\mu_2(E')$ are the attenuation coefficients for the first and the second material correspondingly. The value $R(E')$, defined by the formula

$$R(E') = \frac{|\ \mu_1(E') - \mu_2(E')\ | \cdot 100}{min\{\mu_1(E'), \mu_2(E')\}},$$

is a relative difference (in percent) of the coefficients $\mu_1(E)$ and $\mu_2(E)$ at point $E = E'$, where the absorption coefficients of the first and the second materials are equal: $\mu_{a1}(E') = \mu_{a2}(E')$. The value $R(E')$ shows a difference between the first and the second materials, although they form a poorly visible pair. The names of the chemical elements are presented at the standard chemical notation. The integer number before an element symbol is the atomic number of the element.

The table 2 illustrates our results of artificial creation of poorly visible media, which can be used for masking. It contains examples of two- and three- component mixtures imitating water by energy absorption coefficient for X-ray energy $E = 3$ Kev; w_1, w_2, w_3 are the weight fractions of the components in the mixture, μ_{a0} and μ_a are the energy absorption coefficients for water and the corresponding mixture. The value $R_a(E)$, defined by the formula

Table 1.

The first material	The second material	E'(Kev)	$R(E')$ %
23.V	67.Ho	5.743286	20.48807
23.V	Glass lead	5.647754	20.40850
21.Sc	30.Zn	4.622838	15.82411
21.Sc	57.La	4.781256	14.74743
21.Sc	28.Ni	4.873199	14.62867
21.Sc	29.Cu	5.629836	12.15189
Calcium fluoride	Cesium iodide	4.207931	11.32200
53.I	Calcium fluoride	4.410897	10.64556
45.Rh	82.Pb	4.561324	9.294229
46.Pd	82.Pb	4.881924	8.000983
24.Cr	Calcium sulfate	4.890664	6.827892
Photographic emulsion	Polyvinyl chloride	2.951463	6.771972
12.Mg	Polyvinyl chloride	2.997591	6.452643
81.Tl	82.Pb	5.167814	6.261912
40.Zr	56.Ba	5.764408	5.828532
Concrete, ordinary	Polyvinyl chloride	2.933423	5.736948
21.Sc	Concrete, barite	5.978658	5.523025
Glass, borosilicate	Polyvinyl chloride	2.988139	5.504924

$$R_a(E) = \frac{\mid \mu_a(E) - \mu_{a0}(E) \mid \cdot 100}{min\{\mu_a(E), \mu_{a0}(E)\}},$$

is a relative difference (in percent) of the coefficients $\mu_{a0}(E)$ and $\mu_a(E)$. At both examples the weight fractions w_i were chosen to provide equality $\mu_a(E) = \mu_{a0}(E)$ at the point $E = 3$Kev. The less is $R_a(E)$, the better is the quality of imitation. One can see that use of three-component mixtures allows one to achieve a better conformity between $\mu_a(E)$ and $\mu_{a0}(E)$.

We hope that such new notions as "the indicator of heterogeneity "and" the invisible and poorly visible medium" will play an important role in radiation tomography. Unfortunately, at present, their practical applications are substantiated only for soft X-ray produced by monoenergetic sources. Therefore, we intend to study much more general cases of a transport equation and boundary conditions. The preliminary plan of investigation consists in the following items, which are partially implemented.

Table 2.

Material	Components of the mixture and their weight fractions	Energy (Kev)	$R_a(E)$ %
WATER	Cobalt	1.0000	14.3688
	$w_1 = 0.1236889$	1.5000	9.35584
	Beryllium	2.0000	5.69579
	$w_2 = 0.8763111$	3.0000	0.00000
		4.0000	4.84221
		5.0000	9.16675
		6.0000	13.1577
WATER	Cobalt	1.0000	1.52430
	$w_1 = 0.0285948$	1.5000	0.40729
	Lithium tetraborate	2.0000	0.04928
	$w_2 = 0.3991014$	3.0000	0.00000
	Beryllium	4.0000	0.37082
	$w_3 = 0.5723038$	5.0000	0.87662
		6.0000	1.41366

- Consideration of the Compton effect for description of a more general kind of scattering. This phenomenon becomes essential for rather high level of energy, for example, for γ-ray. In such case we have to use the following form of the indicatrix

$$k(r, \omega \cdot \omega', E, E') = \sigma(r, \omega \cdot \omega', E, E')\delta\left(1 - \omega \cdot \omega' - \frac{E_0}{E} + \frac{E_0}{E'}\right), \quad (9)$$

  where the δ-function of Dirac is multiplied by an ordinary numerical function σ; E_0 is energy of a rest electron, E' and E are the levels of energy respectively before and after an act of scattering. Presence of the δ-function in (9) requires us to use a mathematical technique different from used earlier. Now we research only a forward problem for the indicatrix given by (9). D.S. Anikonov and D.S. Konovalova proved theorems of uniqueness and existence of a solution of a certain boundary value problem[19]. The feature of this result consists in absence of any inequalities of the type: $\mu(r, E) \geq \mu_s^*(r, E)$, which was required earlier (see above).

- Study of additional boundary conditions on the interface ∂G_0 for description of reflection and refraction. Such effects correspond to low levels of energy, in particular, for visible light. I.V. Prothorax researched the forward and inverse problem and created an algorithm concerning this case[20].

- Consideration of a transport equation and boundary conditions in vector forms for description of photon polarization. Earlier, similar models were investigated without full mathematical foundation. We continue this research and prove correctness of a certain forward problem. Also, we intend to develop a qualitative theory of the transport equation solution. A.E. Kovtanyuk and a post-graduated student O.I. Sorochinskaya prepare the article for publication: *"investigation of a boundary value problem for a transport equation in a vector form"*.

- Introduction and verification of the notion "poorly visible media" for strong X-ray and γ-ray when the dependence of energy is essential. We try to prove the hypothesis about importance of the dual coefficient of absorption analogously to the same importance of the absorption coefficient in the monoenergetic case.

- Modification of the indicator $\mathrm{Ind}(r)$ for a more general case of a transport equation and for a case using less known data. This part of research is aimed to creation of a tomography locator. It is being implemented by D.S. Anikonov and a post-graduated student A.N. Rohzko, mainly by computer methods.

4 Final Remark

In the Russian Far East the sciences, directed to a usage of natural resources, prevail over others. Nevertheless, the scientific direction, pointed in the title, takes a marked place, at least, among mathematical, physical and technical sciences in the FEB RAS. We think that this place would be more significant because of an essential part of the applied problems can be connected with inverse problems in a wide sense. A lack of mutual understanding between different specialists from neighboring sciences put obstacles on this way. However that may be, the little progress takes place.

In this paper we mentioned about all similar results, known to us. The list of the references contains only 20 items which seems to be too short for a review. This number can be easily increased due to the references within each paper of our list.

Acknowledgments

This work was supported by the Russian Foundation for Basic Research, project No. 01-01-00128.

References

1. V.A. Akulichev, V.V.Bezotvetnykh , S.I.Kamenev, E.V.Kuz'min , Yu.N.Morgurov , A.V.Nuzhdenko and S.I.Penkin , *Pribory i tekhnica experimenta* **6**, 112–115 (2000)(in Russian).
2. V. A. Akulichev, V. V. Bezotvetnykh, S. I. Kamenev, E. V Kuz'min, Yu. N. Morgunov and A. V. Nuzhdenko, *Acoustic Tomography of Dynamic Processes in the Water Medium in the Shelf Zone of the Japan Sea,* Doklady Akademii Nauk **381(8)**, 234–246(2001) (in Russian).
3. Akulichev V.A., Bezotvetnykh V.V., Kamenev S.I., Kuz'min E.V. and Nuzhdenko A.V., *Acoustical Oceanography, Proceedings of the Institute of Acoustics* **23(2)**, 314–320 (2001).
4. Yu.N.Kulchin, O.B.Vitrik, R.V.Romashko, Yu.S.Petrov, O.V.Kirichenko and O.T.Kamenev, *Tomography methods for vector field study by using space-distributed fiber optic sensors with integral sensitivity,* Fiber and Integrated Optics **17(1)**, 75–84(1998).
5. G.V. Alekseev , *Europ. J. Appl. Math.,* (1998), V. 9, pp. 589–605.
6. Alekseev G.V. and Komarov E.G., *Nonlinear inverse problems of active sound control in two-dimensional waveguides,* Doklady Akademii Nauk **358(1)**, 27–31 (1998) (in Russian).
7. G.V. Alekseev, A.S. Panasyuk and V.G. Sinko, *Inverse problems of active control of acoustic fields in three-dimensional waveguides,* J. Inv. and Ill-Posed Problems **7(5)**, 409–426 (1999).
8. A. Yu. Chebotarev,*Subdifferential inverse problems for stationary systems of Navier – Stokes type,* J. Inv. and Ill-Posed Problems **3(4)**, 268–278 (1995).
9. A.Yu. Chebotarev, *Subdifferential inverse problems for evolution Navier–Stokes systems,* J. Inv. and Ill-Posed Problems **8(3)**,243–254(2000).
10. Sergienko V.I, Avramenko V.A. and Gluschenko V.Yu., *J. Ecotech. Res.* **3(2)**, 81–93(1997).
11. Sergienko V.I., Avramenko V.A., Gluschenko V.Yu., Zheleznov V.V., Marinin D.V. and Chervonetzkyi D.V., *IAAE reports, Korea* (1998).
12. Avramenko V.A., Golikov A.P., Zheleznov V.V., Sergienko V.I., Kaplun E.V., Marinin D.V. and Sokolnitskaya T.A., *International Symposium on Radiation Safety Management: Proc. Int. Symp.-Daejeon, Korea,*

254-265 (2001).
13. A.I.Alexanin and A.V.Kazansky , *Proc. OCEANS-94 OSATES*, 13–16 *Sept.* Brest, France **2**, 11.412–11.417(1994).
14. A.I.Alexanin,M.G.Alexanina , E.E.Herbek and O.Ryabev, *Proc. OCEANS' 98*, 28 *Sept.* – 1 *Oct.*, France **2**, 1000–1005(1998).
15. D.S.Anikonov, A.E.Kovtanyuk and I.V. Prokhorov , *Using the transport equation in tomography* (Publ. House "Logos", Moscow, 3–223(2000)) (in Russian).
16. Anikonov D.S., Kovtanyuk A.E. and Prokhorov I.V. *Transport equation and tomography* (Publ. House VSP, The Netherlands, pp. viii+208 (2002)).
17. D.S.Anikonov, *Integro-differential indicator of nonhomogeneity in tomography problem*, J. Inv. and Ill-Posed Problems **7(1)** , 17–59 (1999).
18. D.S. Anikonov , V.G.Nazarov and I.V.Prokhorov , *Visible and Invisible Media in Tomography*, Doklady Mathematics **56(3)**, 955–958 (1997).
19. D.S. Anikonov and D.S.Konovalova , *Kinetic transport equation in the case of Compton's scattering* , to appeare in Siberian Math. J. (in Russian).
20. I.V.Prokhorov , Differential Equations **36(6)**, 943–948(2000).

INCORPORATION OF UNCERTAINTY IN INVERSE PROBLEMS

H.T. BANKS

Center for Research in Scientific Computation, North Carolina State University, Raleigh, NC 27695

E-mail: htbanks@eos.ncsu.edu

Motivated by examples from biology, electromagnetics and composite materials, we formulate a generic inverse problem for estimation of parameters represented as random variables. A general theoretical framework is outlined and applied to a model for the HIV infection pathway.

1 Introduction

In this lecture we present a summary of some of our experiences in the Industrial Applied Mathematics Program (IAMP-see www.ncsu.edu/crsc/iamp.html) at North Carolina State University. This program involves a number of projects (23 projects in 2000-2001) with industrial and nonacademic lab research groups. These projects incorporate certain common elements, especially with regard to sources of uncertainty.

In particular, we find that uncertainty is important in several distinct aspects including: *i*). data acquisition and related (sensor) observation error; *ii*). modeling intra-and inter-individual variability in systems and data. These aspects arise in diverse applications including inverse problems involving composite materials, biological systems, and electromagnetic imaging.

There are a number of approaches and tools critical to these efforts. These include:

i. Sensitivity analysis (with respect to parameters, geometry, etc.);
ii. "Dispersion" - type modeling and random parameters (to treat intra-individual variability);
iii. Random (stochastic) parameters and mechanisms (to treat inter-individual variability in aggregate data and observations) in the context of "mixed effects"/"mixing distributions" modeling; and
iv. Model reduction techniques including reduced order dynamic models for complex systems (e.g., Proper Orthogonal Decomposition on damages - see Banks et al[10], etc.).

A central feature of each effort is an inverse or parameter estimation problem which can be succinctly stated as follows.

Generic Inverse Problem: One is given a set of data $\hat{d} = \{d_i\}$ corresponding to (perhaps partial) observations, $Cy(t_i;\pi)$, of the state y. The state dynamics are given by a parameter (π) dependent system

$$\frac{dy}{dt} = f(t, y(t), \pi) \tag{1}$$

where f can represent ordinary, functional, or partial differential equations and where π is a (possibly vector valued) random variable. In a least squares setting, the problem is to minimize

$$J(\pi, \hat{d}) = \sum_i \mid Cy(t_i;\pi) - d_i \mid^2 \tag{2}$$

over $\pi \in \Pi =$ subject to (1), where Π is a given family of *random variables* and C is an observation operator.

This generic problem, of course, includes as special cases the usual problems with constant R.V.'s (i.e., the usual vector or function space parameters that are not dependent on any measure or variable of uncertainty).

In the next section we mention briefly several problems that can be formulated in this generic framework.

2 Examples

2.1 PBPK Models for TCE in Fat Cells

In Albanese et al[1], Banks and Potter[13], Banks and Potter[14], Potter[22], the authors consider models for the distribution of trichloroethylene (a cleaning solvent) in adipose tissue. The site of interest (the fat tissue "compartment" in the body) actually consists of millions of cells with varying size, residence time, vasculature and geometry. The resulting models which are nonlinear partial differential equations entail "axial-dispersion" type adipose tissue compartments to embody uncertain physiological heterogeneities in a single organism (rat). This models intra-individual variability due to the disperse nature of the micro structure in an individual. Most data (often even *in vitro* data) available for use in model fitting, however, involve aggregate data from a collection of individuals. This inter-individual variability is treated by viewing parameters (including intra-individual dispersion parameters) as random variables. One then attempts to estimate the associated distributions from aggregate data (multiple rat data) which also contains uncertainty (noise).

2.2 Thermally Conductive Composite Adhesives

Design methodology for composite adhesives is investigated in Bihari[19], Banks and Bihari[7] where the question of interest focuses on how to produce an adhesive with enhanced thermal conductivity. The objects of interest are epoxies and gels (with low thermal conductivity) filled with (highly conductive) particles, such as diamond dust, carbon, and aluminum. To pursue such studies, one employs models for heat transport (effective conductivity) in heterogeneous materials. Data from multiple samples are used to estimate parameters related to thermal properties of the composite materials. Also of importance in design of such materials are sensitivity analysis with respect to particle conductivity, geometry, etc. along with homogenization techniques. Both intra- and inter- individual variability (due to varying particle sizes and non-uniform material samples) must be treated in these studies. This again results in parameters treated as random variables. Another aspect of uncertainty plays a role here since one is interested in properties of materials with randomly generated physical location of particles (i.e., uncertainty in location of particles) with distributions of particle size (uncertainty in particle size).

2.3 Electromagnetic Imaging of Dielectric Materials

The monograph Banks et al[9] focuses on inverse problems involving electromagnetic detection. It initiates the development of computational techniques for pulsed microwave interrogation of targets to determine dielectric properties and geometry. Motivating applications include remote interrogation of military targets and non invasive medical diagnostics. Subsequent efforts (Banks and Raye[15], Banks and Raye[16], Banks and Raye[17]) involve the use of acoustic interfaces as reflecting devices in this developing technology. However, in practice the targets are usually heterogeneous, complex materials that are to be characterized by their polarization and conductivity properties. Therefore, in this class of problems, intra-individual variability arises due to multiple mechanisms at the basis of the polarization and conductivity. This in turn leads to a class of interesting inverse problems that can be considered in the generic framework that is the focus of our lecture here. Specifically, the model equations (Maxwell) are of the form (see Banks et al[9])

$$\varepsilon_r \frac{\partial^2 E}{\partial t^2} + \frac{1}{\varepsilon_0}\frac{\partial^2 \mathcal{P}}{\partial t^2} + \frac{\sigma}{\varepsilon_0}\frac{\partial E}{\partial t} - c^2 \frac{\partial^2 E}{\partial z^2} = J$$

with polarization $\mathcal{P}$ given in very general form by

$$\mathcal{P}(t,z) = \int_0^t g(t-s,z)E(s,z)dz.$$

For a heterogeneous material, the polarization $\mathcal{P}$ involves a "mixture" of mechanisms, represented by models such as
the Debye model:

$$\frac{dP_1}{dt} + \frac{1}{\tau}P_1 = \frac{\varepsilon_0}{\tau}(\varepsilon_s - \varepsilon_\infty)E,$$

the Lorentz model:

$$\frac{d^2P_2}{dt^2} + \frac{1}{\tau}\frac{dP_2}{dt} + \omega_0^2 P_2 = \varepsilon_0 \omega_p^2 E$$

as well as higher order models and mechanisms. Hence the polarization $\mathcal{P}$ is made up of a *distribution* of the P_i's from such a family of differential equation systems. Thus, one seeks to estimate a random variable π defined on a *family of differential equation* side constraints to the Maxwell system (i.e., the sample space is a family of differential equations!).

2.4 Modelling of the HIV Infection Pathway

The final example we present will also be used in the next section to outline and explain the application of a theoretical framework for the generic inverse problem introduced in Section 1. The HIV infection pathway, as depicted in Figure 1, is quite complex and entails a number of steps:

i. viral entry into target (uninfected cell);
ii. reverse transcription of viral RNA into DNA;
iii. transport of newly made DNA into the nucleus;
iv. integration of viral DNA into chromosome;
v. production of viral RNA and protein; and
vi. creation of new virus from newly synthesized RNA molecules and proteins.

An important feature of numerous models (see Banks et al[8], Bortz et al[21] and the references therein) which appears to agree with these biological processes are the *intracellular delays*:

- τ_1: the time that a newly (acutely infected) cell takes to become a productively infected cell, and
- $\tau_1+\tau_2$: the time that an acutely infected cell takes to become a chronically infected cell.

Typical model variables include V, the infectious viral population count, A the number of acutely infected cells, C the number of chronically infected cells, T the number of uninfected target cells, and $X = A + C + T$, the total cell population count. Some models (especially those involving treatment and control) also entail immune response variables.

Models that account for intracellular delays usually involve systems of equations of the form (generally nonlinear)

$$\frac{dV}{dt} = -cV(t) + n_a A(t-\tau) + n_c C(t) - n_{vt} V(t) T(t)$$

where τ is a production delay which actually is distributed across the population of cells. That is, one should write

$$\frac{dV}{dt} = -cV(t) + n_a \int_0^\infty A(t-\tau) k(\tau) d\tau + n_c C(t) - n_{vt} V(t) T(t)$$

where k is a probability density to be estimated from aggregate data.

Even if k is given, these systems are nontrivial to simulate; this requires development of fundamental techniques Banks[5], Banks and Kappel[11] for both simulation and estimation. To be more precise, a specific model (see Banks et al[8]) is given by:

$$\begin{aligned}
\dot{V}(t) &= -cV(t) + n_A \int_0^r A(t-\tau) d\pi_1(\tau) + n_C C(t) - p(V,T) \\
\dot{A}(t) &= (r_v - \delta_A - \delta X(t)) A(t) - \gamma \int_0^r A(t-\tau) d\pi_2(\tau) + p(V,T) \qquad (3) \\
\dot{C}(t) &= (r_v - \delta_C - \delta X(t)) C(t) + \gamma \int_0^r A(t-\tau) d\pi_2(\tau) \\
\dot{T}(t) &= (r_u - \delta_u - \delta X(t)) T(t) - p(V,T) + S
\end{aligned}$$

where

$$C(t) = \mathcal{E}_2\{C(t;\tau)\} = \int_0^r C(t;\tau) d\pi_2(\tau),$$

A is the number of acute cells, $V(t) = V_A(t) + V_c(t)$,

$$V_A(t) = \mathcal{E}_1\{V_A(t;\tau)\} = \int_0^r V_A(t;\tau) d\pi_1(\tau)$$

is the number of virions at time t produced by acutely infected cells, and $V_c(t)$ is the number of virions at time t that have been produced by chronically infected cells. Here π_1 is the probability distribution for the delay from acute infection to viral production, and π_2 is the probability distribution for the delay from acute infection to chronic infection, T is the number of cells, X is the total (infected + uninfected) number of cells.

In the next section we present a theoretical framework and indicate how it can be applied to treat this example.

3 A Theoretical Framework

In order to develop a theoretical basis for the generic inverse problem of Section 1, one must develop some topological notions for a measure space over which to minimize the cost functional (2). For this we rely on some basics of probability theory as summarized in Banks and Bihari[6] (see also Billingsley[20]).

We define $\Pi = \Pi(\mathcal{T}) = \{\pi = (\pi_1, \pi_2) : \pi_i$ are probability measures on $\mathcal{T} = [0, r]\}$. Then $(\Pi(\mathcal{T}), \rho)$ is a metric space with Prohorov metric ρ. It is a complete metric space and is compact since $\mathcal{T}$ is compact. The Prohorov metric is not intuitively defined, but convergence in this metric can be stated in several simple and readily usable forms. They are *equivalent* and are given by:

i. $\rho(\pi^k, \pi) \to 0$;
ii. $\int_{\mathcal{T}} g d\pi^k \to \int_{\mathcal{T}} g d\pi$ for all $g \in C(\mathcal{T})$;
iii. $\pi^k[A] \to \pi[A]$ for all Borel $A \subset \mathcal{T}$ with $\pi[\partial A] = 0$.

For details on Prohorov metric and an approximation theory, see Banks and Bihari[6], Billingsley[20].

Once one has a topology on the space of random variables used as parameters, one can readily develop a general theoretical framework that includes the HIV models of Section 2.4. We consider a parameter (π) dependent functional differential equation (FDE) system:

$$\frac{dx}{dt} = f(t, x_t, \pi), \qquad \pi \in \Pi,$$
$$x_0 = \phi \quad \text{where} \quad x_t(\theta) = x(t - \theta), 0 \leq \theta \leq r.$$

One then assumes (or in the case of the HIV system (3) of Section 2.4, *argues*) that $(t, \psi, \pi) \to f(t, \psi, \pi)$ is

a. continuous from $[0, T] \times C[0, r] \times \Pi$ to R^n, and
b. locally Lipschitz in ψ.

Then by continuous dependence on "parameters" results for FDE's (an extension of standard ordinary differential equation results to FDE's with general vector space parameters, see Banks[2], Banks[3] for the basic ideas), one obtains that $\pi \to x(t; \pi)$ is continuous from Π to R^n for each t. This yields

$$\pi \to J(\pi, \hat{d}) = \sum_i | Cx(t_i; \pi) - d_i |^2$$

is continuous from Π to R^1, where $\Pi = (\Pi(\mathcal{T}), \rho)$, is compact and ρ is the Prohorov metric.

Then the general theory of Banks and Bihari[6] can be followed to obtain existence and stability for inverse problems (continuous dependence with respect to data of solutions of the inverse problem). Moreover, an approximation theory is obtained that can be used as a basis for computational methods.

We can also obtain results for "method stability under approximation" (see Banks[4], Banks and Kunisch[12], Banks et al[18]). To briefly summarize these, let $\mathcal{T}_M = \{\tau_j^M\} \subset \mathcal{T}$ be such that $\cup_M \mathcal{T}_M$ is dense in $\mathcal{T}$ and define

$$\Pi^M(\mathcal{T}) = \{\pi_M \in \Pi : \pi_M = \textstyle\sum_{j=1}^M p_j \delta_{\tau_j^M}, \tau_j^M \in \mathcal{T}_M, p_j \in R, p_j \geq 0, \sum p_j = 1\}.$$

Let $\hat{d} = \{d_i\}, \quad \hat{d}^k = \{d_i^k\}$ be sets of data (observations) such that $\hat{d}^k \to \hat{d}$ as $k \to \infty$. Define

$$\Pi_M^*(\hat{d}^k) = \{\text{set of minimizers for} \quad J^k(\pi) = J(\pi, \hat{d}^k) \quad \text{over} \quad \Pi^M(\mathcal{T})\},$$

and

$$\Pi^*(\hat{d}) = \{\text{set of minimizers for} \quad J(\pi, \hat{d}) \quad \text{over} \quad \Pi(\mathcal{T})\}.$$

Let *dist* (A, B) be the Hausdorff distance between sets A and B. Then we have

Theorem 1 *We have* dist $(\Pi_M^*(\hat{d}^k), \Pi^*(\hat{d})) \to 0$ *as* $M \to \infty, \hat{d}^k \to \hat{d}$, *so that solutions depend continuously on data and approximate problems are "method stable".*

The above theory can be applied to the HIV model (3) given in Section 2.4. To see this, let $x = (V, A, C, T)$ and observe that the π dependent terms in right sides of the system have the form $\int_0^r x(t-\tau)d\pi(\tau)$, so that continuity with respect to π in the Prohorov metric is readily established, i.e., by the equivalence of (*i.*) and (*ii.*) above, we have immediately

$$\int_0^r x(t-\tau)d\pi^k(\tau) \to \int_0^r x(t-\tau)d\pi(\tau) \quad \text{whenever} \quad \rho(\pi^k, \pi) \to 0.$$

The data is such that the d_i are observations for $A(t_i) + C(t_i) + T(t_i)$, so that the remainder of the conditions needed for the theory to be applicable are readily satisfied.

In Banks et al[8], we used experimental data to estimate the delays in the HIV models of Section 2.4 in the context of such a framework as described here. In these inverse problem calculations in Banks et al[8] we used numerical approximation methods for the FDE's (both discrete delays and continuous probability density functions were used). The approximation methods were spline-based as developed in Banks[5] and Banks and Kappel[11].

In the results reported in Banks et al[8], we estimated p of the nonlinear term $p(V,T)$, and Dirac measures $\pi_1 = \delta_{\tau_1}$ and $\pi_2 = \delta_{\tau_1+\tau_2}$ as well as continuous probability density functions associated with the delays from acute infection to viral production and from acute infection to chronic infection. Extremely good fits to the experimental data were obtained.

4 Concluding Remarks

We close with several summary comments.

1. Randomness (uncertainty) is ubiquitous in inverse and estimation problems, whether due to uncertainty in modeling, intra- and inter-individual variability in aggregate data for populations, etc. or combinations of these.
2. Important applications include: biology (PBPK models, HIV cellular infection models); materials (design of modern composites); electromagnetic interrogation (medical diagnostics, remote detection) as well as numerous others.
3. Successful efforts require combining deterministic and probabilistic modeling, and theoretical and computational ideas (mixing distributions, random effects).
4. Both theoretical and computational challenges are significant!!

Some initial efforts have been made, but much is yet to be done - see the program on Inverse Problem Methodology in Complex Stochastic Models (www.samsi.info) for Fall, 2002 at the new institute, the Statistical and Applied Mathematical Sciences Institute (SAMSI).

Acknowledgements

This research was supported in part by the U.S. Air Force Office of Scientific Research under grant AFOSR-F49620-1-00-0026.

References

1. R.A. Albanese, H.T. Banks, M.V. Evans and L.K. Potter, *PBPK models for the transport of trichloroethylene in adipose tissue,* CRSC-TR01-03(Jan., 2001); to appear in Bull. Math Biology.
2. H.T. Banks, *Necessary conditions for control problems with variable time lags,* SIAM J. Control **6** , 9–47(1968).

3. H.T. Banks, *Variational problems involving functional differential equations,* SIAM J. Control **7** , 1–17(1969).
4. H.T. Banks, *On a variational approach to some parameter estimation problems,* LCDS Technical Report #85-14(May, 1985); in *Distributed Parameter Systems* ed. F. Kappel, et. al. (Springer Lecture Notes in Control and Info. Sci. **75**, 1–23(1985)).
5. H.T. Banks, *Identification of nonlinear delay systems using spline methods,* in *Proc. Intl. Conf. on Nonlinear Phenomena in Math. Sciences* Ed. V. Lakshmikantham (Academic Press , 47–55(1982)).
6. H.T. Banks and K.L. Bihari, *Modeling and estimating uncertainty in parameter estimation,* CRSC-TR99-40, NCSU(Dec., 1999); Inverse Problems **17** , 1–17(2001).
7. H.T. Banks and K.L. Bihari, *Analysis of thermal conductivity in composite adhesives,* CRSC-TR01-20, NCSU(August, 2001); Numerical Functional Analysis and Optimization, submitted.
8. H.T. Banks, D.M. Bortz and S.E. Holte, *Incorporation of variability into the modeling of viral delays in HIV infection dynamics,* CRSC-TR01-25(Sept., 2001); Math Biosciences, submitted.
9. H.T. Banks, M.W. Buksas and T. Lin, *Electromagnetic Material Interrogation Using Conductive Interfaces and Acoustic Wavefronts,* SIAM Frontiers in Applied Mathematics, Vol. FR21, Philadelphia (2000).
10. H.T. Banks, M.L. Joyner, B. Wincheski and W.P. Winfree, *Nondestructive evaluation using a reduced-order computational methodology,* ICASE Tech Rep. 2000-10, NASA Langley Res. Ctr.(March 2000); Inverse Problems **16** , 929–945(2000).
11. H.T. Banks and F. Kappel, *Spline approximations for functional differential equations,* J. Differential Equations, **34**, 496–522 (1979).
12. H.T. Banks and K. Kunisch, *Estimation Techniques for Distributed Parameter Systems,* (Birkhaüser, Boston, 1989).
13. H.T. Banks and L.K. Potter, *Well-posedness results for a class of toxicokinetic models,* CRSC-TR01-18(July, 2001); Discrete and Continuous Dynamical Systems, submitted.
14. H.T. Banks and L.K. Potter, *Model predictions and comparisons for three toxicokinetic models for the systemic transport of TCE,* CRSC-TR01-23(August, 2001); to appear in Mathematical and Computer Modeling.
15. H.T. Banks and J.K. Raye, *Computational methods for nonsmooth acoustic systems,* CRSC-TR01-02, NCSU(January, 2001); to appear in Computational and Applied Mathematics.
16. H.T. Banks and J.K. Raye, *Computational methods for nonsmooth acoustic systems arisin gin an electromagnetic hysteresis identification prob-*

lem, to appear in *Proceedings of the* 18th *ASME Biennial Conference on Mechanical Vibration and Noise* (Pittsburgh, PA, Sept. 9-12, 2001).

17. H.T. Banks and J.K. Raye, em Well-posedness for systems representing electromagnetic/acoustic wavefront interaction, CRSC-TR01-34(December, 2001); ESAIM: Control, Optimization and Calculus of Variations, submitted.
18. H.T. Banks, R.C. Smith and Y. Wang, *Smart Material Structures: Modeling, Estimation and Control* (Masson/John Wiley, Paris/Chichester,1996); .
19. K.L. Bihari, *Analysis of Thermal Conductivity in Composite Adhesives,* Ph.D. Thesis, NCSU(August, 2001).
20. B. Billingsley, *Convergence of Probability Measures,* (Wiley, New York, 1968).
21. D. Bortz, R. Guy, J. Hood, K. Kirkpatrick, V. Nguyen and V. Shimanovich, *Modeling HIV infection dynamics using delay equations,* in 6th *CRSC Industrial Math Modeling Workshop for Graduate Students,* NCSU (July, 2000), CRSC-TR00-24 (Oct., 2000).
22. L.K. Potter, *Physiologically Based Pharmacokinetic Models for the Systemic Transport of Trichloroethylene,* Ph.D. Thesis, NCSU(August, 2001).

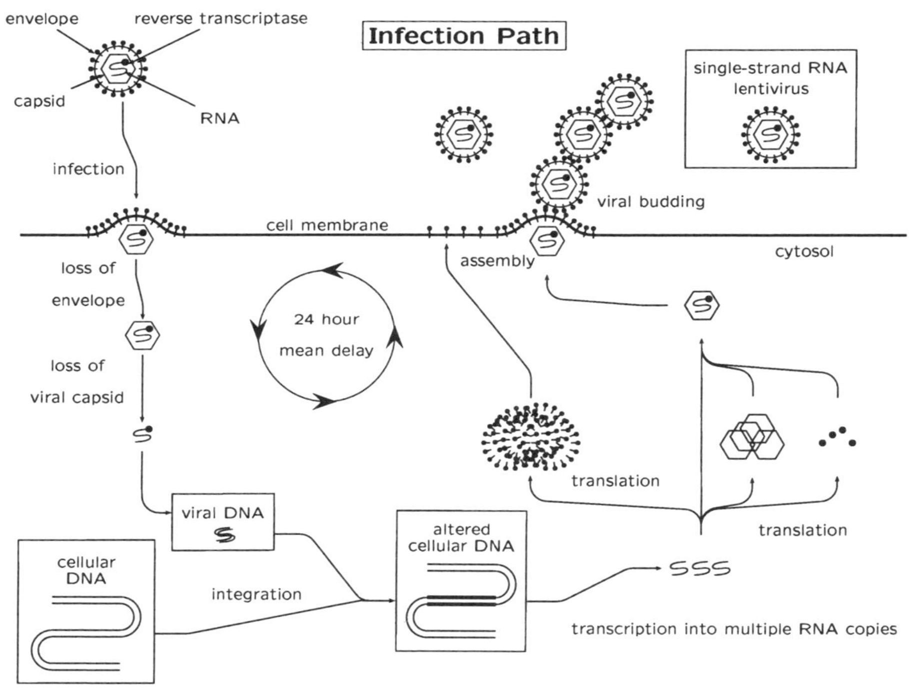

Figure 1. HIV Infection Pathway

INVERSE AND OPTIMAL DESIGN PROBLEMS IN DIFFRACTIVE OPTICS

GANG BAO

Department of Mathematics, Michigan State University, East Lansing, MI 48824-1027, USA

E-mail: bao@math.msu.edu

Consider a time-harmonic electromagnetic plane wave incident from the top on a periodic structure in $\mathbf{R}^3$. The periodic structure separates two regions. Above the structure, the index of refraction k is assumed to be a fixed constant. Below the structure is a perfectly reflecting material. Given the incident field, the inverse diffraction problem is to determine the periodic structure or the shape of the interface from the scattered field. The optimal design problem in this context is to create grating profiles which yield some specified diffraction patterns. In this paper, recent significant progress on the characterization of uniqueness and stability for the inverse problem is discussed. Optimal design of diffractive structures is also addressed. The paper is concluded with discussions on future research directions in the field.

1 Introduction

Consider a time-harmonic electromagnetic plane wave incident on a periodic structure in $\mathbf{R}^3$. The structure is the surface of some perfectly reflecting material or conductor. Above the structure, the index of refraction k is assumed to be a fixed constant. The inverse diffraction problem is to determine the periodic structure or the shape of the interface from the scattered field. In this paper, issues on uniqueness and stability for this inverse diffraction problem are studied. This work is motivated by the study of optimal design problems of gratings where one wishes to design a grating (or periodic) structure that generates some specified scattered field. Recent results on the design problem are also discussed.

The scattering theory in periodic structures has many applications in micro-optics, where periodic structures are often called *diffraction gratings.* A good introduction to the problem of electromagnetic diffraction through periodic structures, along with some earlier numerical methods, can be found in Petit [36]. We refer to Chen and Friedman [24], Nédélec and Starling[35], Dobson[26], Abboud[1], Bao[6-8], Bao and Dobson[13], Bao and Yang[18] for recent existence and uniqueness results, and numerical approximations of solutions by using either integral equation methods or variational approaches. An up-to-date survey on the mathematical modelling of diffractive optics and other related topics may be found in Bao, Cowsar, and Masters[12]. We also refer the reader to the book of Colton and Kress [25] and references therein for the general theory of inverse scattering problems in general (nonperiodic) structures.

2 The Diffraction Problem

The electromagnetic wave propagation is governed by the time harmonic Maxwell's equations (time dependence $e^{-i\omega t}$):

$$\nabla \times E - i\omega\mu H = 0 \,, \tag{1}$$

$$\nabla \times H + i\omega\epsilon E = 0 \,, \tag{2}$$

where E and H denote the electric and magnetic fields, respectively, μ is the magnetic permeability which is assumed to be a fixed positive constant everywhere, and ϵ is the dielectric coefficient.

Assume that a plane wave is incident on a periodic (grating) surface which separates two homogeneous media. The grating is called biperiodic if the surface is periodic in two orthogonal directions. Throughout, we mainly focus on the two-dimensional geometry, *i.e.*, the structure is assumed to be periodic in the x_1 direction and invariant in the x_2 direction.

Let the scattering profile (object) in one period be described by the curve $\Gamma = \{(x_1, x_2, x_3) : x_3 = f(x_1)\}$ with a periodic function f of period $\Lambda > 0$. For convenience, the function f is supposed to be sufficiently smooth, for example of $C^{1,1}$. The space below Γ is filled with some perfectly reflecting material (a conductor).

Let $\Omega = \{(x \in \mathbf{R}^3 : x_3 > f(x_1)\}$ be filled with a material in such a way that its index of refraction $k = \omega\sqrt{\epsilon\mu}$ is a fixed constant. Here ω is the angular frequency. In addition, it is assumed throughout that the index of refraction k satisfies: $Re(k) > 0$ and $Im(k) \geq 0$. The case $Im(k) > 0$ accounts for materials which absorb energy. Suppose that a plane wave is incident on Γ from the top. We then have the following diffraction problem: Given the incident field u_I and the periodic structure, one wishes to predict the behavior of the outgoing reflected waves. Note that since the medium underneath is a conductor, it does not support any transmitted wave.

In the two-dimensional case, there are two fundamental polarizations: TE (transverse electric) and TM (transverse magnetic). In the TE polarization case, *i.e.*, the electric field vector E is assumed to point to the x_2 axis. In other words, $E = u\vec{x}_2$, where $u = u(x_1, x_3)$ is a scalar function. Similarly, in the TM case, the magnetic field $H = u\vec{x}_2$. For the two-dimensional geometry, the Maxwell equations can be further simplified. Let $u_I = e^{i\alpha x_1 - i\beta x_3}$ be the incident plane wave. Here $\alpha = ksin\theta$, $\beta = kcos\theta$, and $-\pi/2 < \theta < \pi/2$ is the incident angle. From the Maxwell equations (1) and (2), it is straightforward to deduce the following Helmholtz equation:

$$(\Delta + k^2)u = 0 \text{ in } \Omega, \tag{3}$$

$$u|_\Gamma = 0\,, \tag{4}$$

where the homogeneous Dirichlet boundary condition (4) comes from the TE polarization assumption and the assumption that the material is a conductor. Note that for TM polarization, the perfect conductor assumption would imply the homogeneous Neumann boundary condition

$$\frac{\partial u}{\partial n} = 0\,.$$

Because of the physics, we seek for quasiperiodic solutions to this problem, *i.e.*, the solution u such that $ue^{-i\alpha x_1}$ is Λ-periodic for every x_3. It is evident that to completely specify the boundary value problem, we need to impose a radiation condition in the x_3 direction. The radiation condition is the boundedness of the scattered fields as x_3 tends to infinity. More precisely, we insist that u is composed of bounded outgoing plane waves plus the incident wave u_I.

Let T be a fixed constant such that $T > max\{f(x_1)\}$. We next present a transparent boundary condition on $x_3 = T$ which may be derived by a combination of the fundamental solution and the periodicity of the solutions. It allows us to reduce the scattering problem to a bounded domain. Let u be the quasiperiodic solution that solves the scattering problem (3) and (4). Then there exists a pseudodifferential operator B of order one [6,26], such that

$$\frac{\partial u}{\partial n}|_{x_3=T} = B(u|_{x_3=T}) - 2i\beta e^{-i\beta T + i\alpha x_1}\,. \tag{5}$$

For the direct scattering problem, questions on existence and uniqueness are well understood, see for example Chen and Friedman[24],Bao[6],Dobson[26] , Nédélec and Starling[35]. Basically, the following general result holds.

Theorem 1 *There is possibly a sequence of frequencies ω_j with $\omega_j \to +\infty$, such that the scattering problem (3), (4), and (5) specified above has a unique quasiperiodic solution provided that $\omega \neq \omega_j$ for any $j = 1, 2, \ldots$.*

Since k is a fixed constant, for simplicity, we always assume that the direct scattering problem has a unique solution in this paper. In general, because of Theorem 1, this may be arranged by perturbing k or ω slightly.

Suppose that u (quasiperiodic) solves the scattering problem (3), (4) and (5) for a given incident plane wave u_I. The inverse problem can be stated as follows: Determine $f(x_1)$ from the knowledge of $u(x_1, T)$ or the trace of u.

3 Uniqueness for the Inverse Problem

Suppose that for a given incident plane wave u_I, $u_j(x_1, x_3)$ $(j = 1,2)$ is Λ-quasiperiodic and solve the scattering problem (3), (4), and (5) with respect to the profiles $f_j(x_1)$, where the functions f_1 and f_2 are Λ-periodic. Let $T > max\{f_1(x_1), f_2(x_1)\}$ be a fixed constant. Denote $h = max\{f_1(x_1), f_2(x_1)\} - min\{f_1(x_1), f_2(x_1)\}$.

We are ready to state a uniqueness result for the inverse problem.

Theorem 2 *(Bao [5]) Assume that* $u_1(x_1, T) = u_2(x_1, T)$. *Assume further that one of the following conditions is satisfied:*

i). k has a nonzero imaginary part;

ii). k is real and h satisfies $k^2 < 2[h^{-2} + \Lambda^{-2}]$.

Then $f_1(x_1) = f_2(x_1)$.

When k has a nonzero imaginary part, a global uniqueness result was proved by Bao [4] and by Ammari [3] in the biperiodic case. However, in general, global uniqueness may not be possible when k is real. This is evident in the simplest case with a plane wave incident on a flat surface. In this case, the solution of the scattering problem can be written down explicitly. The nonuniqueness is obvious since the scattering fields will remain the same when one moves the flat surface up or down in certain multiples of the wavelength.

In the case with real k corresponding to the dielectric medium, one can only prove a local uniqueness theorem. In this case, our uniqueness theorem indicates that any two surface profiles are identical if they generate the same scattering fields (or patterns) and the area in between the two profiles are sufficiently small. Moreover, the smallness of the area is characterized explicitly in terms of a condition which relates the index of refraction k, the period, and the maximum of the difference in height allowed for the two profiles. The proof may be given by an application of Holmgren's uniqueness theorem and some unique continuation argument. A crucial step is to estimate the first eigenvalue of the Dirichlet Laplacian. In fact, by estimating the eigenvalue, one can get precise idea on how close the profiles need to be for uniqueness to hold.

Global uniqueness of the inverse problem in the dielectric medium case by using a finite number of incident waves have been proved in Bao [10], Hettlich and Kirsch [29]. For the 3-D biperiodic problem, a local uniqueness theorem has been obtained by Bao and Zhou [19] where the model and proofs are much more technical. The idea of Bao and Zhou [19] should also yield a local uniqueness result in the TM case.

In Kirsch[32], Kirsch proved a uniqueness theorem by a similar approach as for the general inverse scattering problem in Kirsch and Kress[33]. The main idea was to prove by using many incident waves the denseness of a set of special solutions. Other related results on inverse diffraction problems may be found in Borovikov[22],Bao[9].

4 Stability for the Inverse Problem

In applications, it is impossible to make exact measurements. Thus stability results are crucial in the reconstruction of profiles. This is particularly the case here. In fact, the Rayleigh diffraction theory indicates that the scattered wave may be expressed away from the interface as infinite sum of plane waves, where only a finite number of the plane waves are propagating modes and the rest are exponentially damped. In the far-field, only the propagating modes are detectable. Thus, the measurements are not exact but may be fairly close to the exact boundary values of the solution.

Let first introduce some notations. For any two domains D_1 and D_2 in $\mathbf{R}^2$, denote by $d(D_1, D_2)$ the Hausdorff distance between them. Denote $D = \{x;\ f(x_1) < x_3 < T\}$, and a sequence of domains $D_h = \{x;\ f(x_1) + h\sigma_h(x_1)\nu(x_1) < x_3 < T\}$ for any $0 < h < h_0$, where $\nu(x_1)$ is the normal to $\Gamma = \{x_3 = f(x_1)\}$. Assume also that the boundary $\Gamma_h = \{x_3 = f(x_1) + h\sigma_h(x_1)\nu(x_1)\}$ is periodic of the same period Λ and is of C^2. Further, the function σ_h satisfies $|\sigma_h(x_1)| \le C$. Furthermore, for h_0 is sufficiently small, the sequence of domains is assumed to satisfy that

$$C_1 h \le d(D, D_h) \le C_2 h\ ,$$

where C_1 and C_2 are positive constants.

For the fixed incident plane wave u_I, assume that u and u_h solve the scattering problem with respect to periodic structures Γ and Γ_h, respectively. Then we have the following local stability result.

Theorem 3

$$d(D_h, D) \le C||u_h|_{x_3=T} - u|_{x_3=T}||_{H^{1/2}}\ , \tag{6}$$

where the constant C may depend on the family $\{\sigma_h\}$.

The result indicates that for small h, if the boundary measurements are $O(h)$ close to the scattered fields in the $H^{1/2}$ norm, then D_h is $O(h)$ close to D in the Hausdorff distance.

The theorem was proved in Bao and Friedman [16]. Our proof is based on a variational approach and applications of a unique continuation technique.

Actually, in Bao and Friedman[16], local Lipschitz type stability results were obtained for a more general class of inverse diffraction problems in both the TE and TM (transverse magnetic) polarizations.

More recently, by using the technique of material derivatives with respect to the variation of the dielectric coefficient, Elschner and Schmidt have generalized the local stability result to the case of polygonal (grating) interfaces. A global stability result has been obtained by Bruckner, Cheng, and Yamamoto [23] under certain additional assumptions equivalent to the validness of the maximum principle. The stability question becomes much more challenging in the 3-D biperiodic case. Until now, the only result available is a local stability result similar to Theorem 3 proved by Bao and Zhou [19].

Finally, we mention that local stability results for other inverse problems, for example, inverse conductivity problems, were previously obtained in Bellout and Friedman[20], Bellout *et al* [21].

5 Optimal Design

Given the incident field, the optimal design problem concerns the creation of grating profiles that give rise to some specified diffraction patterns. The problem can be posed as a nonlinear least-squares problem. Difficulties arise since the scattering pattern depends on the interface in a very implicit fashion and in general the set over which the function is minimized is neither convex nor closed. The formulation of the design problem is very close to similar problems in elasticity, for which fast and efficient algorithms have recently been developed. Initial progress on the design problem has been made via weak convergence analysis methods by Achdou and Pironneau [2], Dobson [26], and the homogenization theory by Bao and Bonnetier [11] along with the "relaxation" technique of Kohn and Strang [31]. The main idea is to allow the grating profiles to be highly oscillating and to use relaxed formulation of the optimization problem. The crucial step is to determine the relaxed formulation which involves materials and the effective dielectric properties [11]. We refer to Bao *et al* [14,15,17] for additional results on this and related design problems.

Another important direction in optimal design of diffractive optics is to design "resonances". One of the most exciting new developments in diffractive optics involves the integration of a zero-order grating with a planar waveguide to create a resonance. Such structures, known as guided-mode resonance filters, have been demonstrated to yield ultra-narrow bandwidth filters for a selected center wavelength and polarization with $\approx$ 100% reflectance [34]. With such extraordinary potential performance, these "resonant reflectors" have attracted attention for many applications, such as lossless spectral filters with arbitrarily narrow, controllable linewidth, efficient and low-power optical switch elements, 100% reflective narrow-band spectrally selective mirrors, polarization control, high-precision sensors, lasers, and integrated optics. Significant recent progress has been made in Huang [30] for solving an interesting optimal design problem: to determine the structure and the material that give rise to a resonance at some specified wavelength. By using the variational approach, the design process may be formulated as an optimization problem where the diffraction grating and waveguide problems are solved repeatedly.

6 Future Directions

A closely related problem is to determine the periodic (grating) structure ruled on some nonconductive optical material. In this situation, one places optical detectors both above and below the material. The measurements consist of information on the reflected wave and transmitted wave. In the TE case, the model equation takes the same form as Equation (3). However, the boundary condition (4) is no longer valid. Instead, the direct problem may be formulated in a "box" with nonlocal boundary conditions that are similar to (5) on the top and at the bottom. We believe that a local uniqueness theorem for this inverse problem may be proved by modifying the proof of Theorem 3.1. A local stability result was established in Bao and Friedman[16]. No result is available in the biperiodic case.

Another interesting problem concerns global uniqueness for the inverse diffraction problems. In particular, no result is available in the TM (transverse magnetic) case or the biperiodic case. The corresponding inverse problem turns out to be much more difficult. It is not clear whether additional data such as a finite number of incident waves would be sufficient to assure global uniqueness. The difficulty lies in the fact that the first eigenvalue of the Neumann or vector Laplacian does not have the monotone property with respect to the domain or the diameter of the domain. So far, we were only able to prove some local stability results [16] by combining a variational approach and the analytic index theory.

Numerical solution of the design and inverse diffraction problems is of great interest. As one might expect from the local uniqueness and stability results reported here that some a priori knowledge is necessary in order to determine the structure. An ongoing research is to restrict one's attention to a class of curves with certain geometry and then solve the inverse problem by an optimization method.

A significant future direction is to study the inverse and design problems in nonlinear optics. It has been observed that the use of gratings can significantly enhance the nonlinear effects of second harmonic generation in nonlinear optics. The field is widely open. We refer the reader to Bao, Huang, and Schmidt [17] for some references and preliminary results on optimal design of nonlinear gratings.

Acknowledgments

The research of the author was partially supported by the NSF Applied Mathematics Programs grant DMS 0104001, the NSF Western Europe Programs grant INT 98-15798, the Office of Naval Research (ONR) grant N000140210365, and an Intramural Research Grants Program grant of Michigan State University.

References

1. T. Abboud, *Electromagnetic waves in periodic media*, in *Second International Conference on Mathematics and Numerical Aspects of Wave Propagation,* ed. R. Kleinman et al.(SIAM, Philadelphia , 1–9(1993)).
2. Y. Achdou and O. Pironneau, *Optimization of a photocell,* Optimal Control Appl. Meth. **12** , 221–246(1991).
3. H. Ammari, *Uniqueness theorems for an inverse problem in a doubly periodic structure*, Inverse Problems **11** , 823–833(1995).
4. G. Bao, *A uniqueness theorem for an inverse problem in periodic diffractive optics*, Inverse Problems **10**, 335–340 (1994).
5. G. Bao, *An inverse diffraction problem in periodic structures*, in *Proceedings of Third International Conference on Mathematical and Numerical Aspects of Wave Propagation,* Ed. by G. Cohen(SIAM, Philadelphia, 694–704(1995)).
6. G. Bao, *Finite elements approximation of time harmonic waves in periodic structures*, SIAM J. Numer. Anal. **32**,1155–1169(1995).
7. G. Bao, *Numerical analysis of diffraction by periodic structures: TM polarization*, Numer. Math. **75**, 1–16 (1996).
8. G. Bao, *Variational approximation of Maxwell's equations in biperiodic structures*, SIAM J. Appl. Math. **57**, 364–381(1997).
9. G. Bao, *On the relation between the coefficients and solutions for a diffraction problem*, Inverse Problems **14** , 787–798 (1998).
10. G. Bao, *Inverse diffraction by a periodic perfect conductor with several measurements*, in *Inverse Problems in Engineering, Theory and Practice,* Ed. D. Delaunay, Y. Jarny, and K.A. Woodbury(ASME, 297–303, 1998).

11. G. Bao and E. Bonnetier, *Optimal design of periodic diffractive structures*, Appl. Math. Optim. **43** , 103–116 (2001).
12. G. Bao, L. Cowsar, and W. Masters, ed., *Mathematical Modeling in Optical Science*, the SIAM Frontiers in Applied Mathematics, SIAM, Philadelphia (2001).
13. G. Bao and D. Dobson, *On the scattering by biperiodic structures*, Proc. Am. Math. Soc. **128** , 2715–2723(2000).
14. G. Bao, D. Dobson, and J. A. Cox, *Mathematical studies of rigorous grating theory*, J. Opt. Soc. Am. **A 12** , 1029–1042(1995).
15. G. Bao, D. Dobson, and K. Ramdani, *A constraint on the maximum reflectance of rapidly oscillating dielectric gratings*, SIAM J. Control. Opt. **40** , 1858–1866(2002).
16. G. Bao and A. Friedman, *Inverse problems for scattering by periodic structures*, Arch. Rat. Mech. Anal. **132** , 49–72(1995).
17. G. Bao, K. Huang, and G. Schmidt, *Optimal design of nonlinear gratings*, submitted.
18. G. Bao and H. Yang, *A least-squares finite element analysis for diffraction problems*, SIAM J. Numer. Anal. **2** , 665–682(2000).
19. G. Bao and Z. Zhou, *An inverse problem for scattering by a doubly periodic structure*, Trans. Ameri. Math. Soc. **350** , 4089–4103(1998).
20. H. Bellout and A. Friedman, *Identification problems in potential theory*, Arch. Rational Mech. Anal. **101** , 143–160(1988).
21. H. Bellout, A. Friedman, and V. Isakov, *Stability for an inverse problem in potential theory*, Tran. Amer. Math. Soc. **332** , 271–296(1992).
22. I. Borovikov, *Uniqueness of solutions to one inverse diffraction problem*, Differentsial'nye Uravneniya **28** , 827–831(1992).
23. G. Bruckner, J. Cheng, and M. Yamamoto, *An inverse problem in diffractive optics: conditional stability*, Inverse Problems **18** , 415–433(2002).
24. X. Chen and A. Friedman, *Maxwell's equations in a periodic structure*, Trans. Amer. Math. Soc. **323** , 465–507(1991).
25. D. Colton and R. Kress, *Inverse Acoustic and Electromagnetic Scattering Theory* (Springer-Verlag, New York, 1992).
26. D. Dobson, *Optimal design of periodic antireflective structures for the Helmholtz equation*, Euro. J. Appl. Math. **4** , 321–340(1993).
27. J. Elschner and G. Schmidt, *Numerical solution of optimal design problems for binary gratings*, J. Comput. Phys. **146** , 603–626(1998).
28. J. Elschner and G. Schmidt, *Inverse scattering for periodic structures: stability of polygonal interfaces*, Inverse Problems **17** , 1817–1829(2001).
29. F. Hettlich and A. Kirsch, *Schiffer's theorem in inverse scattering theory for periodic structures*, Inverse Problems **13** , 351–361(1997).

30. K. Huang, *Optimal Design of Diffractive Optics*, Ph.D. Thesis (Michigan State Univ., 2002).
31. R. Kohn and G. Strang, *Optimal design and relaxation of variational problems I, II, III*, Comm. Pure Appl. Math. **39** , 113–137, 139–182, 353–377 (1986).
32. A. Kirsch, *Uniqueness theorems in inverse scattering theory for periodic structures*, Inverse Problems **10** , 145–152(1994).
33. A. Kirsch and R. Kress, *Uniqueness in inverse scattering*, Inverse Problems **9** , 285–299(1993).
34. R. Magnusson and S. Wang, *New principle for optical filters*, Appl. Phys. Lett. **61** , 1022–1024(1992).
35. J. C. Nédélec and F. Starling, *Integral equation methods in a quasi-periodic diffraction problem for the time-harmonic Maxwell's equations*, SIAM J. Math. Anal. **22** , 1679–1701(1991).
36. R. Petit, *Electromagnetic Theory of Gratings*, in *Topics in Current Physics, Vol. 22,* ed.R. Petit(Springer-Verlag, Heidelberg, 1980).

THE INVERSE PROBLEM OF OPTION PRICING

VICTOR ISAKOV

Department of Mathematics and Statistics, Wichita State University, Wichita, KS 67260-0033, U.S.A.
E-mail:victor.isakov@wichita.edu

We consider the problem of recovery of the volatility coefficient of the Black-Scholes equation for option prices as functions of time and of stock price. We give most recent results about uniqueness and stability of reconstruction of volatility from market data and discuss relations with stochastic partial differential equations. We suggest two algorithms of numerical reconstruction, using a parametrix and the linearized inverse problem. We give the results of some numerical tests. For simplicity, we handle only European options.

1 The Black-Schools Equation

For any stock price, $0 < s < \infty$, and time , $0 < t < T$, a price u for an option expiring at time T satisfies the following partial differential equation

$$\frac{\partial u}{\partial t} + \frac{1}{2}s^2\sigma^2(s)\frac{\partial^2 u}{\partial s^2} + s\mu\frac{\partial u}{\partial s} - ru = 0 \tag{1}$$

Here, $\sigma(s)$ is the volatility coefficient that satisfies $0 < m < \sigma(s) < M < \infty$, and μ and r are, respectively, the risk-neutral drift and the risk-free interest rate assumed to be constants. The backward in time parabolic equation (1) is augmented by the final condition specified by the payoff of the call option with the strike price K

$$u(s,T) = (s-K)^+ = max(0, s-K),\ 0 < s \tag{2}$$

It is known (Bouchouev *et al* [4]) that there is a unique solution u to (1),(2) satisfying the bound $|u(s,t)| < C(s+1)$.

The inverse problem of option pricing seeks for σ given

$$u(s^*, t^*; K, T) = u^*(K),\ x \in \omega^*. \tag{3}$$

Here s^* is market price of the stock at time t^*, and $u^*(K)$ denote market price of options with different strikes K for a given expiry T.

The Black-Scholes formula (Black *et al* [7]) gives values of σ in the inverse problem provided it is constant (log-normal distribution). However, this assumption is usually not satisfied in real markets. A way to reconcile the difference is to look for variable σ. Even though there was a considerable effort to solve the inverse option pricing problem and many numerical algorithms have been proposed [1,2,3,6,12], neither theory nor convergence properties of these algorithms are satisfactory for practitioners (Bouchouev *et al* [4]).

While time dependence is important for some markets, one can not predict time variable volatility in future from current market data, so we consider only $\sigma = \sigma(s)$.

To obtain our results we will use that the option premium $u(.,.;K,T)$ satisfies the equation dual to the Black-Scholes equation (1) with respect to the strike price K and expiry time T:

$$\frac{\partial u}{\partial T} - \frac{1}{2}K^2\sigma^2(K)\frac{\partial^2 u}{\partial K^2} + \mu K\frac{\partial u}{\partial K} + (r-\mu)u = 0 \tag{4}$$

Given the volatility function $\sigma(s)$ the problem (4) ,(2) has a unique solution satisfying the bound $|u(K,T)| < C(K+1)$.

The substitution

$$y = \ln(\frac{K}{s^*}), \quad \tau = T - t,\ a(y) = \sigma(s^* e^y),\ U(y,\tau) = u(s^* e^y, \tau + t) \tag{5}$$

transforms the equation (4) and the initial data (2) into

$$\frac{\partial U}{\partial \tau} - \frac{12a^2}{(y)}\frac{\partial^2 U}{\partial y^2} + \left(\frac{1}{2}a^2(y) + \mu\right)\frac{\partial U}{\partial y} + (r-\mu)U = 0 \tag{6}$$

$$U(y,0) = s^*(1-e^y)^+, \quad y \in \mathbf{R}$$

$$U(y,\tau^*) = U^*(y), \quad y \in \omega \tag{7}$$

Here ω is the transformed interval ω^* (ω^* in y- variables (5)). Observe that $\tau^* = T - t^*$.

The equations (6) and (7) for functions $U(\tau,y), a(y)$ form the so-called inverse parabolic problem with the final overdetermination. The known uniqueness conditions for this problem (Isakov[9], section 6.2 and Isakov[10], section 9.2) are not satisfied in our particular situation.

2 Uniqueness and Stability Results

First we give available theoretical results.

Theorem 1 *Let ω_0 be a non-empty open subinterval of ω.*

If a if known on ω_0, then the data (7) uniquely determine a on ω.

If ω is bounded and $a = a_1$, $a = a_2$ are two coefficients given outside ω and on ω_0 which generate the final data U_1^*, U_2^*, then

$$|a_2 - a_1|_\lambda(\omega) \leq C|U_2^* - U_1^*|_{2+\lambda}(\omega) \tag{8}$$

where C depends only on the norms of a_1,a_2 in $C^\lambda(\omega)$, on ω, ω_0, and on τ^*.

This result is obtained in Bouchouev *et al* [3]. We outline the ideas of the proof of uniqueness.

A solution $U(y, \tau)$ of the parabolic problem with a time independent coefficient is analytic with respect to $\tau > 0$. Since a is given on ω_0, we can express $\frac{\partial U}{\partial \tau}(s, \tau^*)$ from the final data (7) and the differential equation (6). Since the coefficients of (6) do not depend on τ, $\frac{\partial U}{\partial \tau}$ satisfies the same partial differential equation. Hence we can repeat this step to conclude that all partial derivatives of U with respect to τ are uniquely determined on $\omega_0 \times \{\tau^*\}$. By analyticity, U is uniquely determined on $\omega_0 \times (0, \tau^*)$. Therefore we are given the Cauchy data for U at an endpoint of $\omega \setminus \omega_0$. Then one can apply the Bukhgeim-Klibanov method of proving uniqueness by Carleman type estimates (Isakov[10]).

This result is not satisfactory because the assumption that a is known on ω_0 excludes possibility of an existence theorem. By extending the previous argument, Masahiro Yamamoto and Bouchouev *et al* [4], observed that for infinitely smooth a given outside ω the data (7) uniquely determine a.

3 The Linearized Inverse Option Pricing Problem

Difficulties with the exact (nonlinear) inverse problem and its features (relatively small ω and fast decay of the Gaussian kernel away from the origin) suggest that the linearization around constant volatilities could be useful.

To derive the linearized inverse problem we assume that

$$\frac{1}{2}a^2(y) = \frac{1}{2}\sigma_0^2 + f(y) \tag{9}$$

where f is small. So

$$U = V_0 + V + v.$$

Here V_0 solves (6) with $a = \sigma_0^2$ and v is quadratically small with respect to f_*, while the principal linear term V satisfies the equations

$$\begin{gathered}\frac{\partial V}{\partial \tau} - \frac{1}{2}\sigma_0^2\frac{\partial^2 V}{\partial y^2} + \left(\frac{\sigma_0^2}{2} + \mu\right)\frac{\partial V}{\partial y} + (r - \mu)V = \alpha_0 f,\\ V(y, 0) = 0, \quad y \in \mathbf{R},\\ V(y, \tau^*) = V^*(y), \quad y \in \omega,\end{gathered} \tag{10}$$

where

$$\alpha_0(y,\tau) = s^* \frac{1}{\sigma_0^2\sqrt{4\pi a\tau}} e^{-\frac{y^2}{2\tau\sigma_0^2}+cy+d\tau},$$

$$c = \frac{1}{2} + \frac{\mu}{\sigma_0^2}, \quad d = -\frac{1}{2\sigma_0^2}\left(\frac{\sigma_0^2}{2}+\mu\right)^2 + \mu - r$$

and V^* is the principal linear part of U^*. One can completely justify this linearization by using standard theory of parabolic boundary value problems (Friedman[8], Ladyzenskaja *et al* [11]).

The new substitution

$$V = e^{cy+d\tau}W \tag{11}$$

simplifies (10) to

$$\begin{aligned} \frac{\partial W}{\partial \tau} - \frac{1}{2}\sigma_0^2\frac{\partial^2 W}{\partial y^2} &= \alpha f, \quad 0<\tau<\tau^*, \quad y\in\mathbf{R}, \\ \alpha(\tau,y) &= \frac{s^*}{\sqrt{2\pi\tau}\sigma_0}e^{-\frac{y^2}{2\tau\sigma_0^2}}, \\ W(y,0) &= 0, \quad y\in\mathbf{R}, \\ W(y,\tau^*) &= W^*(y), \quad y\in\omega, \end{aligned} \tag{12}$$

with

$$W^*(y) = e^{-cy-d\tau^*}V^*(y)$$

Let us denote by Af the solution to (12) on ω : $Af(y) = W(y,\tau^*)$, $y \in \omega$.

Lemma 1 *We have*

$$Af(x) = \int_\omega B(x,y;\tau^*)f(y)dy, \quad x\in\omega, \tag{13}$$

where

$$B(x,y;\tau^*) = \frac{s^*}{\sigma_0^2\sqrt{\pi}}\int_{\frac{|x-y|+|y|}{\sigma_0\sqrt{2\tau^*}}}^{\infty} e^{-\tau^2}d\tau.$$

A proof can be obtained by using the Laplace transform with respect to τ, see Bouchouev *et al* [5]

Corollary 1 *The linearized inverse problem implies the following Fredholm integral equation*

$$f(x) - \frac{1}{2\tau^*\sigma_0^2}\int_\omega e^{-\frac{(|x-y|+|y|)^2-x^2}{2\tau^*\sigma_0^2}}(|x-y|+|y|)f(y)dy =$$

$$-\frac{\sqrt{\pi\tau^*}}{\sqrt{2}s^*}\sigma_0^3 e^{\frac{|x|^2}{2\tau^*\sigma_0^2}}\frac{\partial^2}{\partial x^2}W(x,\tau^*),\ x\in\omega \tag{14}$$

Proof: Differentiating the equation $Af = W(,\tau^*)$ on ω, using (13) and the formula $\frac{\partial}{\partial x}|x-y| = sign(x-y)$ we will have

$$-\frac{1}{\sigma_0^3\sqrt{2\pi\tau^*}}\int_\omega e^{-\frac{(|x-y|+|y|)^2}{2\tau^*\sigma_0^2}} sign(x-y)f(y)dy = \frac{\partial}{\partial x}W(x,\tau^*), \quad x\in\omega.$$

Differentiating once more and multiplying by $-\frac{\sigma_0^3\sqrt{\pi\tau^*}}{\sqrt{2}}e^{\frac{|x|^2}{2\tau^*\sigma_0^2}}$ we get the Fredholm integral equation (14).

Theorem 2 *Let $\omega = (-b,b)$. Let θ_0 be the root of the equation $2\theta - e^{-4\theta} = 3$. If*

$$\frac{b^2}{\tau^*\sigma_0^2} < \theta_0, \tag{15}$$

then a solution $f \in L^\infty(\omega)$ to the integral equation (14) and hence to the inverse option pricing problem (10) is unique.

One can check numerically that $1.5012 < \theta_0 < 1.5013$.
We remind that $\|f\|_\infty(\omega)$ is essential supermum of $|f|$ over ω.

Proof: Due to Corollary 1 to prove Theorem 1 it suffices to show uniqueness of solution f of (14), i.e. to assume that the right side is zero and conclude that $f = 0$. To do it we observe that

$$\int_\omega e^{-\frac{(|x-y|+|y|)^2}{2\tau^*\sigma_0^2}}(|x-y|+|y|)dy = e^{-\frac{x^2}{2\tau^*\sigma_0^2}}(\tau^*\sigma_0^2 + x^2)$$

$$-\frac{\tau^*\sigma_0^2}{2}(e^{-\frac{(2b-x)^2}{2\tau^*\sigma_0^2}} + e^{-\frac{(2b+x)^2}{2\tau^*\sigma_0^2}}). \tag{16}$$

This can be verified by direct calculations, using that for $0 < x$, $|x-y|+|y|$ is $2y-x$ when $x<y$, it is x when $0<y<x$, and it is $x-2y$ when $y<0$. For $x<0$, $|x-y|+|y|$ is $2y-x$ when $0<y$, it is $-x$ when $x<y<0$ and it is $x-2y$ when $y<x$.

Returning to uniqueness of f we assume that f is not zero. We can assume that $\|f\|_\infty(\omega) = f(x_0) > 0$ at some $x_0 \in [-b, b]$. From (14) at $x = x_0$ (with zero right side) we have

$$\|f\|_\infty \leq \frac{1}{2\tau^*\sigma_0^2}\int_\omega e^{-\frac{|x_0-y|+|y|)^2-x_0^2}{2\tau^*\sigma_0^2}}(|x_0-y|+|y|)\|f\|_\infty dy$$
$$= \|f\|_\infty(\frac{\tau^*\sigma_0^2+x_0^2}{2\tau^*\sigma_0^2} - \frac{1}{4}(e^{-\frac{(2b-x_0)^2-x_0^2}{2\tau^*\sigma_0^2}} + e^{-\frac{(2b+x_0)^2-x_0^2}{2\tau^*\sigma_0^2}}))$$

if we use (16).

We will show that

$$g(x) = \frac{x^2}{\tau^*\sigma_0^2} - \frac{1}{2}\left(e^{\frac{2b(x-b)}{\tau^*\ sigma_0^2}} + e^{-\frac{2b(x+b)}{\tau^*\sigma_0^2}}\right) < 1, \quad -b \leq x \leq b. \tag{17}$$

Then the previous inequality yields

$$\|f\|_\infty < \|f\|_\infty\left(\frac{1}{2}+\frac{1}{2}\right) = \|f\|_\infty,$$

and hence $\|f\|_\infty(\omega) = 0$.

To prove (17), by a careful elementary analysis (of g, g', g'') one shows that maximum of g on $\overline{\omega^*}$ is at $x = b$. Then

$$g(b) = \frac{b^2}{\tau^*\sigma_0^2} - \frac{1}{2}\left(1 + e^{-\frac{4b^2}{\tau^*\sigma_0^2}}\right) < 1,$$

if $2\frac{b^2}{\tau^*\sigma_0^2} - e^{-\frac{4b^2}{\tau^*\sigma_0^2}} < 3$. Since the function $2\theta - e^{-4\theta}$ is increasing, the last inequality holds when $\theta < \theta_0$, where $2\theta_0 - e^{-4\theta_0} = 3$.

This completes a short version of the proof.

A complete proof is given in Bouchouev *et al* [5].

By Lemma 1 the linearized inverse option pricing problem implies the integral equation

$$Af(x) = F(x),\ x \in \omega = (-b, b) \tag{18}$$

where F is a given function.

The integral equation (18), Theorem 2 and simple properties of integral operators imply the stability estimate $\|f\|_\infty(\omega) \leq C\|F''\|_\infty(\omega)$, similar to (8).

It is clear from known properties of solutions to parabolic problems (and it can be seen from the equation (14)) that $f \in L^2(\omega)$ implies that $u(,\tau^*) \in H_{(2)}(\omega)$. So the operator A maps $L^2(\omega)$ into the Sobolev space $H_{(2)}(\omega)$ and from Lemma 1 and Corollary 1 it follows that the range of A has the codimension not greater than 2 in $H_{(0)}(\omega)$. In Bouchouev *et al*[5] we show that it is exactly 2. At present we do not know an exact description of the range of A.

4 Numerical Algorithm and its Testing

Practitioners need fast and reliable algorithms.

The numerical algorithms in Avellaneda *et al*[1] and Lagnado *et al*[12] result from a regularized $L_2(\omega)$ (least -squares) matching of the market data (3) and the equations (1),(2). Since the minimized cost functional is not convex, these algorithms do not guarantee convergence and since computations of a solution to the direct problem is quite difficult in any event convergence is very slow.

In Bouchouev *et al*[3] we proposed another algorithm based on use of first two terms of fundamental solution to the Black-Scholes equation built from the well-known parametrix by a standard scheme. Practically, convergence was fast, but again this method lacks a rigorous justification, and hence it hardly can be considered as a reliable one.

In Bodurtha *et al*[6] a (formal) linearization approach was considered, but even a solution of the linearized inverse problem was relatively slow.

Based on the theory described in section 3 we designed a very fast and justified algorithm, proposing to solve numerically the integral equation (18) with a simplified (due to Lemma 1) kernel. We consider the interval $\omega = (-1, 1)$, $s^* = 20$ and we let $\mu = 0$ and $r = 0.05$. On this interval we will use uniformly distributed grid points. Observe that we are solving numerically a linear inverse problem, using the data generated by the original nonlinear problem (6). Of course, it generates data errors, due to the linearization. Otherwise, the data are (numerically) exact.

The direct problem (6) was solved numerically by the finite differences method (the Crank-Nicholson scheme with 80 grid points on the interval $(-1.5, 1.5)$ with artificial zero (Dirichlet) boundary conditions at $y = -1.5$ and $y = 1.5$).

The integral operator (18) is discretized by using standard tables for the error function $errfc$ at uniform grid points $x_1, ..., x_{54}$. The points x_j are the measurements points. Their collection coincides with the points $y_1, ..., y_{54}$.

We considered 5 examples, where we let $\sigma_0 = 1$ and we will use different observation times $\tau^* = 0.1, 0.3, 0.5$ and 0.7. As perturbations $f(y)$ of constant volatility we will take functions $f_1(y) = 0.3y$, $f_2(y) = 0.3y_2$, $f_3(y) = 0.5y^2 - 0.25y$, $f_4(y) = 0.3sin(2\pi y)$, and $f_5(y) = 0.3sin(4\pi y)$. The functions f_4, f_5 are oscillating functions and they are not typical for financial problems. We included them to test how robust is the numerical algorithm. Observe that perturbations can reach $0.3-0.7$ of the magnitude of the unperturbed constant coefficient.

The reconstruction was near perfect on the whole ω when τ^* is$0.5; 0.7$. This is with agreement with the condition (15), which in these examples simplifies to

$$0.6667 < \tau^*$$

The greater is τ^*, the better is the reconstruction on the whole interval $(-1, 1)$, so $\tau^* = 0.5$ or 0.7 correspond to the recovered volatilizes closest to the given ones. On the other hand, for smaller time $\tau^* = 0.1$ the reconstructed f starts to deteriorate near endpoints (on the intervals $(-1, -0.6)$ and $(0.7, 1)$ on figure 1). For $\tau^* = 0.3$ the deterioration is visible, but not as strong.

In Bouchouev *et al* [5] we give more details and illustrating figures.

5 Open Problems and Future Research

Uniqueness in the original (nonlinear) problem is open. Even in the linearized case it is not clear that the condition (15) is necessary for uniqueness. The inverse option pricing problem is a particular case of the more general inverse diffusion problem which has a probabilistic interpretation. We are not aware of any uniqueness results about recovery of diffusion rate from probability of distribution at a fixed moment of time. The described method can be applied at least to a linearized version of this inverse probabilistic problem.

The proposed reconstruction algorithm is expected to perform very well when volatility is not changing fast with respect to stock price s and is changing very slow with respect to time. Sudden and dramatic changes of market situations most likely can not be properly described by our model and more generally by the Black-Scholes equation. Probably, a minor modification of the proposed model (replacing $\mathbf{R}$ in (6) by a finite interval) can eliminate difficulties with existence theorem and generate even better numerical algorithms. We did not test the algorithm on real market data, but we can not see any problem with that. Observe, that to find continuous f the data F must be at least twice differentiable on ω, so the real market data are in need of a proper interpolation, minimizing the size of second derivatives of F. A choice of an appropriate smoothing interpolation and an intensive numerical testing will be a subject of future work.

For simplicity we considered only European options. We hope to adjust the linearization technique to American and more complicated options, which are in particular described by free boundary problems. So far there are actually no results in this very important practical case.

Acknowledgment

The work was in part supported by the NSF grants DMS 98-03397 and DMS 01-04029.

References

1. M. Avellaneda, C. Friedman, R. Holmes and L. Sampieri, *Calibrating volatility surfaces via relative entropy minimization*, Appl. Math. Finance **4**, 37–64(1997).
2. H. Berestycki, J. Busca and I. Florent, *An inverse parabolic problem arising in finance*, C.R. Acad. Sci. Paris **331**, 965–969 (2000).
3. I. Bouchouev and V. Isakov, *The inverse problem of option pricing*, Inverse Problems **13**, L1–L7 (1997).
4. I. Bouchouev and V. Isakov, *Unique ness, stability, and numerical methods for the inverse problem that arises in financial markets*, Inverse Problems **15**, R95–R116 (1999).
5. I. Bouchouev, V. Isakov and N. Valdivia, *Recovery of volatility coefficient by linearization*, (2001), submitted.
6. J.N. Bodurtha and M. Jermakyan, *Non-Parametric Estimation of an Implied Volatility Surface*, J. Comput. Finance **2(4)**, Summer (1999).
7. F. Black and M. Scholes, *The pricing of options and corporate liabilities*, J. Political Econ. **81**, 637–659 (1973).
8. A. Friedman, *Partial differential equations of parabolic type* (Prentice-Hall, 1964).
9. V. Isakov *Inverse Source Problems* (American Mathemathical Society, Providence, Rhode Island, 1990).
10. V. Isakov, *Inverse Problems for PDE* (Springer-Verlag, New York, 1998).
11. O.A. Ladyzenskaja, V.A. Solonnikov and N.N.Uraltseva, *Linear and quasilinear equations of parabolic type* (Academic Press, New York-London, 1969).
12. R. Lagnado and S. Osher, *A technique for calibrating derivation of the security pricing models: numerical solution of the inverse problem*, J. Comput. Finance **1**, 13–25 (1997).

AN OUTLINE OF ADAPTIVE WAVELET GALERKIN METHODS FOR TIKHONOV REGULARIZATION OF INVERSE PARABOLIC PROBLEMS

STEPHAN DAHLKE

Fachbereich Mathematik, Philipps–Universität Marburg, Germany

PETER MAAß

Zentrum für Technomathematik, Universität Bremen, Germany,
E-mail:pmaass@math.uni-bremen.de

In this paper, we discuss some ideas how adaptive wavelet schemes can be applied to the treatment of certain inverse problems. The classical Tikhonov–Phillips regularization produces a numerical scheme which consists of an inner and an outer iteration. In its normal form, the inner iteration can be interpreted as a boundedly invertible operator equation which can be handled very efficiently by using a stable wavelet basis. This general framework is illustrated by an application to the inverse heat equation.

1 Introduction

Due to its theoretical challenges and its practical importance for many industrial applications the theory of regularization methods for inverse problems has gained increasing interest in the mathematical community over the last two decades. Excellent introductions to this field can be found e.g. in [12,14,16]. In this article we aim at presenting a framework for adaptive Tikhonov regularization and its realization by adaptive wavelet methods for parabolic differential equations. Moreover, in order to highlight the main ideas we will only consider inverse problems with a linear or an affine linear operator, e.g., parameter estimation problems for heat transfer equations. Hence we consider a compact operator A between Hilbert spaces X and Y and a corresponding operator equation

$$Ax = y\ , \tag{1}$$

where x is the searched for function and y denotes perfect data, however we assume that only some observed data y^δ with a known error bound $||y - y^\delta|| \leq \delta$ is given.

Tikhonov–Phillips regularization of such an ill–posed problem is achieved by replacing the linear equation (1) by the minimization problem

$$\begin{aligned} &\textit{find } x_\alpha^\delta \in X \textit{ which minimizes} \\ &T_\alpha(x) \;=\; \| Ax - y^\delta \|_Y^2 \;+\; \alpha \, \|x\|_X^2 \;. \end{aligned} \tag{2}$$

The idea of Tikhonov–Phillips regularization (2) is to control the influence of the data error in the regularized solution x_α^δ by adding a penalty term. The unique minimizer of (2) is given as the unique solution of the regularized normal equation

$$(A^*A \,+\, \alpha\, I)\, x_\alpha^\delta \;=\; A^* y^\delta \;. \tag{3}$$

Early results on the convergence of Tikhonov regularization methods were usually entirely based in function spaces, the additional influence of an appropriate discretization of the operator was hardly mentioned. For some exceptions see, e.g., [19,20,21]. However, any numerical scheme for solving inverse problems by Tikhonov regularization depends on at least two parameters (regularization parameter α, a parameter determining the discretization of the operator) and a stopping rule. Characterizing a numerical scheme for operator equations as adaptive usually refers to a nonlinear dependence of these ingredients on the given data y^δ. In this sense, any a posteriori stopping rule leads to an adaptive scheme. In this paper, we address adaptive schemes in a stronger sense: we analyze methods where the regularization parameter and the discretization spaces depend on the unknown solution and are chosen adaptively during the solution procedure without using a priori information. More precisely, we will consider the following framework for Tikhonov regularization:

- given data: $A, y^\delta, \delta, 0 < q < 1, \alpha_0$;
- outer iteration for determining the regularization parameter: choose iteratively $\alpha_n = q^n \alpha_0$, for each α_n determine a critical level of approximation $\epsilon = \epsilon(\alpha_n, \delta, y^\delta)$ for the solution. This parameter has to be chosen, such that the over all scheme realizes optimal convergence rates;
- inner iteration for determining the minimizer $x_{\alpha,\Lambda_\epsilon}^\delta$ of (3): $x_{\alpha,\Lambda_\epsilon}^\delta$ will be determined by suitable wavelet Galerkin approximations of the forward operator A^*A, these wavelet approximations will be chosen adaptively by using local a posteriori error estimates and an appropriate refinement strategy.

The paper is organized as follows. Section 2 contains the description of a model problem, which describes a parameter estimation problem for a heat equation. Section 3 deals with the approximation requirements of the outer iteration and the resulting adaptive approximation levels $\epsilon = \epsilon(\alpha, \delta, y^\delta)$. Finally Section 4 analyzes how to construct an adaptive wavelet Galerkin method which realizes the required levels of approximation.

2 A model problem

In this paper, we just aim at outlining a general approach for adaptive Tikhohonov regularization via wavelet discretizations. Hence we will not present any numerical results. However, in order to focus our ideas we will introduce a simple model problem, which serves as motivation for the subsequent sections. We do not present any new results in this section, to the contrary the content is rather classical and elementary, see, e.g., [23,25]. Since we want to merge results from inverse problems and wavelet analysis, which have developed some conflicting notations and which sometimes even give different meanings to the same expressions, we would like to introduce some basic concepts in detail.

We consider inverse heat problems, the underlying differential equation is hence given by

$$u_t = \operatorname{div}\{\sigma \nabla u\}$$

on $x \in \Omega$, $t \in [0,T]$, where $\Omega \subset \mathbb{R}^2$ denotes a region with piecewise smooth boundary $\Gamma = \partial\Omega$.

The construction of wavelet Galerkin methods and their convergence properties have only recently been analyzed successfully, these results will be described in Section 4.

The inverse problems we consider will differ in terms of the given and/or the measured data: initial data $\mu = u(\cdot, 0)$; boundary data $a(x,t) = u(x,t)$ for $x \in \Gamma$, $t \in [0,T]$; observation at a fixed time instant $g(x) = u(x,T)$, observation on an interior region $b(x,t) = u(x,t)$ for $x \in \tilde{\Omega} \subset \Omega$, $t \in [0,T]$.

Let us first consider the standard inverse heat problem:

$$\text{given data: } a, g \text{ , searched for quantity: } \mu \text{ .}$$

For this model problem the forward operator $A = A(\mu)$ is defined as follows: For a fixed a let L denote the solution operator of the parabolic problem

$$u_t = \operatorname{div}\{\sigma \nabla u\} \text{ for } x \in \Omega$$

with initial data μ and boundary values a, i.e.,

$$L(\mu)(x,t) = u(x,t) \text{ for } x \in \Omega,\ t \in [0,T] \text{ .}$$

Then

$$A(\mu)(x) = L(\mu)(x,T) \quad , \tag{4}$$

which leads to the formal description of the operator equation for the inverse problem

$$A(\mu) = g \ .$$

In order to allow the modelling of measurement error, A is considered as a mapping from $L_2(\Omega) \longrightarrow L_2(\Omega)$.

For non-zero boundary data a, the operator A is nonlinear. However, introducing $u^{\#}$ and $g^{\#} = u^{\#}(\cdot, T)$, where $u^{\#}$ denotes the solution with zero initial and non-zero boundary data, i.e.,

$$u_t = \mathrm{div}\{\sigma \nabla u\} \text{ for } x \in \Omega \ \ , u(\cdot, 0) = 0 \ , \ a(x,t) = u(x,t) \text{ for } x \in \Gamma, \ t \in [0,T] \ ,$$

leads to an affine decomposition

$$A(\mu) = \tilde{A}\mu + g^{\#} \ ,$$

where $\tilde{A}$ is the linear operator, which solves

$$u_t = \mathrm{div}\{\sigma \nabla u\} \text{ for } x \in \Omega \ \ , \ u(\cdot, 0) = \mu \ , \ 0 = u(x,t) \text{ for } x \in \Gamma, \ t \in [0,T] \ ,$$

and restricts the solution to its values at time T. Hence by combining the originally measured data g with the particular solution $g^{\#}$ via

$$\tilde{g} = g - g^{\#}$$

leads to a linear inverse problem $\tilde{A}\mu = \tilde{g}$.

A similar affine decomposition also holds for the inverse problem posed by

$$\text{given data: } b, \text{ searched for quantities: } (\mu, a) \ .$$

In all these cases including many variations, we are finally lead to consider an exponentially ill-posed linear operator equation.

3 A framework for adaptive Tikhonov regularization

We consider Tikhonov regularization for solving a linear operator equation (1), i.e., we consider

$$x_\alpha^\delta = (A^*A + \alpha I)^{-1} A^* y^\delta \quad , \tag{5}$$

where $\|y - y^\delta\| \leq \delta$ and A is a compact operator between Hilbert spaces X, Y

$$A : X \to Y \quad .$$

Now let us incorporate an adaptive Galerkin discretization of $(A^*A + \alpha I)$ in (5). I.e., we fix an approximation tolerance ϵ and construct an index set Λ_ϵ such that the corresponding approximate solution $x^\delta_{\alpha,\Lambda_\epsilon}$ satisfies a guaranteed error estimate

$$||x^\delta_\alpha - x^\delta_{\alpha,\Lambda_\epsilon}|| \leq \text{const.}\ \frac{\epsilon}{\sqrt{\alpha}}\ . \tag{6}$$

An adaptive scheme, which realizes this condition will be described in Section 4.

The choice of α and ϵ determines the approximation properties of $x^\delta_{\alpha,\Lambda_\epsilon}$. So far we have discussed the solution of (2) for a fixed value of α. Let us now discuss how to determine a suitable value of α. We will choose α according to a discrepancy principle of the form (or some modification thereof)

$$\|Ax^\delta_{\alpha,\Lambda_\epsilon} - y^\delta\| = \tau\delta + \sigma\epsilon, \tag{7}$$

where $\tau > 1$ and σ sufficiently large, for a precise statement see Theorem 3.1. This still describes an idealized situation: in practice one never aims at solving (7) precisely, one rather chooses α from a sequence of test parameters and determines $\alpha_N \in \{\alpha_n = q^n\alpha_0 |\ n = 0, 1, 2, ...\}$, for a fixed $0 < q < 1$ by requiring

$$\|Ax^\delta_{\alpha_N,\Lambda_\epsilon} - y^\delta\| \leq \tau\delta + \sigma\epsilon, \tag{8}$$

$$\|Ax^\delta_{\alpha_n,\Lambda_\epsilon} - y^\delta\| > \tau\delta + \sigma\epsilon \quad \text{for} \quad n < N\ . \tag{9}$$

Hence the overall algorithm for computing $x^\delta_{\alpha,\Lambda_\epsilon}$ requires to solve $(N+1)$ operator equations of type (5).

Of course the number of iterations N is a priori unknown. Thus an efficient procedure for obtaining sparse approximations of $(A^*A+\alpha I)$ in connection with a reliable strategy for selecting the approximation level ϵ will greatly reduce the numerical cost of the algorithm. Our main objective in this section is to determine an approximation level $\epsilon(\delta, \alpha)$ such that $x^\delta_{\alpha,\Lambda_\epsilon}$ exhibits optimal convergence rates. Note that the approximation level $\epsilon(\delta, \alpha)$ may change with α during the search process for the optimal regularization parameter α_N. This will later be used to choose coarser approximations for larger values of α.

As usual we assume that the generalized solution x^+ lies in the range of $(A^*A)^\nu$, that is,

$$x^+ = (A^*A)^\nu v, \quad \|v\| \leq \varrho\ . \tag{10}$$

Moreover we restrict ourselves to smoothness assumptions of the order

$$0 < \nu \leq \frac{1}{2}\ ,$$

since – for this choice of discrepancy principle – higher order regularity of x^+ does not further improve the convergence rate of $\|x^\delta_{\alpha,\Lambda_\epsilon} - x^+\|$. This is consistent with the theory of a posteriori parameter selection for classical Tikhonov regularization since – even when using the exact operator A – applying a discrepancy functional of type (7) limits optimal convergence rates to the range $0 < \nu \leq 1/2$. However, it should be mentioned that alternative discrepancy principles lead to optimal convergence rates for a larger range of smoothness assumptions, i.e. for $\nu \leq 1$, see [12].

To avoid unnecessary notation we furthermore assume that

$$\overline{\text{range}\,(A)} = Y, \quad \|y^\delta\| > \delta, \quad \|A\| \leq 1 \ . \tag{11}$$

The starting for this investigation is a basic estimate which reveals the three error contributions in estimating $\|x^\delta_{\alpha,\Lambda_\epsilon} - x^+\|$. This result is a small adaptation of previously published standard estimates, see, e.g.,[19,21].

Lemma 3.1 *Let x^+ be the generalized solution of $Ax = y$ and let $x^\delta_{\alpha,\Lambda_\epsilon}$ be defined by the discretized version of* (5). *Assume that $\|y - y^\delta\| \leq \delta$ and that x^+ obeys* (10). *Then,*

$$\|x^\delta_{\alpha,\Lambda_\epsilon} - x^+\| \leq \frac{\delta}{2\sqrt{\alpha}} + \frac{\epsilon\|x^+\|}{\sqrt{\alpha}} + \alpha^\nu c_{\nu,\alpha}(v)$$

where

$$c^2_{\nu,\alpha}(v) = \sum_{n\geq 0} \left\{ \frac{\alpha^{1-\nu}\sigma_n^{2\nu}}{(\sigma_n^2 + \alpha)} \langle v, u_n \rangle \right\}^2 \leq \{(1-\nu)^{1-\nu}\nu^\nu \varrho\}^2 \ .$$

In connection with the modified discrepancy principle (8) this result gives an optimal convergence rate.

Theorem 3.1 *If $\epsilon = O(\delta^p \alpha^q)$, with $0 < p, q,\ p + q = 1$, and if α is chosen by the modified discrepancy principle* (8) *with $\tau > 2/q$, $\sigma > 9\|x^+\|/4q$, then*

$$\|x^\delta_{\alpha,\Lambda_\epsilon} - x^+\| = O(\delta^{2\nu/(2\nu+1)}) \ .$$

The above theorem shows that we can e.g. choose $p = q = 1/2$ and still obtain optimal convergence rates. Such a choice is preferable for large values of α which is the case in the beginning of our iterative search for the optimal regularization parameter.

Optimal convergence rates cannot be achieved in general if $p + q < 1$.

4 Wavelet Galerkin methods for operator equations

In recent years, much effort has been spent to design efficient numerical schemes based on wavelets. The most far-reaching results were obtained for operator equations of the form

$$\mathcal{A}u = f^\delta, \tag{12}$$

where $\mathcal{A} : H \to H'$ is a linear operator from a Hilbert space H into its normed dual H'. In our applications, H will typically be a Sobolev space H^t on some domain $\Omega \subset \mathbb{R}^d$ or on a closed manifold. We assume that $\mathcal{A}$ is *boundedly invertible* so that

$$\|\mathcal{A}v\|_{H'} \sim \|v\|_H, \ v \in H \tag{13}$$

holds. This setting fits perfectly to the normal equation (5) arising in the inner iteration, i.e., to the problem

$$x_\alpha^\delta = (A^*A + \alpha I)^{-1} A^* y^\delta \quad , \tag{14}$$

since, as already stated above, $\mathcal{A} = (A^*A + \alpha I)$ is boundedly invertible on $L_2(\Omega)$. More precisely, the operator norm of $\mathcal{A}^{-1}$ is bounded by $||\mathcal{A}^{-1}|| \leq \alpha^{-1}$. The right hand side $f^\delta = A^*y^\delta$ satisfies $||f^\delta - f|| \leq ||A^*|| \ \delta$.

Before we discuss later on the specific problems arising in the numerical treatment of (14), let us briefly recall the basic numerical concepts. We are especially interested in *adaptive* schemes, and we shall focus on numerical algorithms based on *wavelets*, i.e., the basis functions are taken from a family $\Psi = \{\psi_\lambda, \lambda \in \mathcal{J}\}$ satisfying the following fundamental assumptions:

- Ψ induces *norm equivalences* for a whole scale of Sobolev spaces, $\| \sum_{\lambda\in\mathcal{J}} d_\lambda \psi_\lambda \|_{H^s} \sim (\sum_{\lambda\in\mathcal{J}} 2^{2|\lambda|s} |d_\lambda|^2)^{1/2}, \ s_0 \leq s \leq s_1$;
- ψ_λ possesses the *cancellation property* $|\langle v, \psi_\lambda\rangle| \lesssim 2^{-|\lambda|m} |v|_{H^m(\mathrm{supp}\psi_\lambda)}$;
- the wavelets are *local* in the sense that $\mathrm{diam}(\mathrm{supp}\psi_\lambda) \sim 2^{-|\lambda|}, \lambda \in \mathcal{J}$.

Nowadays, several constructions of bases satisfying these assumptions are available [4,7,8,9]. Our goal is to develop a suitable Galerkin scheme to approximate the solution of (14). Therefore we consider subspaces of the form

$$S_\Lambda := \{\psi_\lambda : \ \lambda \in \Lambda\}, \qquad \Lambda \subset J, \tag{15}$$

and project our problem onto these spaces, i.e., the Galerkin approximation u_Λ is defined by

$$\langle \mathcal{A}u_\Lambda^\delta, v\rangle = \langle f^\delta, v\rangle \ , \qquad v \in S_\Lambda. \tag{16}$$

In an adaptive scheme, the goal is always to find a *possibly small* set $\Lambda \subset J$ such that the actual error is below some given tolerance. In principle, such a scheme consists of the following three steps:

- compute the current Galerkin approximation u_Λ;
- estimate the error $\|u^\delta - u^\delta_\Lambda\|$ in some suitable norm, with $u^\delta = \mathcal{A}^{-1}f^\delta$;
- add wavelets if necessary which yields a new index set $\hat{\Lambda}$.

For the second step, one clearly needs an *a posteriori error estimator* since the exact solution u is unknown, and for the third step one has to develop a suitable *refinement strategy* so that the whole algorithm converges. In the wavelet setting, an error estimator can be easily constructed by employing assumption (13), norm equivalences, and Galerkin orthogonality:

$$\|u - u^\delta_\Lambda\| \le \|u - u^\delta\| + \|u^\delta - u^\delta_\Lambda\| \quad ,$$

where the first term is controlled by the Tikhonov regularization and the second term gives rise to the error estimator via

$$\begin{aligned} \|u^\delta - u^\delta_\Lambda\|_{H^t} &\sim \|\mathcal{A}(u^\delta - u^\delta_\Lambda)\|_{H^{-t}} \sim \|f^\delta - \mathcal{A}u^\delta_\Lambda\|_{H^{-t}} \\ &= \|r_\Lambda\|_{H^{-t}} \sim \left(\sum_{\mathcal{J}\backslash\Lambda} 2^{-2t|\lambda|}|\langle r_\Lambda, \psi_\lambda\rangle|^2\right)^{1/2} . \end{aligned} \tag{17}$$

In our example for the inverse heat problem we have $\mathcal{A} : L_2(\Omega) \to L_2(\Omega)$, i.e. $t = 0$. From (17), we observe that the current error can be estimated by computing the wavelet coefficients of the *residual* $r_\Lambda = f^\delta - \mathcal{A}u^\delta_\Lambda$. Intuitively, the *residual weights* $\rho_\lambda := 2^{-t|\lambda|}|\langle r_\Lambda, \psi_\lambda\rangle|$ serve as local error indicators. Therefore a suitable refinement strategy can be derived by adding those wavelets which produce large entries in the expansion of the residual, i.e., we define the new index set $\hat{\Lambda}$ in such a way that

$$\left(\sum_{\lambda\in\hat{\Lambda}\backslash\Lambda} 2^{-2t|\lambda|}|\langle r_\Lambda, \psi_\lambda\rangle|^2\right)^{1/2} \ge \beta \left(\sum_{\lambda\in\mathcal{J}\backslash\Lambda} 2^{-2t|\lambda|}|\langle r_\Lambda, \psi_\lambda\rangle|^2\right)^{1/2} \tag{18}$$

for some suitable parameter β. However, this strategy is not directly numerically realizable since catching the bulk of the residual requires knowing *all* its wavelet coefficients. Nevertheless, in [6], it was shown that a judicious variant of this idea exploiting the cancellation property of wavelets indeed leads to an implementable and convergent algorithm, i.e., given a tolerance ϵ, the adaptive scheme produces a final index set $\tilde{\Lambda}_\epsilon$ such that

$$\|u^\delta - u^\delta_{\tilde{\Lambda}_\epsilon}\| \le \epsilon \tag{19}$$

by using only information on the given data. Moreover, in [5], subtle generalizations have been derived which yield asymptotically optimal schemes in the sense that (within a certain range) the convergence rate of best N–term approximation is achieved at a computational expense which stays proportional to the number $N = |\Lambda_\epsilon|$ of degrees of freedom. Furthermore, in [1], a first efficient numerical realization is documented.

As already stated above, we suggest to use this strategy for the numerical treatment of the basic problem (14),

$$x_\alpha^\delta = (A^*A + \alpha I)^{-1} A^* y^\delta \quad . \tag{20}$$

Clearly this problem fits perfectly into the framework described above. However, as explained in detail in [5,6], the design of an implementable refinement strategy requires some *compressibility* properties of the underlying operator. For the special operators considered here, this issue will be further analyzed in the near future. Moreover, for an efficient implementation, the problem remains how to compute the entries of the associated stiffness matrix

$$(\mathcal{A}_\Lambda)_{\lambda,\lambda'} := \langle \mathcal{A}\psi_{\lambda'}, \psi_\lambda \rangle = \langle A\psi_{\lambda'}, A\psi_\lambda \rangle + \alpha \langle \psi_{\lambda'}, \psi_\lambda \rangle \tag{21}$$

and of the right–hand side

$$(A^* y^\delta)_\lambda = \langle y^\delta, A\psi_\lambda \rangle. \tag{22}$$

Fortunately, the adjoint operator A^* is not needed, but nevertheless the task is nontrivial since the operator A is induced by the forward problem (4), i.e., it is given as a parabolic equation. We intend to solve this problem with another fully adaptive scheme as we shall now explain. Following the basic investigations in [2,3], we treat our parabolic equation as an abstract Cauchy problem

$$\begin{aligned} u'(t) + \mathcal{B}u(t) &= 0, \quad t \in (0, T], \\ u(0) &= u_0. \end{aligned} \tag{23}$$

Usually, this problem is treated by the *method of lines.* Discretization in space first leads to a block system of ordinary differential equations. However, as already outlined in [2,3], for an adaptive approach the other discretization sequence, *first time then space*, which is classically known as the *method of Rothe* [24] seems to be preferable. Then (23) is viewed as an ordinary differential equation in some suitable Hilbert space which, due to stability reasons, is solved by an implicit scheme with time-step control. Then, in each step, a certain elliptic subproblem has to be solved. However, since these subproblems are boundedly invertible in the sense of (13), they can again be efficiently discretized by employing the well–known adaptive wavelet algorithm. Clearly, the convergence and efficiency of this strategy has to be analyzed in detail. This will be performed in the near future.

Acknowledgments

This research was partially supported by the Deutsche Forschungsgemeinschaft (DFG), Grant Da360/4–1 and the Bundesministerium für Bildung, Wissenschaft, Forschung und Technologie under grant number BMBF-03-MSM1HB.

References

1. A. Barinka, T. Barsch, P. Charton, A. Cohen, S. Dahlke, W. Dahmen, and K. Urban, Adaptive wavelet schemes for elliptic problems – Implementation and numerical experiments, *SIAM J. Scientific Comp.* **23(3)**, 910–939 (2001).
2. F. Bornemann, An adaptive multilevel approach to parabolic equations I. General theory and 1D implementations, *Impact Comput. Sci. Engrg.* **2**, 279–317 (1990).
3. F. Bornemann, An adaptive multilevel approach to parabolic equations II. Variable–order time discretization based on a multiplicative error correction, *Impact Comput. Sci. Engrg.* **3**, 93–122 (1991).
4. C. Canuto, A. Tabacco, and K. Urban, The wavelet element method, part II: Realization and additional features in 2d and 3d, *Appl. Comp. Harm. Anal.* **8**, 123-165 (2000).
5. A. Cohen, W. Dahmen, and R. DeVore, Adaptive wavelet methods for elliptic operator equations – Convergence rates, *Math. Comp.* **70**, 22–75 (2001).
6. S. Dahlke, W. Dahmen, R. Hochmuth, and R. Schneider, Stable multiscale bases and local error estimation for elliptic problems, *Appl. Numer. Math.* **23**, 21–47 (1997).
7. W. Dahmen and R. Schneider, Composite wavelet bases for operator equations, *Math. Comput.* **68**, 1533–1567 (1999).
8. W. Dahmen and R. Schneider, Wavelets on manifolds I: Construction and domain decomposition, *SIAM J. Math. Anal.* **31**, 184-230 (1999).
9. W. Dahmen and R. Schneider, Wavelets with complementary boundary conditions — function spaces on the cube, *Res. in Math.* **34**, 255–293 (1998).
10. V. Dicken and P. Maaß, Wavelet–Galerkin methods for ill–posed problems, *J. Inv. and Ill-posed Probl.* **4(3)**, 203–222 (1996).
11. H.W. Engl, Discrepancy principles for Tikhonov regularization of ill–posed problems leading to optimal convergence rates, *J. Opti. Theory Appl.* **52**, 209–215 (1987).

12. H.W. Engl, M. Hanke, and A. Neubauer, *Regularization of Inverse Problems*, Kluwer, Boston, (1996).
13. H. Gfrerer, An a posteriori parameter choice for ordinary and iterated Tikhonov regularization of ill–posed problems leading to optimal convergence rates, *Math. Comp.* **49**, 507–522 (1987).
14. C.W. Groetsch, *The Theory of Tikhonov Regularization for Fredholm Equations of the First Kind*, Pitman, Boston (1984).
15. J.T. King and A. Neubauer, A variant of finite-dimensional Tikhonov regularization with a–posteriori parameter choice, *Computing* **40**, 91–109 (1988).
16. A.K. Louis, *Inverse und schlechtgestellte Probleme*, Teubner, Stuttgart (1989).
17. A.K. Louis, P. Maaß, and A. Rieder, *Wavelets – Theorie und Anwendungen*, Teubner, Stuttgart (1994). English version: Wiley, Chichester.
18. P. Maaß and R. Ramlau, Wavelet accelerated regularization methods for hyperthermia treatment planning, *Int. J. Imag. Sys. and Tech.*, **7**, 191–199 (1996).
19. P. Maaß and A. Rieder, Wavelet–accelerated Tikhonov–regularisation with applications, in "Inverse Problems in Medical Imaging and Nondestructive Testing", eds. H.W. Engl, A.K. Louis, and W. Rundell, Springer, Wien, New York, pp. 134–159 (1997).
20. A. Neubauer, An a posteriori parameter selection choice for Tikhonov regularization in Hilbert scales leading to optimal convergence rates, *SIAM J. Numer. Anal.* **25**, 1313–1326 (1988).
21. A. Neubauer, An a posteriori parameter choice for Tikhonov regularization in the presence of modeling error, *Appl. Num. Math.* **4**, 507–519 (1988).
22. S.V. Pereverzev, Optimization of projection methods for solving ill–posed problems, *Computing* **55** (1995).
23. J. Reinhardt, On a sideways parabolic equation, *Inverse Problems* **13**, 297–309 (1997).
24. E. Rothe, Zweidimensionale parabolische Randwertaufgabe als Grenzfall eindimensionaler Randwertaufgaben, *Math. Ann.* **102**, 650–670 (1930).
25. M. Yamamoto and J. Zou, Simultaneous reconstruction of the initial temperature and heat radiation coefficient, *Inverse Problems* **17**, 1181–1202 (2001).

ESTIMATION OF DISCONTINUOUS SOLUTIONS OF ILL-POSED PROBLEMS BY REGULARIZATION FOR SURFACE REPRESENTATIONS: NUMERICAL REALIZATION VIA MOVING GRIDS

ANDREAS NEUBAUER

Institut für Industrie mathematik, Johannes-Kepler-Universität, A-4040 Linz, Austria

E-mail: neubauer@indmath.uni-linz.ac.at

In this paper we discuss the numerical realization of a new regularization method, *regularization for surface representations*, which is well-suited for ill-posed problems with discontinuous solutions: this realization is essentially based on *moving grids*. After describing the method we present several numerical examples showing that this combination with moving grids is a powerful tool to identify discontinuities in two-dimensional problems.

1 Introduction

In this paper we study the estimation of discontinuous solutions of linear or nonlinear ill-posed problems

$$F(f) = g \tag{1}$$

from noisy measurements g^δ of g satisfying $\|g^\delta - g\| \leq \delta$, where

$$F : \mathcal{D}(F)(\subset X) \to Y$$

and X and Y are Hilbert spaces.

Tikhonov regularization is well known to stabilize ill-posed problems [3]. In this method an exact solution of (1) is approximated by a minimizer f_α^δ of the functional

$$\|F(f) - g^\delta\|^2 + \alpha p(f - f^*)\,, \tag{2}$$

where f^* is an initial guess of the exact solution and $p(\cdot)$ is a properly chosen penalty term. Usually one uses the penalty term

$$p(f - f^*) = \|f - f^*\|^2\,. \tag{3}$$

However, this is not appropriate for ill-posed problems with discontinuous solutions, since it has a smoothing effect on the regularized solutions.

Using the bounded variation norm in (2) as penalty term has turned out to be an effective regularization method [1,2,10]. A major drawback of this approach, however, is that this norm is not differentiable.

Neubauer and Scherzer [9] introduced a new approach for regularizing problems with discontinuous solutions, *regularization for curve representations.* The essence of this method is to replace a discontinuous function by its continuous graph. This allows a combination with the usual Tikhonov regularization in Hilbert spaces ((2), (3)) and, therefore, all the results about convergence from the general theory on nonlinear Tikhonov regularization are applicable. The method was successfully applied to one-dimensional parameter estimation problems by Kindermann and Neubauer [5].

Kindermann and Neubauer [7] generalized the method to two-dimensional problems, *regularization for surface representations*, and applied it to the linear problem of deblurring images. The idea of this method is as follows:

Let f be a C^1-function, then its graph $G_f := \{(x, y, f(x, y)) \in \mathbb{R}^3\}$ defines a surface in $\mathbb{R}^3$. Let $(a(u,v), b(u,v), c(u,v))$ be an equivalent parameterization of G_f, then f can be recovered by $f(x,y) = c((a,b)^{-1}(x,y))$. It was shown that discontinuous functions f that are in a certain subset of the functions of bounded variation may be parameterized by smooth parameterizations. To guarantee that the parameterized surface may be interpreted as the graph of a function, we restricted ourselves to parameterizations

$$a(u,v) = a(u) \qquad\qquad b(u,v) = b(v) \tag{4}$$

satisfying $(a, b, c) \in \mathcal{D}$ with

$$\begin{aligned} \mathcal{D} &:= \{(a,b,c) \in X \,:\, a(1) = 1 = b(1),\ \dot{a} \geq 0 \text{ a.e.}, \dot{b} \geq 0 \text{ a.e.}\}\,, \\ X &:= \{(a,b,c) \in H^1[0,1] \times H^1[0,1] \times H^1_0(\Omega) \,:\, a(0) = 0 = b(0)\}\,. \end{aligned} \tag{5}$$

This method, which was successfully applied to two-dimensional parameter estimation problems [6], is still quite restrictive: due to the special choice of surface parameterizations, discontinuities in solutions $f = c(a^{-1}(x), b^{-1}(y))$ are only allowed to occur on lines parallel either to the x- or the y-axis.

The general case of parameterizations was treated from a theoretical point of view in the PhD-Thesis of Stefan Kindermann [4]. He showed that a general parameterization, $f((a(u,v), b(u,v)) = c(u,v)$, exists for all functions of bounded variation with compact support. He also gave conditions on F that guarantee that the Tikhonov regularized solutions converge to the exact one.

Let us consider the two-dimensional linear integral equation

$$g(s,t) = Ff(s,t) := \int_\Omega k(s,t,x,y) f(x,y)\, d(x,y)\,, \quad (s,t) \in \Omega\,, \tag{6}$$

where $\Omega = [0,1]^2 \subset \mathbb{R}^2$ is the unit square and $k \in L^2(\Omega^2)$. Note that $F : L^2(\Omega) \to L^2(\Omega)$ is compact. The reformulation of this equation in terms of (a, b, c) yields

$$g(s,t) = G(a,b,c)(s,t) := \int_{\Omega} k(s,t,a(u,v),b(u,v))c(u,v)|J(a,b)(u,v)|\, d(u,v) \tag{7}$$

with

$$J(a,b)(u,v) = a_u(u,v)b_v(u,v) - a_v(u,v)b_u(u,v)\,.$$

This is now a nonlinear integral equation with respect to a and b. G is at least well defined for those $a, b, c \in H^1(\Omega)$ that are parameterizations of functions of bounded variation with compact support [4].

This problem is stabilized via nonlinear Tikhonov regularization, where we can use the standard penalty term, i.e., we are looking for minimizers $(a_\alpha^\delta, b_\alpha^\delta, c_\alpha^\delta)$ of the functional

$$\|G(a,b,c) - g^\delta\|_{L^2}^2 + \alpha(\|\nabla a\|_{L^2}^2 + \|\nabla b\|_{L^2}^2 + \|\nabla c\|_{L^2}^2)\,. \tag{8}$$

For $(a, b) : \Omega \to \Omega$ to be admissible in the sense that the parameterized surface may be interpreted as the graph of a function, it is necessary that (a, b) is one to one and onto. Besides some boundary conditions this means that $J(a, b)$ is not allowed to change sign (except for a set of measure zero). For the special parameterization (4) this was guaranteed through the conditions in (5). These conditions were also easy to check in the numerical realization. If, in the general case, we used bilinear finite element functions for the approximation of a, b, and c, (8) would turn into a nonlinear minimization problem with respect to the coefficients of a, b, and c governed by nonlinear constraints in each node of the finite element grid of Ω. This is much too involved.

Therefore, we will present a numerical realization based on moving grids which is much faster and yields excellent results.

The method of moving grids is described in the next section. In Section 3 we combine it with regularization for surface representations. Finally, numerical results are presented in Section 4.

2 Moving Grids: The Deformation Method

Moving grids have been developed for the numerical solution of time dependent partial differential equations. A drawback of using fixed grids occurs when the solution of the PDE exhibits large variations due to, for example, shock waves or moving fronts. Due to its static feature the fixed grid is unable to efficiently and accurately resolve such variations. One can improve this by using adaptive grids. The idea is to generate the grids such that nodes will be concentrated in regions where the solution changes rapidly and fewer grid points are used in regions where negligible changes occur.

There are essentially two strategies for grid adaption: *local refinement* and *moving grids*. In local refinement nodes are inserted where and when they are needed. This method is flexible and easy to conform the boundary. However, the solver and data structure have to be modified after insertion or deletion of nodes. For moving grids the total number of nodes and the connectivity between them is fixed. The nodes are redistributed where and when they are needed. There are different methods to move the nodes around. We only describe the so called *traditional deformation method*, which is based on a result from differential geometry. It provides direct control over the cell size of the adaptive grid and the node velocities are directly determined. It turned out that it is also efficient for the numerical calculation of discontinuous solutions of ill-posed problems in combination with regularization for surface representations. The description follows the lines in Liao *et al* [8].

Suppose that u satisfies

$$u_t = L(u)\,, \tag{9}$$

where L is a differential operator defined on a physical domain $\Omega \subset I\!R^2$. The idea is now to construct a transformation $\phi : \tilde{\Omega} \times [0,T] \to \Omega$ which moves a fixed number of grid points on Ω to adapt to the numerical solution. Of course ϕ must be one to one and onto. It is well known that if the Jacobian determinant of a transformation ϕ is positive in $\tilde{\Omega}$, then ϕ is one to one in all of $\tilde{\Omega}$. This ensures that the grid will not fold onto itself. Therefore, the deformation method constructs ϕ such that $\det \nabla\phi = m(\phi, t)$, where m is a positive monitor function. This assures precise control over the cell size relative to the fixed initial grid.

Suppose that the solution of (9) has been computed at time step $t = t_{k-1}$ and a preliminary computation has been done at time level $t = t_k$. Assume that we have some positive error estimator $e(\eta, t)$ at the time steps t_{k-1} and t_k. Let us define the monitor function

$$m(\eta, t) = \frac{\gamma(t)}{e(\eta, t)}\,,$$

where $\gamma(t)$ is a positive scaling parameter so that

$$\int_\Omega \frac{1}{m(\eta, t_k)} d\eta = |\tilde{\Omega}| \tag{10}$$

holds. Note that m is small in regions where the error is large and large in regions where the error is small. We then seek a transformation

$\phi : \tilde{\Omega} \times [t_{k-1}, t_k] \to \Omega$ such that

$$\begin{aligned} \det \nabla\phi(\xi, t) &= m(\phi(\xi,t),t)\,,\ \xi \in \tilde{\Omega}\,,\ t_{k-1} \le t \le t_k\,, \\ \phi(\xi, t_{k-1}) &= \phi_{k-1}(\xi)\,, \qquad \xi \in \tilde{\Omega}\,, \end{aligned} \tag{11}$$

where ξ is a grid node of an initial grid and $\phi_{k-1}(\xi)$ represents the coordinates of the node at $t = t_{k-1}$. In addition we specify that $\phi(\xi) \in \partial\Omega$ for all $\xi \in \partial\tilde{\Omega}$. Note that (10) is necessary for (11) to be true and that (11) ensures that the size of the transformed cells will be proportional to m, i.e., the grid will be appropriately condensed in regions of high errors and stretched in regions of small errors.

The solution of (11) can be done efficiently via the following two steps: the first step is to find a vector field $\rho(\eta, t)$ satisfying

$$\begin{aligned} \operatorname{div} \rho(\eta, t) &= -\frac{\partial}{\partial t}\left(\frac{1}{m(\eta,t)}\right), \eta \in \Omega\,,\ t_{k-1} \le t \le t_k\,, \\ \langle\, \rho(\eta,t), n(\eta)\,\rangle &= 0\,, \qquad\qquad \eta \in \partial\Omega\,,\ t_{k-1} \le t \le t_k\,, \end{aligned}$$

where n denotes the outward normal to $\partial\Omega$. The vector field ρ can be calculated by solving for w in the scalar Poisson equation (for fixed t)

$$\Delta w(\eta, t) = -\frac{\partial}{\partial t}\left(\frac{1}{m(\eta,t)}\right), \ \eta \in \Omega\,, \qquad \frac{\partial w}{\partial n} = 0\,,\ \eta \in \partial\Omega\,, \tag{12}$$

and then setting $\rho = \nabla w$. Note that the constraint

$$\int_\Omega \frac{\partial}{\partial t}\left(\frac{1}{m(\eta,t)}\right) d\eta = 0$$

for the Neumann boundary condition is satisfied due to (10).

The second step is to solve for $\phi(\xi, t)$ from the ODE system (for fixed $\xi \in \tilde{\Omega}$)

$$\begin{aligned} \frac{d}{dt}\phi(\xi,t) &= m(\phi(\xi,t),t)\nabla w(\phi(\xi,t),t)\,,\ \ t_{k-1} \le t \le t_k\,, \\ \phi(\xi, t_{k-1}) &= \phi_{k-1}(\xi)\,. \end{aligned} \tag{13}$$

It was shown in Semper and Liao [11] that this two step procedure actually yields the solution of (11).

3 Numerical Realization

In this section we want to show how a combination of regularization for surface representations with moving grids for adapting the finite element grid yields a fast and effective method to identify discontinuities in two-dimensional problems.

From our numerical calculations for the special parameterization (4) we saw that, as expected, the grid is much finer where the solution is steep [6,7]. The idea is now to replace the minimization problem (8) by a sequence of problems where the minimization is done only with respect to c and afterwards (a, b) is adjusted via moving grids according to an appropriate monitor function.

Obviously, in our case the time t in the moving grid method only plays an artificial role: it is somehow a homotopy parameter connecting the initial grid at stage $t = 0$ with the final one at e.g. $t = 1$. Let us assume that (a_k, b_k, c_k) denotes the parameterization at iteration step k. Then a proper choice for the monitor function is such that

$$\begin{aligned} m(\phi_k(\xi), 0) &= \det \nabla_\xi \phi_k(\xi)\,, & \xi \in \tilde{\Omega}\,, \\ m(\eta, 1) &= \frac{1}{\gamma(1 + \beta|\nabla_\eta f_k(\eta)|^2)^{\frac{1}{2}}}\,, & \eta \in \Omega\,, \end{aligned} \tag{14}$$

where $f_k(\eta) = c_k(\phi_k^{-1}(\eta))$, $\phi_k(\xi) = (a_k(\xi), b_k(\xi))$, $\gamma > 0$ is a constant to satisfy (10), and $\beta > 0$ is a constant for deciding how much the steepness should influence the cell size. In the following we assume that $\Omega = \tilde{\Omega} = [0, 1]^2$.

Since we want to avoid the calculation of the inverse of ϕ_k we propose the following way of calculating ϕ_{k+1}:

$$\phi_{k+1}(\xi) := \phi(\xi, 1) \qquad \text{and} \qquad \phi(\xi, t) := \phi_k(d(\xi, t)) \tag{15}$$

According to (11) we obtain

$$\begin{aligned} \det \nabla\phi(\xi, t) &= \det \nabla\phi_k(d(\xi, t)) \det \nabla d(\xi, t) \\ &= m(\phi(\xi, t), t) \;=\; m(\phi_k(d(\xi, t)), t)\,, \end{aligned}$$

and hence $d : \tilde{\Omega} \times [0, 1] \to \tilde{\Omega}$ satisfies

$$\begin{aligned} \det \nabla d(\xi, t) &= \tilde{m}(d(\xi, t), t)\,,\ \xi \in \tilde{\Omega}\,,\ 0 \le t \le 1\,, \\ d(\xi, 0) &= \xi\,, \qquad\qquad \xi \in \tilde{\Omega}\,, \end{aligned} \tag{16}$$

where

$$\tilde{m}(\xi, t) := \frac{m(\phi_k(\xi, t)}{\det \nabla\phi_k(\xi)}\,.$$

Note that (16) looks like (11) with ϕ and m replaced by d and $\tilde{m}$, respectively. As monitor function $\tilde{m}$, taking into account (14), we have chosen

$$\tilde{m}(\xi, t) := \frac{1}{(1 - t) + t\gamma \det \nabla_\xi \phi_k(\xi)(1 + \beta|\nabla_\eta f_k(\phi_k(\xi))|^2)^{\frac{1}{2}}}\,.$$

The advantage of this choice will be seen later. Together with the chain rule we obtain that

$$\tilde{m}(\xi,t) = \frac{1}{(1-t)+t\gamma D(a_k,b_k,c_k,\beta)^{\frac{1}{2}}(\xi)} \tag{17}$$
$$D(a,b,c,\beta) := (a_u b_v - a_v b_u)^2 + \beta((b_v c_u - b_u c_v)^2 + (a_u c_v - a_v c_u)^2)$$

Since we have to satisfy (10), γ in (17) is chosen such that

$$1 = \gamma \int_\Omega D(a_k,b_k,c_k,\beta)^{\frac{1}{2}}(\xi)\, d\xi \,.$$

For our numerical calculations we approximate (a_k, b_k, c_k) by bilinear finite element functions on a uniform grid, i.e.,

$$a_k = \sum_{i,j=0}^{n} a_{ij}^k \varphi_{ij}\,, \quad b_k = \sum_{i,j=0}^{n} b_{ij}^k \varphi_{ij}\,, \quad c_k = \sum_{i,j=0}^{n} c_{ij}^k \varphi_{ij}\,, \tag{18}$$

where φ_{ij} are the bilinear finite element functions satisfying $\varphi_{ij}(\frac{\bar{i}}{n},\frac{\bar{j}}{n}) = \delta_{i\bar{i}}\delta_{j\bar{j}}$. Besides the restriction that (a_k, b_k) is one to one we need that (a_k, b_k) maps the boundary of $\Omega = [0,1]^2$ onto itself. This is achieved if the nodes of the grid, (a_{ij}^k, b_{ij}^k) satisfy

$$a_{0j} = 0 = b_{i0}\,, \qquad a_{nj} = 1 = b_{in}\,, \qquad 0 \le i,j \le n\,. \tag{19}$$

We obtain an update for (a_k, b_k) via (15), (16). We know from the last section that the solution of (16) can be computed by a two-step procedure:

According to (12), as first step we have to solve the scalar Poisson equation

$$\Delta w(\xi) = 1 - \gamma D(a_k,b_k,c_k,\beta)^{\frac{1}{2}}(\xi)\,,\ \xi \in \tilde{\Omega}\,, \qquad \frac{\partial w}{\partial n} = 0\,,\ \eta \in \partial\tilde{\Omega}\,. \tag{20}$$

Note that due to the special choice of the monitor function in (17), the function w does not depend on t and, hence, has to be calculated only once for each iteration step. We do this by looking for the weak solution of this equation using bilinear finite element functions on the same uniform grid as chosen in (18), i.e.

$$w = \sum_{i,j=0}^{n} w_{ij}\varphi_{ij}\,. \tag{21}$$

Note that then the Neuman boundary condition is trivially satisfied if

$$w_{i0} = w_{i1}\,,\ w_{in-1} = w_{in}\,, \quad w_{0j} = w_{1j}\,,\ w_{n-1j} = w_{nj}\,, \qquad 0 \le i,j \le n\,.$$

We obtain a linear system of equations where the $(n-1)^2 \times (n-1)^2$ matrix is sparse. It has a blocktridiagonal structure and each block is again tridiagonal.

For the calculation of the right hand side of this system we approximate $D(a_k, b_k, c_k, \beta)^{\frac{1}{2}}(\xi)$ via bilinear spline interpolation. Note that due to our choice of (a_k, b_k, c_k) in (18) the function $D(a_k, b_k, c_k, \beta)^{\frac{1}{2}}(\xi)$ is only piecewise continuous, where discontinuities occur on the grid lines $(\frac{i}{n}, v)$ and $(u, \frac{j}{n})$. Using bilinear spline interpolation we approximate it by a continuous function. As coefficients of the splines we take the average of all possible values of $D(a_k, b_k, c_k, \beta)^{\frac{1}{2}}(\xi)$ (up to four) at the nodes $\xi = (\frac{i}{n}, \frac{j}{n})$. The solution of the corresponding system is calculated using an SSOR method.

The next step is now to solve the ODE system (13) for fixed node $\xi = (\frac{i}{n}, \frac{j}{n})$, i.e.

$$\begin{aligned} \frac{d}{dt} d(\xi, t) &= \tilde{m}(d(\xi, t), t) \nabla w(d(\xi, t)), \quad 0 \le t \le 1, \\ d(\xi, 0) &= \xi, \end{aligned} \tag{22}$$

with $\tilde{m}$ as in (17). This system is solved with the classical fourth order Runge Kutta method. Since ∇w for w as in (21) is not continuous, we calculate ∇w approximately using central difference quotients, where we take care that nodes at the boundary do not leave the boundary (see (19)).

Due to (15), we obtain the new grid points via

$$(a_{ij}^{k+1}, b_{ij}^{k+1}) = \phi_k(d((\tfrac{i}{n}, \tfrac{j}{n}), 1)) . \tag{23}$$

Although from the theory of the last section a folding of the grid is not possible, it can happen that due to numerical approximations for very small cells $\det \nabla \phi_{k+1}$ becomes negative. If this is the case we smooth the nodes by the following simple averaging procedure

$$a_{ij}^{k+1} := \tfrac{1}{9} \sum_{\mu,\nu=-1}^{1} a_{i+\mu\, j+\nu}^{k+1}, \qquad b_{ij}^{k+1} := \tfrac{1}{9} \sum_{\mu,\nu=-1}^{1} b_{i+\mu\, j+\nu}^{k+1}. \tag{24}$$

The formula above is used for interior nodes. For boundary nodes the formula is adjusted in accordance with (19).

Summarizing we end up in the following algorithm as realization of regularization for surface representations using moving grids:

Algorithm:

Step 0: As initial grid choose $a_{ij}^0 := \frac{i}{n}$, $b_{ij}^0 := \frac{j}{n}$, i.e. a uniform grid, and set $k = 0$.

Step 1: Minimize the functional

$$\|G(a_k, b_k, c) - g^\delta\|_{L^2}^2 + \alpha \|\nabla c\|_{L^2}^2, \qquad \alpha > 0,$$

with respect to c (with c as in (18)). The solution is denoted by c_k.

Step 2: Update (a_k, b_k) using moving grids, i.e.:

(a) Solve (20) with an SSOR method as described above.

(b) Solve (22) with the classical fourth order Runge Kutta method.

(c) Compute the coordinates of the new grid via (23).

(d) Adjust the nodes according to (24) if necessary.

Step 3: If the change in (a_k, b_k) was large enough, goto Step 1, otherwise Stop. The final iterate is denoted by $(a_\alpha^\delta, b_\alpha^\delta, c_\alpha^\delta)$.

Note that in this algorithm $\det \nabla\phi$ is always positive. It is not possible as for the approach with the special parameterization (4) that $\det \nabla\phi$ vanishes (see (5)). However, the cell size can become arbitrarily small and hence the regularized solutions $f_\alpha^\delta = c_\alpha^\delta((a_\alpha^\delta, b_\alpha^\delta)^{-1})$ can be arbitrarily steep.

4 Numerical Results

In the last section we described the numerical realization of regularization for surface representations using moving grids for adapting the finite element grid. We will now present several examples where we applied our algorithm to two-dimensional linear integral equations (6). The operator G of the transformed equation, given in (7), is linear with respect to c. Therefore, the minimization problem in Step 1 of our algorithm can be computed explicitly by solving the linear system

$$(A + \alpha B)c = r\,, \tag{25}$$

where

$$\begin{aligned} A &:= [\langle T_k\varphi_{ij}, T_k\varphi_{\bar{i}\bar{j}}\rangle]\,, \qquad & B &:= [\langle \nabla\varphi_{ij}, \nabla\varphi_{\bar{i}\bar{j}}\rangle]\,, \\ r &:= [\langle T_k\varphi_{ij}, g^\delta\rangle]\,, & T_k &:= G(a_k, b_k, \cdot)\,. \end{aligned} \tag{26}$$

The solution c is the coefficient vector used in (18) to express c_k.

We have considered deblurring problems where the kernel of the integral equation (6) is given by

$$k(s,t,x,y) = \tilde{k}(|(s,t)-(x,y)|)\,, \qquad \tilde{k}(r) = \begin{cases} (1-\frac{1}{\rho}r^2)^4 & \text{if } r^2 \le \rho\,, \\ 0 & \text{else}\,. \end{cases} \tag{27}$$

For ρ we have chosen the values $\frac{1}{100}$ and $\frac{1}{10}$. The bilinear finite element functions in (18) were chosen with $n = 20$. The regularization parameter in Step 1 of our algorithm was chosen to be $\alpha = 10^{-9}$.

The most time consuming step in each iteration is to build the full (441×441)-matrix A (more than 95% of the CPU time). A speed-up of the program could, therefore, be achieved via parallelization. Nevertheless, this algorithm was already 20 times faster than the one we used to calculate the regularized solutions with the special parameterization (4) [7].

Example 1: As kernel of the integral equation we used (27) with $\rho = \frac{1}{100}$. The exact solution is given by

$$f(x,y) := \chi_{B_{0.2}(0.6,0.5)} \cdot$$

For this solution we were able to compute the right hand side g exactly. Figure 1 shows the regularized solution calculated with our algorithm: the left picture shows the initial grid, the picture in the middle the grid after 10 iterations, and the right picture shows the graph of the 10th iterate. As we can see the location of the discontinuities is identified quite well after only 10 iterations.

In Figures 2 we show that the method is robust with respect to data noise. We added random noise to the exact data such that δ was equal to 5% of the norm of the exact data.

Example 2: In this example we treat Example 2 from Kindermann and Neubauer [7]; $\rho = \frac{1}{10}$ in (27). The exact solution is given by

$$f(x,y) := 10\chi_{B_{0.2}(0.5,0.3)} \cdot$$

Also for this solution we were able to compute the right hand side g exactly. Figure 3 shows the comparison of regularization for surface regularization with the special parameterization (4) and BV-regularization. Since the exact solution has its discontinuities along a circle, we have not been able to identify the location of the discontinuities with the special parameterization [7]. In Figure 4 we show the result obtained with the new approach in our algorithm. One can see quite well where the discontinuities are. The result is not as good as in Example 1. The reason for this is that the radius of the point spread function (the kernel of the integral equation) is three times larger. A better result could be possibly obtained by taking a finer quadrature rule for the calculation of the integrals occurring in the elements of the matrix A (see (26)). However, this would also increase the CPU time. A comparison of Figures 3 and 4 shows that our algorithm yields better results than BV-regularization.

Example 3: In this example we treat Example 3 from Kindermann and Neubauer [7]; $\rho = \frac{1}{10}$ in (27). The exact solution is given by

$$f(x,y) := 8\chi_{B_{0.1}(0.2,0.5)} + 10\chi_{B_{0.15}(0.8,0.3)} + 5\chi_{B_{0.2}(0.5,0.7)} \,.$$

As for Example 2 we show the comparison of regularization for surface regularization with the special parameterization (4) and BV-regularization as well as the result obtained with our algorithm in Figures 5 and 6, respectively. Similar arguments hold as in Example 2. In Figure 6 black circles indicate where the real locations of the discontinuities are. One can see that $n = 20$ is too small to be able to identify the locations accurately enough. An increase of n in (18), however, would increase the needed storage and the CPU-time by a quite a bit. Note that the storage requirement and the number of operations for computing A grows with $O(n^4)$, solving the linear system in (25) even grows with $O(n^6)$.

Example 4: In this example the kernel of the integral equation was chosen as in Example 1, i.e., we used (27) with $\rho = \frac{1}{100}$. The exact solution is given by

$$f(x,y) := 8\chi_{A_1} + 10\chi_{A_2} + 5\chi_{A_3} \,,$$

where A_1, A_2, and A_3 are areas bounded by straight lines as follows:

$$\begin{aligned} A_1 &:= \{(x,y) \in I\!R^2 : 70x - 30y + 9 \geq 0\,, 2x + 10y - 9 \leq 0\,, \\ &\qquad 90x - 70y + 7 \leq 0\,, 4x + 10y - 7 \geq 0\}\,, \\ A_2 &:= \{(x,y) \in I\!R^2 : 3x + 10y - 3 \geq 0\,, 8x - 10y + 2 \geq 0\,, 10x + 4y - 5 \leq 0\}\,, \\ A_3 &:= \{(x,y) \in I\!R^2 : 7x + 10y - 10 \leq 0\,, 10x + 3y - 10 \leq 0\,, \\ &\qquad 5x + 10y - 6 \geq 0\,, 10x - 5y - 4 \geq 0\}\,. \end{aligned}$$

The right hand side g was approximated by bilinear splines on a uniform grid with nodes $(\frac{i}{100}, \frac{j}{100})$, $0 \leq i,j \leq 100$. The node values were calculated by using a very fine quadrature rule in (6), namely a 4-point Gauss quadrature rule on subintervals of length $\frac{1}{100}$ which is five times finer than the one used in the identification algorithm.

Figure 7 shows the regularized solution calculated with our algorithm: the left picture shows the initial grid, the picture in the middle the grid after 10 iterations, and the right picture shows the graph of the 10th iterate.

Black lines in the picture in the middle indicate the locations of the discontinuities. As one can see, these locations are identified quite well except for the sharp corners. Better results could be obtained by increasing n (compare Example 3).

We think that all the examples above show quite well that regularization for surface representations in combination with moving grids is a good alternative to bounded variation regularization. Since the location of the discontinuities is found rather fast one could possibly get a faster method by the following hybrid approach: first identify the location of the discontinuities using our algorithm and then use a standard regularization method to identify the function values having the knowledge where the discontinuities are.

These ideas and the application of this new approach to parameter estimation problems as well as theoretical aspects of convergence of these finite-dimensional regularized solutions will be future work.

Acknowledgments

The author gratefully acknowledges travel support by the Austrian Science Foundation Funds under grant SFB F013/F1317.

References

1. R. Acar and C. Vogel, *Analysis of bounded variation penalty methods for ill-posed problems*, Inverse Problems **10**, 1217–1229 (1994).
2. G. Chavent and K. Kunisch, *Regularization of linear least squares problems by total bounded variation*, ESAIM: Control, Optimisation and Calculus of Variations **2** , 359–376 (1997).
3. H. W. Engl, M. Hanke, and A. Neubauer, *Regularization of Inverse Problems* (Kluwer Academic Publishers, Dordrecht ; Boston , 1996).
4. S. Kindermann, *Regularization of Ill-Posed Problems with Discontinuous Solutions by Curve and Surface Representations* (PhD thesis, Johannes Kepler Universität Linz, October 2001).
5. S. Kindermann and A. Neubauer, *Identification of discontinuous parameters by regularization for curve representations*, Inverse Problems **15**, 1559–1572 (1999).
6. S. Kindermann and A. Neubauer, *Estimation of discontinuous parameters of elliptic partial differential equations by regularization for surface representations*, Inverse Problems **17(4)**, 789–803(2001).
7. S. Kindermann and A. Neubauer, *Regularization for surface representations of discontinuous solutions of linear ill-posed problems*, Numer.

Funct. Anal. Optim. **22**, 79–105 (2001).

8. G. Liao, F. Liu, G. dela Pena, D. Peng, and S. Osher, *Level set based deformation methods for adaptive grids*, J. Comp. Phys. **159**, 103–122(2000).
9. A. Neubauer and O. Scherzer, *Regularization for curve representations: uniform convergence for discontinuous solutions of ill-posed problems*, SIAM J. Appl. Math. **58**, 1891–1900 (1998).
10. L. I. Rudin, S. Osher, and E. Fatemi, *Nonlinear total variation based noise removal algorithms*, Phys. **D 60**, 259–268 (1992).
11. B. Semper and G. Liao, *A moving grid finite element method using grid deformation*, Numer. Methods of PDEs **11**, 603–615 (1995).

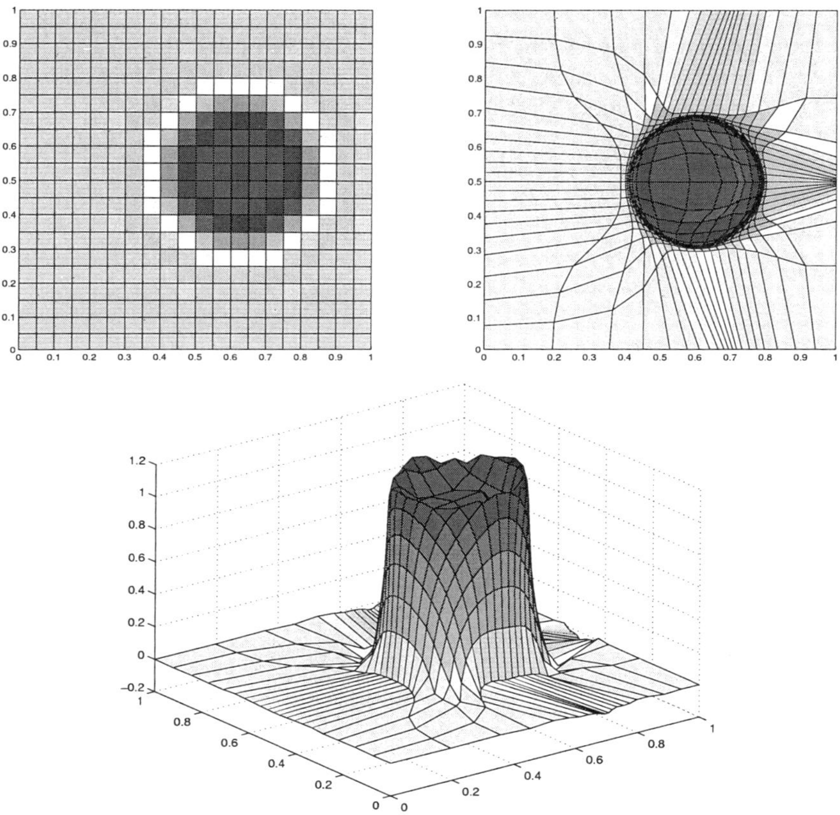

Figure 1. Initial grid, grid and graph of 10th iterate for Example 1.

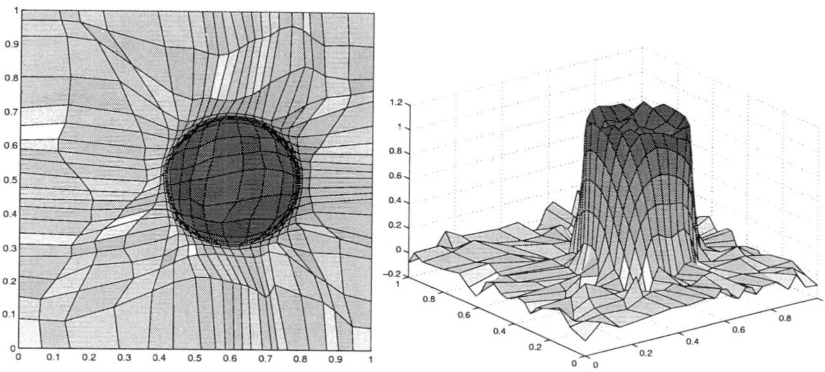

Figure 2. Grid and graph of 10th iterate for perturbed data in Example 1.

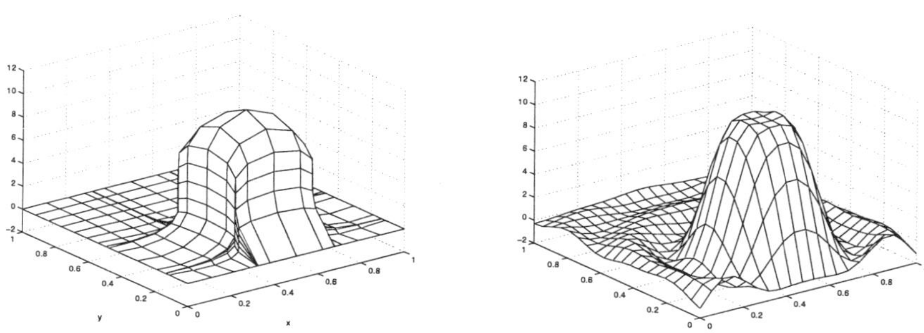

Figure 3. Surface- and BV-Regularization for Example 2.

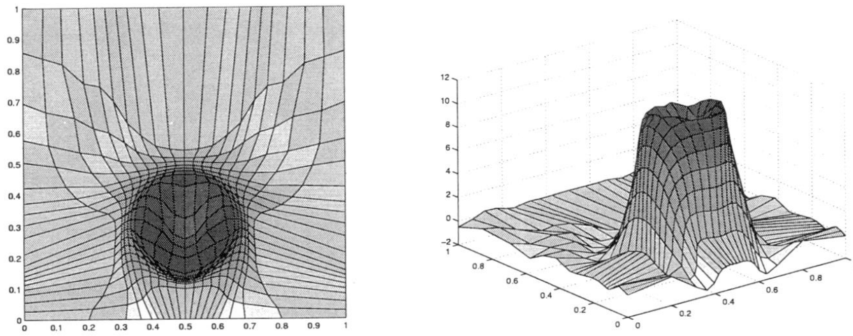

Figure 4. Grid and graph of 10th iterate for perturbed data in Example 2.

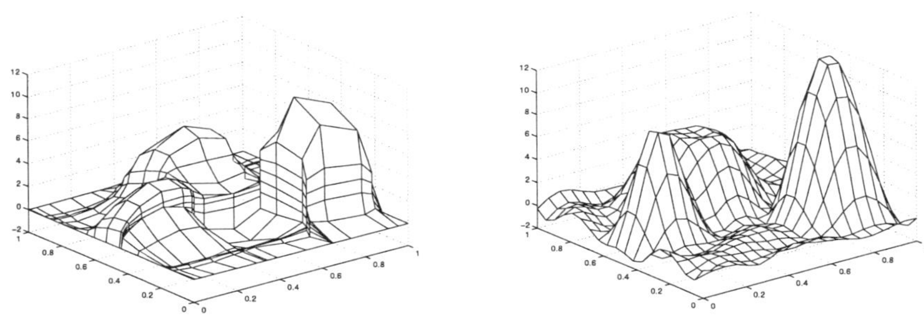

Figure 5. Surface- and BV-Regularization for Example 3.

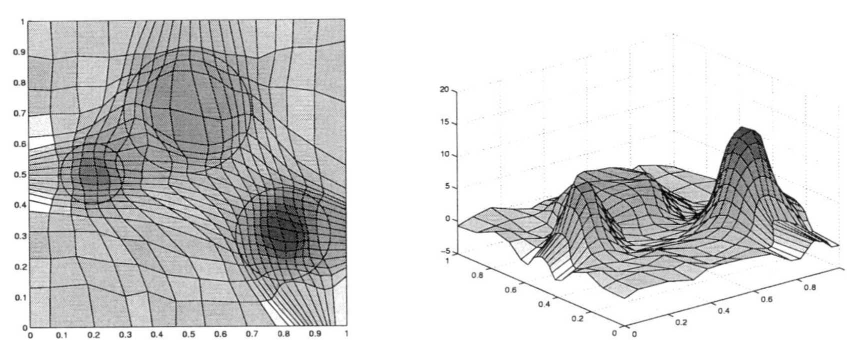

Figure 6. Grid and graph of 10th iterate for perturbed data in Example 3.

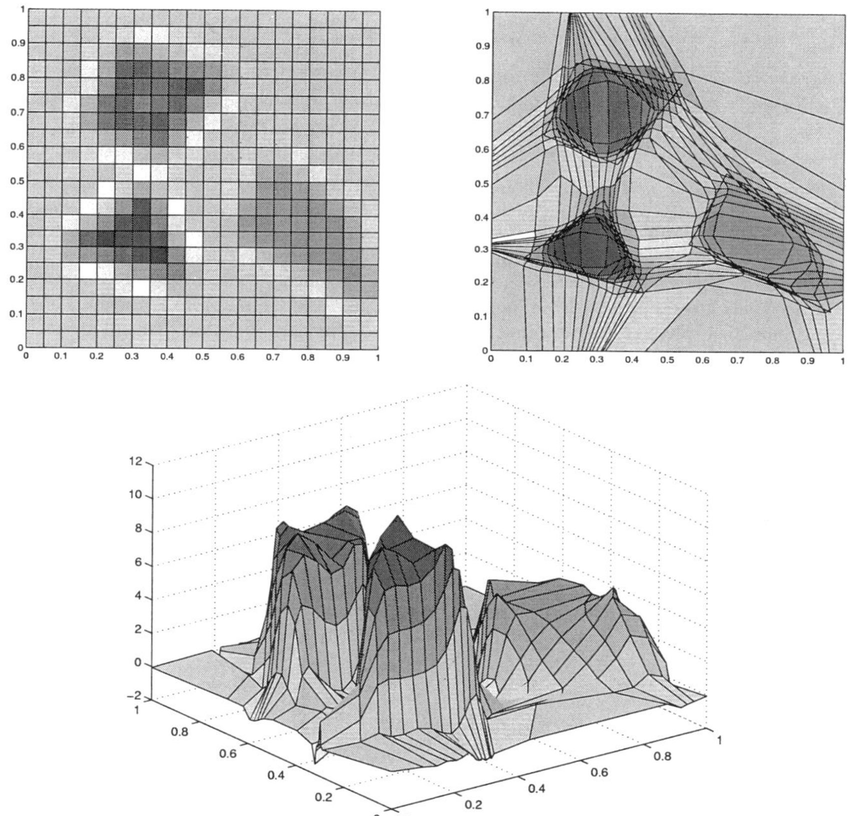

Figure 7. Initial grid, grid and graph of 10th iterate for Example 4.

INVERSE THEORY: SOLVING OR UNDERSTANDING MODELS?

P.C. SABATIER

Laboratoire de Physique Mathématique et Théorique, Université des Sciences et Techniques du Languedoc, Place Eugéne Bataillon 34095 Montpellier, France
E-Mail: sabatier@lpm.univ-montp2.fr

Inverse problems are the problems where information is extracted from results. Thus they may be studied either to get a way for deriving approximate solutions of the generally illposed problems of measurements or as a way of understanding the model which generated the ill posed problem and in some cases to modify its design. They may even be studied similarly on purely mathematical concepts. Hence their interaction with the model is important as well in applied mathematics, mathematical physics, or pure mathematics, and the various aspects complete each other. In this lecture, some progress in this understanding of the problems will be illustrated on examples, so as to introduce those given in the conference.

1 Introduction

In this introductory lecture, we try to draw the main features of Inverse Problems in their interaction with modelling natural phenomena. Inverse Theory can be defined as the Art of reconstructing *parameters from* measurements, for a given *model, previously created to interpret, directly or indirectly, available data.*

HENCE

a) There is always *a priori information*

b) One cannot escape the question : to what extent the *model* may be, and should be, *modified* ?

We shall see on the way that these points are managed somewhat differently in *Mathematical Physics and in* Applied Mathematics.

Throughout the paper, it is good to keep in mind the scheme of Figure 1, where direct and inverse problem are shown, according to the definitions :

Direct Problem : M maps the set C of all possible parameters into the set E of all possible results of measurements.

Regularized Solution of the Inverse Problem : $\overline{M}$ maps the set E into the subset of admissible parameters and is continuous for reasonable topologies in E and C.

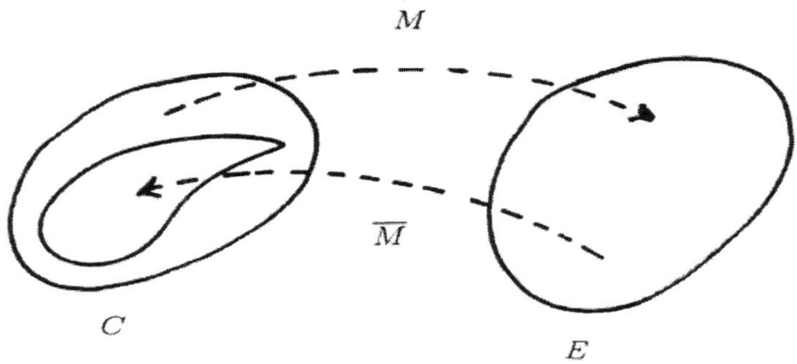

Figure 1.

2 Relation between inverse problems analysis and the mathematical model.

Analysis of inverse problems creates a dialog between defining the model and deriving solutions. We wish to review cases where a new preparation or a modification of the model, either by mathematical or by physical intrusions or both may be a significant element in our game.

A first glance shows these cases and suggests what points of view are different in applied mathematics and in mathematical physics :

a) **Applied Mathematics**

Emphasis is on deriving approximate solutions of relatively *narrow* problem, e.g. getting images from data.

Model or a priori information are questioned almost *only* for numerical analysis or algorithmic purposes.

Algorithms should be *adapted* to the needs of practitioners.

A priori information is in most cases used to put parameters in a compact subset or to seek those for which a *trade off* between several properties is got by the minimum of a cost *Functional.*

b) **Mathematical Physics**

Emphasis is on deriving *global solutions*, exact or approximate, of more general problems.

Model can be questioned and *partially modified.*

All methods of applied mathematics can be used but also *analytic methods* (even if it is necessary doing approximations before) when they give a global analysis of the problem.

The priori information that relies on *physical properties* is privileged.

3 Putting Physical Information into the Model.

Curiously, it often turns out that a better management of inverse problems is obtained by putting physical information in the model :

a) All regularization or filtering techniques do it, not always obviously, by putting down random perturbations ("noise") and/or setting rough physical requirements ; simple examples are shown if C^o is made compact or the cost functional is a convex combination of a "misfit" and a "parameter reliability".

I will not say more on these points, since many if not most other lectures will show them[1].

b) However, special physical features of parameters are often difficult to impose, (e.g. the piecewise continuous character of topographic lines). There exist attempts to do it in some algorithms and in the so-called mathematical morphology[2], but it still remains open problems.

c) What about long range information, to be or not to be inserted into a local model ? 200 years of coastal engineering answered to this question : no !

Example : A slide tsunami Inverse Problem is still modelled[3] (and the design of a harbour too), with absorbing *boundary conditions*, or Sommerfeld-like asymptotics.

In more extended coupled cases, the *patchwork approach* may represent the effects of other sites on a given one by a few parameters, on which each "local inverse problem solution" depends, and which are eventually derived by iterations. Green's functions ("*influence*") are particularly useful.

Example : get V from Neumann Dirichlet operator data at the surface $z = 0$ of a domain D where :

$$\Delta u + [K - V(z)]u = 0 \qquad (-h < z < 0) \tag{1}$$

$$du/dn = 0 \qquad (z = -h) \tag{2}$$

and two "sites" A_1 and A_2 only show non constant $V(z)$. (See Reference[4]). Also, some weakly nonlinear large size inverse problems can be treated as a set of coupled local problems by using *domain decomposition* methods.

d) Solving may be easier if the model contains more physical features.

   Examples: Many electromagnetic Inverse Problems (GPR soundings etc), in papers [5,6] where *Kleinman, Lesselier, Van den Berg,* replace academic "hard shape" scattering problems by medium scattering problems, (target shape corresponding there to steep variations of the descriptive parameter). Standard optimization techniques can then be used, (with nice results).

After others, the last issue of "Inverse Problems" shows several cases[7] where using physics to introduce approximate methods yields better analysis and less ill-posed features in the inverse problems structure. For instance, instead of a simple use of the usual optimization methods, Engl[8] recently tried to optimize an optical waveguide with variable cross section, whose losses increase but some other properties improve if length is shorter. The classical approach starts with discretization and expansion along modes and then uses standard optimization techniques. It is heavier and less stable than solving the nonlinear inverse problem which gives the shape from a given physical invariant of the problem (a preserved physical quantity along the guide).

Notice finally that unusual physical measurements may be suggested by mathematicians because they feel the inverse problem can be well managed : thus was introduced the inverse nodal problem (Hald and J. Mc Laughlin [9]).

4 Representation of Information in the Model

Many Inverse Problems are ill-posed, and data cannot completely determine the model parameters. Hence, it is essential to see that the information we may get from data is represented in the model. As a matter of fact, it is only after several years that scientists understood how a model *should be designed, or modified,* in order to give the most useful representation of data informative content *by no way a trivial point* :

a) using "descriptive" parameters and "imaging" them is almost always the method ; hence, in 2d or 3d, graphs, contours and their various "*fuzzy*" aspects are supposed to show data information and its uncertainties.

b) Assume data are M weighted linear moments of $s(x)$ on an interval I ; Backus and Gilbert[10] showed that weights can be combined to yield at any point x^0 of I a kernel $L\left(x, x^0\right)$ that "looks like" a "smooth" delta function of $x - x^0$, whose width a appraises the *nonuniqueness* of the inverse problem, since we can derive from data the weighted average :

$$\overline{s}\left(x^0\right) = \int L\left(x, x^0\right) s(x) dx \tag{3}$$

hence the method, extensible to weakly nonlinear cases, represents information by an average solution and by the "*resolving length*" a. A byproduct is the method of smooth delta source, which was used in water waves generation problems, and other boundary conditions problems, e.g. recently Cauchy problem (Hon and Wei[?]).

c) In very much under determined problems, the solutions cannot be represented any longer by "average values and uncertainties". One should redefine the model as a set of *well-posed questions* (**WPQ**) whose answer gives either some essential descriptive properties of solutions or a property that enables decision *(decisive modelling)*. Example of the simplest gravimetry model[12]:

$$y_m = \sum_{n=1}^{N} A_{mn}\, x_n \Longleftrightarrow \mathbf{y} = \mathbf{A}\mathbf{x} \tag{4}$$

$$\mathbf{x} : 0 \leq x_n \leq a \qquad n = 1, 2, \cdots N \tag{5}$$

the "parameter" $\mathbf{x}$ is a solution if it satisfies Eqs. (4), (5) where $\mathbf{y} = \{y_m\}\,(m = 1, 2, \cdots M)$ is the "data".

For one data $\mathbf{y}$, the set of solutions $\mathbf{x}$ is convex ; typical cases : M few hundreds, N few thousands ! *Descriptive modelling* may be done out of WPQ on bounds for the center of mass or other moments, or on solutions achieving an extreme property, e.g. body of minimum contrast, or two values only etc ; *Decisive modelling* may be done e.g. with WPQ on bounds for the depth of holes unable or "unlikely" to give any more information.

5 What about Algorithms and Their Own Modelling ?

I am sure that the present workshop will yield several answers to this question, so that I only give one remark :

The aim is to encompass more and more problems (three dimensional scattering, transport, financial, meteorology, etc) : it is necessary to adapt the mathematical description (e.g. topologies or distances in C and E) and the mathematical models : for instance for overpassing secondary minima of functionals. Physics has given sometimes the solution ; examples of "simulated annealing", "genetic algorithms" etc [13]. I have noticed recent progresses of a "deterministic" competitor, the "method" of Heavy Ball with friction[14]. The idea is that the cost functional defines a surface where a heavy ball can move, with inertia, and energy dissipation by friction, so that when it reaches a local minimum, it can still move out and explore a few of them. Some more refinements of the dynamical system suggest that one may build from it algorithms for global optimization.

6 Cases Where Data Sets That Can Be Collected Do Not Determine the Model.

a) Particularly important in **inverse scattering**, these cases may correspond to *transparent potentials*, to *hidden details*, to *stealth targets*, etc. They may correspond to physical properties whose representation is definitely lost by the model : in this case, no mathematical trick for *restoring uniqueness* is truly relevant.

The best example is that of Schrödinger scattering[15] on the line, where the so called Jost solutions of

$$Df =: \left(-\frac{\partial^2}{\partial x^2} + V(x)\right) f(k,x) = k^2 f(k,x) \tag{6}$$

$$V \in L_1'(\mathbb{R}) : \int_{-\infty}^{+\infty} (1+|x|)\,|V(x)|\,dx < \infty \tag{7}$$

$$\exp\left[\mp ikx\right] f^{\pm}(k,x) \rightarrow 1 \qquad (x \rightarrow \pm\infty,\ k \in \mathbb{R}) \tag{8}$$

(where $\pm$ correspond according to their position)
define reflection and transmission coefficient by

$$T(k)f_-(k,x) = f_+(-k,x) + R(k)f_+(k,x) \tag{9}$$

R, and sometines T, can be "measured" for all k.

WANTED, V. *Answer* : $R(k)$, with the eigenvalues $k = i\kappa_n$ and the normalizing constants of the eigenfunctions of operator D determine V, and T, but $R(k)$ only, **does not** ! Hence transparent potentials, with $R = 0$, exist as nonvanishing functions !

b) In "*data assimilation*", solutions of inverse problems from data are used for a dynamical modelling[16]. As a simple example, we may describe the evolution of a state vector X by the equation

$$dX/dt = F(X,t)\,;\; X(0) = U\,;\; U \in E \tag{10}$$

where E is a convenient space of functions. The solution being $X = C(U,t)$, and observations being those of $X(t)$, the control parameter U can be recalibbrated from time to time by minimizing the control functional :

$$J(U) = \int \|C(U,t) - X_{obs}(t)\|^2 \, dt \tag{11}$$

Difficulties increase when there are coupled phenomena *with different time scales*, e.g. in climatology, where several "*regimes*" should be identified, and a unique strategy is impossible to keep.

c) As a simpler example of the last point, let us think to water waves in ocean. Short ones, and swell, may be Fourier-analyzed, and if internal long waves propagate, one may see them for instance by their effect on the electromagnetic waves backscattered by surface waves, which can be discarded from the analysis. However these are extreme favorable situations ; in the general case of a broad spectrum, in particular near the coast, Fourier analysis does not fit because nonlinearly propagating signals have *motion invariants* that are not kept between two points of measurement by this analysis. A specific analysis must then be made[17] . Curiously it has been a by product of the "exact analysis" of the Schrödinger inverse problem "on the line" we saw above. It can be summarized as so : let us "guess" that the "nonlinear" propagation of these long waves follow the Korteveg de Vries equation (KdV) :

$$\frac{\partial \eta}{\partial t} + c_0 \frac{\partial \eta}{\partial x} + \beta \frac{\partial^3 \eta}{\partial x^3} + \alpha \eta \frac{\partial \eta}{\partial x} = 0 \tag{12}$$

If it did it according to the linearized KdV equation (LKdV) :

$$\frac{\partial \eta}{\partial t} + c_0 \frac{\partial \eta}{\partial x} + \beta \frac{\partial^3 \eta}{\partial x^3} = 0 \tag{13}$$

we would process this "signal" by Fourier transforming it :

$$\widetilde{\eta}(k,t) = \int_{-\infty}^{+\infty} e^{ikx}\, \eta(x,t) dx \tag{14}$$

showing from Eq. (13) the so-called dispersion relation

$$\omega(k) = k\left(c_0 - \beta k^2\right) \tag{15}$$

that relates (14) to (13). The signal processing would then achieved along the scheme :

$$\eta(x,0) \quad \longrightarrow \quad \widetilde{\eta}(k,0) \tag{16}$$

$$(\downarrow LKdV) \qquad\qquad \left(\frac{\partial}{\partial t} + i\omega\right)\widetilde{\eta} = 0 \Big\downarrow \tag{17}$$

$$\eta(x,t) \quad \longleftarrow \quad \widetilde{\eta}(k,t) \tag{18}$$

where the horizontal arrow in Eq. (16) stands for making the FT (14), i,e, solving a direct problem, the vertical arrow after Eq. (17) stands for going from $\widetilde{\eta}(k,0)$ to $\widetilde{\eta}(k,t)$ by solving the trivial linear differential equation, and the horizontal arrow in Eq. (18) stands for making the inverse $F.T.$ of $\widetilde{\eta}$, i.e. solving an inverse problem associated to the linear operator $-\frac{\partial^2}{\partial x^2}$, whose spectrum ($\mathbb{R}$) is a motion invariant :

$$-\frac{\partial^2}{\partial x^2} e^{ikx} = k^2 e^{ikx} \tag{19}$$

Exactly in the same spirit, the KdV equation (13) an be processed by following the scheme

$$\eta(x,0) \quad \longrightarrow \quad \{R^+(k,0)\,;\, i\kappa_n,\, c_n(0)\} \tag{20}$$

$$(\downarrow KdV) \qquad \left(\frac{\partial}{\partial t} + i\omega\right)\begin{cases} R^+ \\ c_n \end{cases} = 0 \Big\downarrow \tag{21}$$

$$\eta(x,t) \quad \longleftarrow \quad \widetilde{\eta}(k,t) \tag{22}$$

where $\omega(k)$ is given again by (15), the motion invariant spectrum is that of the Schrödinger operator (6), made of $\mathbb{R}$ and the discrete eigenvalues κ_n, the direct problem in (6.15) and the inverse one in (6.17) are those of the Schrödinger problem on the line. This nonlinear processing was applied successfully[17] to real data (satellite observations). As it is well known, this nonlinear generalization of Fourier transform is called an **inverse scattering transform**, and it suggests an extreme purely mathematical case of modelling

7 An Extreme Case of Modelling : Inverse Transforms and Applications

Through an *Inverse Transform*, a couple of functions are associated by solving a couple direct-inverse problem. During years, it looked a miracle that such a machinery enables one solving interesting classes of *nonlinear p.d.e.*, for special classes of *boundary conditions*. Now there exists ways of *modelling* these problems where every step is transparent! We present here one of them, which enabled us recently to get new results. One can call it : the "Elbow Scattering" model for solving Korteveg de Vries Equation.

a) The nonlinear KdV equation

$$\partial V/\partial t + \frac{1}{4}V''' - \frac{3}{2}VV' = 0 \tag{23}$$

where prime is for x-derivative, is the existence condition of a double differential for the solution $\mathbf{F}$ of two linear (Lax) eqs :

$$\partial\mathbf{F}(k,x,t)/\partial x = \mathbf{M}(k,x,t)\mathbf{F}(k,x,t) \tag{24}$$

$$\partial\mathbf{F}(k,x,t)/\partial t = \mathbf{N}(k,x,t)\mathbf{F}(k,x,t) \tag{25}$$

where the matrices $\mathbf{M}, \mathbf{N}, \mathbf{F}$ are given as so :

$$\mathbf{M} = \begin{pmatrix} 0 & 1 \\ -k^2 & 0 \end{pmatrix} + \begin{pmatrix} 0 & 0 \\ V & 0 \end{pmatrix} =: \mathbf{M}_0 + \mathbf{V} \tag{26}$$

$$\mathbf{N} = k^2\mathbf{M}_0 + \left(\frac{1}{2}k^2\mathbf{V} + \mathbf{W}_0\right) =: \mathbf{N}_0 + \mathbf{W} \tag{27}$$

$$\mathbf{F} = \begin{pmatrix} f \\ f' \end{pmatrix} \tag{28}$$

$$\mathbf{W}_0 = \begin{pmatrix} V_1 & V_0 \\ V_2 & -V_1 \end{pmatrix} \tag{29}$$

$$V_0 = \frac{1}{2}V \qquad V_1 = -\frac{1}{4}V' \qquad V_2 = \left(\frac{1}{2}V^2 - \frac{1}{4}V''\right) \tag{30}$$

We also need the notations :

$$A = W_0^2 + \frac{\partial W_0}{\partial t} = \left(V_1^2 + V_0 V_2\right) I + \frac{\partial W_0}{\partial t} \tag{31}$$

$$B = \begin{pmatrix} B_1 & 0 \\ B_3 & B_1 \end{pmatrix} \quad \begin{cases} B_1 = \frac{3}{4}V^2 - \frac{1}{4}V'' \\ B_3 = \frac{1}{2}\partial V/\partial t \end{cases} \tag{32}$$

b) Exactly as we defined the Schrödinger scattering problem on the line, we can define a scattering problem in the x, t plane, and the Scattering Solutions F (see Fig. 2)

This is possible thanks to the working assumptions :

Basic assumption : $V(x,t)$ goes to 0 for fixed x or for fixed t if the other variable goes to infinite ;

Virtual assumption : in addition, V goes to zero at infinity in all directions of the two "virtual" quadrants II and IV. The support of this additional assumption is that only solitary waves of positive velocity x/t exist for KdV.

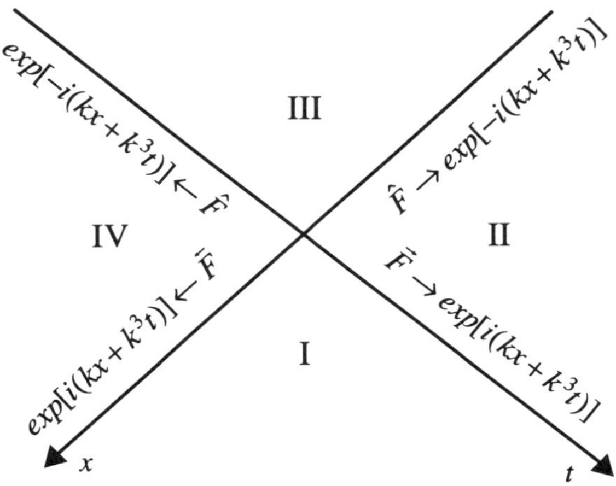

Figure 2.

Global scattering coefficients exist since matrices **M** and **N** are zero trace : the space of solutions of the couple of Lax equations is two dimensional

$$\overleftarrow{F}(k,x,t) = \eta(k)\overrightarrow{F}(k,x,t) + \xi(k)\overrightarrow{F}(-k,x,t) \tag{33}$$

$$\overrightarrow{F}(k,x,t) = \eta(-k)\overleftarrow{F}(k,x,t) - \xi(k)\overleftarrow{F}(-k,x,t) \tag{34}$$

By the way, notice the unitarity (V is real) :

$$\eta(k)\eta(-k) - \xi(k)\xi(-k) = 1 \tag{35}$$

and the symmetries

$$k \in \mathbb{R}, \text{ real } V \ : \ \eta(-k) = [\eta(k)]^* \tag{36}$$

$$\xi(-k) = [\xi(k)]^* \tag{37}$$

As for the Virtual Assumption, it yields

$$\widehat{F}(k,x,t) = \eta(k)\overleftarrow{F}(-k,x,t) - \xi(-k)\overleftarrow{F}(k,x,t) = \overrightarrow{F}(-k,x,t) \tag{38}$$

$$\overset{V}{F}(k,x,t) = \eta(-k)\overrightarrow{F}(-k,x,t) + \xi(-k)\overrightarrow{F}(k,x,t) = \overleftarrow{F}(-k,x,t) \tag{39}$$

c) *Axes translations* : Since η and ξ are space and time invariants, modifications of scattering coefficients on fixed x or fixed t axes are simply obtained by refixing the asymptotic behavior of $\mathbf{F}$: the transmission coefficient remains the invariant $1/\eta$ and the reflection coefficient is ξ/η times the phase factor $\exp[ikx]$ or $\exp[ik^3t]$. Hence both the t-translations, which describe the KdV motion, are isospectral, and the x-translations are too.

d) It is therefore trivial to derive the solution $V(x,t)$ obtained at $t > 0$ from $V(x,0)$: this is the basic "inverse scattering", known from Kruskal[18] (1967), and obtained by using Marchenko-Faddeyev theory of the Schrödinger inverse problem on the line.

Let us recall the steps of this classical method without comment :

$$V(x,0) \in L_1'(\mathbb{R}) \tag{40}$$

$$\overleftarrow{F}(k,x,0) = \begin{pmatrix} 1 \\ ik \end{pmatrix} e^{ikx} -$$

$$\int_x^{\infty} dy \begin{pmatrix} \cos k(y-x) & -\frac{\sin k(y-x)}{k} \\ k\sin k(y-x) & \cos k(y-x) \end{pmatrix} \mathbf{V}(y,0)\overleftarrow{F}(k,y,0) \tag{41}$$

$$\text{Asymptotics of } \overleftarrow{F}(k,x,0) \Rightarrow \eta(k) \text{ and } \xi(k) \tag{42}$$

$$\eta(k) = 0 \Rightarrow \text{ discrete spectrum } k = i\kappa_1,\, i\kappa_2, \cdots i\kappa_N \tag{43}$$

$$\int_{-\infty}^{+\infty} \left|\overleftarrow{F}(i\kappa_p,x,0)\right|^2 dx = \left(s_p^+\right)^2 \tag{44}$$

$$S(z,t) = \frac{1}{2\pi}\int_{-\infty}^{+\infty} dk \frac{\xi(-k)}{\eta(k)} e^{i\left(kz+2k^3t\right)} - \sum_p \left(s_p^+\right)^{-2} e^{-\kappa_p z + 2\kappa_p^3 t} \tag{45}$$

$$\kappa(x,y,t) = S(x+y,t) + \int_x^{\infty} \kappa(x,z,t)S(z+y,t)dz \tag{46}$$

$$V(x,t) = -2\frac{\partial}{\partial x}\kappa(x,x,t) \tag{47}$$

$$\overleftarrow{F}(k,x,t) = e^{i\left(kx+k^3t\right)}\begin{pmatrix} 1 \\ ik-\kappa(x,x,t)\end{pmatrix} + \int_x^\infty K(x,y,t)e^{i\left(ky+k^3t\right)}dy \tag{48}$$

$$K(x,y,t) = \begin{pmatrix} \kappa(x,y,t) \\ \frac{\partial}{\partial y}\kappa(x,y,t)\end{pmatrix} \tag{49}$$

From (49), (48) and (47) everything is known !

(e) The presentation given above shows that it is also possible to symmetrize the problem and to derive a solution $V(x,t)$ obtained at $x>0$ from $V, V', V''\,(0,t)$(Sabatier[19]).

$\overrightarrow{F}(k,0,t)$ can be derived by solving the Volterra equation

$$\overrightarrow{F}(k,0,t) = \begin{pmatrix} 1 \\ ik \end{pmatrix} e^{ik^3t} -$$

$$\int_t^\infty du \begin{pmatrix} \cos k^3(u-t) & -\frac{\sin k^3(u-t)}{k} \\ k\sin k^3(u-t) & \cos k^3(u-t)\end{pmatrix} \mathbf{W}(k,0,u)\overrightarrow{F}(k,0,u) \tag{50}$$

For analytic properties in $\lambda = k^3$, we need transforming 2-vectors as $G(k)$ into 6-vectors as $\Gamma(\lambda)$ and 2,2 matrices $a(k)$ into 3,3 matrices $A(\lambda)$ by

$$G(k) \Rightarrow \mathbf{G}(k) = \begin{pmatrix} G(k) \\ G(jk) \\ G\left(j^2k\right)\end{pmatrix} \Rightarrow \Gamma(\lambda) = \frac{1}{3}T(k)\mathbf{G}(k) \tag{51}$$

$$a(k) \Rightarrow \mathbf{a}(k) = \begin{pmatrix} a(k) & 0 & 0 \\ 0 & a(jk) & 0 \\ 0 & 0 & a\left(j^2k\right)\end{pmatrix} \Rightarrow A(\lambda) = T\,\mathbf{a}\,T^{-1} \tag{52}$$

$$T(k) = \begin{pmatrix} 1 & 1 & 1 \\ k^{-1} & j^2k^{-1} & jk^{-1} \\ k^{-2} & jk^{-2} & j^2k^{-2}\end{pmatrix} \quad T^{-1}(k) = \frac{1}{3}\begin{pmatrix} 1 & k & k^2 \\ 1 & jk & j^2k^2 \\ 1 & j^2k & jk^2\end{pmatrix} \tag{53}$$

By means of this transformation, analyticity properties in the λ plane can be shown for the transforms of F and the scattering coefficients :

$$\eta(-k) \Rightarrow C(\lambda) \qquad -\xi(-k) \Rightarrow D(\lambda) \tag{54}$$

$$R(\lambda) = [C(\lambda)]^{-1} D(\lambda) \tag{55}$$

$$F(k, \cdots) \Rightarrow H(\lambda, \cdots) \tag{56}$$

Coefficients $\eta(k)$ and $\xi(k)$, and their ratio, $r(k)$, are determined at $x = 0$ by the large negative t behavior of $\mathbf{F}(k, 0, t)$. At $x > 0$, r receives the phase factor $\exp(2ikx)$. Hence we derive $R(\lambda, x)$ and :

$$S(x, v) = \frac{1}{2\pi} \begin{pmatrix} i & 0 & 0 \\ 0 & 1 & 0 \\ 0 & 0 & -i \end{pmatrix} \int_{-\infty}^{+\infty} d\lambda \, e^{i\lambda v} R(\lambda, x) \begin{pmatrix} i & 0 & 0 \\ 0 & 1 & 0 \\ 0 & 0 & -i \end{pmatrix} \tag{57}$$

which is the "reflexive" ingredient of a "*Marchenko-like*" equation for the 6-vector valued function $L(x, t, u)$:

$$L(x, t, u) = S(x, t + u) \begin{pmatrix} -P_0(x, t) \\ P_1 \\ 0 \end{pmatrix} + \int_t^{\infty} dv \, S(x, u + v) L(x, t, v) + \widetilde{G}_x(t, u) \tag{58}$$

where, omitting the fixed variable x in $\widetilde{G}$

$$\widetilde{G}(t, u) = \frac{1}{2\pi} \int_{-\infty}^{+\infty} \left\{ \begin{pmatrix} i & 0 & 0 \\ 0 & 1 & 0 \\ 0 & 0 & -i \end{pmatrix} [C(\lambda)]^{-1} \overset{V}{H}(\lambda, t) - e^{-i\lambda t} \begin{pmatrix} -P_0 \\ P_1 \\ 0 \end{pmatrix} \right\} e^{i\lambda u} d\lambda \tag{59}$$

$$P_0 = \begin{pmatrix} 1 \\ -\frac{1}{2} \int_t^{\infty} B_1(x, u) du \end{pmatrix} \quad P_1 = \begin{pmatrix} 0 \\ 1 \end{pmatrix} \tag{60}$$

In the most regular cases (virtual assumption), the Fourier transform can be dispatched in the remainder of the equation as a *finite number of discrete spectrum contributions*, and the Equation can be solved, giving L and hence $V(x, t)$ as :

$$\mathbf{L} = \begin{pmatrix} L_0 \\ L_1 \\ L_2 \end{pmatrix} \; ; \quad \frac{d}{dt} L_2(x,t,t) = \left(\frac{1}{2} B\, P_0\right)(x,t) \Rightarrow V(x,t) \tag{61}$$

$$\overrightarrow{F}(k,x,t) = \mathbf{q}(ik) \cdot \mathbf{P}(x,t) e^{i(kx+k^3 t)} + \int_t^\infty du \mathbf{q}(ik) \cdot \mathbf{L}(x,t,u) e^{i(kx+k' t)} \tag{62}$$

$$\mathbf{q}(k) = \begin{pmatrix} I \\ kI \\ k^2 I \end{pmatrix} \quad \mathbf{P} = \begin{pmatrix} P_0 \\ P_1 \\ 0 \end{pmatrix} \quad \mathbf{L} = \begin{pmatrix} L_0 \\ L_1 \\ L_2 \end{pmatrix} \tag{63}$$

and again, everything can be derived from the three last equations.

8 Extreme opposite case : inverse problem solved by a physical device.

In *time reversal acoustics*, a signal is recorded by transducers, time reversed, then retransmitted to the medium. Using the *time reversal cavity* (TRC), or *mirror* (TRM) is connected to the inverse source problem and gives ways for analysing inverse scattering problems[20].

A TRC is a 2d transducer array that samples the wavefield ; a receiving amplifier, a storage memory and a programmable transmitter are able to *synthesize a time reversed version of the stored signal. Surrounding the source*, a TRC reconcentrates the signal into it, solving the inverse problem.

Using a TRM and time windowing separates waves reflected by a target according to their velocities and make possible concentrating on their apparent generation points ("*secondary sources*") in the target. Hence the *first reflected echo* (which determines the contour of a rigid target), the *elastic echos* (which correspond to various elastic modes) can be analyzed.

9 Conclusions

Made for describing a natural system, a first *model* defines the inverse problem.

"*Physical*" information is then or, should be, a basis of regularization methods, but the remarks below must be made first.

There are problems *not too much ill posed*, where admissible solutions x exist for each data y, known with possible errors δ, and the set $\{x\}$ may be represented by one solution exact or approximate, x^0, depending continuously on y, and known with a possible resolving distance η. This case is the one convenient for regularization and algorithms.

There are problems *more ill posed*, but where each solutions set can be related to a *new "physical" parameter*, whose inclusion modifies the original model. Mathematical tricks that regularize or restore uniqueness are often artificial[21] and may be misleading.

There are problems *so much underdetermined* that no solution imaging makes sense : one must substitute to the descriptive model a set of (hopefully) *well posed questions* whose answer enables one decision : "*decisive modelling*".

Algorithms are mathematical constructions, but some very sophisticated ones follow models issued from natural sciences, e.g., in the random case, simulated annealing and genetic algorithms, or dynamical systems in the deterministic case.

Models of mathematical physics can sometimes be constructed, that reduce a problem of analysis to solving inverse problems. Example : inverse scattering transforms. Sometimes one may be able to design a set of physical devices for solving an inverse problem!

General remark : most of our remarks are related as well to exact inverse problems with exact solutions and to applied inverse problems with regularized, or approximate solutions.

References

1. H.W. Engl, M. Hanke, and A. Neubauer, *Regularisation of Inverse Problems* (Kluwer, Dordrecht, 1996).
2. J.P.Cocquerez and S.Philipp, *Analyse d'images : filtrage et segmentation*(Masson, 1995).
3. S. Assier-Rzadkiewicz, P. Heinrich, P.C. Sabatier, B. Savoye and J.F. Bourillet, *Numerical modelling of a landslide-generated tsunami: The 1979 Nice event*, Pure and Applied Geophysics **157**, 1707–1727 (2000).
4. P.C.Sabatier, *A patchwork approach to problems with boundary measurements*, Inverse Problems **11**, 1233–1245 (1995); in *Inverse Problems in Geophysical Applications*, ed. H. Engl, W. Rundell, D. Colton, and A. Louis (SIAM, Philadelphia, 27–42(1997)).
5. L. Souriau, B. Duchêne, D. Lesselier, and R.E. Kleinman, *Modified gradient approach to inverse scattering for binary objects in stratified media L Souriau*, Inverse Probl. **12(4)**, 463–481 (1996).
6. G. Dassios and R. Kleinman, *Low Frequency Scattering* (Oxford S. Publications, Clarendon Press, 2000).
7. Determination of point wave sources by boundary measurements A El

Badia and T Ha-Duong Inverse Problems **17(4)**, August (2001).
8. A El Badia and T Ha-Duong,*Determination of point wave sources by boundary measurements* **17(4)**, 1127–1140 (2001).
9. O.H. Hald and J.R. Mc Laughlin, *Inverse Nodal Problems : Finding the Potential from Nodal Lines*, Memoir of the American Mathematical Society **119** , 572 (1996).
10. G. Backus and F. Gilbert, Geophys. J.R. Astron. Soc. **13**, 247–276(1967) : *ibid.* **16**, 169-205 (1968) ; G. Backus, Proc. Natl. Acad. Sci. USA **65**, 1-105, 281-287 (1970).
11. Y.C. Hon and T. Wei, *Backus-Gilbert algorithm for the Cauchy problem of the Laplace equation*, Inverse Problems **17**, 261–271 (2001).
12. P.C. Sabatier, *Inverse and III-Posed Problems* , ed. H.W. Engl and C.W. Groetsch (Academic, Boston, 1987).
13. P.J.M. Van Laarhoven and E.H.L. Aarts, *Simulated Annealing, Theory and Practice* (Kluwer Academic, Dordrecht, 1992).
14. H. Attouch, X. Goudou, P. Redont, *The heavy ball with friction method*, Communications in Contemporary Mathematics **2(1)**, 1–34 (2000).
15. K. Chadan and P.C. Sabatier, *Inverse Problems in Quantum Scattering Theory*, 2nd ed (Springer, New York, 1989).
16. P. Courtier and O. Talagrand, Tellus **42A**, 531–549 (1990).
17. A.R. Osborne, in *Nonlinear Topics in Ocean Physics*, ed. A.R. Osborne (North Holland, Amsterdam, 669–698(1991)); *Nonlinear Ocean Waves and the Inverse Scattering Transform*, in *SCATTERING*,ed. Roy Pike and Pierre Sabatier(Academic Press 637–666(2001)).
18. F. Calogero and A. Degasperis, *Spectral Transform and Solitons* (North Holland, Amsterdam, 1982).
19. P.C. Sabatier, *Elbow scattering and inverse scattering applications to LKdV and KdV*, J. Math. Phys. **41**, 414-436 (2000); A.S. Fokas and B. Pelloni, Proc. R. Soc. London, Ser. A **454**, 645–657,to be published.
20. M. Fink and Claire Prado, *Acoustic time-reversal mirrors*, Inverse Problems **17**, R1–R38 (2001).
21. P.C. Sabatier, J. Inverse and Ill-posed Problems **4**, 707–717 (1996).
22. P.C.Sabatier, *Past and Future of Inverse Problems*, J.Math.Phys. **41**, 4082-4124 (2000).
23. A. S. Fokas, *On the integrability of linear and nonlinear partial differential equations*, J. Maths. Phys. **41**, 4188–4237 (2000).
24. M.J. Ablowitz and P.A. Clarkson, *Solitons, Nonlinear Evolutions Equations and Inverse Scattering* (Cambridge U.P., Cambridge, 1991); in *the almost annual proceedings of the meeting " Needs"* (Nonlinear Evolution Equations and Dynamical Systems), existing since 1980 (World Scientific,

Singapore)
25. R.E. Kleinman and P.M. Van der Berg, *Two-dimensional location and shape reconstruction*, Radio Sci. **29(4)**, 1157–1169 (1994).
26. D. Goldberg, *Genetic Algorithms in Search, Optimization and Machine Learning* (Addison-Wesley, Reading, MA, 1989).
27. R. Cerf., *Une thorie asymptotique des algorithmes gntiques (synthse)*, Ph. D. Thesis, Montpellier Technical Report **94-04** (1994).
28. V. Richard, Ph. D. thesis, Montpellier (1983).

Section II

Theoretical Aspects

INVERSE BOUNDARY VALUE PROBLEMS FOR SYSTEMS OF PARTIAL DIFFERENTIAL EQUATIONS.

GREGORY ESKIN AND JAMES RALSTON

Department of Mathematics, UCLA, Los Angeles, CA 90095-1555, USA.
E-mail: eskin@math.ucla.edu

We describe the main results and the ideas of the proofs in the papers of Eskin[2], Eskin and Ralston[4] (see References). In addition, we simplify the construction of asymptotic solutions in Eskin[2], using the results of Eskin and Ralston[4], and we simplify the proof of estimate (21) that was given in Eskin and Ralston[4].

1 The Schrödinger Equation with an External Yang-Mills Potential.

Let Ω be a smooth bounded domain in $\mathbf{R}^n$, $n \geq 3$. Consider the Dirichlet problem, $u = f$ on $\partial\Omega$, for a system of differential equations of the form

$$-\Delta u - 2i\Sigma_{k=1}^{n} A_k(x)\frac{\partial u}{\partial x_k} + Bu = 0,\ x \in \Omega, \tag{1}$$

where $u = (u_1, \ldots, u_m)$, and A_k, $k = 1, \ldots, n$, and B are smooth $m \times m$ matrix functions. We assume that Ω is such that the Dirichlet problem has a unique solution $u \in H_1(\Omega)$ for every $f \in H_{\frac{1}{2}}(\partial\Omega)$. Let Λ be the Dirichlet-to-Neumann operator,

$$\Lambda f = \frac{\partial u}{\partial \hat{n}} + iA \cdot \hat{n} u \text{ on } \partial\Omega,$$

where $\hat{n}$ is the exterior unit normal to $\partial\Omega$ and u is the solution of (1) with $u = f$ on $\partial\Omega$. The inverse boundary value problem is to find the coefficients in (1) given Λ. When one rewrites (1) in the form

$$(-i\frac{\partial}{\partial x} + A(x))^2 u + V(x)u = 0 \tag{2}$$

with $A = (A_1, .., A_n)$ and $V = B - \Sigma_{k=1}^{n} A_k^2 + i\Sigma_{k=1}^{n} \frac{\partial A_k}{\partial x_k}$, it becomes the time-independent Schrödinger equation for a particle under the influence of the Yang-Mills potential (A, V). We say that two Yang-Mills potentials $(A^{(1)}, V^{(1)})$ and $(A^{(2)}, V^{(2)})$ are gauge equivalent if there exists a smooth, invertible matrix function $g(x)$ on $\overline{\Omega}$ such that

$$A^{(2)} = g^{-1}A^{(1)}g - ig^{-1}\frac{\partial g}{\partial x} \text{ and } V^{(2)} = g^{-1}V^{(1)}g. \tag{3}$$

The following theorem was proven in Eskin[2].

Theorem 1 *Let $L^{(j)}u = 0$, $j = 1,2$, be two Schrödinger equations of the form (2) in $\Omega \subset \mathbf{R}^n$, $n \geq 3$, with Yang-Mills potentials $(A^{(j)}, V^{(j)})$, $j = 1,2$, and let $\Lambda^{(j)}$, $j = 1,2$, be their Dirichlet-to-Neumann operators. Assume that Ω is convex. Then $\Lambda^{(1)} = \Lambda^{(2)}$ if and only if $(A^{(1)}, V^{(1)})$ and $(A^{(2)}, V^{(2)})$ are gauge equivalent.*

The proof of Theorem 1 is based on the method of complex exponential solutions with a large parameter that was introduced in Sylvester and Uhlmann[7]. Let μ,ν and l be pairwise orthogonal vectors in $\mathbf{R}^n$ with $|\mu| = |\nu| = 1$. We set $\theta = \mu + i\nu$, and for $\tau >> 0$ we set $\zeta = l/2 + (\tau^2 - |l|^2/4)^{1/2}\mu$ and $\delta = \zeta + i\tau\nu$. We look for solutions of (2) of the form $u = v\exp(ix \cdot \delta)$. Hence v must satisfy

$$L_\delta v =_{\text{def.}} (-i\frac{\partial}{\partial x} + \delta)^2 v - 2iA(x) \cdot (\frac{\partial}{\partial x} + i\delta)v + B(x)v = 0. \tag{4}$$

In order to solve systems of the form (4) or, more generally, inhomogeneous systems

$$L_\delta v = f \text{ in } \Omega_0, \tag{5}$$

where $\overline{\Omega} \subset \Omega_0$, we will need solutions of matrix equations of the form

$$i\theta \cdot \frac{\partial C(x,\theta)}{\partial x} = \theta \cdot A(x)C(x,\theta), \ x \in \Omega_0, \tag{6}$$

where C is an invertible $m \times m$ matrix function. One can show that, when $A(x)$ is extended to a function of compact support in $\mathbf{R}^n$, there may be no solution of (6) which tends to the identity matrix as $|x| \to \infty$. However, there are always solutions which grow polynomially. The following lemma was proven in Eskin[2].

Lemma 1 *There exists an invertible matrix function $C(x,\theta)$, solving (6) and depending smoothly on (x,θ) on the domain $\overline{\Omega_0} \times \{\theta = \mu + i\nu : \mu \cdot \nu = 0, \ |\mu| = |\nu| = 1\}$. This solution is not unique, but it can be chosen to satisfy $C(x, e^{i\omega}\theta) = C(x,\theta)$.*

Using Lemma 1, we can prove (see Eskin[2]):

Lemma 2 *For any $f \in L^2(\Omega_0)$ there is a $v(\tau) \in H^2(\Omega_0)$ such that for τ sufficiently large $v(\tau)$ solves (5) in Ω, and*

$$||v||_{H^l(\Omega_0)} \leq \frac{C}{(1+\tau)^{1-l}}||f||_{L^2(\Omega_0)}, \ l = 0,1,$$

where C is independent of τ and f.

To prove Lemma 2 we proceed as follows. Let $C(x, \mu + i\nu)$ be as in Lemma 1, and let $c(x, D, \tau)$ be the pseudo-differential operator with symbol $C(x, \frac{\xi'+\zeta}{|\xi'+\zeta|} + i\nu)\chi$, where $\xi' = \xi - (\xi \cdot \nu)\nu$ and χ is a suitable cutoff function in (ζ, ξ, τ). We look for v in the form

$$v = c(x, D, \tau)E(\tau)g, \text{ where } E(\tau) = (-i\frac{\partial}{\partial x} + \delta)^{-2} \text{ and } g \in L^2(\Omega_0). \quad (7)$$

Substituting (7) into (5) one gets $g + T(\tau)g = f$, where the norm of $T(\tau)$ goes to zero as $\tau \to \infty$. Therefore $I + T(\tau)$ is invertible for τ large. Note that the proof of Lemma 2 in Eskin[2] is a generalization of the method in Eskin and Ralston [3] which treated the (scalar) case of electro-magnetic potentials. Lemma 2 is used in Eskin[2] to prove the following:

Lemma 3 *For every vector of polynomials, $p(z)$, in the complex variable $z = \theta \cdot x$ there is a solution $v(\tau)$ of (4) satisfying*

$$\lim_{\tau \to \infty} v(\tau) = C(x, \theta)\Pi^+(C^{-1}p), \quad (8)$$

where C is the matrix function from Lemma 1 and Π^+ is a Toeplitz projection.

Using (7) to construct the leading order term in $v(\tau)$ made both the proof of Lemma 3 and its applications somewhat complicated in Eskin[2]. In Eskin and Ralston [4] we found a way to construct such solutions more simply with explicit higher order asymptotics as $\tau \to \infty$, and this lead to simpler proofs. We will give this construction in section 2.

Now we can complete the proof of Theorem 1. Following the strategy of Sylvester and Uhlmann[7], we can use the assumption $\Lambda^{(1)} = \Lambda^{(2)}$, Green's formula and Lemma 3 to derive integral identities involving $C^{(j)}$ and $(A^{(j)}, V^{(j)})$, $j = 1, 2$. Then arguments involving $\overline{\partial}$-equations in the parameters in these identities (see §5 and §6 in Eskin[2]) lead to the proof of Theorem 1.

2 Construction of Complex Exponential Solutions.

To construct solutions of (1) of the form $u = v\exp(ix \cdot \delta)$ we will proceed as follows. Substituting $u = v\exp(ix \cdot \delta)$ into (1), one sees that v must satisfy

$$L_\delta v =_{\text{def.}} -\Delta v - 2i\delta \cdot \frac{\partial v}{\partial x} - 2iA \cdot (i\delta v + \frac{\partial v}{\partial x}) + Bv = 0. \quad (9)$$

We will construct v in the form $v = \Sigma_{k=0}^{n} v_k + \tilde{v}_n$, where $v_k = O(\tau^{-k})$ and is explicit modulo solutions of (6), and $\tilde{v}_n$ is $O(\tau^{-n-1})$. We have

$$\delta = \tau\theta + \frac{l}{2} + O(\tau^{-1}) =_{\text{def.}} \tau\theta + \delta' \text{ and}$$

$$L_\delta =_{\text{def.}} -2i\tau\theta \cdot \frac{\partial}{\partial x} + 2\tau\theta \cdot A + M_{\delta'}.$$

To solve (9) modulo terms of order τ^{-n} we require

$$2\tau(i\theta \cdot \partial_x - \theta \cdot A)v_k = M_{\delta'} v_{k-1}$$

for $k = 0, \ldots, n$ with $v_{-1} = 0$. We set $v_0 = C_0(x, \theta)p(x \cdot \theta)$, where C_0 is a solution of (6) with the properties described in Lemma 1 and $p(z)$ is a vector of polynomials in the complex variable z. Since we only require that v satisfy (9) on Ω, we can introduce a cut-off function $\psi \in C_0^\infty(\Omega_0)$, $\psi = 1$ on a neighborhood of Ω, and set

$$v_k = -i(2\tau)^{-1} C_k(x, \theta)(\theta \cdot \partial_x)^{-1}(C_k^{-1}(\cdot, \theta)\psi M_{\delta'} v_{k-1}), \; k = 1, \ldots, n. \tag{10}$$

In (10) the operator $(\theta \cdot \partial_x)^{-1}$ multiplies the Fourier transform by $(i\theta \cdot \xi)^{-1}$, and C_k is again a solution of (6) as in Lemma 1 (in applications so far we have taken $C_k = C_0$, but this is not necessary). Since $M_{\delta'}$ does not increase the order of a term in τ and $i(2\tau)^{-1}(\theta \cdot \partial_x)^{-1}$ adds a factor of τ^{-1}, we have $v_k = O(\tau^{-k})$ and

$$L_\delta(v_0 + v_1 + \cdots + v_n) = M_{\delta'} v_n = O(\tau^{-n})$$

on the neighborhood of Ω where $\psi = 1$. Hence, taking $\psi_0 \in C^\infty(\Omega_0)$ supported in the set where $\psi = 1$ such that $\psi_0 = 1$ on Ω, $v = v_0 + \cdots + v_n + \tilde{v}_n$ will be a solution of (10) in Ω if

$$L_\delta \tilde{v}_n = -\psi_0 M_{\delta'} v_n \tag{11}$$

in Ω_0.

To solve (11) apply Lemma 2. Lemma 2 holds in a more general form (see Eskin and Ralston[4]): if $f \in H^k(\Omega_0)$, $k \geq 0$, then $v \in H^{k+2}(\Omega_0)$, and one has the estimate

$$\|v\|_{H^{k+l}(\Omega_0)} \leq \frac{C}{(1+\tau)^{1-l}} \|f\|_{H^k(\Omega_0)}, \; l = 0, 1.$$

Since v_n is bounded by $C\tau^{-n}$, the solutions given by Lemma 2 will be bounded by $C\tau^{-n-1}$ in $H^k(\Omega_0)$ for $k \geq 0$. Thus the $v(\tau)$ that we have constructed is a solution of (9) whose asymptotics in $H^k(\Omega)$ up to order τ^{-n} are given by the asymptotics of $v_0 + \cdots + v_n$. For the leading term in the asymptotics we have

$$v(\tau) = C_0(x,\theta)p(\theta \cdot x) + O(\tau^{-1}), \tag{12}$$

where $O(\tau^{-1})$ means bounded by $C\tau^{-1}$ in $H^k(\Omega), k \geq 0$. Since the limit of $v(\tau)$ no longer involves a Toeplitz projection, one no longer needs one of the arguments (Lemma 5.1) in Eskin[2].

3 The Equations of Isotropic Elasticity.

The construction presented in section 2 can be used to construct solutions of the system of isotropic elasticity as well. Using subscripts for derivatives, this system is given by

$$\Sigma_{j=1}^3 (\lambda w_{x_j}^j)_{x_k} + \Sigma_{j=1}^3 (\mu(w_{x_j}^k + w_{x_k}^j))_{x_j} = 0, \; k = 1,2,3, \tag{13}$$

where $w = (w^1, w^2, w^3)$ is the deformation of an elastic body with "Lamé parameters" $\lambda(x)$ and $\mu(x)$. Let w be the solution of the Dirichlet problem for (13) with $w = h$ on $\partial\Omega$. Then the inverse boundary value problem for this system is to recover λ and μ from the Dirichlet-to-Neumann map

$$\Lambda(h)^k = \Sigma_{j=1}^3 (\lambda w_{x_j}^j)\hat{n}^k + \Sigma_{j=1}^3 \mu(w_{x_j}^k + w_{x_k}^j)\hat{n}^j, \; k = 1,2,3.$$

There is as yet no proof that Λ determines λ and μ. Partial results are given in Nakamura and Uhlmann[6],and Eskin and Ralston [4]. The system (13) is not in the form (1). However, Ang, Ikehata, Trong and Yamamoto in Ang *et al* [1] show that $w = \mu^{-1/2}u + \mu^{-1}\nabla f - f\nabla\mu^{-1}$ will satisfy (1) when the 4-vector (u, f) satisfies the system

$$\Delta \begin{pmatrix} u \\ f \end{pmatrix} + V_1(x) \begin{pmatrix} \nabla f \\ \nabla \cdot u \end{pmatrix} + V_0(x) \begin{pmatrix} u \\ f \end{pmatrix} = \begin{pmatrix} 0 \\ 0 \end{pmatrix}. \tag{14}$$

Here

$$V_1(x) = \begin{pmatrix} -2\mu^{1/2}\nabla^2\mu^{-1} & -\mu^{-1}\nabla\mu \\ 0 & \frac{\lambda+\mu}{\lambda+2\mu}\mu^{1/2} \end{pmatrix}$$

and $\nabla^2 f$ denotes the Hessian matrix $\partial^2 f/\partial x_j \partial x_k$. The matrix V_0 is a complicated expression in λ, μ and their derivatives, but it vanishes when μ is constant. Since (14) does have the form (1), the method above can be used to construct solutions $(u, f) = \exp(i\delta \cdot x)(r, s)$ of (13) with prescribed asymptotics as $\tau \to \infty$.

Suppose that one has two elastic bodies occupying the region Ω with Lamé parameters $(\lambda^{(j)}, \mu^{(j)})$. $j = 1, 2$. Let $\Lambda^{(j)}$, $j = 1, 2$, be the corresponding Dirichlet-to-Neumann maps. Using Green's formula, one can verify that $\Lambda^{(1)} = \Lambda^{(2)}$ is equivalent to

$$\begin{aligned} H(w^{(2)}, w^{(1)}) &= \text{def.} \int_\Omega [(\lambda^{(2)} - \lambda^{(1)})(\nabla \cdot \overline{w^{(2)}})(\nabla \cdot w^{(1)}) \\ &\quad + \frac{1}{2}(\mu^{(2)} - \mu^{(1)})\Sigma_{1\le j,k\le 3}(\overline{w_k^{(2),j} + w_j^{(2),k}})(w_k^{(1),j} + w_j^{(1),k})]dx \\ &= 0 \end{aligned} \tag{15}$$

for all solutions to $L^{(1)}w^{(1)} = 0$, $L^{(2)}w^{(2)} = 0$, where now $L^{(j)}w^{(j)} = 0$ is the system (13) with $\lambda = \lambda_j$ and $\mu = \mu_j$. We can use (15) in the following way. The method of section 2 yields solutions of (14) with prescribed asymptotics. For the problem corresponding to $L^{(1)}$ we use these with $\delta^{(1)} = \delta$ as defined earlier, but for the problem corresponding to $L^{(2)}$ we set $\delta^{(2)} = \overline{\delta} - l$. Taking $w^{(j)} = \mu_j^{-1/2}u^{(j)} + \mu_j^{-1}\nabla f^{(j)} - f^{(j)}\nabla\mu_j^{-1}$, we substitute $w^{(1)}$ and $w^{(2)}$ into (15). When we collect the terms of each order in τ, this gives

$$0 = H(w^{(2)}, w^{(1)}) = \tau^2 H_2 + \tau^1 H_1 + H_0 + \cdots + \tau^{-N}H_{-N} + O(\tau^{-N-1}),$$

where each H_j is independent of τ. How far one can continue this expansion depends on the choice of n in section 2 . Note that each H_j must vanish when $\Lambda^{(1)} = \Lambda^{(2)}$. In particular, one has

$$H_2 = \int_\Omega e^{il\cdot x}(\theta \cdot \overline{r}_0^{(2)}, \overline{s}_0^{(2)})V(x,\theta)(\theta \cdot r_0^{(1)}, s_0^{(1)})^t dx = 0, \tag{16}$$

where

$$V(x,\theta) = \begin{pmatrix} (\lambda_1 + \mu_1 - \lambda_2 - \mu_2)\frac{(\mu_1\mu_2)^{1/2}}{(\lambda_1+2\mu_1)(\lambda_2+2\mu_2)} & 2(\mu_2^{-1} - \mu_1^{-1})\mu_2^{-1/2}\theta\cdot\partial_x b_2 \\ 2(\mu_2^{-1} - \mu_1^{-1})\mu_1^{-1/2}\theta\cdot\partial_x b_1 & 2(\mu_2^{-1} - \mu_1^{-1})(b_1 a_1 + b_2 a_2) \end{pmatrix}.$$

The functions a_j and b_j, $j = 1, 2$ are given by

$$a_j = (\theta\cdot\partial_x)^2\mu_j^{-1} \quad \text{and} \quad b_j = \frac{\mu_j}{2}\frac{\lambda_j + \mu_j}{\lambda_j + 2\mu_j}.$$

The functions $(r_0^{(j)}, s_0^{(j)})$ are vector solutions of the following version of (6)

$$-2\theta\cdot\partial_x\begin{pmatrix} r_0 \\ s_0 \end{pmatrix} = V_1\begin{pmatrix} s_0\theta \\ \theta\cdot r_0 \end{pmatrix}. \tag{17}$$

One can analyze H_2 by the techniques that were used for the Yang-Mills case in Eskin[2]. However, here that does not lead to the conclusion that

$\lambda_1 = \lambda_2$ and $\mu_1 = \mu_2$. Instead one arrives at the identity (Theorem 1.3 of Eskin and Ralston[4])

$$b_1^{1/2}(\theta\cdot\partial_x)^2(b_1^{-1/2}) - b_1(\theta\cdot\partial_x)^2(\mu_1^{-1}) = b_2^{1/2}(\theta\cdot\partial_x)^2(b_2^{-1/2}) - b_2(\theta\cdot\partial_x)^2(\mu_2^{-1}).$$

Since this holds identically in θ, it is equivalent to a system of five partial differential equations satisfied by $(\lambda_1, \mu_1, \lambda_2, \mu_2)$. However, even in the case that λ_1 and μ_1 are constant, these equations have solutions with λ_2 and μ_2 nonconstant.

A more readily useful result is the following (Theorem 2 of Eskin and Ralston[4]):

Theorem 2 *Let*

$$||f||_\alpha^2 = \int_\Omega |f|^2 e^{2\alpha|x|^2}dx.$$

Then, if $\Lambda^{(1)} = \Lambda^{(2)}$, *for* $\alpha > \alpha_0$, *one has*

$$||\lambda_1 + \mu_1 - \lambda_2 - \mu_2||_\alpha \leq \frac{C}{\alpha}||\mu_1 - \mu_2||_\alpha$$

with C *independent of* α.

This result comes from $\overline{\partial}$-equations that arise when one follows the method of §6 in Eskin[2]. Theorem 2 can be used to deduce uniqueness for constrained forms of the inverse problem. For instance, if one is given either that $\lambda_1 = \lambda_2$ or that $\mu_1 = \mu_2$, and that the Dirichlet-to-Neumann maps are equal, then both Lamé parameters must be equal.

The second way that one can use (16) is to choose θ as a function of $l/|l|$ keeping $\theta(l/|l|)\cdot l = 0$. This makes (16) (and each of the other equations in the sequence $\{H_j = 0\}_{j=2}^{-\infty}$) equivalent to a pseudo-differential equation of the form

$$P(x,D)(\lambda_2 - \lambda_1) + Q(x,D)(\mu_2 - \mu_1) = 0. \tag{18}$$

To prove uniqueness for the inverse boundary value problem we would like to use (18) to bound $\lambda_2 - \lambda_1$ in terms of $\mu_2 - \mu_1$ or to bound $\mu_2 - \mu_1$ in terms of $\lambda_2 - \lambda_1$. Boundary determination for this inverse problem (see Nakamura and Uhlmann[5]) implies that $\lambda_2 - \lambda_1$ and $\mu_2 - \mu_1$ vanish to infinite order on $\partial\Omega$. Nonetheless, (18) does not imply any estimates of this kind without more information on the operators P and Q. Note that the construction in section 2 makes (r, s) a function of θ, so that it contributes to the symbols of P and Q,and, since the solutions of (17) are not explicit in general, one usually does not know what P and Q are. However, (18) becomes useful when one assumes that $\nabla\mu_j$ is small in C^∞. In that case all entries in V_1 except $(V_1)_{44}$ become small and one can solve (17) explicitly modulo small terms. In fact (17) has a unique matrix solution tending to the identity as $|x| \to \infty$ and given by

$$C_{00}(x,\theta) = \begin{pmatrix} I_{3\times 3} & 0 \\ \varphi\theta & 1 \end{pmatrix}, \tag{19}$$

modulo small terms, where $\varphi = (\theta \cdot \partial_x)^{-1}(-\psi b\mu^{-1})$. Choosing

$$p = (\mathrm{Re}\{\theta(l/|l|)\}, 0)$$

in (12), one gets

$$(r^{(j)}, s^{(j)}) = (\mathrm{Re}\{\theta(l/|l|)\}, 0) + (\tilde{r}_0^{(j)}, \tilde{s}_0^{(j)}) + O(\tau^{-1}), \tag{20}$$

where $\tilde{r}_0^{(j)}, \tilde{s}_0^{(j)}$ are symbols of order zero and $\tilde{r}_0^{(j)}$ is small. Since $\theta \cdot \mathrm{Re}\{\theta\} = 1$, (20) makes $P(x,D)$ in (18) simply multiplication by $(\mu_1\mu_2)^{1/2}(\lambda_1 + 2\mu_1)^{-1}(\lambda_2 + 2\mu_2)^{-1}$ plus an operator of order zero whose norm can be made arbitrarily small by taking $\nabla\mu_j$, $j = 1, 2$, sufficiently small in $C^k(\Omega)$. Thus (18) implies

$$||\lambda_2 - \lambda_1||_{H^k(\Omega)} \le C_k ||\mu_2 - \mu_1||_{H^k(\Omega)} \tag{21}$$

for all k, when $\nabla\mu_j$, $j = 1, 2$, is sufficiently small.

Uniqueness for the inverse boundary value problem when $\nabla\mu_j$ is small will follow from (21) if we can find another estimate bounding a norm of $\mu_2 - \mu_1$ by a small constant times a norm of $\lambda_2 - \lambda_1$. One can get this estimate by using $H_0 = 0$ in the following way. The contributions to H_0 come from the expansion of (r, s) in τ up to order τ^{-2}, and are therefore quite complicated in general. Nonetheless, using (19) and taking $p = (0, 0, 0, 1)$ in (12), we find that (17) has the solution $(r_0, s_0) = (0, 0, 0, 1) + (\tilde{r}_0, \tilde{s}_0)$, where $(\tilde{r}_0, \tilde{s}_0)$ and its derivatives can be made arbitrarily small by taking $\nabla\mu_j$, $j = 1, 2$, sufficiently small in C^k-norm. Moreover, checking further, one sees that the higher order terms in the expansion of (r, s) that one constructs using (10) are also small under these hypotheses. When one uses this choice of (r, s) in the construction of $w^{(1)}$ and $w^{(2)}$, one gets

$$\begin{aligned} H_0 = \int_\Omega & e^{il\cdot x}[(\lambda_1 - \lambda_2)A(x, \theta(l/|l|), l) \\ & + (\mu_1 - \mu_2)[(2\mu_1\mu_2)^{-1}|l|^4 + B(x, \theta(l/|l|), l)]]dx, \end{aligned}$$

where A and B are small symbols in l of orders two and four respectively. Thus we have (18) with $||P(x,D)(\lambda_2 - \lambda_1)||_{L^2(\Omega)}$ bounded by a small constant times $||\lambda_2 - \lambda_1||_{H^2(\Omega)}$ and

$$Q(x,D)(\mu_2 - \mu_1) = (2\mu_1\mu_2)^{-1}(\Delta)^2(\mu_2 - \mu_1) + R(x,D)(\mu_2 - \mu_1),$$

where $||R(x, D)(\mu_2 - \mu_1)||_{L^2(\Omega)}$ is bounded by a small constant times $||\mu_2 - \mu_1||_{H^4(\Omega)}$. Thus we have

$$||\mu_2 - \mu_1||_{H^4(\Omega)} \leq \varepsilon(||\mu_2 - \mu_1||_{H^4(\Omega} + ||\lambda_1 - \lambda_2||_{H^2(\Omega)}), \tag{22}$$

where ε can be taken arbitrarily small when $\nabla\mu_j,\ j = 1, 2$ is sufficiently small in $C^k(\Omega)$. Combining (21) and (22), we have the following result.

Theorem 3 *Given that λ_j, μ_j and μ_j^{-1}, $j = 1, 2$, belong to a bounded set, B, in $C^k(\Omega)$ for k sufficiently large, there is an $\varepsilon(B) > 0$ such that $||\nabla\mu_j||_{C^{k-1}(\Omega)} < \varepsilon(B)$, $j = 1, 2$, implies $(\lambda_1, \mu_1) = (\lambda_2, \mu_2)$, if $\Lambda^{(1)} = \Lambda^{(2)}$.*

This is the main result of Nakamura and Uhlmann[6] and it is also Theorem 1 of Eskin and Ralston[4]. The first version of Eskin and Ralston[4] (July 2001) contains Theorem 3 with additional hypothesis $||\nabla\lambda_j||_{C^{k-1}(\Omega)} < \varepsilon(B)$, $j = 1, 2$. We did not notice that our proof did not use this hypothesis until we received a preprint of Nakamura and Uhlmann[6] (November 2001). We are grateful to G.Nakamura and G.Uhlmann for sending this to us.

References

1. D.D.Ang, M.Ikehata, D.D.Trong and M.Yamamoto, *Unique continuation for a stationary isotropic Lamé system with variable coefficients,* CPDE **23** , 371–376(1998).
2. G.Eskin, *Global uniqueness in the inverse scattering problem for the Schrödinger operator with external Yang-Mills potentials,* Commun. Math. Phys. **222**, 503–531(2001).
3. G. Eskin and J.Ralston, *Inverse scattering problem for the Schrödinger equation with magnetic potential at a fixed energy,* Commun. Math. Phys. **173**, 199–224(1995).
4. G.Eskin and J.Ralston, *On the inverse boundary value problem for linear isotropic elasticity,* to appear in Inverse Problems.
5. G. Nakamura and G.Uhlmann *Inverse problems at the boundary for an elastic medium,* SIAM J. Math. Anal. **26**, 263–279(1995).
6. G.Nakamura, and G.Uhlmann *Correction to: Global uniqueness for an inverse boundary problem arising in elasticity,* Inventiones Mathematica **118**, 457–474 (1994), preprint November 2001.
7. J.Sylvester and G.Uhlmann *A global uniqueness theorem for an inverse boundary value problem,* Ann. Math.**125**, 153–169(1987).

UNIQUENESS IN THE TWO-DIMENTIONAL INVERSE GRAVIMETRY PROBLEM : CASE OF VARIABLE COEFFICIENT

[1]SUNGWHAN KIM AND [2]MASAHIRO YAMAMOTO

Department of Mathematical Sciences
The University of Tokyo
3-8-1 Komaba, Meguro, Tokyo 153 Japan
E-mail : [1] *sungwhan@ms.u-tokyo.ac.jp*
[2] *myama@ms.u-tokyo.ac.jp*

We consider an inverse problem of identifying the support D of a source term in an elliptic equation

$$-\Delta u(x) + q(x)\chi_D(x) = 0, \quad x \in R^2 \qquad \text{and} \qquad u(x) = O(\ln|x|) \quad \text{as} \quad |x| \to \infty.$$

Here q is a given positive function and χ_D is the characteristic function of a bounded domain D. By using a Carleman estimate, we prove the global uniqueness in this inverse problem within convex hulls of sums D's of polygons.

1 Introduction

In this paper, we are interested in finding a sum D of a finite number of bounded domains in R^n, $n \geq 2$, from the gradient of its exterior gravitational potential u, which is a solution to the Poisson equation

$$-\Delta u(x) + q(x)\chi_D(x) = 0, \qquad x \in R^n \tag{1}$$

and

$$u(x) = \begin{cases} O(\ln|x|) & \text{as} \quad |x| \to \infty, \quad n = 2 \\ O(|x|^{2-n}) & \text{as} \quad |x| \to \infty, \quad n \geq 3, \end{cases} \tag{2}$$

where $\chi_D = \chi_D(x)$ is the characteristic function of D.

We are mainly concerned with the uniqueness: let u_1 and u_2 be the gravitational potentials that correspond to domains D_1, $D_2 \subset \Omega$ respectivly. Here Ω is a known bounded domain. Then can we conclude $D_1 = D_2$ from

$$\nabla u_1 = \nabla u_2 \qquad \text{on } \partial\Omega \text{ ?}$$

In the case where q is a non-zero constant, this inverse problem has been studied by [1], [9] and [10]. If D_j are star-shaped with respect to their centres of gravity or D_j are convex in x_n-direction, then the uniqueness is true [9]. To convert integrals in D_j into integrals on ∂D_j, in addition to integration by parts, [9] used an observation that if v is harmonic, then so is $x \cdot \nabla v + 3v$. The result in [9] (Chapter 3, §3.1) yields the uniqueness also for polygons under some constraints. However that method is difficult to be applied to the case where q depends on the whole components of x. In Section 3.4 in [9], a counterexample for the uniqueness is given, in the case where D is x_n-convex and q is given, positive and Hölder continuous.

As for related inverse problems of determining piecewise continuous $\gamma = \gamma(x)$ in $\nabla \cdot (\gamma \nabla u) = 0$ in Ω, we can refer [3], [8], [10] and [13]. The case where $\gamma(x) = 1 + k\chi_D(x)$ and k is constant, in particular, has been studied by many researchers. If D_1 and D_2 have a common piece of the boundaries and satisfy other condition (called i-contact), then the uniqueness is true (Friedman and Isakov [3]). Furthermore [3] established the uniqueness for convex polygons ($n = 2$) and convex polyhedra ($n = 3$), under an extra condition

$$\text{diam } D < \text{dist } (D, \partial D). \tag{3}$$

Seo [13] removes (3) to prove the uniqeness within polygons (not necessarily convex) by two boundary measurements. As for the case where k is not constant, however, the uniqueness is still open.

Now our main task in this paper is to prove the global uniqueness within sums D's of polygons for variable coefficient $q(x)$. To the authors' best knowledge, there are no results on the global uniqueness for variable q, even within sums D's of polygons. In this paper, for simplicity, we have to be restricted to the two-dimensional case, that is, $n = 2$.

The paper is organized as follows:

Section 2: Main results

Section 3: Non-existence of H^1-solution

Section 4: Proof of the main result.

2 Main results

Let $\Omega \subset R^2$ be a bounded domain with C^2 boundary $\partial\Omega$ and let D_1, D_2 be sums of a finite number of polygons such that $\overline{D_1}, \overline{D_2} \subset \Omega$:

$$D_1 = \cup_{n=1}^{N_1} D_n^{(1)} \quad \text{and} \quad D_2 = \cup_{n=1}^{N_2} D_n^{(2)},$$

where $D_n^{(1)}$, $D_n^{(2)}$ are polygons and the $\overline{D_n^{(1)}}, \overline{D_n^{(2)}} \subset \Omega$, $\overline{D_n^{(1)}} \cap \overline{D_m^{(1)}} = \overline{D_n^{(2)}} \cap \overline{D_m^{(2)}} = \emptyset$ if $n \neq m$.

Let $u_j \in H^1(R^2)$, $j = 1, 2$, be the weak solution to

$$-\Delta u_j(x) + q(x)\chi_{D_j}(x) = 0, \qquad x \in R^2 \tag{4}$$

and

$$u_j(x) = O(\ln |x|) \quad \text{as} \quad |x| \to \infty. \tag{5}$$

Here $q \in C^2(R^2)$ and $q > 0$ on $\overline{\Omega}$.

Then we can prove (e.g., [4], [11]) that

$$u_j \in H^2(\Omega) \cap C^1(\Omega). \tag{6}$$

For $D \subset R^2$, we denote the convex hull (i,e., the smallest convex set containing D) by $co(D)$. Now we are ready to state our main results.

Theorem 1 If

$$\nabla u_1 = \nabla u_2 \qquad \text{on} \quad \partial\Omega, \tag{7}$$

then $co(D_1) = co(D_2)$.

When we do not assume the convexity, we do not know comprehensive uniqueness results. However, for polygonal domains D_1, D_2, we apply an argument similar to Kim and Yamamoto [14] to prove the following two uniqueness results :

Theorem 2 Assume that D_1 and D_2 are polygons and that a line segment $\overline{A_0B_0} \subset \partial D_1 \cap \partial D_2$ lies on $\partial_{\text{out}}(D_1 \cup D_2)$, where $\partial_{\text{out}}(D \cup E) = \{x \in \partial(D \cup E) \mid$ there exists a continuous curve in $\Omega \setminus \overline{(D \cup E)}$ joining x with some point of $\partial\Omega\}$. Then $\nabla u_1 = \nabla u_2$ on $\partial\Omega$ yields $D_1 = D_2$.

Theorem 3 Assume that D_1 and D_2 are polygons such that there exist two independent vectors $\vec{a}$ and $\vec{b}$ such that all the edges of D_1 and D_2 are parallel to $\vec{a}$ or $\vec{b}$. Then $\nabla u_1 = \nabla u_2$ on $\partial\Omega$ yields $D_1 = D_2$.

3 Non-existence of an H^2- solution to a Cauchy problem

In this section, we will show a proposition about some non-existence of an H^2- solution to a Cauchy problem of the Laplace equation. This proposition plays an essential role in proving our main theorems.

Proposition 1 By D, let us denote the interior of a triangle $\triangle AOB$ that has three vertices O (the origin), A, $B \in R^2$, and by Γ, the union of the edges $\overline{OA}$ and $\overline{OB}$ of $\triangle AOB$.

Let $G \in H^1(D)$, $\partial_{x_2} G$ be bounded and G be strictly positive along the edge $\overline{OA}$ or $\overline{OB}$. Then there exists no solution $v \in H^2(D)$ to

$$\Delta v = G \quad \text{in} \quad D \quad \text{and} \quad v = |\nabla v| = 0 \quad \text{on} \quad \Gamma.$$

The proof of Proposition 1 is lengthy and, by the page limitation, we omit the details (see Kim and Yamamoto [14]). However, we will show one essential tool, a Carleman estimate, for the proof, which we will use similarly to an inverse "hyperbolic" problem (Imanuvilov and Yamamoto [7]). Now we will state only the necessary Carleman estimate.

For $\beta > 0$, we define the functions $\psi = \psi(x_1, x_2)$ and $\varphi = \varphi(x_1, x_2)$ by

$$\psi(x_1, x_2) = x_1 + \beta x_2^2 \qquad \text{and} \qquad \varphi(x_1, x_2) = e^{-\lambda\psi(x_1, x_2)}$$

with a parameter $\lambda > 0$. Moreover we introduce an elliptic operator in the following form

$$Pv = \Delta v + \alpha \partial_{x_1} \partial_{x_2} v,$$

where a constant α satisfies

$$|\alpha| < 2.$$

We set $\nabla = (\partial_{x_1}, \partial_{x_2})$.

Proposition 2 Let $Q := (0, R) \times (-T, T)$ be an open rectangle in R^2. Then there exists $\lambda_0 > 0$ such that for all $\lambda > \lambda_0$ there exist constants $s_0 = s_0(\lambda) > 0$ and $C = C(s_0, \lambda_0, R, T)$ such that

$$\int_Q (s|\nabla v|^2 + s^3 v^2) e^{2s\varphi} \; dx \le C \int_Q |Pv|^2 e^{2s\varphi} \; dx$$

for all $s > s_0(\lambda)$, provided that

$$\begin{cases} Pv \in L^2(Q), \qquad v \in H^1(Q) \\ v(0,\cdot) = v(R,\cdot) = 0 \qquad \text{in } L^2(-T,T) \\ \partial_{x_1} v(0,\cdot) = \partial_{x_1} v(R,\cdot) = 0 \qquad \text{in } H^{-\frac{1}{2}}(-T,T) \\ v(\cdot,T) = v(\cdot,-T) = 0 \qquad \text{in } L^2(0,R) \\ \partial_{x_2} v(\cdot,T) = \partial_{x_2} v(\cdot,-T) = 0 \qquad \text{in } H^{-\frac{1}{2}}(0,R). \end{cases}$$

For the proof of Proposition 2, we can refer to Theorem 8.3.1 in [6] for example.

4 Proof of Theorem 1

Let us define $v := u_1 - u_2$ in R^2 and denote by $\mathcal{D}$ the connected component of $R^2 \setminus (D_1 \cup D_2)$ containing $\partial\Omega$. By (4), (5) and (7) we have

$$\Delta v = 0 \qquad \text{in} \quad R^2 \setminus \overline{\Omega}, \tag{8}$$

$$\nabla v = 0 \qquad \text{on} \quad \partial\Omega, \tag{9}$$

and

$$v(x) = C \ln |x| \qquad \text{as} \quad |x| \to \infty \quad \text{for some constant } C. \tag{10}$$

Then it follows from Hopf's Lemma that

$$v \equiv 0 \qquad \text{in} \quad R^2 \setminus \overline{\Omega}. \tag{11}$$

Since $R^2 \setminus \overline{\Omega} \subset \mathcal{D}$ and v is harmonic in $\mathcal{D}$, (11) implies

$$v \equiv 0 \qquad \text{in} \quad \mathcal{D}. \tag{12}$$

Assume contrarily that $co(D_1) \neq co(D_2)$. Then, since $co(D_1)$ and $co(D_2)$ are convex polygons, there exists a vertex O of $co(D_1)$ such that $O \in \Omega \setminus \overline{co(D_2)}$ or a vertex O of $co(D_2)$ such that $O \in \Omega \setminus \overline{co(D_1)}$. Without loss of generality, we may assume the former case. Henceforth $\triangle OAB$ means the interior of the triangle with the vertices O, A and B. Then, since $O \in \Omega \setminus \overline{co(D_2)}$, we can take a sufficiently small triangle $\triangle OAB$ such that

$$\overline{OA} \cup \overline{OB} \subset \partial(co(D_1)) \quad \text{and} \quad \triangle OAB \subset co(D_1) \setminus \overline{co(D_2)}. \tag{13}$$

Since $\Omega \setminus (\overline{co(D_1) \cup co(D_2)}) \subset \mathcal{D}$, we have

$$\overline{OA} \cup \overline{OB} \subset \overline{\mathcal{D}}. \tag{14}$$

Then there exists a polygon $D_k^{(1)}$ for some integer $1 \leq k \leq N_1$ such that O is a vertex of $D_k^{(1)}$. Since $co(D_1)$ is convex in a neighborhood of O, so is $D_k^{(1)}$. Therefore O is a convex vertex of $D_k^{(1)}$. By $co(D_1) \supset D_k^{(1)}$, we can take $\triangle OA'B'$ such that

$$\overline{OA'} \cup \overline{OB'} \subset \partial D_k^{(1)} \quad \text{and} \quad \triangle OA'B' \subset \triangle OAB. \tag{15}$$

Hence it follows from $\triangle OAB \subset co(D_1) \setminus \overline{co(D_2)}$ that $\triangle OA'B' \subset co(D_1) \setminus \overline{co(D_2)}$. Moreover, by (13), (14) and (15), we see that $(\overline{OA'} \cup \overline{OB'}) \subset \overline{\mathcal{D}}$. Therefore, by (12), the function v satisfies

$$\begin{array}{ll} \Delta v = q \quad \text{in} \quad \triangle OA'B', & \\ v = |\nabla v| = 0 \quad & \text{on} \quad \overline{OA'} \cup \overline{OB'}. \end{array} \tag{16}$$

The condition (6) means that the function v is a H^2-solution of (16) and the assumptions of Proposition 1 are satisfied. Hence a contradiction occurs, and so we can conclude that $co(D_1) = co(D_2)$.

References

[1] G. Anger, Inverse Problems in Differential Equations, Plenum Publ., New York, 1990

[2] A. Friedman, Detection of mines by electric measurements, SIAM J.Appl. Math., **47**(1987), 201-212

[3] A. Friedman and Isakov V., On the uniqueness in the inverse conductivity problem with one measurement, Indiana Univ. Math. J., **38**(1989), 563-579

[4] D. Gilbarg and N.S. Trudinger, Elliptic Partial Differential Equations of Second Order, Springer-Verlag Berlin Heidelberg, 1977

[5] F. Hettlich and W. Rundell, Recovery of the support of a source term in an elliptic differential equation, Inverse Problem, **13**(1997), 959-976

[6] L. Hörmander, Linear Partial Differential Operators, Springer-Verleg Berlin, 1963

[7] O.Y. Imanuvilov and M. Yamamoto, Global Lipschitz stability in an inverse hyperbolic problem by interior observations, Inverse Problems, **17**(2001), 717-728

[8] V. Isakov, On uniqueness of recovery of a discontinuous conductivity coefficient, Commun. Pure Appl. Math., **41**(1988), 865-877

[9] V. Isakov, Inverse Source Problems, American Mathemathical Society, Providence, Rhode Island, 1990

[10] V. Isakov, Inverse Problems for Partial Differential Equations, Springer-Verleg, Berlin, 1998

[11] O.A. Ladyzhenskaya and N.N. Ural'tseva, Linear and Quasilinear Elliptic Equations, Academic Press, New York, 1968

[12] H. Kang, K. Kwon and K. Yun, Recovery of an inhomogeneity in an elliptic equation, Inverse Problems, **17**(2001), 25-44

[13] J.K. Seo, On the uniqueness in the inverse conductivity problem, J. Fourier Anal. Appl., **2**(1996), 227-235

[14] S. Kim and M. Yamamoto, Uniqueness in identification of the support of a source term in an elliptic equation, Preprint Series, UTMS 2001-30, University of Tokyo, Graduate School of Mathematical Sciences, 2001

INVERSION OF DISCONTINUOUS ANISOTROPIC CONDUCTIVITIES

D. LESNIC
Department of Applied Mathematics, University of Leeds, Leeds, LS2 9JT, UK
E-mail: amt5ld@amsta.leeds.ac.uk

An inverse problem is considered to identify the geometry of discontinuities in a conductive material $\Omega \subset R^d$ with anisotropic conductivity $(I + (K - I)\chi_D)$ from Cauchy data measurements taken on the boundary $\partial\Omega$, where $D \subset \Omega$, K is a symmetric and positive definite tensor not equal to the identity I and χ_D is the characteristic function of the domain D. The previous results of Ikehata[1] for estimating the size of the inclusion D are proved and applied to several examples. Further, we develop an integral representation of the solution and we propose an efficient boundary integral method in conjunction with a least-squares constrained minimization procedure to detect the anisotropic inclusion D.

1 Introduction

We consider the inverse conductivity problem which requires the determination of an anisotropic object D contained in a domain Ω, from measured electric voltage, ϕ, and electric current flux, $\frac{\partial \phi}{\partial n}$, on the boundary $\partial\Omega$. Uniqueness results of identifying D when K is known can be found in Ikehata[1], Isakov[2] and Ikehata *et al.*[3]. However, their proofs do not tell us how to reconstruct the anisotropic inclusion D and the purpose of this paper is to develop such a numerical algorithm. The plan of the paper is as follows. The mathematical formulation of the inverse problem under investigation is presented in section 2, whilst in section 3 the previous results of Ikehata[1] for estimating the size of the inclusion D are proved and applied to several examples. Further, in section 4, we develop an integral representation of the solution and we propose an efficient boundary integral method in conjunction with a least-squares constrained minimization procedure to detect the anisotropic inclusion D. Finally, section 5 presents conclusions and future work.

2 Mathematical Formulation

The problem can be formulated mathematically as follows. Let Ω be a bounded domain of R^d, $d \geq 2$, with Lipschitz boundary and D a subdomain compactly contained in Ω. The anisotropic conductivity tensor B of the domain D is non-dimensionalised with respect to the isotropic constant conductivity tensor $kI \neq B$, $k > 0$, of the domain $\Omega - D$. Assuming the physical requirement for B to be symmetric and positive definite, we obtain $K = B/k \neq I$ to be a symmetric and positive tensor (matrix), representing the non-dimensional anisotropic conductivity of the inclusion D, whilst the medium $\Omega - D$ becomes isotropic with conductivity I. Then the refraction (transmission, conjugate) problem for the electrical potential ϕ is given by

$$\nabla \bullet ((I + (K - I)\chi_D)\nabla\phi) = 0, \quad \text{in } \Omega \tag{1}$$

$$\alpha \frac{\partial \phi}{\partial n} + \beta\phi = h, \quad \text{on } \partial\Omega \tag{2}$$

subject to refraction conditions related to the continuity of the voltage ϕ and its current flux densities $(\partial\phi/\partial n^-)$ and $(K\nabla\phi) \bullet \underline{n}^+$ across the interface ∂D, where n, n^- and n^+ are the outward unit normals to the boundaries $\partial\Omega$, $\partial(\Omega - D) - \partial\Omega$ and ∂D, respectively. In Eq. (2), $h \in L^2(\partial\Omega)$ is a given function and $\alpha = 0$, $\beta = 1$, i.e. Dirichlet condition, or $\alpha = 1$, $\beta = 0$, i.e. Neumann condition. Note that in the Neumann case we further require the conditions

$$\int_{\partial\Omega} h \, dS_\Omega = \int_{\partial\Omega} \frac{\partial \phi}{\partial n} \, dS_\Omega = \int_{\partial\Omega} \phi \, dS_\Omega = 0. \tag{3}$$

It is well-known that the direct problem of finding $\phi \in H^1(\Omega)$ satisfying (1) and (2), when K and D are known, has a unique solution, see Ladyzhenskaya[4] (p.197). Assuming that K is known, the inverse conductivity problem requires finding D from the knowledge of the Dirichlet-to-Neumann map if $\alpha = 0$, $\beta = 1$, or from the Neumann-to-Dirichlet map if $\alpha = 1$, $\beta = 0$, (for finitely many data h on $\partial\Omega$). In the following section we estimate the size of the unknown inclusion D from one boundary measurement.

3 Size Estimation of the Inclusion

Let us consider for simplicity the Dirichlet boundary condition case, i.e. $\alpha = 0$, $\beta = 1$, and redefine h in (2) by $f \in H^{1/2}(\partial\Omega)$. We can define the Dirichlet-to-Neumann map $\Lambda_D(f) = \frac{\partial \phi}{\partial n} := g \in H^{-1/2}(\partial\Omega)$, via Green's formula as

$$< \Lambda_D f, q >= \int_{\partial\Omega} g \, q \, dS_\Omega = \int_\Omega ((I + (K - I)\chi_D)\nabla\phi) \bullet \nabla w \, d\Omega \tag{4}$$

where $q \in H^{1/2}(\partial\Omega)$ and w is any function in $H^1(\Omega)$ with $w_{|\partial\Omega} = q$. Let $\phi_0 \in H^1(\Omega)$ be the unique solution of the direct problem with $K = I$, i.e.

$$\begin{aligned} \nabla^2\phi_0 &= 0, \quad \text{in } \Omega \\ \phi_0 &= f, \quad \text{on } \partial\Omega. \end{aligned} \tag{5}$$

Then we can define the Dirichlet-to-Neumann map $\Lambda_0(f) = \frac{\partial \phi_0}{\partial n} := g_0 \in H^{-1/2}(\partial\Omega)$, via Green's formula as

$$< \Lambda_0 f, q >= \int_{\partial\Omega} g_0 q \, dS_\Omega = \int_\Omega \nabla\phi_0 \bullet \nabla w \, d\Omega \tag{6}$$

where $q \in H^{1/2}(\partial\Omega)$ and w is any function in $H^1(\Omega)$ with $w_{|\partial\Omega} = q$. In particular, we have

$$< \Lambda_0 f, f >= \int_{\partial\Omega} g_0 f \, dS_\Omega = \int_\Omega | \nabla\phi_0 |^2 \, d\Omega \tag{7}$$

and

$$< \Lambda_D f, f >= \int_{\partial\Omega} g f \, dS_\Omega = \int_\Omega ((I + (K - I)\chi_D)\nabla\phi) \bullet \nabla\phi \, d\Omega. \tag{8}$$

Since K is symmetric and positive definite there exists $\delta > 0$ such that

$$K\underline{x} \bullet \underline{x} \geq \delta \mid \underline{x} \mid^2, \quad \forall \underline{x} \in R^d. \tag{9}$$

From Eqs. (8) and (9) we obtain

$$< \Lambda_D f, f > \geq \int_{\Omega - D} | \nabla\phi |^2 \, d(\Omega - D) + \delta \int_D | \nabla\phi |^2 \, dD \geq 0. \tag{10}$$

We also have that $< \Lambda_0 f, f >= \int_\Omega | \nabla\phi_0 |^2 \, dD \geq 0$ and thus $< \Lambda_0 f, f >> 0$ if and only if $f \not\equiv constant$, or equivalently if and only if $< \Lambda_D f, f >> 0$.

The following proposition and theorems can be found in Ikehata[1], but without proof. We find it useful to give these proofs. First we have the following Lemmas which were proved in Ikehata[1].

Lemma 1 *Let C_1 and C_2 be symmetric and positive definite tensors and u_1, $u_2 \in H^1(\Omega)$ be the unique solutions of*

$$\begin{aligned} \nabla \bullet (C_j \nabla u_j) &= 0, \quad in\,\Omega \\ u_j = \phi_j &\in H^{1/2}(\partial\Omega), \quad on\,\partial\Omega. \end{aligned} \tag{11}$$

Then the following identity holds:

$$\begin{aligned} &\int_\Omega [(C_1 \nabla(u_1 - u_2)) \bullet \nabla(u_1 - u_2) + ((C_2 - C_1)\nabla u_2) \bullet \nabla u_2] \, d\Omega \\ &=< \Lambda_{C_1}(\phi_1) - \Lambda_{C_2}(\phi_2), \phi_1 - \phi_2 > + < (\Lambda_{C_2} - \Lambda_{C_1})(\phi_2), \phi_1 > \end{aligned} \tag{12}$$

where

$$< \Lambda_{C_j}(\phi_1), q >:= \int_\Omega (C_j \nabla u_i) \bullet \nabla w \, d\Omega, \quad i, j = 1, 2 \tag{13}$$

for $q \in H^{1/2}(\partial\Omega)$ and w any function in $H^1(\Omega)$ with $w_{|\partial\Omega} = q$.

Remark 1 *The proof of (12) makes use of Alessandrini's identity, namely,*

$$< (\Lambda_{C_2} - \Lambda_{C_1})(\phi_2), \phi_1 >= \int_\Omega ((C_2 - C_1)\nabla u_2) \bullet \nabla u_1 \, d\Omega \tag{14}$$

Lemma 2 *Let C_j, u_j, $j = 1, 2$ be as in Lemma 1. Then*

$$\int_\Omega ((I - C_2^{-1} C_1)\nabla u_1) \bullet (C_1 \nabla u_1) \, d\Omega \leq < \Lambda_{C_1}(\phi_1) - \Lambda_{C_2}(\phi_2), \phi_1 - \phi_2 > \\ + < (\Lambda_{C_2} - \Lambda_{C_1})(\phi_2), \phi_1 > . \tag{15}$$

Now we can formulate the following proposition.

Proposition 1 *Assume that f is not a constant function which is equivalent to $< \Lambda_0 f, f >> 0$. Then*

$$\int_D ((I - K^{-1})\nabla \phi_0) \bullet \nabla \phi_0 \, dD \leq \frac{< \Lambda_0 f, f >}{< \Lambda_D f, f >} < (\Lambda_D - \Lambda_0) f, f >, \tag{16}$$

$$< (\Lambda_D - \Lambda_0) f, f > \leq \int_D ((K - I)\nabla \phi_0) \bullet \nabla \phi_0 \, dD. \tag{17}$$

Proof: To establish the first inequality, take in Lemma 1, $C_1 = I$, $\phi_1 = f$, $C_2 = I + (K - I)\chi_D$, $\phi_2 = (1 + t)f$, with t an arbitrary real number. Then the solutions of (11) are $u_1 = \phi_0$ and $u_2 = (1 + t)\phi$. Using the inequality (15) from Lemma 2 we obtain

$$\begin{aligned} \int_D ((I - K^1)\nabla \phi_0) \bullet \nabla \phi_0 \, dD &\leq -t < \Lambda_0 f - (1 + t)\Lambda_D f, f > \\ &\quad +(1 + t) < (\Lambda_D - \Lambda_0) f, f > \\ &= < \Lambda_D f, f > \left(t + \frac{< (\lambda_D - \Lambda_0) f, f >}{\Lambda_D f, f >} \right)^2 \\ &\quad + \frac{< \Lambda_0 f, f >}{\Lambda_D f, f >} < (\Lambda_D - \Lambda_0) f, f > . \end{aligned} \tag{18}$$

Remark that since f is not a constant, from Eq.(10) we have that $< \Lambda_D f, f >> 0$. Putting $t = -\frac{<(\lambda_D - \Lambda_0) f, f>}{<\Lambda_D f, f>}$ in (18) we obtain the inequality

(16). To establish the second inequality, take in Lemma 1, $C_1 = I + (K-I)\chi_D$, $C_2 = I$, $\phi_1 = \phi_2 = f$ in which case $u_1 = \phi$ and $u_2 = \phi_0$. Using (7) and (8) we obtain

$$< (\Lambda_0 - \Lambda_D)f, f > = \int_\Omega ((I + (K-I)\chi_D)\nabla(\phi - \phi_0)) \bullet \nabla(\phi - \phi_0)\, d\Omega$$
$$- \int_\Omega ((K-I)\nabla\phi_0) \bullet \nabla\phi_0\, d\Omega. \tag{19}$$

Since the first term in the right-hand side of (19) is non-negative we obtain the inequality (17).

Theorem 1 *Let $f \in H^{1/2}(\partial\Omega)$ such that $inf_\Omega \mid \nabla\phi_0 \mid > 0$. Then we have the following estimates for the size of D.*

(i) If $(K-I)\chi_D$ is positive definite, then so is $I - K^{-1}\chi_D$ and

$$min\{a^{-1}\}(sup_\Omega \mid \nabla\phi_0 \mid)^{-2} < (\Lambda_D - \Lambda_0)f, f > \leq \mid D \mid$$
$$\leq max\{b^{-1}\}(inf_\Omega \mid \nabla\phi_0 \mid)^{-2} \frac{< \Lambda_0 f, f >}{< \Lambda_D f, f >} < (\Lambda_D - \Lambda_0)f, f > \tag{20}$$

where a is an eigenvalue of $(K-I)\chi_D$ and b is an eigenvalue of $I - K^{-1}\chi_D$.

(ii) If $(K-I)\chi_D$ is negative definite, then so is $I - K^{-1}\chi_D$ and

$$max\{-a^{-1}\}(inf_\Omega \mid \nabla\phi_0 \mid)^{-2} < (\Lambda_0 - \Lambda_D)f, f > \geq \mid D \mid$$
$$\geq min\{-b^{-1}\}(sup_\Omega \mid \nabla\phi_0 \mid)^{-2} \frac{< \Lambda_0 f, f >}{< \Lambda_D f, f >} < (\Lambda_0 - \Lambda_D)f, f > . \tag{21}$$

Proof: Consider only the case (i) when $(K-I)$ is positive definite. We have that $I - K^{-1} = (K-I)[(K-I)^{-1} - K^{-1}](K-I)$, so $(K-I)$ positive definite implies that $(I - K^{-1})$ is positive definite. Then

$$min\{b\} \mid D \mid inf_\Omega \mid \nabla\phi_0 \mid^2 \leq \int_D ((I - K^{-1})\nabla\phi_0) \bullet \nabla\phi_0\, dD,$$
$$\int_D ((K-I)\nabla\phi_0) \bullet \nabla\phi_0\, dD \leq max\{a\} \mid D \mid sup_\Omega \mid \nabla\phi_0 \mid^2 \tag{22}$$

and the inequality (20) follows from (16) and (17).

Remark 2 *If $n = 2$, the condition $inf_\Omega \mid \nabla\phi_0 \mid > 0$ is ensured if one chooses the Dirichlet data f to be nonconstant and satisfying the property that f is monotonic on $\partial\Omega_1$ and on $\partial\Omega_2$, separately, where $\partial\Omega$ is decomposed into two disjoint arcs $\partial\Omega_1$ and $\partial\Omega_2$, see Alessandrini and Magnanini[5]. If we replace the Dirichlet condition $\phi \mid_{\partial\Omega} = f$ with the Neumann condition $\frac{\partial\phi}{\partial n} \mid_{\partial\Omega} = g$, then the monotonicity conditions should be replaced by the conditions $g \not\equiv 0$, $g \geq 0$ on $\partial\Omega_1$, $g \leq 0$ on $\partial\Omega_2$. Further, in any dimensions, the condition $inf_\Omega \mid \nabla\phi_0 \mid > 0$ is ensured if one chooses, for example, $g(\underline{x}) = < \underline{\nu}, \underline{n}(\underline{x}) >$, where $\underline{\nu}$ is a constant vector, see Kang et al [6]. Also, boundary conditions of the mixed type can be considered, see Alessandrini and Isakov[7]. Needless to say, we can obtain similar results to those of Theorem 1 even in the case where K depends on $\underline{x}$.*

Theorem 2 *Let λ be a nonzero eigenvalue of $(K - I)\chi_D$ and $\underline{a}$ the corresponding normalized eigenvector. Set the Dirichlet boundary data $f(\underline{s}) = \underline{a} \bullet \underline{s}$ for $\underline{s} \in \partial\Omega$. Then*

$$\begin{aligned} &\tfrac{\lambda}{\lambda+1} \mid D \mid \leq \tfrac{<\Lambda_0 f, f>}{<\Lambda_D f, f>} < (\Lambda_D - \Lambda_0) f, f >, \\ &< (\Lambda_D - \Lambda_0) f, f > \leq \lambda \mid D \mid . \end{aligned} \tag{23}$$

Proof: Since $K = I + (K - I)$ is positive definite we have that $\lambda + 1 > 0$. Also $(K - I)\underline{a} = \lambda\underline{a}$ and hence $(I - K^{-1})\underline{a} = \frac{\lambda}{\lambda+1}\underline{a}$. Since ϕ_0 is the solution of problem (5) with $f(\underline{s}) = \underline{a} \bullet \underline{s}$ on $\partial\Omega$, we obtain that $\phi_0(\underline{s}) = \underline{a} \bullet \underline{s}$ in Ω and thus $\nabla\phi_0 = \underline{a}$. Using that $\mid \underline{a} \mid = 1$, from (16) and (17), and $f \not\equiv$ *constant*, we obtain that

$$\frac{\lambda}{\lambda+1} \mid D \mid = \int_D ((I - K^{-1})\underline{a}) \bullet \underline{a}\, dD \leq \frac{< \Lambda_0 f, f >}{< \Lambda_D f, f >} < (\Lambda_D - \Lambda_0) f, f > \tag{24}$$

and

$$< (\Lambda_D - \Lambda_0) f, f > \leq \int_D ((K - I)\underline{a}) \bullet \underline{a}\, dD = \lambda \mid D \mid . \tag{25}$$

Remark 3

(i) From Theorem 2, one can estimate the size $\mid D \mid$ in terms of $< (\Lambda_D - \Lambda_0) f, f >$.

(ii) We do not need to assume that $(K - I)$ is positive or negative definite.

(iii) We do not assume any regularity on D, only that is a Lebesgue measurable subset of Ω.

(iv) $< \Lambda_0 f, f >= \int_\Omega | \nabla\phi_0 |^2 \, d\Omega = \int_\Omega | \underline{a} |^2 \, d\Omega =| \Omega |$ *and* $< \Lambda_D f, f >= \int_{\partial\Omega} (\underline{a} \bullet \underline{s}) g(\underline{s}) \, dS_\Omega := T > 0.$

(v) $< \Lambda_D f, f >=< \Lambda_0 f, f >$ *if and only if* $| D |= 0$. *So if* $T =| \Omega |$ *then there is no inclusion.*

The estimates (23) give

$$T- | \Omega |\leq \lambda | D |\leq \frac{(\lambda+1) | \Omega |}{T}(T- | \Omega |). \tag{26}$$

It is interesting to look at the case when $T- | \Omega |= \frac{(\lambda+1)|\Omega|}{T}(T- | \Omega |)$. This can happen if and only if $T =| \Omega |$ (in which case $| D |= 0$) or $T = (\lambda+1) | \Omega |$ (in which case $| D |=| \Omega |$).

Example: Let Ω be the unit circle in 2-D and consider an orthotropic material with conductivity $K = \begin{bmatrix} 1 & 0 \\ 0 & 2 \end{bmatrix}$. Then $\lambda = 1$, $\underline{a} = \begin{bmatrix} 0 \\ 1 \end{bmatrix}$, $| \Omega |= \pi$, $f(\underline{s}) = \underline{a} \bullet \underline{s} = y$ and $T = \int_{\partial\Omega} y g(x, y) dS_\Omega = \int_0^{2\pi} \sin(\theta) g(\cos(\theta), \sin(\theta)) d\theta$. In this case the estimate (26) gives

$$T - \pi \leq| D |\leq \frac{2\pi}{T}(T - \pi) \tag{27}$$

so a necessary condition for the existence of a solution is $2\pi \geq T \geq \pi$. Of course the Neumann data (current flux) g cannot be taken arbitrarily once the Dirichlet data f already been given. However assuming that a solution exists, We can then consider various choices of the current flux g satisfying $\int_0^{2\pi} g(\cos(\theta), \sin(\theta)) d\theta = 0$ as follows.

(i) $g = \sin(\theta)$, then $T = \int_0^{2\pi} \sin^2(\theta) d\theta = \pi =| \Omega |$ and hence $| D |= 0$, i.e. there is no inclusion, as expected since in this case $\phi = \phi_0 = y$.

(ii) $g = \cos(\theta)$, then $T = \int_0^{2\pi} \sin(\theta)\cos(\theta) d\theta = 0$ and hence f would be a constant which is a contradiction. In this case no size estimation can be established.

(iii) $g = \frac{3}{2}\sin(\theta)$, then $T = \int_0^{2\pi} \frac{3}{2}\sin^2(\theta) d\theta = \frac{3\pi}{2}$ and the estimates (27) give $\frac{\pi}{2} \leq| D |\leq \frac{2\pi}{3}$.

Let $K = \begin{bmatrix} k_{11} & k_{12} \\ k_{12} & k_{22} \end{bmatrix}$ be a positive definite matrix, i.e. $k_{11}k_{22} - k_{12}^2 > 0$.

The eigenvalues of $(K - I)$ are given by $\lambda_{1,2} = \frac{(k_{11}+k_{22}-2) \pm \sqrt{(k_{11}-k_{22})^2 + 4k_{12}^2}}{2}$. Note that since $K \neq I$ we have that $\lambda_1 \neq \lambda_2$ and that an eigenvalue can be zero if and only if $k_{11}k_{22} - k_{12}^2 = k_{11} + k_{22} - 1$. In particular in the orthotropic case, i.e. $k_{12} = 0$, $\lambda_{1,2} = \frac{(k_{11}+k_{22}-2)\pm|k_{11}-k_{22}|}{2}$ can be zero if and only if $k_{11} + k_{22} = 1 + k_{11}k_{22}$, i.e. when $k_{11} = 1$ or $k_{22} = 1$. So we have the possibility to choose better experiments for an orthotropic inclusion for which $k_{11} \neq 1$ and $k_{22} \neq 1$. Let us therefore consider $K = \begin{bmatrix} 2 & 0 \\ 0 & 3 \end{bmatrix}$. Then $\lambda_1 = 1$ and $\lambda_2 = 2$ with the corresponding normalized eigenvectors for $(K - I)$ given by $\underline{a}_1 = \begin{bmatrix} 1 \\ 0 \end{bmatrix}$ and $\underline{a}_2 = \begin{bmatrix} 0 \\ 1 \end{bmatrix}$. Then we can experiment with $f_1(\underline{s}) = \underline{a}_1 \bullet \underline{s} = x$, or $f_2(\underline{s}) = \underline{a}_2 \bullet \underline{s} = y$, which give $T_1 = \int_0^{2\pi} \cos(\theta)g(\cos(\theta), \sin(\theta))d\theta$ and $T_2 = \int_0^{2\pi} \sin(\theta)g(\cos(\theta), \sin(\theta))d\theta$, respectively. From the estimates (27) we can choose the experiment. However, in general there is no control between g and D. All one can do in practice is to calculate T_1 and T_2 and according to (27) decide which experiment is the better, i.e. which estimates are the sharper.

4 Boundary Integral Formulation

The refraction model under investigation, given by Eqs. (1) and (2) and the corresponding transmission conditions, can be recast in the more convenient form by defining $\phi = \phi_1$ in $\Omega - D$ and $\phi = \phi_2$ in D, where ϕ_1 and ϕ_2 satisfy

$$\nabla^2 \phi_1 = 0, \quad \text{in } \Omega - D \tag{28}$$

$$\alpha \frac{\partial \phi_1}{\partial n} + \beta \phi_1 = h, \quad \text{on } \partial\Omega \tag{29}$$

$$\nabla \bullet (K \nabla \phi_2) = 0, \quad \text{in } D \tag{30}$$

$$\phi_1 = \phi_2, \quad \frac{\partial \phi_1}{\partial n^-} = -(K\nabla\phi_2) \bullet \underline{n}^+ = -\frac{\partial \phi_2}{\partial n^*}, \quad \text{on } \partial D \tag{31}$$

where $\underline{n}^* = K\underline{n}^+$. Let us assume now that Ω and D are simply connected and have C^2 boundary. Then $\phi_1 \in H^2(\Omega - D) \cap C(\overline{\Omega})$ and $\phi_2 \in H^2(D)$, see Ladyzenskaia[4] (p.198), and thus Green's formula is applicable.

Prior to this study, Kang and Seo[8] and Ki and Sheen[9] developed an integral representation for the isotropic case. However, their approach cannot be extended to the anisotropic case so easily. Instead, the boundary integral methods of Duraiswami *et al.* [10] and Lesnic[11] for the isotropic case can be extended to the anisotropic case as follows. Let G and G_K be fundamental solutions of the Laplace equation and anisotropic Laplace Eq.(1) in R^d, respectively,

$$G(\underline{x},\underline{\xi}) = \begin{cases} -\frac{1}{2\pi} ln(r),\ d=2 \\ \\ \frac{1}{4\pi r}, \quad d=3 \end{cases} \tag{32}$$

$$G_K(\underline{x},\underline{\xi}) = \begin{cases} -\frac{|K^{-1}|^{1/2}}{2\pi} ln(R),\ d=2 \\ \\ \frac{|K^{-1}|^{1/2}}{4\pi R}, \quad d=3 \end{cases} \tag{33}$$

where $r = |\ \underline{x} - \underline{\xi}\ |$, $|\ K^{-1}\ |$ is the determinant of the inverse matrix K^{-1} and the geodesic distance R is defined by $R^2 = K^{-1}(\underline{x} - \underline{\xi}) \bullet (\underline{x} - \underline{\xi})$.

Considering, for simplicity, the Dirichlet boundary condition in (29), i.e. $\phi_1(x) = f(x)$ for $x \in \partial\Omega$, and applying the interface transmission condition (31) we obtain the following integral representation formulae:

$$\eta(\underline{x})\phi_1(\underline{x}) = \int_{\partial\Omega} \left[G(\underline{x},\underline{x'}) \frac{\partial \phi_1}{\partial n}(\underline{x'}) - f(\underline{x'}) \frac{\partial G}{\partial n(\underline{x'})}(\underline{x},\underline{x'}) \right] dS_\Omega -$$

$$\int_{\partial D} \left[G(\underline{x},\underline{x'})(K\nabla\phi_2(\underline{x'})) \bullet \underline{n}^+(\underline{x'}) + \phi_2(\underline{x'}) \frac{\partial G}{\partial n^-(\underline{x'})}(\underline{x},\underline{x'}) \right] dS_D,$$

$$x \in \overline{\Omega - D} \tag{34}$$

$$\eta(\underline{x})\phi_2(\underline{x}) = \int_{\partial D} \left[G_K(\underline{x},\underline{x'}) \frac{\partial \phi_2}{\partial n^*}(\underline{x'}) - \phi_2(\underline{x'}) \frac{\partial G_K}{\partial n^*(\underline{x'})}(\underline{x},\underline{x'}) \right] dS_D,\ x \in \overline{D} \tag{35}$$

where $\eta(\underline{x}) = 1$, if $\underline{x} \in \Omega - \partial D$ and $\eta(\underline{x}) = 0.5$ if $\underline{x} \in \partial\Omega \cup \partial D$.

Analytical solutions to (34) and (35) are in general not feasible and therefore some form of numerical approximation given by the boundary element method, see Chang *et al* [12], has to be performed. In two-dimensions, i.e. $d = 2$, the boundary $\partial\Omega$ is discretized uniformly into M constant boundary elements in a counterclockwise sense, and the boundary ∂D into M constant boundary elements in both a counterclockwise and clockwise sense. Then applying Eq.(34) at the nodes on $\partial\Omega \cup \partial D$ and Eq.(35) at the nodes on ∂D, results in a system of $3M$ nonlinear equations with $5M$ unknowns, say $A(\underline{x})\underline{u} = \underline{b}$, where $\underline{u}$ contains the unspecified values of $\phi_1 = \phi_2$ on ∂D, $\partial\phi_1/\partial n$ on $\partial\Omega$ and $\partial\phi_2/\partial n^*$ on ∂D, and $\underline{x} = (x_j, y_j)$ for $j = \overline{1, M}$ are the two-dimensional Cartesian coordinates of the boundary element nodes on ∂D. For a given initial guess of $\underline{x}$ this system of equations becomes linear and it can be solved to determine the (calculated) current flux data $\partial\phi_1/\partial n$ on $\partial\Omega$. We can then minimize

$$Objf(\underline{x}) := \| \left(\frac{\partial \phi_1}{\partial n}\right)_{calculated} - \left(\frac{\partial \phi_1}{\partial n}\right)_{measured} \|^2. \tag{36}$$

However, even by minimizing the functional (36), the above system of equations is still underdetermined having $3M$ equations with $4M$ unknowns. Additional information is therefore necessary in order to account for the ill-conditioned nature of the discretized inverse problem. Such constraints (additional information) may include:

(i) $\underline{x} \in \Omega$, such that the unknown object D is always contained in Ω.

(ii) The inclusion in (36) of penalty regularizing terms such as $\lambda_1\|\underline{x}\|^2$, $\lambda_2\|\underline{x}'\|^2$ or $\lambda_3\|\underline{x}''\|^2$, where λ_1, λ_2, $\lambda_3 > 0$ are regularization parameters which may allow for continuous, C^1 or C^2-boundaries ∂D, which also stabilize the numerical solution.

(iii) The boundary ∂D is the union of two disjoint graphs of functions, say $y^1 = y^1(x)$, $y^2 = y^2(x)$, $x \in [a, b]$, such that the number of unknowns is reduced by M, with only the components y_j for $j = \overline{1, M}$ needed to be recovered.

All these situations will be numerically investigated in a future work following the lines of Lesnic[11] for an isotropic conductivity. Moreover, the recent reconstruction methods of Ikehata[13,14] can also be considered numerically using the boundary element method proposed. So far, preliminary numerical studies showed that elliptical inclusions can be uniquely retrieved from a single boundary measurement, but the theoretical proof still remains a conjecture.

5 Conclusions

In this paper the inverse conductivity problem which requires the determination of the location, size and/or non-dimensional anisotropic conductivity, K, of a circular inclusion D contained in a domain Ω from measured electric voltage, ϕ, and electric current flux, $\frac{\partial \phi}{\partial n}$, on the boundary $\partial\Omega$ has been investigated. The proofs of the theorems quoted in Ikehata[1] as exercises for the reader have been provided and various examples have been discussed. Furthermore, a boundary integral representation has been developed and a boundary integral method combined with a constrained minimization procedure have been setup for a future numerical implementation.

References

1. M. Ikehata, *Size estimation of inclusion*, J. Inv. Ill-Posed Problems **6**, 127–140 (1998).
2. V. Isakov, *Commun. Pure Appl. Math.* **41**, 865 (1988).
3. M. Ikehata *et al*, *Appl. Anal.* **72**, 17 (1999).
4. O.A. Ladyzhenskaya, *The Boundary Value Problems of Mathematical Physics* (Springer-Verlag, Berlin, 1985).
5. G. Alessandrini and R. Magnanini, *Elliptic Equations in Divergence Form, Geometric Critical Points of Solutions, and Stekloff Eigenfunctions*, SIAM J. Math. Anal. **25**, 1259–1268(1994).
6. Hyeonbae Kang, Jin Keun Seo and Dongwoo Sheen, *The Inverse Conductivity Problem with One Measurement: Stability and Estimation of Size*, SIAM J. Math. Anal. **28**, 1389–1405(1997).
7. G. Alessandrini and V. Isakov, *Rend. Istit. Mat. Univ. Trieste* **28**, 351 (1996).
8. H. Kang and J.K. Seo, *Identification of domains with near-extreme conductivity: global stability and error estimates*, Inverse Problems **15**, 851–867(1999).
9. H. Ki and D. Sheen, *Numerical inversion of discontinuous conductivities*, Inverse Problems **16**, 33–47(2000).
10. R. Duraiswami *et al*, Eng. Anal. Boundary Elem. **22**, 13 (1998).
11. D. Lesnic, *A numerical investigation of the inverse potential conductivity problem in a circular inclusion*, Inverse Problems in Engineering **9(1)**,1–17,(2001).
12. Y.P. Chang *et al*, Int. J. Heat Mass Transfer **16**, 1905 (1973).
13. M.Ikehata, *Reconstruction of inclusion from boundary measurements*, J. Inv. Ill-Posed Problems **10**, 37–66(2002).
14. M.Ikehata,*On reconstruction in the inverse conductivity problem with one measurement* , Inverse Problems **16**, 785–793(2000).

ON STABILITY ESTIMATE FOR A BACKWARD HEAT TRANSFER PROBLEM

JIJUN LIU
Department of Mathematics, Nanjing Normal University
Department of Applied Mathematics, Southeast University
Nanjing, 210096, P.R.China
E-mail: jjliu@seu.edu.cn

The author considers an inverse problem for 1-D heat transfer problem with variable coefficients and the Robin boundary. Our aim is to determine the initial temperature distribution from the noisy data measured at some final time $T > 0$. We establish a stability estimate and the uniqueness for this inverse problem, under a-prior knowledge for the solution. Furthermore, a regularization scheme, as well as the convergence rate, is proposed based on this stability result.

1 Introduction

Let $\Omega = (0,1), Q_T = \Omega \times (0,T]$. Consider the 1-d parabolic system

$$\begin{array}{l} \partial_t u = \partial_x(a(x,t)\partial_x u) - q(x,t)u, \quad (x,t) \in Q_T, \\ -\partial_x u(0,t) + hu(0,t) = 0, \quad t > 0, \\ \partial_x u(1,t) + Hu(1,t) = 0, \quad t > 0, \\ u(x,0) = f(x), \quad x \in [0,1], \end{array} \tag{1}$$

where $h, H \geq 0$ are two known constants. Let $\beta \in (0,1)$ be a constant. For the coefficients $a(x,t), q(x,t)$, we assume that

(H1). $a(x,t) \geq a_0 > 0, a(x,t), q(x,t), a_x(x,t) \in C^{\beta,\frac{\beta}{2}}(\overline{Q}_T)$, then from the standard theory of linear parabolic equation[8], we know there exists unique solution $u(x,t) \in C^{2+\beta,1+\frac{\beta}{2}}(\overline{Q}_T)$ for $f(x) \in D(f)$ with

$$D(f) = \{\phi(x) : \phi(x) \in C^{2+\beta}(\overline{\Omega}), -\phi'(0) + h\phi(0) = 0, \phi'(1) + H\phi(1) = 0\}.$$

Moreover, there exists a constant $C_1 = C_1(a, q, h, H, Q_T)$ such that

$$|u|^{(2+\beta)}_{\overline{Q}_T} \leq C_1 |f|^{(2+\beta)}_{\overline{\Omega}} \tag{2}$$

holds for all $f(x) \in D(f)$. Here we apply the standard Sobolev space $C^{2k+\beta,k+\frac{\beta}{2}}(\overline{Q}_T)$ for $k = 0,1$ and $C^{2+\beta}(\overline{\Omega})$. That is,

$$C^{2k+\beta,k+\frac{\beta}{2}}(\overline{Q}_T) = \{u : u \in C^{2k,k}(\overline{Q}_T), H_{\beta,\frac{\beta}{2}}(D_t^r D_x^s u) < +\infty, 2r + s = 2k\}$$

with the norm

$$|u|_{\overline{Q}_T}^{(2+\beta)} = \sum_{0\le 2r+s\le 2k} \max_{\overline{Q}_T} |D_t^r D_x^s u| + \sum_{2r+s=2k} H_{\beta,\frac{\beta}{2}}(D_t^r D_x^s u)$$

and

$$C^{2+\beta}(\overline{\Omega}) = \{\phi(x) : \phi \in C^2(\overline{\Omega}), H_\beta(D^2\phi) < +\infty\}$$

with the norm

$$|\phi|_{\overline{\Omega}}^{(2+\beta)} = \sum_{l\le 2} \max_{\overline{\Omega}} |D^l\phi| + \sum_{l=2} H_\beta(D^l\phi),$$

where

$$H_{\beta,\frac{\beta}{2}}(u) = \sup\left\{\frac{|u(x,t)-u(y,s)|}{|x-y|^\beta + |t-s|^{\beta/2}} : x,y \in \Omega, t,s \in (0,T), (x,t) \neq (y,s)\right\},$$

$$H_\beta(\phi) = \sup\left\{\frac{|\phi(x)-\phi(y)|}{|x-y|^\beta} : x,y \in \Omega, x \neq y\right\}.$$

It is obvious from these definitions that $u(x,t) \in C^{2+\beta,1+\frac{\beta}{2}}(\overline{Q}_T)$ implies $u(x,T) \in C^{2+\beta}(\overline{\Omega})$. Therefore (1) defines a map

$$\mathbf{K} : \mathbf{K}f(x) = u(x,T). \tag{3}$$

from $f(x) \in D(f)$ to $u(x,T) \in C^{2+\beta}(\overline{\Omega}) \subset L^2(\Omega)$.

In this paper, we also assume

(H2). $q(x,t), -\partial_t a(x,t), -\partial_t q(x,t)$ are continuous and nonnegative.

The problem considered in this paper is the so-called backward heat transfer problem. That is, we want to determine the initial temperature $u_0(x,0) = f_0(x)$ approximately from the final temperature measured at time $T > 0$. More precisely, assume that the exact solution to (1) corresponding to $f(x) = f_0(x) \in D(f)$ at time $t = T$ is

$$u_0(x,T) = g_0(x), \tag{4}$$

and that we have measured the approximate value $g_\delta(x)$ of $g_0(x)$ in the sense

$$||g_\delta - g_0||_{L^2(\Omega)} \le \delta, \tag{5}$$

we want to determine $f_\delta(x)$, the approximation of $f_0(x)$, from $g_\delta(x)$. Of course, the approximation $f_\delta(x)$ is not unique. However, all the approximate functions should satisfy $f_\delta(x) \to f_0(x)$ as $\delta \to 0$, if these approximations are reasonable.

It is well-known this inverse problem is ill-posed. Therefore, in order to get the approximate solution $f_\delta(x)$, some regularization should be applied.

However, the choice of regularized parameter and the estimate of convergence rate are very difficult for general regularization scheme. For backward heat transfer problems, there have been some works related to this topic. The logarithmic convexity method[6], which is studied by Payne.L.E. as early as in 1975, seems to play an important role still in recent years. In 1996, T.I.Seidman considered a 2-d backward heat problem with Dirichlet boundary condition and got the estimate $\|u_0(\cdot,t) - u_\delta(\cdot,t)\|_{L^2(\Omega)} \leq C\delta^{t/T}$, where $u_\delta(\cdot,t)$ is some regularized solution which is constructed by filtering the large eigenvalues[7]. However, this estimate is true but nonsense at $t = 0$, which implies we do not know whether or not $u_\delta(x,0) \to f_0(x)$ as $\delta \to 0$, to say nothing of the convergence rate.

In 2000, for general operator equation, a new strategy was proposed for choosing the Tikhonov regularized parameter and get the estimate of convergence rate, based on some conditional stability assumptions on the solution[2]. This technique has been applied to treat many inverse problems and has achieved great success[1,5]. Especially, this technique is applied to treat a 2-d backward heat transfer problem with Dirichlet boundary and the variable coefficients depending only on the spatial variable[5]. For the regularized solution, the author got a Holder type estimate for $\|u_\delta(\cdot,t) - u_0(\cdot,t)\|$ for $0 < t < T$, where the time-independence of variable coefficients and the Dirichlet boundary play a very important role[5]. However, the Holder type estimate does not work at initial time $t = 0$, which implies one can not get the convergence estimate on the recovery of initial temperature $u_0(x,0)$.

In this paper, we consider the inverse problem constituted by (1) and (3), which is a model problem for general higher-dimensional model. Since the boundary is of Robin type, we apply the technique proposed by V.Isakov before[4] to establish the logarithmic convexity estimate, which is more difficult than that in the Dirichlet boundary. Then we construct the regularized solution and give an estimate on the convergence rate of $\|u_\delta(\cdot,0) - f_0(\cdot)\|$. Our result is a little weaker than logarithmic type, due to the time-dependance of our variable coefficients. The method proposed in this paper can be generalized to higher-dimensional problems with a little complicated computation.

2 Conditional Stability

Firstly, we establish the conditional stability for recovering $f(x)$. Define $p(t) = \|u(t)\|^2$ and $F(t) = \ln p(t)$, where

$$\|u(t)\| = \left(\int_0^1 u^2(x,t)dx\right)^{1/2},$$

then we have

Lemma 1 *Let $a(x,t), q(x,t)$ satisfy the above conditions (H1) and (H2). For the solution to (1), it holds*

$$\|u(t)\| \leq \|u(0)\|^{1-t/T} \|u(T)\|^{t/T} \tag{6}$$

for all $0 \leq t \leq T$.

Remark 1 *This fact tells us that the solution to (1) depends continuously on the final value $g(x)$ in L^2-norm sense for all $0 < t < T$, if we restrict the solution to (1) in some function space in which $\|u(0)\| \leq m$ for known constant $m > 0$.*

Proof: Firstly, by simple computation, we get

$$\begin{aligned} p'(t) &= 2\langle u, \partial_t u\rangle = 2\int_0^1 u(x,t)\partial_t u(x,t)dx \\ &= 2[u(1,t)a(1,t)u_x(1,t) - u(0,t)a(0,t)u_x(0,t)] - 2\int_0^1 (au_x^2 + qu^2)dx \\ &= \psi(t) - 2\int_0^1 (au_x^2 + qu^2)dx, \end{aligned} \tag{7}$$

from (1), where we set

$$\psi(t) = -2[Hu^2(1,t)a(1,t) + hu^2(0,t)a(0,t)].$$

Moreover, it follows from (H2) that

$$\begin{aligned} p''(t) &= \psi'(t) - 2\int_0^1 [2au_x\partial_t u_x + \partial_t a u_x^2 + 2qu\partial_t u + \partial_t q u^2]dx \\ &\geq \psi'(t) - 4\int_0^1 [au_x\partial_t u_x + qu\partial_t u]dx \\ &= \psi'(t) - 4au_x\partial_t u\,\big|_{x=0}^{x=1} + 4\int_0^1 (\partial_t u)^2 dx \geq 4\int_0^1 (\partial_t u)^2 dx, \end{aligned} \tag{8}$$

since

$$\psi'(t) - 4au_x\partial_t u\,\big|_{x=0}^{x=1} = -2[Hu^2(1,t)\partial_t a(1,t) + hu^2(0,t)\partial_t a(0,t)] \geq 0.$$

Hence we get from the above estimates that $(\ln p(t))'' > 0$ which implies that

$$\ln p(\theta t_1 + (1-\theta)t_2) \leq \theta \ln p(t_1) + (1-\theta)\ln p(t_2)$$

for any $0 \leq \theta \leq 1$ and $t_1, t_2 \in [0, T]$. Now we take $\theta = t/T$ and $t_1 = T, t_2 = 0$ in this estimate, then the above inequality generates our result immediately.

Let the admissible set for the initial function $f(x)$ be

$$\mu_m = \{f(x) : f \in D(f), |f|_{\Omega}^{(2+\beta)} \leq m\}$$

for some known constant $m > 0$. Then the following result is obvious from Lemma 1 due to the fact $f \in \mu_m$ implies $\|f\|_{L^2(\Omega)} \leq m$.

Lemma 2 *For $f(x) \in \mu_m$, the solution $u(x,t)$ to (1) satisfies*

$$\|u(t)\| \leq m^{1-t/T} \|u(T)\|^{t/T} = m\epsilon^{t/T} \tag{9}$$

for all $0 \leq t \leq T$, where we set $\epsilon = \|g\|/m$.

The above estimate is true at $t = 0$ but nonsense. Our main result, the relation between $u(x, 0)$ and $g(x)$ is given by

Theorem 1 *For the solution $u(x,t)$ to (1) corresponding to $f(x) \in \mu_m$, it follows*

$$\|u(0)\|^2 \leq -C_1 m^2 T \frac{1 - \ln \frac{-C_1 T}{\ln \epsilon}}{\ln \epsilon} \tag{10}$$

for $\epsilon > 0$ small enough.

Proof: Firstly, expanding $\|u(t)\|^2$ at $t = 0$ says

$$\|u(t)\|^2 = \|u(0)\|^2 + 2\langle u(\cdot, \xi), \partial_t u(\cdot, \xi)\rangle t$$

for some $0 < \xi < t$, which implies

$$\begin{aligned} \|u(0)\|^2 &= \|u(t)\|^2 - 2\langle u(\cdot, \xi), \partial_t u(\cdot, \xi)\rangle t \\ &\leq \|u(t)\|^2 + 2\|u(\cdot, \xi)\| \|\partial_t u(\cdot, \xi)\| t \\ &\leq m^{2(1-t/T)} \|g\|_{L^2(\Omega)}^{2t/T} + 2m^{1-\xi/T} \|g\|_{L^2(\Omega)}^{\xi/T} \|\partial_t u(\cdot, \xi)\| t \end{aligned} \tag{11}$$

for all $t \in [0, T]$ from Lemma 2. On the other hand, it follows from (2) that

$$\|\partial_t u(\cdot, \xi)\| \leq C_1 |f|_{\Omega}^{(2+\beta)} \leq C_1 m = M.$$

Therefore the above estimate leads to

$$\begin{aligned} \|u(0)\|^2 &\leq m^2 \epsilon^{2t/T} + 2mM\epsilon^{\xi/T} t \\ &\leq m^2 \epsilon^{2t/T} + 2mMt \end{aligned} \tag{12}$$

for $\epsilon > 0$ small enough. By elementary computation, we get

$$\min_{t\in[0,T]} \left\{m^2\epsilon^{2t/T} + 2mMt\right\} = -mMT\frac{1-\ln\frac{-MT}{m\ln\epsilon}}{\ln\epsilon},$$

which completes our proof due to $M = C_1 m$.

Remark 2 *Under a restrictive condition $u(x,0) = f(x) \in \mu_m$, this theorem asserts that $u(x,0)$ depends on the final value $u(x,T) = g(x)$ in a weak topology, that is, we measure both $u(x,0)$ and $u(x,T)$ by $L^2(\Omega)$-norm, rather than by $C^{(2+\beta)}(\overline{\Omega})$-norm. This is due to the ill-posedness of our inverse problem. We do not know weather the L^2-norm can be improved in our stability result.*

Now the conditional stability for our inverse problem (1) can be obtained from the above lemmas immediately:

Theorem 2 *Let $u_i(x,t)$ solve (1) with $f = f_i \in \mu_m$ for $i = 1,2$. Then*

$$\|(u_1 - u_2)(0)\|^2 \le -4m^2T\frac{1-\ln\frac{-C_1T}{\ln\epsilon_0}}{\ln\epsilon_0}, \tag{13}$$

where $g_i(x) = u_i(x,T), \epsilon_0 = \frac{1}{2m}\|g_1 - g_2\|$.

From this stability estimate, the uniqueness of the backward heat transfer problem up to the initial time $t = 0$ can also be obtained.

Theorem 3 *Let $u_i(x,t)$ solve (1) with $f(x) = f_i(x) \in \mu_m$ for $i = 1,2$. Then $f_1(x) = f_2(x)$ in $C^{2+\beta}(\Omega)$ if $g_1(x) = g_2(x)$ in $L^2(\Omega)$.*

Proof: It is obvious from (13) that

$$\|f_1 - f_2\|^2_{L^2(\Omega)} = \int_\Omega |f_1(x) - f_2(x)|^2 dx = 0$$

for $g_1(x) = g_2(x)$ in $L^2(\Omega)$, which means $f_1(x) = f_2(x)$ in $C(\overline{\Omega})$ due to $f_1(x) - f_2(x) \in C^{2+\beta}(\overline{\Omega}) \subset C(\Omega)$. So we get $f_1(x) = f_2(x)$ in $C^{2+\beta}(\overline{\Omega})$ for $f_1(x), f_2(x) \in C^{2+\beta}(\overline{\Omega})$.

Remark 3 *If the coefficients in the heat equation does not depend on time variable, then the uniqueness of backward heat transfer equation is obvious under the assumption that the solution exists, since the solution can be expressed in terms of the eigenfunctions of forward heat problem[4]. However, in our problem, this representation is general impossible due to the time dependance of coefficients. So the uniqueness is not obvious. In this sense, our* *stability estimate is very important both in the uniqueness and in the regularization scheme for this inverse problem. The other possible way to get the uniqueness for the back-ward parabolic equation with the Robin boundary condition in Q_T can be obtained by a classical way*[3]. *However, the uniqueness for $u(x,0)$ is not obtained there.*

3 Regularization and Convergence Analysis

In this section, we will apply our stability result to establish a regularization scheme so that we can determine the approximation of $f_0(x)$ from the noisy data $g_\delta(x)$. Moreover, we also give the estimate of convergence rate.

Assume that the exact final temperature $g_0(x) = u_0(x,T)$ is generated from some initial temperature $f_0(x) = u_0(x,0) \in D(f)$ from system (1). Now if we get the measured data $g_\delta(x)$ of $g_0(x)$ with the error level $\delta > 0$ in the sense of (5), we want to find the initial temperature distribution $f_0(x)$ approximately.

Suppose that we have a-prior knowledge of the exact solution $f_0(x)$, say $f_0(x) \in \mu_m$. This means we know the bound of $|u_0(x,t)|_{Q_T}^{(2+\beta),1+\frac{\beta}{2}}$ from the estimate (2). Furthermore, define a functional

$$F_\alpha^\delta(f) = \|Kf - g_\delta\|_{L^2(\Omega)}^2 + \alpha\left(|f|_{\overline{\Omega}}^{(2+\beta)}\right)^2 \tag{14}$$

over $D(f)$.

Theorem 4 *For any $C_0^2 > m^2+1$, there exists an approximate minimizer $f_\delta(x)$ for functional $F_{\delta^2}^\delta(f)$ over $D(f)$ which satisfies*

$$F_{\delta^2}^\delta(f_\delta) \le C_0^2\delta^2, \tag{15}$$

$$\|Kf_\delta - Kf_0\|_{L^2(\Omega)} \le (C_0+1)\delta. \tag{16}$$

Proof: It is obvious that

$$\begin{aligned} F_{\delta^2}^\delta(f_0) &= \|Kf_0 - g_\delta\|_{L^2(\Omega)}^2 + \delta^2\left(|f_0|_{\overline{\Omega}}^{(2+\beta)}\right)^2 \\ &\le \|g_0 - g_\delta\|_{L^2(\Omega)}^2 + m^2\delta^2 \\ &\le \delta^2 + m^2\delta^2 = (m^2+1)\delta^2, \end{aligned} \tag{17}$$

which implies $\{f : F_{\delta^2}^\delta(f) \le C_0^2\delta^2\} \ne \emptyset$ due to $C_0^2 > m^2+1$. Hence (15) is proven. From this inequality we also know that $f_\delta \in D(f)$ satisfies

$$|f_\delta|_{\overline{\Omega}}^{(2+\beta)} \le C_0, \tag{18}$$

$$\|Kf_\delta - g_\delta\|_{L^2(\Omega)} \le C_0\delta. \tag{19}$$

Therefore we get

$$\|Kf_\delta - Kf_0\|_{L^2(\Omega)} \le \|Kf_\delta - g_\delta\|_{L^2(\Omega)} + \|g_\delta - Kf_0\|_{L^2(\Omega)} \le (C_0+1)\delta.$$

So we get (16). The proof is complete.

Now $u_\delta(x,t)$, the approximate value of $u_0(x,t)$ can be construct from $f_\delta(x)$ by solving a forward heat problem. That is,

Theorem 5 *For $f_\delta(x) \in D(f)$ generated in the above theorem, we solve the forward heat conduction problem (1) with $f(x) = f_\delta(x)$ to get the approximate temperature $u_\delta(x,t)$. For such a approximate solution, it holds*

$$\|(u_\delta - u_0)(t)\|_{L^2(\Omega)} \le 2(m+2)^2\delta^{t/T} \tag{20}$$

for all $0 < t < T$, while at $t = 0$, it holds that

$$\|(u_\delta - u_0)(0)\|^2_{L^2(\Omega)} \le 4C_1(m+1)^2T\frac{1-\ln\frac{C_1T}{\ln\frac{1}{\delta}}}{\ln\frac{1}{\delta}} \tag{21}$$

for $\delta > 0$ small enough.

Proof: The proof can be completed from Theorem 4and Theorem 2, by taking $u_1 = u_\delta$ and $u_2 = u_0$ respectively.

Firstly, we fix $C_0 = m+1$ for certainty. Then (18) says $f_\delta \in \mu_{(m+1)}$ which means $f_\delta - f_0 \in \mu_{2(m+1)}$ due to $f_0 \in \mu_m \subset \mu_{(m+1)}$. Now (9) in Lemma 2tells us

$$\|(u_\delta - u_0)(t)\| \le (2(m+1))^{1-t/T}\|Kf_\delta - Kf_0\|^{t/T}. \tag{22}$$

for $0 < t < T$. Now inserting (16) into this estimate leads to (20).

For (21), it is obvious from Theorem 2that

$$\|(u_\delta - u_0)(0)\|^2 \le -4C_1(m+1)^2T\frac{1-\ln\frac{-C_1T}{\ln\epsilon_1}}{\ln\epsilon_1}, \tag{23}$$

where $\epsilon_1 = \frac{1}{2(m+1)}\|Kf_\delta - Kf_0\| \le \frac{m+2}{2(m+1)}\delta < \delta$ due to (16). For small $\epsilon_1 > 0$, it is easy to verify that

$$\frac{d}{d\epsilon_1}\left(\frac{1-\ln\frac{-C_1T}{\ln\epsilon_1}}{\ln\epsilon_1}\right) = \frac{\ln\frac{-C_1T}{\ln\epsilon_1}}{\epsilon_1(\ln\epsilon_1)^2} < 0,$$

therefore we get

$$\|(u_\delta - u_0)(0)\|^2 \le -4C_1(m+1)^2T\frac{1-\ln\frac{-C_1T}{\ln\delta}}{\ln\delta},$$

which complete the proof of (21) immediately.

Remark 4 *The only information in our method is the up bound of the exact solution $u_0(x,t)$ at $t = 0$. The constant m is not difficult to get in many cases. Further, our estimate gives the error bound by m and δ explicitly. From the convergence rate, we know that $u_\delta(x,t)$ converges to $u_0(x,t)$ fast near $t = T$ and slowly near $t = 0$. This is reasonable from the physics background. Especially, our estimate on $\|(u_\delta - u_0)(0)\|$ is a little weaker than $\frac{1}{\ln\frac{1}{\delta}}$ from (21), due to the fact $1 - \ln\frac{C_1 T}{\ln\frac{1}{\delta}} \to +\infty$.*

Acknowledgments

The author would like to give his thanks to Prof. J.Cheng for the useful discussions on this paper. This work is also partly supported by the Science Foundation at Southeast University (No.9207011148).

References

1. J.Cheng and G.Nakamura, *Stability for the inverse potential problem by finite measurement on the boundary*, to appear in Inverse Problems.
2. J.Cheng and M.Yamamoto, *The global uniqueness for determining two convection coefficients from Dirichlet to Neumann map in two dimensions*, Inverse Problems **16(3)**, L25–L30 (2000).
3. A. Friedman, *Partial Differential Equations of Parabolic Type* (Prentice-Hall, Inc., 1964).
4. V.Isakov, *Inverse Problems for Partial Differential Equations* (Springer-Verlag, New York, 1998).
5. J.J.Liu, *Determination of temperature field from backward heat transfer problem*, Communications of Korea Mathematical Socity **16(3)**, 371–384(2001).
6. L.E.Payne, *Improperly Posed Problems in Partial Differential Equations* (Regional Conference Series in Applied mathematics, SIAM, Philadelphia, 1975).
7. T.I.Seidman, *Optimal filtering for the backward heat equation*, SIAM, J. Numer. Anal. **33(1)**, 162–170(1996).
8. Qixiao Ye, Zhengyuan Li, *An Introduction to Reaction-Diffusion Equations* (Science Press, Beijing, 1994).

AN EXISTENCE FOR AN INVERSE PROBLEM FROM COMBUSTION THEORY AND ITS NUMERICAL SIMULATION

YICHEN MA, QI CHEN AND GENJUN YING
Science College, Xi'an Jiaotong Univ., Xi'an, 710049
E-mail: ycma@xjtu.edu.cn

GONGSHENG LI
Department of math. and phys., Zibo Univ., Zibo city, Shandong, 255000
E-main: ligs@ns2.zbu.net.cn

In this paper we are concerned with a quasilinear parabolic equation with homogeneous Cauchy and non-homogeneous Neumann conditions arising from combustion theory. By using the Schauder fixed point theorem and Green function of the second homogeneous boundary value problem, we give a local existence result to the solution of an inverse problem defined on a semi-infinite space. Numerical simulation results show that the proposed numerical algorithm is efficient and applicable.

1 Introduction

Inverse problem for a parabolic equation is an important research field of inverse problem. In particular the inverse problems concerning nonlinear parabolic equations are challenging. For example: the determination of the diffusion parameter. Since the 80's, many scientists are interested in determining the nonlinear right-hand term of the quasilinear parabolic equation. The list of researchers includes J.R. Cannon[1,2], P.C. Ducheutrau[2] etc. In the paper[3], S. Gatti gave the local existence proof on the solution (u,a) to the following inverse problem, where u is the thermal profile and $\Omega_T^1 = (-\infty, 0) \times (0, T)$.

$$\begin{cases} \partial_t u(x,t) - \partial_x^2 u(x,t) + R(u(0,t))\partial_x u(x,t) \\ \qquad\qquad = a(u(x,t)) + f(x,t), \qquad (x,t) \in \Omega_T^1 \\ u(x,0) = u_0(x), \qquad x \le 0 \\ \partial_x u(0,t) = p(t) + F(t, u(0,t)), \qquad 0 \le t \le T \\ u(0,t) = h(t) \\ a(\tau) = a_0(\tau) \end{cases} \tag{1}$$

The physical model describes a semi-infinite one-dimensional space of homogeneous solid propellent burning in a vessel at an uniform pressure. We assume that the propellent is adiabatic except at the burning surface. Two external sources act on the propellent: p(t) is the part deposited at the surface of an external radiant flux originating from a continuous wave source concentrated at the burning surface, while $f(x,t)$ is the remaining part of the flux distributed volumetrically along the propellent. The function $R(\phi)$ is the burning rate described by the pyrolysis law(Arrhenius law, see DeLuce[4,5] for details). S. Gatti considered that the data of the inverse problem are the surface temperature at $x = 0$. By means of the Schauder fixed-point theorem, he proved that, for a sufficiently small T, there exists at least one solution to the inverse problem (1).

A similar problem was studied by Lorenzi and Paparoni in A. Lorenzi [6,7] for a semi-linear parabolic equation on a bounded domain. In this paper, we assume that $(x,t) \in \Omega_T = (0,\infty) \times (0,T)$. The additional measured data are

$$u(x_0, t) = \theta(t) \tag{2}$$

where $(x_0, t) \in \Omega_T, x_0 > 0$ is a fixed point. We then obtain the following equation which will be discussed later in this paper:

$$\begin{cases} \partial_t u(x,t) - \partial_x^2 u(x,t) + R(u(0,t))\partial_x u(x,t) = f(u(x,t)), (x,t) \in \Omega_T \\ u(x,0) = 0, \qquad x \geq 0 \\ \partial_x u(0,t) = F(t, u(0,t)), \qquad 0 \leq t \leq T \end{cases} \tag{3}$$

It was firstly proposed by J.R. Cannon that the finding of $f(u)$ in the equation (3) with condition (2) is an inverse problem for a quasilinear parabolic equation with the special boundary condition. For a general third boundary condition, non-local boundary condition and the complex boundary condition[1], the inverse problems become very difficult to be solved. In this paper, we prove an existence result based on the condition that the $R(u(0,t))$ is sufficient small. We also give the numerical algorithm and examples.

2 Assumptions

According to the theory of the parabolic equation[8,9], when source term $f(x,t)$, initial data and boundary data satisfy some proper conditions, the direct problem (3) is well-posed, i.e. there exists a unique solution. Let's introduce the spaces and norms.

$S = C^{0,\alpha} \times C^1 = \{(f,\mu), f \in C^{0,\alpha}, \mu \in C^1\}, \qquad ||(f,\mu)||_S = ||f||_{C^{0,\alpha}} + ||\mu||_{C^1}$

So $|u(.,t)|_1 = \sup_{x>0} |u_x(x,t)|, t \in [0,T]$. The admissible set of f is

$$Y = \{f \in C(D), ||f||_C \leq E, f(0) = 0, |f(u) - f(v)| \leq L|u - v|\}$$

where L, E are positive constants. Define $|f|_\alpha = sup_{u,v \in D} \dfrac{|f(u) - f(v)|}{|u - v|^\alpha}$.

To obtain the expression of the solution of equation (3), the Green function[10] on the parabolic equation with the second homogeneous boundary value condition is

$$K(x,y;t) = \frac{1}{2\sqrt{\pi t}}[exp^{(-\frac{(x-y)^2}{4t})} + exp^{(-\frac{(x+y)^2}{4t})}]. \tag{4}$$

We assume that $\mu(t)=u(0,t)$.When f,μ are known, we treat $R(u(0,t))\partial_x u(x,t)$ as a part of right term of (3). Thus the solution of (3) is

$$\begin{aligned} u(x,t) = &- \int_0^t K(x,0;t-\tau)F(\tau,\mu(\tau))d\tau \ + \int_0^t\int_0^\infty K(x,y;t-\tau)f(u(y,\tau))dyd\tau \\ &+ \int_0^t\int_0^\infty K(x,y;t-\tau)(-R(\mu(\tau))u_x(y,\tau))dyd\tau \end{aligned} \tag{5}$$

Substituting $x = x_0, x = 0$ into (3) and (5) respectively, we get

$$\begin{cases} \overline{f}(\theta(t)) = \theta'(t) + \int_0^t K_{xx}(x_0,0,t-\tau)F(\tau,\mu(\tau))d\tau \\ \qquad - \int_0^t\int_0^\infty K_{xx}(x_0,y;t-\tau)f(u(y,\tau))dyd\tau \\ \qquad + \int_0^t\int_0^\infty K_{xx}(x_0,y;t-\tau)R(\mu(\tau))u_x(y,\tau)dyd\tau \\ \qquad +R(\mu)u_x(x_0,t) \\ \overline{\mu}(t) \quad = - \int_0^t K(0,0;t-\tau)F(\tau,\mu(\tau))d\tau \\ \qquad + \int_0^t\int_0^\infty K(0,y;t-\tau)f(u(y,\tau))dyd\tau \\ \qquad - \int_0^t\int_0^\infty K(0,y;t-\tau)R(\mu(\tau))u_x(y,\tau)dyd\tau \end{cases} \tag{6}$$

For $(f,\mu) \in S$, define:

$$T(f,\mu) = (T_1,T_2)(f,\mu)$$

where

$$\begin{aligned} T_1(f,\mu) &= \overline{f}(\theta(t)) \\ &= \theta'(t) + \int_0^t K_{xx}(x_0,0;t-\tau)F(\tau,\mu(\tau))d\tau \\ &\quad - \int_0^t\int_0^\infty K_{xx}(x_0,y;t-\tau)f(u(y,\tau))dyd\tau \\ &\quad + \int_0^t\int_0^\infty K_{xx}(x_0,y;t-\tau)R(\mu(\tau))u_x(y,\tau)dyd\tau \\ &\quad +R(\mu)u_x(x_0,t) \end{aligned} \tag{7}$$

and

$$\begin{aligned} T_2(f,\mu) &= \overline{\mu}(t) \\ &= - \int_0^t K(0,0;t-\tau)F(\tau,\mu(\tau))d\tau \\ &\quad + \int_0^t\int_0^\infty K(0,y;t-\tau)f(u(y,\tau))dyd\tau \\ &\quad - \int_0^t\int_0^\infty K(0,y;t-\tau)R(\mu(\tau))u_x(y,\tau)dyd\tau \end{aligned} \tag{8}$$

Assume that $F(t,\mu), f(x,t), R(\mu)$ satisfy (9),(10) and (11) respectively

$$\begin{cases} F(0,0) = 0; \\ |F_t|, |F_\mu|, |F_{t\mu}|, |F_{\mu\mu}| \le C_F \end{cases} \tag{9}$$

$$\begin{cases} |f(u)-f(v)|_1 \le C_f|u-v|_1; \\ |f(u)-f(v)|_1 = \sup_{y\neq x} \dfrac{|(f(u)-f(v))(y,\tau)-(f(u)-f(v))(x,\tau)|}{|y-x|}; \\ |u-v|_1 = \sup_{x\in R^+} |u_x(y,\tau)-v_x(x,\tau)|, \tau\in[0,T] \end{cases} \quad (10)$$

$$\begin{cases} R(0)=0; \\ |R_\mu|, |R_t| \le C_R \end{cases} \quad (11)$$

Suppose the admissible set of (f,μ) is

$$S_{cd} = \{(f,\mu)\in S, ||(f,\mu)||_S \le E\}. \quad (12)$$

where $E>0$ is a constant.

3 Lemmas and the Estimations

Without loss of generality, suppose $(f,\mu)\in S_{cd}$, $0<t_2<t_1<T$, from the properties of Green function (4) and the assumption of (9),(10) and (11), it is not difficult to prove the following lemmas:

Lemma 1 *Let $||f||_\infty \le E, ||\mu||_\infty \le E$. If $F(\tau,\mu(\tau)), R(\mu(\tau))$ satisfy properties (9), (10), (11) and T is sufficient small, then*

(1) $$\begin{cases} ||u_x||_\infty \le d_2C_F(T+||\mu'||_\infty T)+d_2||f||_\infty c_1(0)T^{\frac{1}{2}}, \\ d_2=(1-c_1(0)T^{1/2}||R||_\infty)^{-1}=\zeta_1(T) \end{cases}$$

(2) $$\begin{cases} ||u_{xx}||_\infty \le d_1C_F(2/\sqrt{\pi})(1+||\mu'||_\infty)T^{1/2}+d_1c_2(\alpha)||u_x||_\infty^\alpha|f|_\alpha T^{\alpha/2} \\ d_1=(1-||R||_\infty c_2(1)T^{1/2})^{-1}=\zeta_2(T) \end{cases}$$

Lemma 2 *(K. Yosida[13]) For $0<\alpha\le 1$,$C^{0,\alpha}(D)$ is compactly imbedded in $C(D)$, and $T>0$,$C^1[0,T)$ is compactly imbedded in $C[0,T)$.*

Lemma 3 *If $I=\int_0^t K_{xx}(x_0,0;t-\tau)F(\tau,\mu(\tau))d\tau$, then*

(1) $|I| \le C_Fc(x_0)(1+||\mu'||_\infty)t^2$

(2) $|I(t_1)-I(t_2)| \le C_Fc(x_0)(1+||\mu'||_\infty)\frac{1}{2}(t_2^2+t_1+t_2)|t_1-t_2|$

Lemma 4 *For fixed $x \in (0,\infty)$, then*

$$|u_x(x,t_1) - u_x(x,t_2)| \le c_u|t_1 - t_2|^{1/2}$$

where c_u is a bounded positive constant for $\forall t_1, t_2 \in (0,T)$.

Remark 1 *When $||R||_\infty$ is sufficient small, d_1, d_2 is positive and tend to 1, as $T \to 0$, so $\zeta_1(T)$, and $\zeta_2(T)$ tend to 0, when $T \to 0$.*

Lemma 5 *If $f(u(x,t)), R(\mu(t))$ satisfy properties (10) (11), then*

$$\mu'(t) = -\frac{1}{\sqrt{\pi}} \int_0^t (t-\tau)^{-1/2}(F_\tau + F_\mu \mu')d\tau$$

$$+ \int_0^t \int_0^\infty K_{xx}(0,y;t-\tau) f(u(y,\tau))dyd\tau$$

$$- \int_0^t \int_0^\infty K_{xx}(0,y;t-\tau) R(\mu(\tau))u_x(y,\tau)dyd\tau$$

$$+f(\mu(t)) - R(\mu(t))F(t,\mu(t))$$

4 Existence of the Inverse Problem

Theorem 1 *Let $\theta(t)$ and $F(t,\mu(t))$ satisfy*

(1) $\theta(0) = \theta'(0) = 0; u_x(x_0,t) = 0, t \in [0,T]$;

(2) $\theta(t) \in C^{1,\alpha}([0,T])$;

(3) $F(t,\mu(t))$ satisfies (9).

Then there exists T^*, such that T maps S_{cd} to S_{cd}, when $t \in [0,T^*]$, where

$$S_{cd} = \{(f,\mu), ||(f,\mu)||_S \le E\}.$$

For given $(f_1,\mu_1) \in S_{cd}$, define the series

$$f_{n+1} = T_1[f_n,\mu_n]; \quad \mu_{n+1} = T_2[f_n,\mu_n] \tag{13}$$

According to theorem 1, we have $\{(f_n,\mu_n)\} \in S_{cd}$. If $0 < \alpha < 1/2$, from lemma 2 there is subsequence $\{(f_n,\mu_n)\}$ strongly converging in $C^{0,\alpha}(D) \times C^1[0,T]$.

Since formula (7)

$$\begin{aligned} f_{n+1} = \theta'(t) &+ \int_0^t K_{xx}(x_0,0;t-\tau)F(\tau,\mu_n(\tau))d\tau \\ &- \int_0^t\int_0^\infty K_{xx}(x_0,y;t-\tau)f_n(u^n(y,\tau))d\tau \\ &+ \int_0^t\int_0^\infty K_{xx}(x_0,y;t-\tau)R(\mu_n(\tau))u_x^n(y,\tau)dyd\tau \\ &+ R(\mu_n(t))u_x^n(x_0,t) \end{aligned} \tag{14}$$

$$\begin{aligned} \mu_{n+1} = &- \int_0^t K(0,0;t-\tau)F(\tau,\mu_n(\tau))d\tau \\ &+ \int_0^t\int_0^\infty K(0,y;t-\tau)f_n(u^n(y,\tau))dyd\tau \\ &- \int_0^t\int_0^\infty K(0,y;t-\tau)R(\mu_n(\tau))u_x^n(y,\tau)dyd\tau \end{aligned} \tag{15}$$

It follows from (5)

$$\begin{aligned} u^n(x,t) = &- \int_0^t K(x,0;t-\tau)F(\tau,\mu_n(\tau))d\tau \\ &+ \int_0^t\int_0^\infty K(x,y;t-\tau)f_n(u^n(y,\tau))dyd\tau \\ &+ \int_0^t\int_0^\infty K(x,y;t-\tau)(-R(\mu_n(\tau))u_x^n(y,\tau))dyd\tau \end{aligned} \tag{16}$$

If $(f_n,\mu_n) \to (\hat{f},\hat{\mu}), n \to \infty$, $u_n \to \hat{u}$, $\hat{u}$ satisfies

$$\begin{aligned} \hat{u}(x,t) = - \int_0^t K(x,0;t-\tau)F(\tau,\hat{\mu})d\tau + \int_0^t\int_0^\infty K(x,y;t-\tau)\hat{f}(\hat{u}(y,\tau))dyd\tau \\ + \int_0^t\int_0^\infty K(x,y;t-\tau)(-R(\hat{\mu})\hat{u}_x)dyd\tau \end{aligned}$$

Now we obtain a convergence subsequence and its limit $(\hat{f},\hat{u})$. It is necessary to prove the limit is the solution of the equation (3). Thus

Theorem 2 *Under the conditions of theorem 1, if f_n satisfies property (10), $(\hat{f},\hat{u})$ is the solution of the inverse problem (2)–(3), then*

$$\hat{u}(x,t) = \lim_{n\to\infty} u^n(x,t), \qquad \hat{f}(\hat{u}) = \lim_{n\to\infty} f_n(u^n) \tag{17}$$

We will describe the proof of the theorem 1 and theorem 2 in detail in next section.

5 The Proof of the Theorem 1 and Theorem 2

In this section,we give some lemmas and their proofs.At first,we give some basic properties of the *Green* function $K(x,y;t)$, which will be used in the following proofs.

Property 1 $\int_0^\infty K_x(x,y;t)dy = \int_0^\infty K_{xx}(x,y;t)dy = \int_0^\infty K_{xxt}(x,y;t)dy = \int_0^\infty K_{xt}(x,y;t)dy = 0, \forall x > 0;$

Property 2 $\int_0^\infty K(x,y;t)dy = 1; \int_0^t K_x(x,0;t-\tau)d\tau \le 1;$

Property 3

$$\begin{cases} (1)\ \int_0^t \int_0^\infty |K(x,y;t-\tau)|x-y|^\alpha dyd\tau \le c_0(\alpha)t^{\alpha/2+1}, \alpha \ge 0; \\ (2)\ \int_0^t \int_0^\infty |K_x(x,y;t-\tau)|x-y|^\alpha dyd\tau \le c_1(\alpha)t^{(1+\alpha)/2}, \alpha \ge 0; \\ (3)\ \int_0^t \int_0^\infty |K_{xx}(x,y;t-\tau)|x-y|^\alpha dyd\tau \le c_2(\alpha)t^{\alpha/2}, \alpha > 0; \\ (4)\ \int_0^t \int_0^\infty |K_{xxt}(x,y;t-\tau)|x-y|^\alpha dyd\tau \le c_3(\alpha)\dfrac{t^{\alpha/2-1}}{1-\alpha/2}, 0 < \alpha < 2 \\ (5)\ \int_0^t \int_0^\infty |K_{xt}(x,y;t-\tau)||x-y|^\alpha dyd\tau \le c_5(\alpha)t^{\alpha/2-1/2}, \forall\alpha > 0. \end{cases}$$

where

$$\begin{cases} c_0(s) = \dfrac{2^{s+1}}{\sqrt{\pi}(s+2)} \int_{-\infty}^\infty e^{-t^2}|t|^2 dt; \\ c_1(s) = \dfrac{2^s}{\sqrt{\pi}(s+1)} \int_{-\infty}^\infty e^{-t^2}|t|^{1+s} dt; \\ c_2(s) = \dfrac{2^{s+1}}{\sqrt{\pi}s} \int_{-\infty}^\infty e^{-t^2}|t^2|(t^2+1/2)|dt; \\ c_3(s) = \dfrac{2^s}{\sqrt{\pi}} \int_{-\infty}^\infty e^{-t^2}|t|^s(t^4+3t^2+3/2)dt \\ c_5(s) = \frac{5\cdot 2^{s+1}}{\sqrt{\pi}(s-1)} \int_{-\infty}^\infty e^{-t^2}t^{3+s}dt \end{cases}$$

Property 4 *There exists a constant* $c(x_0)$ *only depending on* $x_0 > 0$*, such that*

$$|K(x_0,0;t)|, |K_x(x_0,0;t)|, |K_{xx}(x_0,0;t)|, |K_{xxt}(x_0,0;t)|, |K_{xt}(x_0,0;t)| \le c(x_0)$$

Lemma 6 *Under the conditions of the theorem 1, if* f_n *satisfies the property (10), and* $||R||_\infty$ *is sufficient small, then*

$$\begin{aligned} ||u_x^n - \overset{\Lambda}{u}_x||_\infty \le d_u[C_F||\mu_n - \overset{\Lambda}{\mu}||_\infty + c_1(0)||f_n - \overset{\Lambda}{f}||_\infty T^{1/2} \\ + c_1(0)L||u^n - \overset{\Lambda}{u}||_\infty T^{1/2} + c_1(0)||R_\mu||_\infty ||u_x^n||_\infty ||\mu_n - \overset{\Lambda}{\mu}||_\infty T^{1/2}] \end{aligned}$$

where $d_u = (1 - T^{1/2}||R||_\infty c_1(0))^{-1} > 0.$

Proof: From formula (5) and the expression of $\stackrel{\Lambda}{u}$

$$
\begin{aligned}
|\stackrel{\Lambda}{u}_x - u_x^n| &\leq |\int_0^t K_x(x,0;t-\tau)[F(\tau,\stackrel{\Lambda}{\mu}) - F(\tau,\mu_n)]d\tau| \\
&+| \int_0^t \int_0^\infty K_x(x,y;t-\tau)[\stackrel{\Lambda}{f}(\stackrel{\Lambda}{u}) - f_n(u^n)]dyd\tau| \\
&+| \int_0^t \int_0^\infty K_x(x,y;t-\tau)[R(\stackrel{\Lambda}{\mu})\stackrel{\Lambda}{u}_x - R(\mu_n)u_x^n]dyd\tau| \\
&= T_0 + T_1 + T_2
\end{aligned}
$$

According to assumption (9) and property 2, we have

$$|F(\tau,\mu_n) - F(\tau,\stackrel{\Lambda}{\mu})| \leq C_F|\mu_n - \stackrel{\Lambda}{\mu},|$$

and

$$\int_0^t K_x(x,0;t-\tau)d\tau \leq 1$$

Thus

$$T_0 \leq C_F||\mu_n - \stackrel{\Lambda}{\mu}||_\infty.$$

Because of $f_n \in Y, \stackrel{\Lambda}{f} \in Y$,

$$T_1 \leq c_1(0)(||f_n - \stackrel{\Lambda}{f}||_\infty + L||u^n - \stackrel{\Lambda}{u}||_\infty)T^{1/2}$$

Similarly

$$
\begin{aligned}
T_2 &\leq |\int_0^t \int_0^\infty K_x(x,y;t-\tau)[R(\stackrel{\Lambda}{\mu})(u_x^n - \stackrel{\Lambda}{u}_x) + (R(\mu_n) - R(\stackrel{\Lambda}{\mu}))u_x^n]dyd\tau| \\
&\leq ||R||_\infty||u_x^n - \stackrel{\Lambda}{u}_x||_\infty c_1(0)t^{1/2} + c_1(0)||R_\mu||_\infty||\mu_n - \stackrel{\Lambda}{\mu}||_\infty||u_x^n||_\infty t^{1/2}
\end{aligned}
$$

If $1 - ||R||_\infty c_1(0)T^{1/2} > 0$, the proof is completed.

Remark 2 *If $||R||_\infty$ is sufficient small, d_u will be positive constant depending on T, and as T tends to 0, d_u tends to 1.*

Lemma 7 *Under the conditions of lemma 6, u^n tends to u, as n tends ∞.*

Proof: Under the formula (5) and the expression of $\stackrel{\Lambda}{u}$, we have

$$
\begin{aligned}
|\stackrel{\Lambda}{u} - u^n| &\leq |\int_0^t \int_0^\infty K(x,y;t-\tau)[R(\stackrel{\Lambda}{\mu})\stackrel{\Lambda}{u}_x - R(\mu_n)u_x^n]dyd\tau| \\
&+| \int_0^t \int_0^\infty K(x,y;t-\tau)[\stackrel{\Lambda}{f}(\stackrel{\Lambda}{u}) - f_n(u^n)]dyd\tau| \\
&+| \int_0^t K(x,0;t-\tau)[F(\tau,\stackrel{\Lambda}{\mu}) - F(\tau,\mu_n)]d\tau| \\
&= T_3 + T_4 + T_5
\end{aligned}
$$

According the conditions of F, R, f and the properties of *Green* function, we get

$$\begin{aligned}T_3 &= \int_0^t\int_0^\infty K(x,y;t-\tau)\{[R(\overset{\wedge}{\mu})-R(\mu_n)]\hat{u}_x+R(\mu_n)[\hat{u}_x-u_x^n]\}dyd\tau\\ &\le ||\hat{u}_x||_\infty||R_\mu||_\infty||\overset{\wedge}{\mu}-\mu_n||_\infty c_0(0)t+||R||_\infty||u_x^n-\hat{u}_x||_\infty c_0(0)t\,,\end{aligned}$$

then

$$T_3 \le \gamma(||\mu_n-\overset{\wedge}{\mu}||_\infty)+||R||_\infty||u_x^n-\hat{u}_x||_\infty t\,,$$

where $\gamma(p,t)=||\hat{u}_x||_\infty||R_\mu||_\infty c_0(0)t\cdot p$, so $\gamma(p,t)\to 0$, as $t\to 0$.

$$T_5 \le \frac{2C_F}{\sqrt{\pi}}||\mu_n-\overset{\wedge}{\mu}||_\infty t^{1/2}\,.$$

Similarly, for T_4, we have

$$T_4 \le L||u^n-\hat{u}||_\infty t+||f_n-\hat{f}||_\infty t\,.$$

Following form the lemma 6 and $\lim_{n\to\infty} f_n=\hat{f}, \lim_{n\to\infty}\mu_n=\overset{\wedge}{\mu}$, when T is small enough, we have

$$\lim_{n\to\infty} u^n(x,t)=u(x,t),(x,t)\in\Omega_T$$

The proof is completed.

Lemma 8 *Under the conditions of lemma 6, if x is fixed, then*

$$|u_{xx}^n-\hat{u}_{xx}|_\infty\to 0, n\to\infty$$

Proof: From formula (6) and the expression of $\hat{u}$

$$\begin{aligned}|\hat{u}_{xx}-u_{xx}^n| &\le |\int_0^t\int_0^\infty K_{xx}(x,y;t-\tau)[R(\overset{\wedge}{\mu})\hat{u}_x-R(\mu_n)u_x^n]dyd\tau|\\ &\quad+|\int_0^t\int_0^\infty K_{xx}(x,y;t-\tau)[\hat{f}(\hat{u})-f_n(u^n)]dyd\tau|\\ &\quad+|\int_0^t K_{xx}(x,0;t-\tau)[F(\tau,\overset{\wedge}{\mu})-F(\tau,\mu_n)]d\tau|\\ &= T_6+T_7+T_8\end{aligned}$$

For $|F(\tau,\mu_n)-F(\tau,\overset{\wedge}{\mu})|\le C_F|\mu_n-\overset{\wedge}{\mu}|$ and property 4, because of $\lim_{n\to\infty}\mu_n=\overset{\wedge}{\mu}$, then

$$T_8\le c(x)C_F||\mu_n-\overset{\wedge}{\mu}||_\infty t\to 0,(n\to\infty)$$

Note $\lim_{n\to\infty} f_n = \overset{\Lambda}{f}$, $|f_n(u^n) - f_n(\overset{\Lambda}{u})| \leq L||u^n - \overset{\Lambda}{u}||_\infty$ and lemma 6, then

$$\begin{aligned} T_7 &\leq |\int_0^t \int_0^\infty K_{xx}(x,y;t-\tau)[f_n(u^n) - f_n(\overset{\Lambda}{u})]dyd\tau| \\ &\quad + |\int_0^t \int_0^\infty K_{xx}(x,y;t-\tau)(f_n(\overset{\Lambda}{u}) - \overset{\Lambda}{f}(\overset{\Lambda}{u})]dyd\tau| \\ &\leq ||f_n - \overset{\Lambda}{f}||_\infty c_2(0) + Lc_2(0)||u^n - \overset{\Lambda}{u}||_\infty \\ &\quad \to 0, (n \to \infty) \end{aligned}$$

Similarly

$$\begin{aligned} T_6 &\leq |\int_0^t \int_0^\infty K_{xx}(x,y;t-\tau)[R(\overset{\Lambda}{\mu})(u_x^n - \overset{\Lambda}{u}_x) + (R(\mu_n) - R(\overset{\Lambda}{\mu}))u_x^n]dyd\tau| \\ &\leq ||R||_\infty ||u_x^n - \overset{\Lambda}{u}_x||_\infty c_2(0) + ||R_\mu||_\infty ||\mu_n - \overset{\Lambda}{\mu}||_\infty || \overset{\Lambda}{u^n}_x ||_\infty c_2(0) \end{aligned}$$

Following lemma 6 and lemma 7, when $n \to \infty$, $T_6 \to 0$. The proof is completed.

The proof of theorem 1

Following from (7) and lemma 1(1), we obtain

$$\begin{aligned} ||\overline{f}||_\infty &\leq ||\theta'||_\infty + C_F c(x_0)(T + ||\mu||_\infty)T \\ &\quad + c_2(\alpha)||u_x||_\infty^\alpha |f|_\alpha T^{\alpha/2} + c_2(1)||u_{xx}||_\infty ||R||_\infty T^{1/2} \\ &\quad + ||R||_\infty ||u_x||_\infty \end{aligned} \tag{18}$$

Substituting the estimations of $||u_x||_\infty, ||u_{xx}||_\infty$ into (18), we get

$$||\overline{f}||_\infty \leq ||\theta'||_\infty + \zeta_f(T) \tag{19}$$

where

$$\begin{aligned} \zeta_f(T) &= C_F c(x_0)(T + ||\mu||_\infty)T \\ &\quad + c_2(\alpha)|f|_\alpha [d_2 C_F(1 + ||\mu'||_\infty)T \\ &\qquad + d_2||f||_\infty c_1(0)T^{1/2}]^\alpha T^{\alpha/2}(1 + d_1 c_2(1)||R||_\infty T^{1/2}) \\ &\quad + ||R||_\infty [d_2 C_F(1 + ||\mu'||_\infty)T + d_2||f||_\infty c_1(0)T^{1/2}] \\ &\quad + c_2(1)||R||_\infty d_1 C_F(2/\sqrt{\pi})(1 + ||\mu'||_\infty)T \end{aligned}$$

For estimating $|\overline{f}|_{C^{0,\alpha}}$, let's estimate $||\overline{f}||_\alpha$. First,

$$\begin{aligned} \overline{f(\theta(t_1))} - \overline{f(\theta(t_2))} &= \theta'(t_1) - \theta'(t_2) + I(t_1) - I(t_2) + J(t_1) - J(t_2) \\ &\quad + K(t_1) - K(t_2) + Q(t_1) - Q(t_2) \end{aligned}$$

where $I(t)$ is defined as in lemma 2,and

$$\begin{aligned} J(t_1) - J(t_2) = & \int_0^{t_1} \int_0^{\infty} K_{xx}(x_0, y; t_1 - \tau) f(u) dy d\tau \\ & - \int_0^{t_2} \int_0^{\infty} K_{xx}(x_0, y; t_2 - \tau) f(u) dy d\tau \end{aligned}$$

$$\begin{aligned} K(t_1) - K(t_2) = & \int_0^{t_1} \int_0^{\infty} K_{xx}(x_0, y; t_1 - \tau) R(\mu) u_x dy d\tau \\ & - \int_0^{t_2} \int_0^{\infty} K_{xx}(x_0, y; t_2 - \tau) R(\mu) u_x dy d\tau \end{aligned}$$

$$Q(t_1) - Q(t_2) = -[R(\mu(t_1)) u_x(x_0, t_1) - R(\mu(t_2)) u_x(x_0, t_2)]$$

(1) Estimation of $|Q(t_1) - Q(t_2)|$:

$$\begin{aligned} Q(t_1) - Q(t_2) &= T_1 + T_2 \\ &= -(R(\mu(t_1)) - R(\mu(t_2))) u_x(x_0, t_1) \\ &\quad - R(\mu(t_2))(u_x(x_0, t_1) - u_x(x_0, t_2)) \end{aligned}$$

From property (11),

$$|R(\mu(t_2)) - R(\mu(t_1))| \leq ||R_\mu(Z)||_\infty |t_1 - t_2|,$$

so

$$|T_1| \leq ||u_x||_\infty ||R_\mu(Z)||_\infty |t_1 - t_2|.$$

Because of lemma 4, $|T_2| \leq ||R||_\infty c_u |t_1 - t_2|^{1/2}$. Thus

$$|Q(t_1) - Q(t_2)| \leq (||u_x||_\infty ||R_\mu(Z)||_\infty |t_1 - t_2|^{1/2} + c_u)|t_1 - t_2|^{1/2}$$

(2) Estimation of $|K(t_1) - K(t_2)|$, From property (11), property 1, property 3 and lemma 4

$$\begin{aligned} |K(t_1) - K(t_2)| &= D_1 + D_2 \\ &\leq | \int_0^{t_2} \int_0^{\infty} [K_{xx}(x_0, y; t_1 - \tau) \\ &\qquad - K_{xx}(x_0, y; t_2 - \tau)] R(\mu) u_x dy d\tau | \\ &\quad + | \int_{t_2}^{t_1} \int_0^{\infty} K_{xx}(x_0, y; t_1 - \tau) R(\mu) u_x dy d\tau | \end{aligned}$$

Thus

$$\begin{aligned} D_2 &\leq ||R||_\infty | \int_{t_2}^{t_1} \int_0^{\infty} K_{xx}(x_0, y; t_1 - \tau)[u_x(y, \tau) - u_x(x_0, \tau)] dy d\tau | \\ &\leq ||R||_\infty ||u_{xx}||_\infty \int_{t_2}^{t_1} \int_0^{\infty} |K_{xx}(x_0, y; t_1 - \tau)||y - x_0| dy d\tau \\ &\leq ||R||_\infty ||u_{xx}||_\infty c_2(1) |t_1 - t_2|^{\frac{1}{2}} \end{aligned}$$

Further estimate D_1

$$\begin{aligned} D_1 &= |\int_{t_2}^{t_1}\int_0^{t_2}\int_0^\infty R(\mu)u_x(y,\tau)K_{xxt}(x_0,y;s-\tau)dyd\tau ds| \\ &\leq ||R||_\infty \int_{t_2}^{t_1}\int_0^{t_2}\int_0^\infty |K_{xxt}(x_0,y;s-\tau)|\frac{|u_x(y,\tau)-u_x(x_0,\tau)|}{|y-x_0|}|y-x_0|dyd\tau ds \\ &\leq ||R||_\infty||u_{xx}||_\infty \int_{t_2}^{t_1}\int_0^{t_2}\int_0^\infty |K_{xxt}(x_0,y;s-\tau)||y-x_0|dyd\tau ds \\ &\leq ||R||_\infty||u_{xx}||_\infty \int_{t_2}^{t_1}\int_0^{s}\int_0^\infty |K_{xxt}(x_0,y;s-\tau)||y-x_0|dyd\tau ds \\ &\leq c_3(1)||R||_\infty||u_{xx}||_\infty| \int_{t_2}^{t_1} s^{\frac{1}{2}-1}ds|| \\ &\leq 2c_3(1)||R||_\infty||u_{xx}||_\infty|t_1-t_2|^{\frac{1}{2}} \end{aligned}$$

(3) Estimation of $|J(t_1)-J(t_2)|$. From property 3 and $u_x(x_0,t)=0$, we obtain

$$\begin{aligned} |J(t_1)-J(t_2)| &= |\int_0^{t_2}\int_0^\infty[K_{xx}(x_0,y;t_1-\tau)-K_{xx}(x_0,y;t_2-\tau)]f(u)dyd\tau| \\ &\quad +|\int_{t_1}^{t_2}\int_0^\infty K_{xx}(x_0,y;t_1-\tau)f(u)dyd\tau| \\ &\leq |\int_{t_1}^{t_2}\int_0^{t_2}\int_0^\infty K_{xxt}(x_0,y;s-\tau)f(u)dyd\tau ds| \\ &\quad +|f|_\beta||u_{xx}||_\infty^\beta \int_0^{t_2}\int_0^\infty K_{xx}(x_0,y;t_1-\tau)|y-x_0|^{2\beta}dyd\tau \\ &\leq \tfrac{2}{(2-\beta)\beta}c_3(2\beta)|f|_\beta||u_{xx}||_\infty^\beta|t_1^\beta-t_2^\beta| \\ &\quad +c_2(2\beta)||u_{xx}||_\infty^\beta|f|_\beta|t_1-t_2|^\beta \\ &\leq \frac{2c_3(2\beta)}{\beta(2-\beta)}|f|_\beta||u_{xx}||_\infty^\beta|t_1-t_2|^\beta \\ &\quad +c_2(2\beta)||u_{xx}||_\infty^\beta|f|_\beta|t_1-t_2|^\beta \end{aligned}$$

When $0<\alpha=\beta<1/2$, we obtain

$$||\overline{f}||_{C^{0,\alpha}} \leq ||\theta||_{C^{1,\alpha}} + \{\phi(T,E)\} \tag{20}$$

where

$$\begin{aligned} \phi(T,E) &= ||u_x||_\infty||R_\mu||_\infty + ||R||_\infty||u_{xx}||_\infty[c_2(1)+2c_3(1)]|t_1-t_2|^{(2-\alpha)/2} \\ &\quad +[c_2(2\alpha)+2\tfrac{c_3(\alpha)}{\alpha(2-\alpha)}]|f|_\alpha||u_{xx}||_\infty^\alpha \end{aligned}$$

Then to estimate $||\overline{\mu'}||_\infty, ||\overline{\mu}||_\infty$. From (8)

$$\begin{aligned} \overline{\mu}(t) &= -\int_0^t K(0,0;t-\tau)F(\tau,\mu(\tau))dyd\tau \\ &\quad +\int_0^t\int_0^\infty K(0,y;t-\tau)f(u(y,\tau))dyd\tau \\ &\quad -\int_0^t\int_0^\infty K(0,y;t-\tau)R(\mu(\tau))u_x(y,\tau)dyd\tau \end{aligned}$$

because of

$$|\int_0^t\int_0^\infty K(0,y;t-\tau)R(\mu(\tau))u_x(y,\tau)dyd\tau| \le ||R_\mu||_\infty||\mu||_\infty||u_x||_\infty c_0(0)t,$$

$$|\int_0^t K(0,0;t-\tau)F(\tau,\mu(\tau))dyd\tau| \le C_F(t+||\mu||_\infty)t^{1/2}\,,$$

and

$$|\int_0^t\int_0^\infty K(0,y;t-\tau)f(u(y,\tau))dyd\tau| \le ||f||t\,,$$

so

$$||\overline{\mu}||_\infty \le C_F(T+||\mu||_\infty)T^{1/2} + ||R_\mu||_\infty||u_x||_\infty||\mu||_\infty c_0(0)T + ||f||_\infty T$$

According to lemma 2

$$\begin{aligned}\overline{\mu'}(t) = &-\frac{1}{\sqrt{\pi}}\textstyle\int_0^t(t-\tau)^{-1/2}(F_\tau+F_\mu\mu')d\tau\\ &+\textstyle\int_0^t\int_0^\infty K_{xx}(0,y;t-\tau)f(u(y,\tau))dyd\tau\\ &-\textstyle\int_0^t\int_0^\infty K_{xx}(0,y;t-\tau)R(\mu(\tau))u_x(y,\tau)dyd\tau\\ &+f(\mu(t))-R(\mu(t))F(\mu(t))\end{aligned}$$

From property 3

$$\begin{cases} |\int_0^t\int_0^\infty K_{xx}(0,y;t-\tau)R(\mu(\tau))u_x(y,\tau)dyd\tau| \le c(\alpha)||u_x||_\infty||R||_\infty t^{1/2}\\ |\int_0^t\int_0^\infty K_{xx}(0,y;t-\tau)f(\mu(\tau))dyd\tau| \le c(\alpha)||u_x||_\infty^\alpha||f||_\alpha t^{\alpha/2}\\ |\int_0^t(t-\tau)^{-1/2}(F_\tau+F_\mu\mu')d\tau| \le C_F(1+||\mu'||_\infty)\frac{2}{\sqrt{\pi}}t^{1/2}\\ |R(\mu(t))| = |R(\mu(t))-R(0)| \le ||R'||_\infty||\mu'||_\infty t\\ |f(\mu(t))| = |f(\mu(t))-f(0)| \le L||\mu'||_\infty t\end{cases}$$

Thus

$$\begin{aligned}||\mu'||_\infty \le &(2C_F/\sqrt{\pi})(1+||\mu'||_\infty)T^{1/2} + c_2(\alpha)||u_x||_\infty^\alpha|f|_\alpha T^{\alpha/2}\\ &+c_2(1)||u_{xx}||_\infty||R||_\infty T^{1/2} + (L+||R_\mu||_\infty)||\mu'||_\infty T\end{aligned}$$

If $||f||_{C^{0,\alpha}} \leq E, ||\mu||_{C^1} \leq E$, under the conditions of the theorem and the formulas above, we get

$$||\overline{f}||_{C^{0,\alpha}} \leq \xi_T^1(E) + ||\theta||_{C^{1,\alpha}} \qquad ||\overline{\mu}||_{C^1} \leq \xi_T^2(E)$$

where

$$\xi_T^1(E) \leq ||u_x||_\infty ||R_\mu||_\infty + 2^{\frac{2-\alpha}{2}} T^{\frac{2-\alpha}{2}} [c_2(1) + 2c_3(1)]||R||_\infty ||u_{xx}||_\infty$$

$$+ [c_2(2\alpha) + 2\frac{c_3(\alpha)}{\alpha(2-\alpha)}]|f|_\alpha ||u_{xx}||_\infty^\alpha$$

$$\xi_T^2(E) = (2C_F/\sqrt{\pi})(1 + ||\mu^{'}||_\infty)T^{1/2} + c_2(\alpha)||u_x||_\infty^\alpha |f|_\alpha T^{\alpha/2}$$

$$+ c_2(1)||u_{xx}||_\infty ||R||_\infty T^{1/2} + (L + ||R_\mu||_\infty)||\mu^{'}||_\infty T$$

$$+ C_F(T + ||\mu||_\infty)T^{1/2} + ||R_\mu||_\infty ||u_x||_\infty ||\mu||_\infty c_0(0)T + ||f||_\infty T$$

So, we have $||T[f,\mu]||_S \leq \xi_T^1(E) + \xi_T^2(E) + ||\theta||_{C^{1,\alpha}}$. For given $E > 2||\theta||_{C^{1,\alpha}}$, we can choice a proper T^*, such that $\xi_T^1(E) + \xi_T^2(E) \leq ||\theta||_{C^{1,\alpha}}$, i.e.$||T[f,\mu]||_S \leq E$.

The proof is completed.

Remark 3 *From (18), it's not difficult to obtain that $\zeta_f(T)$ tends to 0, while $T \to 0$. It follows that lemmas in section 3. Similarly $\Phi(T,E)$ in (20) tends to 0 too, as $T \to 0$. And the $\xi_T^1(E)$ or $\xi_T^2(E)$ tends to 0, as $T or E$ tends to 0. The conclusions described here is important to complete the proof.*

The proof of the theorem 2:

Formula (17) is obtained by lemma 7, $|f_n(u^n) - \hat{f}(\hat{u})| \leq |f_n(u^n) - \hat{f}(u^n)| + |\hat{f}(u^n) - \hat{f}(\hat{u})|$ and the properties of $f_n, \hat{f}$. Because $\hat{u}(x,t)$ satisfies the equation (3), $\hat{u}_t - \hat{u}_{xx} + R(\hat{u}(0,t))\hat{u}_x(x,t) = \hat{f}(\hat{u})$. If $\hat{u}$ also satisfies (2), the theorem is proven. Set $x = x_0$, we obtain

$$\hat{f}(\hat{u}(x_0,t)) = \hat{u}_t(x_0,t) - \hat{u}_{xx}(x_0,t) + R(\hat{u}(0,t))\hat{u}_x(x_0,t)$$
$$f_{n+1}(\theta) = \theta^{'}(t) - u_{xx}^n(x_0,t)\,,$$

then

$$\begin{aligned}\hat{f}(\hat{u}(x_0,t)) - f_{n+1}(\theta) = {} & \hat{u}_t - \theta^{'} + u_{xx}^n(x_0,t) - \hat{u}_{xx} + [R(\hat{u}(0,t)) \\ & - R(u^n(0,t))]\hat{u}_x(x_0,t) \\ & + R(u^n(0,t))[\hat{u}_x(x_0,t) - u_x^n(x_0,t)].\end{aligned}$$

Let $n \to \infty$. From the lemma 6 and lemma 7 and property (11), i.e.

$$||u^n_{xx}(x,t) - \overset{\Lambda}{u}_{xx}||_\infty \to 0, \quad ||\overset{\Lambda}{u}_x(x,t) - u^n_x(x,t)||_\infty \to 0$$
$$||R(\overset{\Lambda}{u}(0,t)) - R(u^n(0,t))||_\infty \le ||R||_\infty ||u^n(x,t) - \overset{\Lambda}{u}||_\infty \to 0$$

we obtain

$$\overset{\Lambda}{f}(\overset{\Lambda}{u}(x_0,t)) - \overset{\Lambda}{f}(\theta) = \overset{\Lambda}{u}_t(x_0,t) - \theta'(t) \tag{21}$$

From the property (10) of f, set $r(t) = \overset{\Lambda}{u}(x_0,t) - \theta(t)$. Using the Lipschitz property of f(i.e. the definition of Y), we get $[\overset{\Lambda}{f}(\overset{\Lambda}{u}(x_0,t)) - \overset{\Lambda}{f}(\theta)]r(t) = [\overset{\Lambda}{u}_t(x_0,t) - \theta'(t)]r(t)$, then $r'(t)r(t) \le Lr^2(t)$. From the Grownwall theorem, we have $r^2(t) \le r^2(0)e^{-2Lt}$, because of $r(0) = \overset{\Lambda}{u}_t(x_0,0) - \theta'(0) = 0$, $\overset{\Lambda}{u}(x_0,t) = \theta(t), \forall t \in [0,T]$. The proof is completed.

6 Numerical Simulation

For given (f_n, μ_n), we can get f_{n+1}, μ_{n+1} from (14)–(15). Following from the theorem 1, the solution of inverse problem (2)–(3) exists, and we know (f_n, u_n) converges to the solution of the problem (2)–(3), because of the theorem 2. So our numerical algorithm is:

1. Set $n = 0$, Given (f_0, μ_0) and the error: ϵ;
2. For the data (f_n, μ_n), from the equation (1.3), u_n is solved;
3. Following from (14)–(15), compute the values f_{n+1}, μ_{n+1};
4. Get the error $e_1 = ||f_{n+1} - f_n||_C, e_2 = ||\mu_{n+1} - \mu_n||_C$, if $max\{e_1, e_2\} \le \epsilon$, algorithm stops; otherwise, $n := n + 1$ and turn to 2.

Example: $T = 0.02$, $\epsilon = 1.0000e - 2$, $R(u(0,t)) = u(0,t)t$, $F(t, u(0,t)) = -u(0,t)$, $f(u) = (-t^3 - 1)u + 2te^{-x}$, the exact solution of (3) is t^2e^{-x}, $\theta(t) = t^2e^{-x_0}$, where $x_0 = \frac{1}{2}$.

References

1. J.R.Cannon, *The one dimensional heat equation* (Addison-Wesley, London, 1984).
2. J.R.Cannon and P.DuCheutrau, *Determining unknown coefficient in a nonlinear heat conduction problem*, SIAM J.AP, **24(3)**, 298–314 (1973).

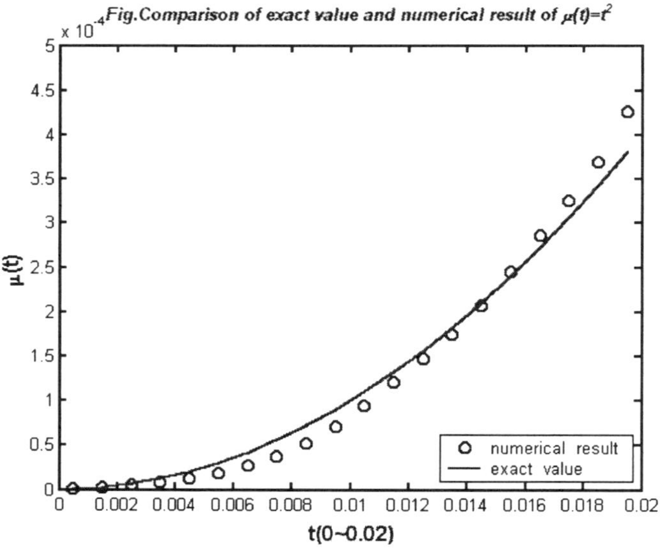

Figure 1. Comparison of exact value and numerical result of $\mu(t) = t^2$

3. S.Gatti,*An existence result for an inverse problem for quasilinear parabolic equation* , Inverse Problems **14**, 53–65 (1998).
4. L.De Luca, *Non-steady burning and combustion of solid propellants,* AIAA Prog. Astronaut. Axonaut **143**, 519–600 (1992).
5. L.De Luca, Extinction theories and experiments, AIAA Prog. Astronaut. Acronaut **90**, 661–694 (1983).
6. A. Lorenzi and E. Paparoni, *An identification problem for a semi-linear parabolic equation,* Ann.Mat.Pura Appl.CLI, 263–287 (1988).
7. A. Lorenzi and E. Paparoni, *An existence theorem for an identification problem related to a semi-linear parabolic equation* (Quaderno No.2, Department of mathematics, University of Milan, 1987).
8. A. Friedman, *P.D.E. of parabolic type* (Prentice-Hall, Englewood Cliffs N J, 1964).
9. Minxin Wang, *Nonlinear parabolic equation* (BeiJing Univ. Press, BeiJing, 1993).
10. Chaohao Gu, Zhongfan Xu etc.., *Mathematical physical equations* (Shanghai Science and technical Press, Shanghai, 1962).
11. V. Isakov, *Inverse problem for partial differential equations* (Springer-

Table 1. The result of $f(u(x,t))$ in the 11th step.

x	t	$f(x,t)$	exact value of $f(x,t)$	error
0.0	0.001000	1.92825e-03	1.99900e-03	3.539044%
1.0	0.003000	2.20356e-03	2.20397e-03	0.018249%
1.0	0.017000	1.19087e-02	1.24016e-02	3.974484%
1.0	0.019000	1.32339e-02	1.38466e-02	4.425226%
2.0	0.009000	2.30594e-03	2.42507e-03	4.912706%
2.0	0.011000	2.97298e-03	2.96100e-03	0.404626%
3.0	0.009000	8.54700e-04	8.92134e-04	4.196033%
3.0	0.011000	1.09432e-03	1.08929e-03	0.461700%
4.0	0.001000	3.66313e-05	3.66130e-05	0.049962%
4.0	0.011000	3.84526e-04	4.00728e-04	4.043144%
4.0	0.013000	4.50116e-04	4.73111e-04	4.860505%
6.0	0.001000	4.95750e-06	4.95503e-06	0.049981%
6.0	0.017000	8.00905e-05	8.35612e-05	4.153532%
6.0	0.019000	8.90598e-05	9.32977e-05	4.542395%
8.0	0.009000	5.76282e-06	6.01115e-06	4.131313%
8.0	0.011000	7.39122e-06	7.33959e-06	0.703513%
9.0	0.001000	2.46819e-07	2.46696e-07	0.049889%
9.0	0.003000	7.37695e-07	7.39348e-07	0.223633%
9.0	0.009000	2.12713e-06	2.21138e-06	3.809753%
9.0	0.011000	2.57118e-06	2.70008e-06	4.774103%
9.0	0.019000	4.42655e-06	4.64502e-06	4.703421%

Verlag, New York, 1998).
12. ShuXing Chen and JiaXin Hong, *Modern methods of P.D.E.* (Fudan Unvi, Press, ShangHai, 1988).
13. K. Yosida, *Functional Analysis* (Springer-Verlag, New York, 1987).

IDENTIFICATION AND CONTROL OF STRONGLY DAMPED NONLINEAR HYPERBOLIC PROBLEMS WITH APPLICATIONS

S. MIGÓRSKI

Jagiellonian University, Faculty of Mathematics, Physics and Computer Science Institute of Computer Science, ul. Nawojki 11, PL-30072 Cracow, Poland
E-mail: migorski@softlab.ii.uj.edu.pl

In this paper we study the optimal control of systems monitored by an abstract second order evolution inclusion with damping. First exploiting a surjectivity result we show that the problem has a global weak solution. Then, for distributed parameter control systems we consider the Bolza optimal control problem. Finally we present a result on the parameter identification problem and we give applications to hemivariational inequalities.

1 Introduction

Let V be a reflexive Banach space, let V^* be its dual and let H be a Hilbert space such that V is densely and compactly embedded in H. Let $A: (0,T) \times V \to V^*$ be a nonlinear operator such that $v \mapsto A(t,v)$ is pseudomonotone, $B \in \mathcal{L}(V,V^*)$ and $N \in \mathcal{L}(H,Y)$, Y being a reflexive Banach space and let $\partial J: (0,T) \times Y \to 2^{Y^*}$ be the Clarke generalized gradient of a function $J(t,\cdot): Y \to \mathbb{R}$. In this paper we consider a class of second order evolution inclusions of the form

$$(\mathrm{P}) \quad \begin{cases} y''(t) + A(t,y'(t)) + By(t) + N^*(\partial J(t,Ny(t))) \ni f(t) & \text{a.e. } t \in (0,T) \\ y(0) = y_0, \quad y'(0) = y_1, \end{cases}$$

where $f \in L^2(0,T;V^*)$, $y_0 \in V$ and $y_1 \in H$.

When J is independent of t and differentiable with derivative J', the problem (P) becomes an evolution equation and it serves as a formulation for systems arising in "smart" materials (see Banks, Smith and Wang [2] and the references therein). In this case the term $N^*(g(Nu))$ with $g(v) = J'(v)$ models a neo-Hookean type stress-strain constitutive law.

The problem (P) appears also in the description of several models of the so-called nonsmooth nonconvex mechanics (cf. Panagiotopoulos [13,14,15] for more information on hemivariational inequalities in nonmonotone contact problems in elasticity, skin effects of viscoelastic materials, nonconvex superpotential constitutive laws, etc.). In Section 2 below we present a simple example of a hemivariational inequality which can be formulated in the form

(P). For this reason sometimes we refer to (P) as to a hemivariational inequality. We refer also to Naniewicz and Panagiotopoulos [11] for the theory of stationary hemivariational inequalities and their applications. Some recent results on the existence and optimal control of evolution hemivariational inequalities can be found in Ochal [12], Migórski and Ochal [10] and Migórski [7,9].

In this note first we show the (global in time) existence of weak solutions for (P). Then we deal with the Bolza optimal control problem for (P). Independently we study the identification problem in which the control appears in the operators governing the process and in the multivalued term. This situation corresponds to the inverse problem when the goal is to identify unknown real stress-strain law in materials. Due to the lack of room the proofs of our results and a detailed discussion will be presented elsewhere.

2 Motivating Example

In this section we present an example leading to nonlinear hyperbolic hemivariational inequality and we recall some definitions.

Example 1 We consider a model of an elastic beam obeying linear Hook's law with Kelvin-Voigt damping. The beam is assumed fixed at $x = 0$ and at $x = l$. From its upper side along the segment $(l_1, l_2) \subset (0, l)$ the beam is adhesively connected with a support. The displacement at time t and position x is given by $y(t, x)$. The action of the adhesive material on the beam is described by a multivalued and nonmonotone relation

$$-f(t,x) \in \partial j(y(t,x)) \text{ on } (0,T) \times (l_1, l_2) \tag{1}$$

with f being the reaction force per unit length due to the gluing material and $j\colon \mathbb{R} \to \mathbb{R}$ denotes a locally Lipschitz function. Balance of forces and moments yields the following differential equation:

$$\rho \frac{\partial^2 y}{\partial t^2} + \frac{\partial^2}{\partial x^2}\left(EI \frac{\partial^2 y}{\partial x^2} + c_D I \frac{\partial^3 y}{\partial x^2 \partial t} \right) = f + g, \tag{2}$$

where ρ denotes the linear mass density, EI and $c_D I$ stand for the stiffness and damping coefficients respectively, and g is the prescribed loading. Along with the equation we consider the initial conditions

$$y(0,x) = y_0(x), \quad \frac{\partial y}{\partial t}(0,x) = y_1(x) \text{ for } x \in (0,l) \tag{3}$$

and the boundary conditions

$$y(t,0) = y(t,l) = 0, \quad \frac{\partial^2 y}{\partial x^2}(t,0) = \frac{\partial^2 y}{\partial x^2}(t,l) = 0 \tag{4}$$

for $t \in (0,T)$. We introduce the Hilbert space $V = H^2(0,l) \cap H_0^1(0,l)$ with the inner product $\langle w, z\rangle = \int_0^l \frac{\partial^2 w}{\partial x^2} \frac{\partial^2 z}{\partial x^2}\, dx$. Next we define the operators A, $B: V \to V^*$ by $\langle Aw, z\rangle = \frac{c_D I}{\rho}\langle w, z\rangle$ and $\langle Bw, z\rangle = \frac{EI}{\rho}\langle w, z\rangle$. Then the equation (2) with the law (1) and the specified boundary and initial conditions (3) and (4) can be written in the form: find $y: (0,T) \to V$ such that

$$\begin{cases} y''(t) + Ay'(t) + By(t) - \frac{1}{\rho} f(t) = \frac{1}{\rho} g(t) \ \text{ a.e. } t \in (0,T) \\ -f(t) \in U^*(\partial j(Uy(t))) \ \text{a.e. } t \in (0,T) \\ y(0) = y_0, \quad y'(0) = y_1. \end{cases} \tag{5}$$

Here the operator $U: L^2(\Omega) \to L^2(\Omega')$ is defined by $Uv = v|_{\Omega'}$, where $\Omega = (0,l)$ and $\Omega' = (l_1, l_2)$. Its adjoint operator $U^*: L^2(\Omega') \to L^2(\Omega)$ is given by

$$(U^*v)(x) = \begin{cases} v(x) & \text{if } x \in \Omega' \\ 0 & \text{otherwise.} \end{cases}$$

Therefore the multivalued relation in (5) is equivalent to the following two conditions

$$\begin{cases} -f(t,x) \in \partial j(y(t,x)) & \text{for } (t,x) \in (0,T) \times \Omega' \\ -f(t,x) = 0 & \text{for } t \in (0,T),\ x \notin \Omega'. \end{cases}$$

The problem analogous to (5) can be considered in the case of Kirchhoff plates (see Panagiotopoulos and Pop [16]). For other examples appearing in the modelling of linear visco-elastic materials (cf. Vol. 1, Chapter 3 of Dautray and Lions [6]), see Ochal [12].

Let Y be a reflexive Banach space and let $T: Y \to 2^{Y^*}$ be a multivalued operator. An operator T is said to be *pseudomonotone* (cf. Browder and Hess [4]) if it satisfies :

a) for every $y \in Y$, Ty is a nonempty, convex and weakly compact set in Y^*;

b) T is u.s.c. from every finite dimensional subspace of Y into Y^* endowed with the weak topology; and

c) if $y_n \to y$ weakly in Y, $y_n^* \in Ty_n$ and $\limsup \langle y_n^*, y_n - y\rangle_Y \le 0$, then for each $z \in Y$ there exists $y^*(z) \in Ty$ such that $\langle y^*(z), y - z\rangle_Y \le \liminf \langle y_n^*, y_n - z\rangle_Y$.

Let $L: D(L) \subset Y \to Y^*$ be a linear densely defined maximal monotone operator. An operator T is said to be *L-pseudomonotone* if and only if a) and b) hold and

d) if $\{y_n\} \subset D(L)$ is such that $y_n \to y$ weakly in Y, $Ly_n \to Ly$ weakly in Y^*, $y_n^* \in T(y_n)$, $y_n^* \to y^*$ weakly in Y^* and $\limsup \langle y_n^*, y_n \rangle_Y \leq \langle y^*, y \rangle_Y$, then $(y, y^*) \in Graph(T)$ and $\langle y_n^*, y_n \rangle_Y \to \langle y^*, y \rangle_Y$.

A single-valued operator $T: Y \to Y^*$ is said to be *pseudomonotone* if for each sequence $\{y_n\} \subseteq Y$ such that it converges weakly to $y_0 \in Y$ and $\limsup \langle Ty_n, y_n - y_0 \rangle_Y \leq 0$, we have $\langle Ty_0, y_0 - y \rangle_Y \leq \liminf \langle Ty_n, y_n - y \rangle_Y$ for all $y \in Y$.

Finally, we recall (see Clarke [5]) that given a locally Lipschitz function $h: E \to \mathbb{R}$, where E is a Banach space, the generalized directional derivative of h at x in the direction v, denoted by $h^0(x; v)$, is defined by $h^0(x; v) = \limsup_{y \to x,\ t \downarrow 0} \frac{1}{t}(h(y + tv) - h(y))$. The generalized gradient of h at x, denoted by $\partial h(x)$, is a subset of a dual space E^* given by $\partial h(x) = \{\zeta \in E^* : h^0(x; v) \geq \langle \zeta, v \rangle_{E^* \times E}$ for all $v \in E\}$.

3 Existence Result

Let $(V, \|\cdot\|)$ be a real reflexive, separable Banach space which is densely and compactly embedded in a Hilbert space H. The dual space of V is denoted by V^* and $|\cdot|$ stands for the norm in H. By $\langle \cdot, \cdot \rangle$ we denote the duality of V and V^*. Given $0 < T < +\infty$, we introduce the following spaces $\mathcal{V} = L^2(0, T; V)$, $\mathcal{H} = L^2(0, T; H)$, $\mathcal{W} = \{v \in \mathcal{V} : v' \in \mathcal{V}^*\}$, $\mathcal{Z} = \{v \in \mathcal{V} : v' \in \mathcal{W}\}$, where $\mathcal{V}^* = L^2(0, T; V^*)$, $\mathcal{H}^* \cong \mathcal{H}$ and the time derivative is taken in the sense of vector valued distributions. It is well known (cf. Zeidler [18]) that $\mathcal{W} \subset C(0, T; H)$ and $\mathcal{W} \subset C(0, T; V)$ continuously and $\mathcal{W} \subset \mathcal{H}$ compactly. Moreover, let Y be a reflexive Banach space.

The multivalued second order evolution equation under consideration is the following: given $y_0 \in V$, $y_1 \in H$ and $f \in \mathcal{V}^*$, find $y \in \mathcal{V}$ such that $y' \in \mathcal{W}$ and

$$\begin{cases} y''(t) + A(t, y'(t)) + By(t) + N^*(\partial J(t, Ny(t))) \ni f(t) \quad \text{a.e. } t \in (0, T) \\ y(0) = y_0, \quad y'(0) = y_1, \end{cases} \tag{6}$$

where $A: (0, T) \times V \to V^*$ is a nonlinear operator, $B: V \to V^*$ and $N: H \to Y$ are bounded linear operators, $\partial J: (0, T) \times Y \to 2^{Y^*}$ is the generalized (Clarke) gradient with respect to the second variable of a locally Lipschitz function $J(t, \cdot): Y \to \mathbb{R}$ and N^* denotes the adjoint operator of N.

We remark that the initial conditions in (6) have a sense since the embeddings $\mathcal{Z} \subset C(0, T; V)$ and $\mathcal{W} \subset C(0, T; H)$ are continuous. We say that $y \in \mathcal{V}$ is a solution of the problem (6) with $y_0 \in V$ and $y_1 \in H$ if $y' \in \mathcal{W}$ and there exists $\xi \in \mathcal{H}$ such that

$$\begin{cases} y''(t) + A(t, y'(t)) + By(t) + \xi(t) = f(t) \quad \text{a.e. } t \in (0,T) \\ y(0) = y_0, \quad y'(0) = y_1, \\ \xi(t) \in N^*(\partial J(t, Ny(t))) \quad \text{a.e. } t \in (0,T). \end{cases} \tag{7}$$

Remark 1 *The problem (6) is equivalent to the following inequality: find $y \in \mathcal{V}$ such that $y' \in \mathcal{W}$ and*

$$\begin{cases} \langle y''(t) + A(t, y'(t)) + By(t) - f(t), v\rangle + J^0(t, Ny(t); Nv) \geq 0 \\ \qquad\qquad\qquad\qquad\qquad\qquad \textit{for a.e. } t \textit{ and all } v \in V \\ y(0) = y_0, \quad y'(0) = y_1, \end{cases}$$

where $J^0(t, z; w)$ is the generalized directional derivative of J at a point $z \in Y$ in the direction $w \in Y$. This justifies the name hemivariational inequality given to problem (6).

We make the following assumptions:

$\underline{H(A)}$: $A: (0,T) \times V \to V^*$ is an operator such that

(i) $t \mapsto A(t, v)$ is measurable on $(0,T)$;

(ii) $v \mapsto A(t, v)$ is pseudomonotone for each t;

(iii) $||A(t, v)||_{V^*} \leq a_1(t) + b_1||v||$ a.e $t \in (0,T)$, for all $v \in V$, $a_1 \in L^2(0,T)$, $a_1 \geq 0$, $b_1 > 0$;

(iv) $\langle A(t, v), v\rangle \geq \beta_1||v||^2$ a.e. t, for all $v \in V$ with $\beta_1 > 0$.

$\underline{H(B)}$: $B: V \to V^*$ is a linear, bounded, positive and symmetric operator.

$\underline{H(N)}$: $N: H \to Y$ is a linear and bounded operator.

$\underline{H(J)}$: $J: (0,T) \times Y \to \mathbb{R}$, $J = J(t, z)$ is measurable in $t \in (0,T)$, locally Lipschitz in $z \in Y$ and for $\zeta \in \partial J(t, z)$ we have $||\zeta||_{Y^*} \leq \overline{c}(1 + ||z||_Y)$ for every $z \in Y$, $t \in (0,T)$ with $\overline{c} > 0$.

$\underline{(H_0)}$: $y_0 \in V$, $y_1 \in H$ and $f \in \mathcal{V}^*$.

$\underline{(H_1)}$: $\beta_1 > 4\beta^2 T \overline{c}\, ||N||^2$, where β is an embedding constant of V into H and $||N|| = ||N||_{\mathcal{L}(H,Y)}$.

Lemma 1 *If hypotheses $H(A)$, $H(B)$, $H(N)$, $H(J)$ and (H_0) hold and y is a solution to (6), then there exists a constant $C > 0$ such that*

$$\|y\|_{C(0,T;V)} + \|y'\|_{\mathcal{W}} \le C\left(1 + \|y_0\| + |y_1| + \|f\|_{\mathcal{V}^*}\right).$$

Lemma 2 *Assume hypotheses $H(N)$ and $H(J)$. Then the multivalued map $R\colon (0,T) \times H \to 2^H$ defined by $R(t,v) = N^*(\partial J(t, Nv))$ has nonempty, convex and weakly compact values in H, $R(t,\cdot)$ is from H into H_{weak} and there is a constant $\tilde{c} > 0$ such that $|R(t,v)| \le \tilde{c}(1 + |v|)$ for all $v \in H$.*

Theorem 1 *Under hypotheses $H(A)$, $H(B)$, $H(N)$, $H(J)$, (H_0) and (H_1), the problem (6) admits at least one solution.*

Proof: We present the main idea of it. First we reduce the order of the problem (6). We consider the operator $K\colon \mathcal{V} \to C(0,T;V)$ defined by $Kv(t) = \int_0^t v(s)\,ds + y_0$ for $v \in \mathcal{V}$. The operator K is bounded and continuous from $\mathcal{V}$ into $C(0,T;V)$. Using K we can write the problem (6) as follows: find $z \in \mathcal{W}$ such that

$$\begin{cases} z'(t) + A(t, z(t)) + B(Kz(t)) + R(t, Kz(t)) \ni f(t) \text{ a.e. } t \in (0,T) \\ z(0) = y_1. \end{cases} \tag{8}$$

Now we can see that $z \in \mathcal{W}$ solves (8) if and only if $y = Kz$ solves (6). Therefore, it is enough to prove the existence of solutions to (8). We consider two cases: first we study the problem (8) with regular initial condition $y_1 \in V$ and then we deal with a general case $y_1 \in H$. In the first case we define the following operators $\mathcal{A}_1\colon \mathcal{V} \to \mathcal{V}^*$, $\mathcal{B}_1\colon \mathcal{V} \to \mathcal{V}^*$ and $\mathcal{R}_1\colon \mathcal{V} \to 2^{\mathcal{V}^*}$ by $(\mathcal{A}_1)(\cdot) = A(\cdot, v(\cdot) + y_1)$, $(\mathcal{B}_1)(\cdot) = B(K(v + y_1)(\cdot))$ and $\mathcal{R}_1 v = \{w \in \mathcal{H} : w(t) \in R(t, K(v + y_1)(t))$ a.e. t for all $v \in \mathcal{V}$, respectively. Here $v + y_1$ is understood as follows $(v + y_1)(\cdot) = v(\cdot) + y_1$. Exploiting the above operators the problem (8) is formulated in the following way

$$\begin{cases} z' + \mathcal{A}_1 z + \mathcal{B}_1 z + \mathcal{R}_1 z \ni f \quad \text{a.e. } t \in (0,T) \\ z(0) = 0. \end{cases} \tag{9}$$

Note that $z \in \mathcal{W}$ solves (8) if and only if $z - y_1 \in \mathcal{W}$ solves (9). Next, introducing the operators $L\colon D(L) \subset \mathcal{V} \to \mathcal{V}^*$ and $T\colon \mathcal{V} \to 2^{\mathcal{V}^*}$ given by $Lz = z'$ with $D(L) = \{z \in \mathcal{W} : z(0) = 0\}$ and $Tz = \mathcal{A}_1 z + \mathcal{B}_1 z + \mathcal{R}_1 z$, respectively, the problem (9) takes the form: find $z \in D(L)$ such that $Lz + Tz \ni f$. In order to establish the existence of a solution to (9), we use a surjectivity result of Papageorgiou, Papalini and Renzacci [17]. To this end, exploiting a result of Berkovits and Mustonen [3] and the Convergence Theorem of Aubin and Cellina [1], we are able to prove that T is a bounded, coercive and L-pseudomonotone operator (cf. the definition in Section 2).

Example 2 Let $\Omega \subset \mathbb{R}^N$ be a bounded domain with a Lipschitz boundary and let $Y = L^2(\Omega; \mathbb{R}^N)$. We consider a function $j: (0,T) \times \Omega \times \mathbb{R}^N \to \mathbb{R}$ such that

$\underline{H(j)}$: $j(\cdot,\cdot,v): (0,T) \times \Omega \to \mathbb{R}$ is measurable for all $v \in \mathbb{R}^N$, $j(t,\cdot,0) \in L^1(\Omega)$, $j(t,x,\cdot): \mathbb{R}^N \to \mathbb{R}$ is locally Lipschitz for all $(t,x) \in (0,T) \times \Omega$ and for all $\xi \in \partial_v j(t,x,v)$ we have $\|\xi\|_{\mathbb{R}^N} \le c\,(1 + \|v\|_{\mathbb{R}^N})$ with $c > 0$.

We define $J: (0,T) \times Y \to \mathbb{R}$ by $J(t,v) = \int_\Omega j(t,x,v(x))\,dx$. It can be verified that if the integrand j satisfies $H(j)$, then the functional J satisfies $H(J)$. Furthermore, we can easily see that if $N = 1$ and $\beta \in L^\infty_{loc}(\mathbb{R})$ is such that $|\beta(s)| \le c(1+|s|)$ for $s \in \mathbb{R}$, then $j(t,x,v) = \int_0^v \beta(s)\,ds$ satisfies the hypothesis $H(j)$.

4 Optimal Control for Hemivariational Inequalities

4.1 Bolza Type Optimal Control Problem

We consider a system described by the following controlled second order evolution inclusion:

$$\begin{cases} y''(t) + A(t,y'(t)) + By(t) + N^*(\partial J(t,Ny(t))) \ni f(t) + C(t)u(t) \text{ a.e. } t \\ y(0) = y_0, \quad y'(0) = y_1, \end{cases} \tag{10}$$

where $y = y(u)$ is the solution corresponding to a control variable $u \in \mathcal{U} = L^2(0,T;X)$, X being the space of controls, C represents a controller and A, B, N, J, f, y_0 and y_1 are as in the previous section. We deal the following Bolza type optimal control problem (CP):

$$\begin{cases} \Phi(y,u) \to \inf, \quad \text{where } \; y \in S(u) \; \text{ and} \\ u(t) \in U(t) \text{ a.e. } t \in (0,T), u(\cdot) \text{ is measurable} \end{cases}$$

and the cost functional is given by

$$\Phi(y,u) = l(y(T),y'(T)) + \int_0^T F(t,y(t),y'(t),u(t))\,dt.$$

We admit the following assumptions:

$\underline{H(C)}$: $C \in L^\infty(0,T;\mathcal{L}(X,H))$ and X is a separable reflexive Banach space.

$\underline{H(\Phi)}$: $l: H \times H \to \mathbb{R}$ is weakly lower semicontinuous; $F: [0,T] \times H \times H \times X \to \mathbb{R} \cup \{+\infty\}$ is a measurable function such that

(i) $F(t, \cdot, \cdot, \cdot)$ is sequentially lower semicontinuous;

(ii) $F(t, y, z, \cdot)$ is convex;

(iii) there exist $M > 0$ and $\psi \in L^1(0,T)$ such that $F(t, y, z, u) \geq \psi(t) - M(|y| + |z| + \|u\|_X)$.

$\underline{H(U)}$: $U: [0,T] \to 2^X \setminus \{\emptyset\}$ is a multifunction such that for all $t \in [0,T]$, $U(t)$ is a closed convex subset of X and $t \to \sup\{\|u\|_X \, : \, u \in U(t)\}$ belongs to L^∞_+.

Theorem 2 *If the hypotheses $H(A)$, $H(B)$, $H(N)$, $H(J)$, (H_0), (H_1), $H(C)$, $H(\Phi)$ and $H(U)$ hold, then the problem (CP) admits an optimal solution.*

For other control problems for systems modeled by (10), a time optimal control problem and a maximum stay control problem, we refer to Ochal [12] and Migorski [8].

4.2 The Identification Problem

We consider the parameter estimation problem for the hemivariational inequality model (6). We state this problem in terms of finding parameters which give the best fit of the parameter dependent solutions of hemivariational inequality to the observation data for response of the system to excitations.

Let the collection of unknown parameters be denoted by p and we assume that it belongs to some admissible parameter set $\mathcal{P}$. Given $p \in \mathcal{P}$ we denote by $S(p)$ the solution set of

$$\begin{cases} y''(t) + A(p)y'(t) + B(p)y(t) + N^*(\partial J(p)(t, Ny(t))) \ni f(t) \quad \text{a.e. } t \in (0,T) \\ y(0) = y_0, \quad y'(0) = y_1. \end{cases} \tag{11}$$

The formulation of the inverse problem is as follows: given a cost functional $F = F(p,y)$, $F: \mathcal{P} \times \mathcal{Z} \to \mathbb{R}$, find $p^* \in \mathcal{P}$ and $y^* \in S(p^*)$ such that

$$F(p^*, y^*) = \inf\{F(p,y) \, : \, p \in \mathcal{P}, \; y \in S(p)\}. \tag{12}$$

We admit the following hypotheses:

$\underline{H(P)}$: $\mathcal{P}$ is a compact subset of a metric spaces of parameters $\widehat{\mathcal{P}}$,

$\underline{H(A)_1}$: for any $p \in \mathcal{P}$, $A(p) \in \mathcal{L}(V, V^*)$, $\langle A(p)v, v\rangle \geq c_1 ||v||^2$ for all $v \in V$ with $c_1 > 0$ independent of p and $p_n \to p$ in $\widehat{\mathcal{P}}$ implies $A(p_n) \to A(p)$ in $\mathcal{L}(V, V^*)$;

$\underline{H(B)_1}$: for any $p \in \mathcal{P}$, $B(p) \in \mathcal{L}(V, V^*)$, $B(p)$ is symmetric and positive and $p_n \to p$ in $\widehat{\mathcal{P}}$ implies $B(p_n) \to B(p)$ in $\mathcal{L}(V, V^*)$;

$\underline{H(J)_1}$: for any $p \in \mathcal{P}$, $J(p)\colon (0,T) \times Y \to \mathbb{R}$ is measurable in $t \in (0,T)$ and locally Lipschitz function in $v \in Y$ such that

(i) $||\xi||_{Y^*} \leq c_2\,(1 + ||v||_Y)$ for $\xi \in \partial J(p)(t, v)$, $v \in H$ with $c_2 \geq 0$,

(ii) if $p_n \to p$ in $\widehat{\mathcal{P}}$, then $\limsup_{n\to\infty} Gr\, \partial J(p_n)(t,\cdot) \subset Gr\, \partial J(p)(t,\cdot)$ in $Y \times Y_{weak}$ topology, for all $t \in (0,T)$,

where $Gr\, \partial J(p)(t,\cdot) = \{(z,w) \in Y \times Y^* \,:\, w \in \partial J(p)(t,z)\}$ stands for the graph of $\partial J(p)(t,\cdot)$.

Theorem 3 *If hypotheses $H(P)$, $H(A)_1$, $H(B)_1$, $H(J)_1$ and (H_1) hold, $(y_0, y_1) \in V \times H$, $f \in \mathcal{V}^*$ and F is lower semicontinuous in $\mathcal{P} \times \mathcal{Z}_{weak}$ topology, then the problem (12) admits a solution.*

We remark that the hypothesis $H(J)_1(ii)$ holds, for example, if $J(p_n)(t,\cdot)\colon Y \to \mathbb{R}$, $n \geq 1$, are locally Lipschitz, equi-lower semidifferentiable, locally equi-bounded and $J(p_n)(t,\cdot) \xrightarrow{\Gamma} J(p)(t,\cdot)$ for all $t \in (0,T)$ (see Theorem 1 of Zolezzi [19]).

In the examples, we may consider the problem of estimating of parameters by fitting data w obtained from displacement, velocity or acceleration measurements at various locations in a body $\Omega \subset \mathbb{R}^N$. This leads to functional $F(p,y) = G(y) + I(p)$, where $I\colon \mathcal{P} \to \mathbb{R}$ is a lower semicontinuous on $\mathcal{P}$ and $G\colon \mathcal{Z} \to \mathbb{R}$ is of the form

$$G(y) = \sum_{i=1}^{r} \left(\left\| y(t_i;p) - w_i^1 \right\|_H^2 + \left\| y'(t_i;p) - w_i^2 \right\|_H^2 \right)$$

or

$$G(y) = \int_{\Sigma_1} \left(|y(x,t) - w^3|^2 + |y'(x,t) - w^4|^2 \right) dxdt$$

$(\Sigma_1 = \Gamma_1 \times (0,T)$, $\Gamma_1 \subset \partial\Omega$, $m(\Gamma_1) > 0)$ subject to $y = y(\cdot;p)$ satisfying (11), where $0 < t_1 < t_2 < \ldots t_r \leq T$ are points of measurements and w_i^1, w_i^2, w^3, w^4 are fixed targets.

Acknowledgments

The research was supported in part by the State Committee for Scientific Research of the Republic of Poland (KBN) under Grants No. 2 P03A 004 19 and 7 T07A 047 18.

References

1. J. P. Aubin and A. Cellina, *Differential Inclusions. Set-Valued Maps and Viability Theory* (Springer, Berlin, New York, Tokyo, 1984).
2. H. T. Banks, R. C. Smith and Y. Wang, *Smart Material Structures: Modeling, Estimation and Control* (Wiley, Chichester, Masson, Paris, 1996).
3. J. Berkovits, V. Mustonen, *Monotonicity Methods for Nonlinear Evolution Equations*, Nonlinear Anal. **27**, 1397–1405 (1996).
4. F. E. Browder and P. Hess, *Nonlinear mappings of monotone type in Banach spaces*, J. Funct. Anal. **11**, 251–294 (1972).
5. F. H. Clarke, *Optimization and Nonsmooth Analysis* (Wiley Interscience, New York, 1983).
6. R. Dautray and J.-L. Lions, *Mathematical Analysis and Numerical Methods for Science and Technology*, Vol.1, *Physical Origins and Classical Methods* (Springer-Verlag, Berlin, 1992).
7. S. Migórski, *Existence and convergence results for evolution hemivariational inequalities*, Topological Methods Nonlinear Anal. **16**, 125–144 (2000).
8. S. Migórski, *Evolution hemivariational inequalities in infinite dimension and their control*, Nonlinear Anal. **47**, 101–112 (2001).
9. S. Migórski, *On existence of solutions for parabolic hemivariational inequalities*, J. Comp. Appl. Math. **129**, 77–87 (2001).
10. S. Migórski and A. Ochal, *Optimal control of parabolic hemivariational inequalities*, J. Global Optim. **17**, 285–300 (2000).
11. Z. Naniewicz and P. D. Panagiopopoulos, *Mathematical Theory of Hemivariational Inequalities and Applications* (Marcel Dekker, Inc., New York, Basel, Hong Kong, 1995).
12. A. Ochal, *Optimal Control of Evolution Hemivariational Inequalities* (PhD Thesis, Jagiellonian Univ., Cracow, Poland, p.63(2001)).
13. P.D. Panagiotopoulos, *Inequality Problems in Mechanics and Applications. Convex and Nonconvex Energy Functions* (Birkhäuser, Basel, 1985).
14. P.D. Panagiotopoulos, *Coercive and semicoercive hemivariational in-*

equalities, Nonlinear Anal. **16**, 209–231 (1991).

15. P.D. Panagiotopoulos, *Hemivariational Inequalities, Applications in Mechanics and Engineering* (Springer, Berlin, 1993).
16. P.D. Panagiotopoulos and G. Pop, *On a type of hyperbolic variational-hemivariational inequalities*, J. Applied Anal. **5**, 95–112 (1999).
17. N.S. Papageorgiou, F. Papalini and F. Renzacci, *Existence of Solutions and Periodic Solutions for Nonlinear Evolution Inclusions*, t Rend. Circ. Mat. Palermo **48**, 341–364 (1999).
18. E. Zeidler, *Nonlinear Functional Analysis and Applications II A/B* (Springer, New York, 1990).
19. T. Zolezzi, *Convergence of Generalized Gradients*, Set-Valued Anal. **2**, 381–393 (1994).

ESTIMATION OF ALL MATHEMATICAL MODEL PARAMETERS AND EXPERIMENT INFORMATIVENESS

M. R. ROMANOVSKI

CAD/CAE Department, POINT Ltd., 79-1-334, Schelkovskoe shosse, 107497, Moscow, Russia

E-mail: mromanovski@netscape.net,

URL: http://mywebpage.netscape.com/mromanovski/IP.htm

The conditions providing the highest achievable informativeness of experiment processing are examined. It is proved that a realization of a single experiment is sufficient to identify all phenomenological properties of a test object described by a superposition of mutually commuting operators. Simultaneous identification both model coefficients and boundary conditions is considered for the first time. The practical meaning of the study consists in the approach development guaranteeing the experiment against unidentifiable states and observation.

1 Introduction

Let us consider the problem of determining of the maximal information on properties of a test object during an interpretation of observation data. The main purpose is to define the maximum number of unknown parameters of an input signal, which is conveyed by a received signal.

For the sake of the following terminology shortness, we will define the amount of data concerned in a sampling as *an informativeness of an observed event*. This will understand as the permissible maximal volume of useful information on input signal components that is contained in a given received signal and can be unambiguously reconstructed during further observation processing. Here we will deal with the qualitative aspect, related to finding the upper bound of the informativeness, as well as with cases of the proper information degeneration.

The question response will be grounded on the condition determination that breaks down the one-to-one correspondence between a direct problem solution and its coefficients. In the theory of inverse problems many authors studied the conditions to identify more than one model parameter [1,2,3,4]. These investigations deal with uniqueness of inverse problem solutions to substantiate a correct mapping of sought functions.

In contrast to these studies, we consider the problem to extend a number of desired quantities as much as possible. A similar viewpoint is directed to practical problems to give ample opportunities of a correct identification of test object properties without numerous measurements.

2 Violation of One-to-One Correspondence

Let us define the permissible upper bound of the informativeness and establish the highest achievable volume of object properties that can be obtained having observation data of a single state function. Consider the following abstract model

$$\sum_{k=1}^{p} a_k L_k u = f, \tag{1}$$

where a_k denotes a phenomenological object's property that accurate within the equivalence describes an input-output mapping, p is a number of model parameters, u is a state function, f is an input function, L_k is an operator.

Equation (1) defines in a general form a known relation between a directly observed state u, an external impact f, and object's properties $a = \{a_k\}_{k=\overline{1,p}}$. This equation with initial-boundary conditions conveys a typical direct problem. In particular, the models of control systems as a rule are added up to Eq. (1), where a_k is the equivalence combination of virgin system parameters. A lot of distributed parameter systems also give rise to Eq. (1).

It is assumed that Eq. (1) has a unique and stable solution u with fixed a and f. Also, the domain of operator $L_a \equiv \sum\limits_{k=1}^{p} a_k L_k$ is supposed to be independent of the sought quantities. The dependence of the boundary conditions on the sought quantities will be considered separately in section 3.

Let us suppose that model (1) is specified as a superposition of mutually commuting operators, i.e., $\forall i, j \in [1,p] : L_i L_j = L_j L_i$. In this case our main result is given in the following theorem[5].

Theorem 1 *To determine all properties of a test object described by Eq. (1), where $a_k = const$ it is necessary and sufficient to perform a single experiment, in which the object state function does not satisfy the linear dependence condition*

$$\sum_{k=1}^{p} \beta_k L_k u^* = 0, \beta_k \in B, \tag{2}$$

where $B = \{\beta_k = const, \exists i, j \in [1,p] : \beta_i, \beta_j \neq 0\}$.

Theorem 1 conveys only the basic possibility for a number of unknowns to be simultaneously identified. It is also necessary to define the properties of the function f^* generating the coefficient invariance. This will allow us to answer the question what happens to inverse problem solutions during the violation of the one-to-one correspondence.

If the operators $\{L_k\}_{k=\overline{1,p}}$ are mutually commuting, then acting on the two parts of Eq. (1) and adding up the terms multiplied by the coefficients we get a linear dependence condition for the image space elements of Eq. (1). Therefore, condition (2) defines the subspace, whose elements being mapped accordingly to (1) with commuting operators retain the linear dependence condition. Further, expressing any term of Eq. (1) from condition (2) we reveal the form of the non-unique subset of unknown quantities as well as the equation whose solution satisfies the linear dependence condition (2). This results in the following corollary.

Corollary 1 *If the solution of Eq. (1) with mutually commuting operators satisfies condition (2), then the equation coefficients belong to the family*

$$\beta_p a_k - \beta_k a_p = b_k, \quad k = \overline{1, p-1} \tag{3}$$

the input function also satisfies the linear dependence condition

$$\sum_{k=1}^{p} \beta_k L_k f^* = 0, \beta_k \in B, \tag{4}$$

and the state u^* is determined as the solution of the equation

$$\sum_{k=1}^{p-1} b_k L_k u^* = \beta_p f^* \tag{5}$$

Hence, family (3) conveys the non-uniqueness character of the mapping from the space of object states into the space of the Eq. (1) coefficients. All its elements generate the only solution u^* that satisfies Eq. (5) with reduced order in respect to the initial equation. One-to-one correspondence is violated but under certain values of f^*, which are to satisfy condition (4). The violation occurs due to the linear dependence of the mathematical model terms. This dependence generates the subspace of direct problem solutions U^*, whose elements correlate with the subset of the sought object properties A^* determining the inverse problem solution accurate to family (3).

In the above case the higher-order operator L_p is singled out, so that family (3) sets the one-parameter dependence of Eq. (1) coefficients relative to a_p. Expressing latter in the terms of family (3) one can prove that another form of one-parameter dependence with the same non-zero values $\beta_k \in B$ but found with different terms of Eq.(2) is equivalent to the initial ambiguity family. If the number of the properties is $p > 2$, then new linearly dependent terms in Eq.(5) can be selected. Hence, there is a two-parameter family and the order of Eq.(1) is reduced once more. As a result, the non-uniqueness subset of Eq.(1) contains coefficient families of one- up to $(p-1)$-parameters.

Note, that condition (4) holds for arbitrary β_k, when model (1) is homogeneous, $f \equiv 0$. Therefore, the one-parameter family from the subset A^* has the form $a_k/a_p = \beta_k$, and the coefficients of a homogeneous equation, the boundary conditions of which do not depend on their values, can be only found within the ration β_k. This commonly known property of equivalence is coupled by the results quoted. As it seems, aside from a one-parameter family, homogeneous Eq.(1) leaves room for two-parameter up to $(p-1)$-parameter family of the model coefficients.

Theorem 1 and Corollary 1 give the complete answer the foregoing question about the useful information contents extracted during the experimental data processing. Namely, a single experiment can provide simultaneous estimation of all phenomenological properties of the test object, if the appearance of the state u^* satisfying the linear dependence condition of the initial equation terms is excluded. It is hence possible to identify the object properties starting from the observation with limited number of measurements.

On the other hand, the results obtained attest that there exists a class of a direct problem solution u^*, completely retained, if the free term is effected according to (4) while the equation coefficients should satisfy (3). For the reason the variance of object properties and characteristics does not change the direct problem solution at every point of its variable domain.

The result obtained conveys the general functional properties of mathematical models that generate the non-unique correspondence between a direct problem solution and its coefficients. The properties' manifestation is determined by specifying certain boundary conditions and external impacts. We will now study their determination based on the next equation

$$a_1 L_1 u + a_2 L_2 u = f \tag{6}$$

that is assumed to be given in a variable domain Q with boundary $\partial Q = G_1 \cup G_2$, on which the solution u satisfies the conditions

$$K_i u|_{G_i} = \varphi_i, i = 1, 2, \tag{7}$$

where $L_{1,2}$ are given linear operators, $K_{1,2}$ are the corresponding linear operators, one of which, for example K_1, determines the Cauchy conditions on the boundary G_1; f and $\varphi_{1,2}$ are known functions. It is assumed that there exists a unique function u satisfying (6) and (7) while the functions f and $\varphi_{1,2}$ are smooth enough so that the values of $L_{1,2}f, K_1 f|_{G_1}, K_2 f|_{G_2}, L_1\varphi_2, L_2\varphi_1$ can be determined. In this case the following theorem holds.

Theorem 2 *The one-to-one correspondence between the coefficients $a_{1,2}$ and the solution of problem (6), (7) breaks down, if and only if its solution is the function $u^* = b^{-1}L_1^{-1}f$, for the existence of which it is necessary and sufficient the adjustment of boundary conditions*

$$K_1\varphi_2|_{G_1} = K_2\varphi_1|_{G_2}, \tag{8}$$

the free term f^* must satisfy the equation

$$\beta L_1 f^* = L_2 f^*, \beta \neq 0 \tag{9}$$

with the conditions

$$\beta K_1 f^*|_{G_1} = bL_2\varphi_1 \tag{10}$$

$$K_2 f^*|_{G_2} = bL_1\varphi_2 \tag{11}$$

and the coefficients are given from the single family

$$a_1 + \beta a_2 = b, \tag{12}$$

where b and β are the parameters of the ambiguity subset.

The theorem proof is carried out similar Romanovskii[6]. From a practical viewpoint the following corollary is important.

Corollary 2 *If the conditions (8) and (9) hold, then the one-to-one correspondence is provided for the homogeneous conditions $L_1\varphi_2 = 0$ and $L_2\varphi_1 = 0$ or $K_1 f^*|_{G_1} = 0$ and $K_2 f^*|_{G_2} = 0$.*

These conditions guarantee the preservation of the one-to-one correspondence, if one designs experimental conditions. Satisfaction of any of the corollary 2 conditions ensures the absence of the unidentifiable class u^* of direct problem solutions. Analysis of the conditions (9)–(11) with terms identically zero gives the following affirmation.

Corollary 3 *If $K_1\varphi_2|_{G_1} = K_2\varphi_1|_{G_2}, K_2 f^*|_{G_2} = 0$, $L_{1,2}f = 0$, $L_1\varphi_2 = 0$, $K_1 f^*|_{G_1}/L_2\varphi_1 = const$, then all the problem (6),(7) solutions are determined as*

$$u^* = \frac{K_1 f^*|_{G_1} - a_2 L_2\varphi_1}{a_1 K_1 f^*|_{G_1}} L_1^{-1} f \tag{13}$$

The existence of the similar kind of solutions means that for every $a \in A$ problem (6), (7) has linearly dependent terms $L_{1,2}u$. In this case the subset of non-uniqueness is the entire original set of coefficients, $A^* \equiv A$, and for the reason it contains an infinite number of families (12). This does not contradict theorem 2, since the infinity of families is associated with different solutions of problem (6), (7) and each of them has the unique family (12). Formulation

$$a_1 \frac{\partial u}{\partial t} = a_2 \frac{\partial^2 u}{\partial x^2} + f, 0 < x < 1, t > 0;$$

$$u|_{t=0} = x(x-1), -\left.\frac{\partial u}{\partial x}\right|_{x=0} = 1, -\left.\frac{\partial u}{\partial x}\right|_{x=1} = -1$$

exemplifies the conditions of corollary 3. Every solution of this direct problem with $f = const \neq 0$ is defined by formula (13). So simulated field $u(x,t)$ does not provide the unique inverse problem solution for any thermal properties $a_{1,2}$ and observation data. This example demonstrates, that the violation of the one-to-one correspondence between the direct problem solution and sought quantities must be kept in mind as an important problem.

The results obtained convey the invariant properties of linear equations. Nonlinear equations have also the violation of the one-to-one correspondence. For example, in the class $u(x,t) \in C^{2,1}$ a solution of the equation

$$a_1(u)\frac{\partial u}{\partial t} = \frac{\partial}{\partial x}\left[a_2(u)\frac{\partial u}{\partial x}\right]$$

that satisfies the condition $\frac{du}{d\varsigma} = \exp\left(\int\limits_0^u \beta(\theta)d\theta\right)$ is invariant relatively transformation $a_1'' = a_1' + \exp\left(\int\limits_0^u \beta(\theta)d\theta\right)\frac{\varsigma_x^2}{\varsigma_t}\left(\beta h + \frac{dh}{du}\right)$, $a_2'' = a_2' + h$, where $h(u)$ is a displacement, the function β satisfies

$$\frac{d\beta}{du} + \beta^2 + \beta\frac{d}{du}\ln\left(\frac{\varsigma_x^2}{\varsigma_t}\frac{a_2^2}{a_1}\right) = \frac{1}{a_2}\frac{da_2}{du}\frac{d}{du}\ln\left(\frac{\varsigma_t}{\varsigma_x^2}\frac{a_1}{\frac{da_2}{du}}\right)$$

and $\varsigma(x,t) = C_0 + C_1 x + C_2 t$ or $\varsigma(x,t) = C_0 + (C_1 + x)/\sqrt{C_3(C_2 - t)}$, C_{0-3} are arbitrary constants. The similar kind of the heat equation solutions are the scaling solutions [7].

Thus, if we want to reconstruct several unknown quantities, then the invariant properties have to be taken into account as an important part of the uniqueness investigation.

Previously, we have studied the informativeness depending on the number of experiments necessary to define all the model coefficients. We are now to see the conditions that added to a discrete set $u_i = u|_{\Xi_i}$ on a measurement design $\{\Xi_i\}_{i=\overline{1,m}}$ to assure the identification of the maximal number of unknown quantities. This question will be analyzed within the previous mathematical framework. The following affirmation is valid.

Theorem 3 *Equation (1) is identifiable as to the parameters* $\{a_k\}_{k=\overline{1,p}}$*, if both the discrete set of observation* $\{u_i\}_{i=\overline{1,p}}$ *and free term* f *exclude the satisfaction of the linear dependence conditions* $\sum\limits_{i=1}^{p} \beta_i L_k u|_{x_i} = 0, k = \overline{1,p}$*,* $\sum\limits_{i=1}^{p} \beta_i f|_{x_i} = 0$ *respectively, where* $\beta_k = const, \exists i,j \in [1,p] : \beta_i, \beta_j \neq 0$.

The result obtained indicates that, generally, there are points at an observation domain, where no information of sought properties can be found against high measurement precision. The existence of such sensor locations is commonly known in the theory of oscillation. Theorem 2 is to confirm that the informativeness degeneration shares a common property of mathematical models. Therefore, it is necessary to expect the existence of such sensor locations for other kind of measurements, which do not ensure, for instance, the identification of thermal properties.

To exemplify the similar situation we consider the mathematical model

$$a_1 \frac{\partial u}{\partial t} = a_2 \frac{\partial^2 u}{\partial x^2} + f, 0 < x < \ell, t > 0; \tag{14}$$

$$u|_{t=0} = u_0,\ u|_{x=0} = v_1,\ u|_{x=1} = v_2 \tag{15}$$

where the coefficients $a_{1,2} = const$ are the sought quantities. For the case $f, u_0, v_{1,2} = const$ the linear dependence conditions, pointed by theorem 3, take place, if the condition

$$\exp\left[-\frac{\overline{a}_2}{\overline{a}_1}\left(\frac{k\pi}{\ell}\right)^2 (t_2 - t_1)\right] = \lambda \frac{\sin x_2 \frac{k\pi}{\ell}}{\sin x_1 \frac{k\pi}{\ell}},\ k = 1, 2, ...,\ \lambda \neq 0$$

is satisfied. Then the sought coefficients $a_{1,2}$ cannot be uniquely determined on a discrete sample $u_i^* = u^*(x_i^*, t_i)|_{i=1,2}$, if, and only if, the measurements are made at any two locations $x_{1,2}^*$ such that $x_1^* + x_2^* = \ell$ and also executed at the same moment, $t_1 = t_2$.

So, we can give the following final answer to the question posed above.

First, it is necessary and sufficient to perform a single experiment satisfying the certain requirements to reveal a maximum volume of information on properties of a test object. In the class of the linear abstract models their state functions and corresponding observation data must meet the conditions of linear independence of the model terms. To hold these conditions in practical situations a number of simple requirements must be fulfilled.

Second, strictly defined boundary conditions and external impacts are necessary to break down the one-to-one correspondence between the state function and its coefficients. There exist models all of whose states are unidentifiable on a whole. Also, there are observation points, where no information of sought properties can be found against high measurement precision.

Summarizing this part of the investigation, the following general consequence can be made. *A single experiment can convey information on all the test object properties accurate within the equivalence and there are conditions to identify them uniquely.*

3 Simultaneous Identification of Model Coefficients and Boundary Conditions

Let us now analyze typical experiments from a viewpoint of the maximal informativeness with absolute minimum of input data. Here we want to extend the question studied and pose the new problem of a reconstruction both model coefficients and boundary conditions grounding on *a sole observation point.*

Consider the direct problem (14), (15). It is required to establish the existence and features of the design $\Xi = \{x_i, t_j\}_{i=1}^{j=\overline{1,n}}$ with one observation point and n measurement times for which the known discrete set (sampling) $u^\delta_{i,j} = \overline{u}(x_i, t_j) + \varepsilon_j$, $i = 1, j = \overline{1,n}$ allows to identify the constant unknowns $a = \{a_{1,2}, v_{1,2}\}$ simultaneously. Here ε denotes a measurement noise, about which we only know its upper bound, $\max|\varepsilon_j| \leq \delta$.

We reduce the problem posed to the determination and analysis of the behavior of estimation errors $\mu_{1,2} = (\overline{a}_{1,2} - a_{1,2})/\overline{a}_{1,2}, \mu_{3,4} = (\overline{v}_{3,4} - v_{3,4})/\overline{v}_{3,4}$, where $\overline{a} = \{\overline{a}_{1,2}, \overline{v}_{1,2}\}$ denotes the actual values of sought quantities. Being grounded on the approach [8] that provides the comprehensive analysis of estimation errors, one can obtain the following system

$$\begin{aligned}\max|\frac{1}{1-\mu_2}F_1(\xi_1,\tau_k) + (2\vartheta_0 - \vartheta_1 - \vartheta_2 + \vartheta_1\mu_3 + \vartheta_2\mu_4)F_2(\xi_1,\tau_k) + \\ +(\vartheta_2 - \vartheta_1 + \vartheta_1\mu_3 - \vartheta_2\mu_4)F_3(\xi_1,\tau_k) - F_1(\xi_1,\tau_k)|_{\mu_{1,2}=0} - \\ -(2\vartheta_0 - \vartheta_1 - \vartheta_2)F_2(\xi_1,\tau_k)|_{\mu_{1,2}=0} - (\vartheta_2 - \vartheta_1)F_3(\xi_1,\tau_k)|_{\mu_{1,2}=0} - \\ -\mu_3\vartheta_1 - (\vartheta_1\mu_3 - \vartheta_2\mu_4)\xi_1 + \frac{\varepsilon_j}{\delta}\Re| = \Re, k = \overline{1,n}\end{aligned}$$

where

$$F_1 = \tfrac{4}{\pi^2}\sum_{k=1}^{\infty}\tfrac{\sin(2k-1)\pi\xi}{(2k-1)^3}\left\{1 - \exp\left[-\tfrac{1-\mu_2}{1-\mu_1}(2k-1)^2\pi^2\tau\right]\right\},$$

$$F_2 = \tfrac{2}{\pi}\sum_{k=1}^{\infty}\tfrac{\sin(2k-1)\pi\xi}{2k-1}\exp\left[-\tfrac{1-\mu_2}{1-\mu_1}(2k-1)^2\pi^2\tau\right],$$

$$F_3 = \frac{2}{\pi} \sum_{k=1}^{\infty} \frac{\sin 2k\pi\xi}{2k} \exp\left[-4\frac{1-\mu_2}{1-\mu_1} k^2\pi^2\tau\right],$$

$$\xi = \frac{x}{\ell}, \tau = \frac{a_2}{a_1\ell^2}t, \Re = \frac{\delta a_2}{f\ell^2}, \vartheta_0 = \frac{u_0 a_2}{f\ell^2}, \vartheta_1 = \frac{v_1 a_2}{f\ell^2}, \vartheta_2 = \frac{v_2 a_2}{f\ell^2}$$

To estimate the unknowns $a = \{a_{1,2}, v_{1,2}\}$ four measurement times $(n = 4)$ for any sensor location ξ_1 should be fixed. The sampling $\{u_{i,j}^{\delta}\}_{i=1}^{j=\overline{1,4}}$ is the minimal volume discrete set of observation for the problem studied.

The desired solution μ_{1-4} does not exist, if the observation is made at one of the points $\xi_1^* = (0, 1/2, 1)$. For the sensor location $\xi_1^* = 0$ or $\xi_1^* = 1$ the estimation errors $\mu_{1,2} \to \infty$. If the measurements are fulfilled in the middle of a specimen, $\xi_1^* = 1/2$, then the inverse problem solution depends on the combination $\vartheta_1 + \vartheta_2$ of the desired temperatures and the roots $\mu_{3,4}$ become arbitrary magnitudes.

Thus, in the class of constant initial-boundary conditions only three points of measurements do not provide the reconstruction both the model coefficients $a_{1,2}$ and boundary temperatures $v_{1,2}$. Any other sensor location ensures the identification of thermal properties and boundary temperatures simultaneously. The further important problem here is a determination of optimal initial-boundary conditions and sensor location to provide the minimal errors μ_{1-4} for fixed $\delta \neq 0$. This problem will be studied in the future.

Let us now consider other commonly known inverse problem with a heat flux loading scheme. We specify the following mathematical model

$$a_1 \frac{\partial u}{\partial t} = a_2 \frac{\partial^2 u}{\partial x^2}, 0 < x < \ell, t > 0; \tag{16}$$

$$u|_{t=0} = u_0, \left.\frac{\partial u}{\partial x}\right|_{x=0} = 0, \left. a_2 \frac{\partial u}{\partial x}\right|_{x=\ell} = \begin{cases} q, & for\, t<t_h \\ 0, & for\, t>t_h \end{cases} \tag{17}$$

In the frame of the informativeness investigation the following question is raised. *Is it possible to reconstruct both the thermal conductivity coefficient a_2 and heat flux q simultaneously being grounded on a single sensor sample?*

The question response will be sought as an experimental design problem solution. Namely, the existence and features of the observation design $\Xi = \{x_i, t_j\}_{i=1}^{j=1,2}$ with one sensor and two measurement times will be analyzed. The errors $\mu_1 = (\overline{a}_2 - a_2)/\overline{a}_2$ and $\mu_2 = (\overline{q} - q)/\overline{q}$ should be sought, where $\overline{a}_2, \overline{q}$ denote the actual values. Being grounded on the approach [6] one can obtain the following estimates of the errors of the inverse problem solution

$$\mu_1 \approx \frac{\pm 2\Re}{\pm 2\Re\left(1 + \frac{t_1}{t_h}\right) + C\left(\frac{\xi^2}{2} - \frac{1}{6}\right)} \tag{18}$$

and $\mu_2^{(C)} \equiv \min \mu_2 = \pm 2\Re/\Phi$, where $\Re = \delta a_2/q\ell$, $\Phi = t_h a_2/a_1\ell^2$, $C < 1$ is a certain parameter.

As is seen, the kind of function (18) is defined by the parameter $\Re > 0$. Other parameter t_1/t_h has the restricted variation because of its limiting in accordance with formulation (16), (17), for which $t_1 \leq t_h$. To guarantee the solution existence the variation of $\Re$ has also to be restricted, $\Re < \Re_{\max}$. Analyzing the asymptotic behavior $\mu_1 \to 1$ at the point $\xi = 1$, we obtain $\Re_{\max} = 1/12$. Hence, the estimation error (18) is depicted as a function with the asymptote at the point

$$\xi^* = \sqrt{\frac{1}{3} \pm \frac{4\Re}{C}\left(1 + \frac{t_1}{t_h}\right)}$$

Here only one of these two magnitudes determines the asymptote location. Their choice is defined from the condition of obtaining of the worst estimation error.

On the interval $|\xi| < \xi^*$ the function μ_1 has the extremum at the point $\xi = 0$. If $|\xi| > \xi^*$, then the estimation error is a steadily decreasing function and in this case $|\mu_1|_{\xi=1}| < |\mu_1|_{\xi=0}|$ for any $\{\Re, \Phi\}$.

So, there exists a such $0 < \xi^* < 1$ depending on the parameters $\{\Re, \Phi\}$ and observation time t_1, for which the function μ_1 increases without a limit. In general, the region near $\xi^* = 1/\sqrt{3}$ conveys the cases of worst sensor locations, where the errors of the inverse problem solution go beyond a permissible bound $\mu_1 > 1$. Otherwise, if the solution with minimal error $\mu_1 < 1$ is sought, then there are two points, $\xi = 0$ and $\xi = 1$, where the desired inverse problem solution attains its minimal error. This means that for fixed experimental conditions $\Re < 1/12$ and $\Phi < 1/6$ there exists an observation design guaranteeing the optimal reconstruction of the quantities $\{a_2, q\}$.

Thus, even the reduction of the observation volume to one sensor sample does not decrease the experiment informativeness. *Being grounded on a single discrete set of observation it is possible to reconstruct not only mathematical model coefficients and also boundary conditions.*

The results obtained indicate the new directions of investigations of mathematical model properties. Coefficient invariance inquiry permits to reveal a number of new identification peculiarities. Analysis of a behavior of inverse problem solutions with fixed non-zero measurement noise and discrete set of observation will give also novel theoretical outcomes.

References

1. J. R. Cannon, and P. DuChateau, *Int. J. Eng. Sci.*, **11**, 783–794 (1973).

2. S. Kitamura and S. Nakagiri, *Identifiability of spatially-varying and constant parameters in distributed systems of parabolic type*, SIAM J. Control and Optimization **15**, 785–802 (1977).
3. L. Carotenuto, and G. Raiconi, *Int. J. Syst. Sci.*, **11**, 1035–1049 (1980).
4. J. Cheng and M. Yamamoto, *The global uniqueness for determining two convection coefficients from Dirichlet to Neumann map in two dimensions*, Inverse Problems **16**, L25–L30 (2000).
5. M. R. Romanovskii, J. Eng. Physics **57**, 1112–1117 (1989).
6. M. R. Romanovskii, J. Eng. Physics **42**, 351–357 (1982).
7. M. R. Romanovskii, J. Eng. Physics **45**, 309–316 (1983).
8. M. R. Romanovskii, Industrial Laboratory **59**, 89–96 (1993).

SCATTERING OF ELASTIC WAVES IN THE HALF-SPACE AND RELATION BETWEEN THE LAX-PHILLIPS THEORY AND THE WILCOX THEORY

MISHIO KAWASHITA
Department of Mathematics, Hiroshima University,
Higashi-Hiroshima, 739-8526, Japan
E-mail: kawasita@math.sci.hiroshima-u.ac.jp

WAKAKO KAWASHITA
Institute of Mathematics, University of Tsukuba,
Tsukuba, Ibaraki, 305-8571 Japan
E-mail: wakako@math.tsukuba.ac.jp

HIDEO SOGA
Faculty of Education, Ibaraki University,
Mito, Ibaraki, 310-8512, Japan
E-mail: soga@mito.ipc.ibaraki.ac.jp

We classify plane waves for the elastic wave equation in the half-space. Employing those classified waves, we construct fundamental expressions in the Lax-Phillips scattering theory, i.e., translation representation, etc. This is accomplished in view of relation between the Lax-Phillips theory and the Wilcox theory. The relation of them is cleared in the abstract setting.

1 Introduction

In elastic bodies there are several kinds of the waves. In the isotropic case we have P-wave (the longitudinal wave), S-wave (the transversal wave), the Rayleigh wave, the evanescent wave, etc. P-wave and S-wave go into the inside of the bodies and are called the body waves. The Rayleigh wave and the evanescent wave are concentrated near the boundaries and called the surface waves.

We are interested in the surface waves. Those waves seem to behave individually and to be influenced by situations of the boundaries. By this observation we are motivated to formulate a scattering theory suitable for the surface waves. One of the aims in the scattering theory is to give a framework to present the limit state of the waves as time t tends to $+\infty$ (or $-\infty$) in the terms of the equation in the free space, and to examine well the waves in the free space, e.g., to get concrete expressions of the waves by something like the Fourier transformation, etc.

We think the half-space is suitable as the free space when taking the surface waves into account . And we want to formulate the scattering theory of the Lax-Phillips type in the free space (the half-space), i.e., to construct the translation representation, to give concrete expressions of the solutions, etc. These are described in section 5. The results in section 5 give a basis on investigation of scattering inverse problems.

On the other hand we know the theory of another type different from the Lax-Phillips one, called the Wilcox type. We extract also characteristic points from the both types, and in view of those points we show in section 4 that the both types can be translated into each other in the abstract setting. This consideration helps us to accomplish our construction of the Lax-Phillips type.

In section 2 we summarize reflection of plane waves in the half-space, and classify the situations of the reflection. This classification is used for expression of the solutions in the half-space stated the later sections. Section 3 is devoted to summarization of the generalized Fourier transformation (the spectral representation). Precise descriptions of proofs of the main results in this note are given in the other paper of the authors [2].

2 Reflection of the Plane Waves

In this section we consider the isotropic elastic wave equation in the half-space $\mathbb{R}^3_+ = \{x = {}^t(x_1, x_2, x_3) : x_3 > 0\}$:

$$(\partial_t^2 - L(\partial_x))u(t,x) = 0 \quad \text{in} \quad \mathbb{R}^1 \times \mathbb{R}^3_+,$$

where $u = {}^t(u_1, u_2, u_3)$ is the displacement vector and $L(\partial_x) = \sum_{i,j=1}^3 a_{ij}\partial_{x_i}\partial_{x_j}$. The coefficients a_{ij} are 3×3-matrices whose (p,q)-components a_{ipjq} are given by $a_{ipjq} = \lambda\delta_{ip}\delta_{jq} + 2\mu(\delta_{ij}\delta_{pq} + \delta_{iq}\delta_{jp})$, where λ and μ are the Lame constants and δ_{ij} are Kronecker's delta. The density is assumed to equal 1.

Plane waves in the whole space $\mathbb{R}^3$ mean the solutions of the form

$$c\, e^{i\sigma(t-\eta x)}\, v \qquad (\sigma > 0,\, c \in \mathbb{C}),$$

where $\eta, v \in \mathbb{R}^3$ are taken to satisfy $\det(I - L(\eta)) = 0$ and $v \in \operatorname{Ker}(I - L(\eta))$. In the case of P-wave, the direction η of the propagation and the direction v of the amplitude are parallel each other, and in the case of S-wave they are perpendicular.

In the half-space $\mathbb{R}^3_+$ we add some waves to the above plane wave (the incident wave) so that a boundary condition is satisfied. Here we impose the

Neumann boundary condition

$$Nu \equiv \sum_{i,j=1}^{3} \nu_i a_{ij} \partial_{x_j} u \,|_{x_3=0} = 0,$$

where ν is outer unit normal vector to the boundary (i.e. $\nu = {}^t(0,0,-1)$).

The added waves are, so called, the reflected waves. We can classify the phenomena of reflection in the following way:

(P) For an incident P-wave, P- and S-waves are reflected.

(SV) For an incident S-wave, P- and S-waves are reflected.

(SH) For an incident S-wave, only S-wave is reflected.

(SVO) For an incident S-wave, S-wave is reflected together with the evanescent wave.

(SVO) is the total reflection. Furthermore we have different wave not associated with the reflection (P) $\sim$ (SVO):

(R) There exists the wave called "the Rayleigh wave".

The evanescent and Rayleigh waves are concentrated exponentially near the boundary, and called the surface waves. Because of the surface waves, formulation of the theory in $\mathbb{R}^3_+$ becomes different from that in $\mathbb{R}^n$, as is described in section 5.

Getting rid of the part $e^{i\sigma t}$ from the above plane waves and the Rayleigh wave, we call the remainder functions "the generalized eigenfunctions" since they satisfy $L(\partial_x)u = -\sigma^2 u$. Dermenjian and Guillot[1] have shown that any data are expressed by superposition of these generalized eigenfunctions, and have developed the scattering theory of the Wilcox type for the equation in the half-space. We shall explain their results later (in section 3).

3 The Generalized Fourier Transformation

The Fourier transformation $\mathcal{F}$ is a powerful tool in the scattering theories. The Fourier transform $\mathcal{F}[f] = \hat{f}$ of f is of the form $\hat{f}(\xi) = \int e^{-ix\xi} f(x)dx$. This is used directly in the case of the d'Alembert equation in $\mathbb{R}^n$. For this transformation, one of the important formulas is the inversion formula $f(x) = (2\pi)^{-n} \int e^{ix\xi} \hat{f}(\xi)d\xi$. This means that f can be reconstructed by superposition of the plane waves $e^{ix\xi}$ (without the time-dependent part).

We can develop the similar operator $\mathcal{F}(\sigma)$ for the elastic equation, and call it the generalized Fourier transformation (or the spectral representation). In this case, $\mathcal{F}(\sigma)$ is an operator-valued function of σ (> 0) such that

$$\mathcal{F}(\sigma)(-Lf) = \sigma^2\mathcal{F}(\sigma)f$$
$$f = (2\pi)^{-n}\int_0^\infty \mathcal{F}(\sigma)^*[\mathcal{F}(\sigma)f]d\sigma, \tag{1}$$

where $\mathcal{F}(\sigma)^*$ is the adjoint operator of $\mathcal{F}(\sigma)$. Furthermore, $\mathcal{F}(\sigma)$ is expressed by the resolvent $(L(\partial_x) + (\sigma \pm i\varepsilon)^2 I)^{-1}$ ($\varepsilon > 0$): There exists the limit of the resolvent as $\varepsilon \to +0$, which is known as the limiting absorption principle, and we have

$$\mathcal{F}(\sigma)^*\mathcal{F}(\sigma) = 2(2\pi)^{n-1}(i\sigma)\{(L(\partial_x)+(\sigma-i0)^2I)^{-1} - (L(\partial_x)+(\sigma+i0)^2I)^{-1}\}.$$

This suggests that the generalized Fourier transformation (the spectral representation) may be constructed concretely by using the solutions of the equation $(L(\partial_x) + \sigma^2 I)u = 0$, e.g., by using the plane waves (the generalized eigenfunctions).

Wilcox[8] accomplished this construction actually for the d'Alembert equation, and established an elegant scattering theory. Dermenjian and Guillot[1] have applied the Wilcox idea to the elastic equation (in the perturbed half-space). For this equation, they have used the plane waves (the generalized eigenfunctions) $\phi_\alpha(x;\sigma,\omega)$ associated with the cases α = P, SV, SH, SVO, R in section 2. ϕ_α for α = P, SV, SH, SVO is of the form

$$\phi_\alpha(x;\sigma,\omega) = \phi_\alpha^i(x;\sigma,\omega) + \phi_\alpha^r(x;\sigma,\omega),\ \ \sigma > 0, \omega \in S_\alpha\ \ (\alpha = \text{P, SV, SH, SVO}),$$

where S_α is the zone associated with the case α ($\cup_\alpha S_\alpha = \{\xi \in \mathbb{R}^3_+; |\xi| = 1\}$), ϕ_α^i is the incident wave of the form

$$\phi_\alpha^i(x;\sigma,\omega) = m_\alpha(\sigma,\omega)e^{i\sigma\eta_\alpha(\omega)x}a_\alpha(\omega) \tag{2}$$

and $\phi_\alpha^r(x;\sigma,\omega)$ is the reflected wave for the incident wave ϕ_α^i. Furthermore, $m_\alpha(\sigma,\omega)$ is a function satisfying $|m_\alpha(\sigma,\omega)| = 1$, and $\eta_\alpha(\omega)$, $a_\alpha(\omega)$ are some vectors satisfying $\det(I - L(\eta_\alpha(\omega)) = 0$ and $[I - L(\eta_\alpha(\omega))]a_\alpha(\omega) = 0$ (α = P, SV, SH, SVO). Let us note that all m_α (α = P, SV, SH, SVO) were taken equal to 1 in Dermenjian and Guillot[1]. The generalized eigenfunction ϕ_R of the Rayleigh wave is the form

$$\phi_R(x;\sigma,\zeta) = \sum_{j=1}^{2} c_j e^{i\sigma\eta_R^j(\zeta)x}a_R^j(\omega,\zeta), \quad \zeta \in S_R,$$

where $S_R = \{\xi' \in \mathbb{R}^2; |\xi'| = 1\}$, c_j are some constants and η_R^j, a_R^j are some vectors satisfying $\det(I - L(\eta_R^j)) = 0$, $(I - L(\eta_R^j))a_R^j = 0$. In detail, see M.

Kawashita, W. Kawashita and Soga[2]. The third component $\eta^j_{R\,3}$ of η^j_R is taken satisfying $\mathrm{Im}[\eta^j_{R\,3}] > 0$, and so $\phi_R(x;\sigma,\zeta)$ decays exponentially as $x_3 \to +\infty$.

The generalized Fourier transformation (spectral representation) is defined in the following way:

$$\begin{aligned} \mathcal{F}(\sigma) &= (\mathcal{F}_P(\sigma), \mathcal{F}_{SV}(\sigma), \mathcal{F}_{SH}(\sigma), \mathcal{F}_{SVO}(\sigma), \mathcal{F}_R(\sigma)), \\ (\mathcal{F}_\alpha(\sigma)f)(\omega) &= c_\alpha(-i\sigma)(f, \phi_\alpha(x;\sigma,\omega))_H, \quad \omega \in S_\alpha, \end{aligned} \tag{3}$$

where c_α are some constants.

The Fourier transformation plays an important role also on the Lax-Phillips theory. Lax and Phillips expressed concretely the solutions in the free space by mean of the Radon transformation (cf. Lax and Phillips[3]):

$$\mathcal{R}f(s,\theta) = \int_{\theta x = s} f(x) dS_x. \tag{4}$$

This operator is much connected with the Fourier transformation: $\mathcal{R}f(s,\theta) = \frac{1}{2\pi}\int e^{is\sigma}\ \hat{f}(\sigma\omega)d\sigma$. Therefore, it is expected that we can derive the various expressions in the Lax-Phillips theory from the results of the Wilcox type. In fact, this expectation is accomplished, and moreover both the settings (the Wilcox and Lax-Phillips types) can be changeable even in the abstract situations. But there are many choices on selection of the above superposition (i.e. selection of the functions $\phi_\alpha(\sigma,\omega)$), and each choice of $\phi_\alpha(\sigma,\omega)$ is corresponding to one translation representation in the Lax-Phillips setting. In the Lax-Phillips theory we are required to choose the representation nicely to have a good property, and so every discussion is not finished by the one of Dermenjian and Guillot[1]. More precise discussion is given later in section 5.

4 Relation Between The Wilcox and Lax-Phillips Theories

In this section we consider the (abstract) wave equation

$$(\frac{d^2}{dt^2} - L)u(t) = 0,$$

where $-L$ is a positive self-adjoint operator on a Hilbert space $\mathcal{H}$. In the Lax-Phillips[3] scattering theory , one of the main assertions is that the solution operator

$$U(t) : {}^t(u(0), \frac{du}{dt}(0)) \to {}^t(u(t), \frac{du}{dt}(t))$$

can be transformed into translation by some two operators $T^{\pm}$: There exist subspaces $D_\pm$ in the space of the data (the energy space H) and unitary operators $T^{\pm}$ from H to $L^2(\mathbb{R}^1_s; N)$ (N is an extra Hilbert space) such that

$$\begin{aligned}&(T^{\pm}U(t)f)(s) = (T^{\pm}f)(s-t),\\&T^{\pm}(D_{\pm}) = \{k(s) : k(s) = 0 \text{ if } \pm s < 0\}\end{aligned} \tag{5}$$

if and only if there exist subspaces $D_{\pm}$ in H satisfying

(i) $U(t)D_{\pm} \subset D_{\pm}$ for any $\pm t > 0$,

(ii) $\cap_{t\in\mathbb{R}} U(t)D_{\pm} = \{0\}$,

(iii) $\cup_{t\in\mathbb{R}} U(t)D_{\pm}$ are dense in H.

T^+ and T^- are called the outgoing and incoming translation representations, and D_+ and D_- are called the outgoing and incoming subspaces. In the Lax-Phillips theory, the scattering operator S is defined by $S = T^+(T^-)^{-1}$, and is desired to contain all the information about the scatterer.

The generator of $U(t)$ is of the form

$$A = \begin{pmatrix} 0 & I \\ -L & 0 \end{pmatrix}$$

and the spectral representation for A means (in the Lax-Phillips sense) a unitary operator $\mathcal{T}$ from H to $L^2(\mathbb{R}^1; N)$ such that

$$\mathcal{T}A = i\sigma\mathcal{T}.$$

We see that this $\mathcal{T}$ is connected with a translation representation T by the equality: $(\mathcal{T}f)(s) = \int e^{-i\sigma s}(Tf)(s)ds$ ($f = {}^t(f_1, f_2) \in H$).

The spectral representation $\mathcal{T}$ in the Lax-Phillips sense can be translated into the one in the Wilcox sense $\mathcal{F}(\sigma)$ (the generalized Fourier transformation in section 3):

Theorem 1 *(i) If we have $\mathcal{T}$ or $\mathcal{F}(\sigma)$, then by this we can make the other.*

(ii) These $\mathcal{T}$ and $\mathcal{F}(\sigma)$ are connected each other by the equality

$$(\mathcal{T}f)(\sigma) = i\sigma\mathcal{F}(\sigma)f_1 + \mathcal{F}(\sigma)f_2 \quad \textit{for } \sigma > 0.$$

This theorem is proved in M. Kawashita, W. Kawashita and Soga[2].

5 Expressions in the Free Space

In this section we consider the isotropic elastic equation in the half-space $\mathbb{R}^3_+$, and construct the fundamental expressions in the Lax-Phillips theory (e.g., the translation representations, etc.). In the whole space $\mathbb{R}^n$ those are described by Lax and Phillips[3] (for the d'Alembert equation), Shibata and Soga[5] (for the elastic equation). The construction in the half-space is fairly different from that in the whole space. This is mainly due to existence of the surface waves (i.e., the Rayleigh and evanescent waves).

If we want only to make a translation representation T (or spectral representation $\mathcal{T}$), then by Theorem 1 we can derive it soon from the generalized Fourier transformation $\mathcal{F}(\sigma)$ (of the Wilcox type) which has been obtained by Dermenjian and Guillot[1]. In the Lax-Phillips theory, we like to construct T with a good property: Lax and Phillips made a translation representation T such that

$$\begin{aligned} &[U(t)f](x) = 0 \quad \text{for all} \;\; (t,x) \text{ with } |x| \leq t\,(>0) \\ &\text{if and only if } (Tf)(s) = 0 \quad \text{for all } s < 0. \end{aligned} \tag{6}$$

This implies that D_+ consists of the data to make the lacuna arise. Lax and Phillips[3] carried out the construction of T with the property (6) very concretely by means of the Radon transformation (4). This concreteness and the property (6) are very useful for further investigations on scattering problems, e.g., the inverse problems (cf. Majda[4], Soga[6,7], etc.).

Thus we hope that our translation representation also has the property (6). As is explained later, however, we cannot get such a representation, and only can obtain the one with partially similar property (cf. Theorems 3 and 4).

Let $\mathcal{F}(\sigma) = (\mathcal{F}_P(\sigma), \mathcal{F}_{SV}(\sigma), \mathcal{F}_{SH}(\sigma), \mathcal{F}_{SVO}(\sigma), \mathcal{F}_R(\sigma))$ be the generalized Fourier transformation (3) defined in section 3. Then, by (3) and (ii) of Theorem 1 we obtain the spectral representation $\mathcal{T}$ and consequently the translation representation T (in the odd dimensional half-space $\mathbb{R}^3_+$, T^- becomes equal to T^+, i.e., $T^+ = T^- = T$):

Theorem 2 *We can obtain a translation representation T with the properties stated (5) which is of the form $T = (T_P, \ldots, T_R)$ and is a unitary operator from H to $L^2(\mathbb{R}^1; N)$ where $N = \oplus_{\alpha\in\Lambda} L^2(S_\alpha)$, $\Lambda = \{P, \ldots, R\}$.*

We can express T by means of the Radon transformation in (4) also. But then we need to employ the following modified one $\tilde{\mathcal{R}}_j$ to deal with the surface waves.

$$\tilde{\mathcal{R}}_j\, g(s,\theta) = \frac{1}{\pi}\int_{\mathbb{R}^3_+} \frac{(\tilde{\eta}_3(\theta)x_3)^{1-j} + (\tilde{\eta}'(\theta)x' - s)^j}{(\tilde{\eta}_3(\theta)x_3)^2 + (\tilde{\eta}'(\theta)x' - s)^2}\, g(x)dx, \quad (j = 0, 1),$$

where $\theta \in S_{SVO}$ (or S_R) and $\tilde{\eta}(\theta) = (\tilde{\eta}'(\theta), \tilde{\eta}_3(\theta))$ is a certain vector satisfying $\det(I - L(\tilde{\eta}(\theta))) = 0$. In detail, see §5 in M. Kawashita, W. Kawashita and Soga[2].

Since we can reconstruct the data f by $\mathcal{F}(\sigma)$ and $\mathcal{F}(\sigma)^*$ (cf. (1)), we can express concretely the solution $u(t,x)$ (or $U(t)f$) by means of the above translation representation T (cf. §5 of Kawashita *et al* [2]). Noting this expression of $U(t)f$, we can decompose $U(t)f$ into two parts:

$$U(t)f = U_B(t)f + U_{SR}(t)f,$$

where $U_B(t)f$ is superposition of the plane body waves associated with the real roots $\tilde{\eta}$ of det $(I - L(\tilde{\eta})) = 0$ and $U_{SR}(t)f$ is the remainder term, i.e., consists of the surface waves.

We choose the functions m_α and η_α in (2) as follows:

$$m_\alpha(\sigma,\omega) = 1 \quad \text{for} \quad \alpha = \text{P, SV, SH}\ ,$$

$$m_{SVO}(\sigma,\omega) = \frac{(1 - 2|\omega'|^2)^2 + 4i(\text{sgn}\ \sigma)|\omega'|^2\omega_3(|\omega'|^2 - c_S^2/c_P^2)^{1/2}}{|(1 - 2|\omega'|^2)^2 + 4i(\text{sgn}\ \sigma)|\omega'|^2\omega_3(|\omega'|^2 - c_s^2/c_p^2)^{1/2}|}$$

$$\eta_P(\omega) = c_P^{-1}\, {}^t(\omega', -\omega_3), \quad \eta_\alpha(\omega) = c_S^{-1}\, {}^t(\omega', -\omega_3) \quad \text{for} \quad \alpha = \text{SH, SV, SVO}$$

where c_P and c_S are the propagation speeds of the P- and S-waves respectively. Then we can see that T has a similar property to (6), but weaker than (6):

Theorem 3 *$f \in H$ belongs to $D_\pm$, i.e.,*

$$(Tf)(s) = 0 \quad \textit{for all}\ \pm s < 0,$$

if and only if the following conditions (i) and (ii) hold:

(i) $\quad \textit{supp}\mathcal{P}\,[U_B(t)f]_1 \subset \{x \in \mathbb{R}^3_+\,; \pm c_S t < |x|\},$

(ii) $\textit{supp}\mathcal{P}\,[U(t)f]_1|_{x_3=0} \subset \{x' \in \partial\mathbb{R}^3_+\,;\ \pm c_R t < |x'|\},$

where ${\mathcal{P}}\,{}^t(x', x_3) = x'$ *and* c_S, c_R *are some constants independent of* f, t *and* x.

It is because of existence of the surface waves to need to choose m_α and η_α in the above way.

In case of the half-space, we cannot make the translation representation T with the exactly same property that (6), which follows from

Theorem 4 *$f \in H$ satisfies the condition*

$$supp\ [U(t)f] \subset \{x \in \mathbb{R}^3_+ ; ct < |x|\} \quad (t > 0)$$

for a constant $c > 0$ if and only if $U_{SR}(t)f = 0$, i.e., $T_{SVO}f = 0$ and $T_R f = 0$.

Lax and Phillips gave a characterization of the value of $(Tf)(s, \theta)$ at any fixed (s, θ) for any f in some class: We choose a certain straight line $\{x(t)\}_{t \in \mathbb{R}}$ in $\mathbb{R}^n$, and then we have $(Tf)(s, \theta) = \lim_{t \to \infty} t^{(n-1)/2}[U(t)f]_1(x(t))$. For our equation we obtain a similar result:

Theorem 5 *For any fixed $(s, \theta) \in S_\alpha$ ($\alpha = P, SV, SH$) set*

$$x_\alpha(t) = c_\alpha(s + t)\check{\theta}, \quad \check{\theta} = {}^t(\theta', -\theta_3),$$

where c_P and $c_{SV} = c_{SH}$ are the propagation speeds of the P- and S-waves respectively. Assume that $(Tf)(s, \theta)$ is sufficiently smooth and decreasing (as $|s| \to \infty$). Then we have

$$(T_\alpha f)(s, \theta) = 4\pi c_\alpha^{3/2} \lim_{t \to -\infty} t[U(t)f]_2(x_\alpha(t)) \cdot a_\alpha(\check{\theta}) \quad for\ \alpha = P,\ SV,\ SH\ ,$$

where a_α is the vector in (2).

The above Theorems 2 $\sim$ 5 are proved by M. Kawashita, W. Kawashita and Soga[2]. For the proofs, see §5 of Kawashita *et al* [2].

Acknowledgments

Mishio Kawashita was Partially supported by Grant-in-Aid for Encouragement of Young Scientists A-09740085 from JSPS, Wakako Kawashita were Partially supported by Grant-in-Aid for Encouragement of Young Scientists A-13740090 from JSPS, and Hideo Soga was Partially supported by Grant-in-Aid for Sci. Research (C) 13640150 from JSPS.

References

1. Y. Dermenjian and J. Guillot, *Scattering of elastic waves in a perturbed isotropic half space with a free boundary.* The limiting absorption principle, Math. Meth. Appl. Sci. **10**, 87–124 (1988).
2. M. Kawashita, W. Kawashita and H. Soga, *Relation between scattering theories of the Wilcox and Lax-Phillips types and a concrete construction of the translation representation*, submmited.

3. P. D. Lax and R. S. Phillips, Scattering theory (Academic Press, New York, 1967).
4. A. Majda, *A representation formula for the scattering operator and the inverse problem for arbitrary bodies*, Comm. Pure Appl. Math. **30**, 165–194 (1977).
5. Y. Shibata and H. Soga, *Scattering theory for the elastic wave equation*, Publ. RIMS Kyoto Univ. **25** , 861–887 (1989).
6. H. Soga, *Singularities of the scattering kernel for convex obstacles*, J.Math. Kyoto Univ. **22**, 729–765 (1983).
7. H. Soga, *Representation of the scattering kernel for the elastic wave equation and singularities of the back-scattering*, Osaka J. Math. **29**, 809–836 (1992).
8. C. H. Wilcox, *Scattering Theory for the d'Alembert Equation in Exterior Domains*, Lect. Notes in Math. **442** (Springer, Berlin, 1975).

FORMULAS FOR RECONSTRUCTING CONDUCTIVITY AND ITS NORMAL DERIVATIVE AT THE BOUNDARY FROM THE LOCALIZED DIRICHLET TO NEUMANN MAP

GEN NAKAMURA

Department of Mathematics, Faculty of Science, Hokkaido University, Sapporo 060-0810, Japan

E-mail: gnaka@math.sci.hokudai.ac.jp

KAZUMI TANUMA

Department of Mathematics, Faculty of Engineering, Gunma University, Kiryu 376-8515, Japan

E-mail: tanuma@math.sci.gunma-u.ac.jp

We consider the problem of determining conductivity of the medium from the measurements of the electric potential on the boundary and the corresponding current flux across the boundary, that is, from the Dirichlet to Neumann map. We give three kinds of formulas for reconstructing conductivity and its normal derivative from the localized Dirichlet to Neumann map. They are the formulas for pointwise reconstruction, reconstruction in a weak form, and reconstruction in the form of Fourier transform. In particular, the normal derivative of the conductivity at the boundary is reconstructed directly from the localized Dirichlet to Neumann map.

1 Introduction

Let $\Omega \in R^n$ $(n \geq 2)$ be a bounded domain with Lipschitz boundary $\partial\Omega$. Physically Ω is considered as an isotropic, static and conductive medium with conductivity $\gamma \in L^\infty(\Omega)$. When an electric potential $f \in H^{1/2}(\partial\Omega)$ is applied to the boundary $\partial\Omega$, the potential u solves the Dirichlet problem

$$\nabla \cdot (\gamma \nabla u) = 0 \quad \text{in} \quad \Omega, \qquad u|_{\partial\Omega} = f. \tag{1}$$

Assume that there is a constant $\delta > 0$ such that $\gamma(x) \geq \delta$ (a.e. $x \in \Omega$). Then, there exists a unique weak solution $u \in H^1(\Omega)$ to (1). Define the Dirichlet to Neumann map $\Lambda_\gamma : H^{1/2}(\partial\Omega) \longrightarrow H^{-1/2}(\partial\Omega)$ by

$$< \Lambda_\gamma f,\ g >= \int_\Omega \gamma \nabla u \cdot \nabla v \, dx \quad (g \in H^{1/2}(\partial\Omega)), \tag{2}$$

where u is the solution to (1), v is any function in $H^1(\Omega)$ satisfying $v|_{\partial\Omega} = g$ and $< \, , \, >$ is the bilinear pairing between $H^{1/2}(\partial\Omega)$ and $H^{-1/2}(\partial\Omega)$. Note that $\Lambda_\gamma f = \gamma \nabla u \cdot n$ when $f \in H^{3/2}(\partial\Omega)$, $\gamma \in C^1(\overline{\Omega})$ and $\partial\Omega$ is C^2, where

n is the unit outer normal to $\partial\Omega$. Hence $\Lambda_\gamma f$ is the current flux across $\partial\Omega$ produced by the potential f on $\partial\Omega$.

The problem of determining conductivity of the medium from the measurements of the electric potential on the boundary and the corresponding current flux across the boundary is expressed as *Inverse Problem* "Determine $\gamma(x)$ from Λ_γ". Since this problem was posed by A.P.Calderon, many results on uniqueness, stability, reconstruction have been proved. Here we give a brief review of some of the previous works on reconstruction. When γ and $\partial\Omega$ are C^∞, using the fact that Λ_γ is a pseudodifferential operator in this case, Sylvester and Uhlmann[9] showed how to recover γ and all of its derivatives on $\partial\Omega$ from the symbol of Λ_γ. When $\partial\Omega$ is Lipschitz continuous, from Λ_γ Nachman[3] recovered γ on $\partial\Omega$ if $\gamma \in W^{1,p}(\Omega)$ with $p > n$ and recovered the first normal derivative of γ on $\partial\Omega$ if $\gamma \in W^{2,p}(\Omega)$ with $p > n/2$.

On the other hand, reconstruction of conductivity from the localized Dirichlet to Neumann map has been studied first by Brown[1]. Reconstruction from the localized Dirichlet to Neumann map means that for $x_0 \in \partial\Omega$, assuming some regularity conditions on $\partial\Omega$ and on the conductivity locally around x_0, we take Dirichlet data f's to be the functions compactly supported in a neighborhood of x_0 on $\partial\Omega$, measure Neumann data $\Lambda_\gamma f$ in that neighborhood and then reconstruct conductivity and its derivatives in that neighborhood.

Under the condition that $\partial\Omega$ is C^2 and $\nabla\gamma$ is continuous locally around $x_0 \in \partial\Omega$, Brown[1] reconstructed γ and its first derivatives at x_0 from the localized Dirichlet to Neumann map. Recently, Nakamura and Tanuma[4] reconstructed the higher order derivatives of γ at $x_0 \in \partial\Omega$ *inductively* according to the regularity which γ and $\partial\Omega$ have around x_0.

In this report, we give three kinds of formulas for reconstructing conductivity and its normal derivative from the localized Dirichlet to Neumann map. They are the formulas for pointwise reconstruction, reconstruction in a weak form, and reconstruction in the form of Fourier transform.

Our standpoint is to reconstruct normal derivative of the conductivity *directly* from the localized Dirichlet to Neumann map. More precisely, when recovering the normal derivative of γ at $x_0 \in \partial\Omega$, we need only some regularity assumption on γ around x_0 and need not any information on the values of γ around x_0. This standpoint is different from the reconstruction methods in Brown[1], Nachman[3] and Nakamura *et al* [4], where they reconstructed conductivity and its normal derivative inductively. For example, when recovering the normal derivative of γ at x_0 they needed to know not only the value $\gamma(x_0)$ but also all the values of γ in a neighborhood of x_0 on $\partial\Omega$ in advance. (There is a recent work[2] on inductive reconstruction by using only the value at x_0.)

Our direct reconstruction of the normal derivative of γ can be done by using two special kinds of Dirichlet data compactly supported in a neighborhood of x_0 on $\partial\Omega$ and Λ_γ. Full proofs are given by Nakamura and Tanuma[6]. In this report we give a brief sketch of the idea for these proofs. We believe that our direct reconstruction formulas are useful also for numerical computations.

Finally we note that there are results on reconstruction of elastic tensor for the isotropic and anisotropic elasticity from the localized Dirichlet to Neumann map (Robertson[8], Nakamura *et al* [7]).

In Section 2, we give a formula which reconstructs conductivity and its normal derivative pointwisely at x_0 and give formulas which reconstruct them as the functions defined in a neighborhood of x_0 (in a weak form and in the form of Fourier transform). Since our reconstruction formulas involve limiting process, we give the estimates for their convergences in Section 3.

2 Reconstruction Formulas

To make the essential point of the problem clear, let us assume that $\partial\Omega$ is flat around $x = 0 \in \partial\Omega$ and that $\Omega, \partial\Omega$ are given by

$$\Omega = \{x_n > 0\},\ \partial\Omega = \{x_n = 0\}$$

locally around $x = 0$, where $x = (x', x_n) = (x_1, \cdots, x_{n-1}, x_n)$. Let $t = (t', 0) = (t_1, \cdots, t_{n-1}, 0)$ be any unit tangent to $\partial\Omega$ at $x = 0$.

The starting point is the following theorem.

Theorem 1 *(Brown[1]) Suppose that $\gamma(x)$ is continuous around $x = 0$. Letting $\eta(x') \in C_0^\infty(R^{n-1})$ satisfy*

$$0 \leq \eta \leq 1, \quad \int_{R^{n-1}} \eta^2\, dx' = 1, \quad supp\, \eta \subset \{|x'| < 1\}, \tag{3}$$

we take

$$\phi_N(x') = e^{\sqrt{-1}\, N x' \cdot t'} \eta(\sqrt{N} x') \tag{4}$$

for any positive integer N. Then

$$\lim_{N \to \infty} N^{\frac{n-3}{2}} < \Lambda_\gamma \phi_N, \overline{\phi_N} >= \gamma(0).$$

Note: In Brown[1] he obtained this formula for more general class of γ which includes piecewise continuous γ and γ in $W^{1,1}(\Omega)$.

For the formulas which reconstruct γ and its derivatives inductively we have referred to Nachman[3], Brown[1], Nakamura *et al* [4]. Here we give the formula in Nakamura *et al* [4].

Theorem 2 *Let* $\chi(x) \in C_0^\infty(R^n)$ *satisfy* $0 \leq \chi \leq 1$ *on* $\{|x| \leq \epsilon\}$, $\chi(x) = 1$ *and* $supp\,\chi \subset \{|x| < 2\epsilon\}$ *for small* $\epsilon > 0$. *Define* $\gamma_k(x)$ $(k = 0, 1, 2, \cdots)$ *by*

$$\gamma_k = 1 - \chi(x) + \chi(x)\Big(\gamma(x', 0) + x_n \partial_{x_n}\gamma(x', 0) + \cdots + \frac{{x_n}^k}{k!}\,\partial_{x_n}^k \gamma(x', 0)\Big).$$

Assume that for $k \geq 1$, $\partial_{x'}^{\alpha'}\partial_{x_n}^{\alpha_n}\gamma$ *is continuous around* $x = 0$ *for any multi-index* (α', α_n) *such that* $|\alpha'| + 2\alpha_n \leq 2k$ *and let* $\phi_N(x')$ *be given by (4). Then,*

$$\lim_{N\to\infty} N^{\frac{n-3}{2}+k} < (\Lambda_\gamma - \Lambda_{\gamma_{k-1}})\phi_N, \overline{\phi_N} >= \frac{1}{2^k}\frac{\partial^k}{\partial x_n^k}\,\gamma(0).$$

If x_0 can be any point in a small open subset Γ of $\partial\Omega$, we can recover γ on Γ by using Theorem 1 and hence Λ_{γ_0} can be defined. Then, we can recover $\frac{\partial\gamma}{\partial x_n}$ on Γ by using Theorem 2 with $k = 1$ and hence Λ_{γ_1} can be defined. Repeating this process, we can obtain the higher order normal derivatives of γ on Γ. In this sense the formula in Theorem 2 is an inductive reconstruction formula.

Now we give our direct reconstruction formulas, which are the main results in this report.

Theorem 3 *(Pointwise Reconstruction (Nakamura and Tanuma[5])). Suppose that* $D_{x'}^{\alpha'} D_{x_n}^{\alpha_n}\gamma$ *is continuous around* $x = 0$ *for any multi-index* (α', α_n) *such that* $|\alpha'| + 2\alpha_n \leq 2$. *Letting* $\eta(x') \in C_0^4(R^{n-1})$ *satisfy (3) we take* $\phi_N(x')$ *in (4) and*

$$\psi_N(x') = e^{\sqrt{-1}\,\frac{N}{2}x'\cdot t'}\,\eta(\sqrt{N}x'). \tag{5}$$

Then,

$$\lim_{N\to\infty} N^{\frac{n-1}{2}}\Big[4 < \Lambda_\gamma\psi_N,\ \overline{\psi_N} > -2 < \Lambda_\gamma\phi_N,\ \overline{\phi_N} >\Big]$$
$$= \frac{\partial}{\partial x_n}\gamma(0) + 3\,\gamma(0)\int_{R^{n-1}} \big(|\nabla_{x'}\eta|^2 - (t'\cdot\nabla_{x'}\eta)^2\big)\,dx'. \tag{6}$$

In this formula, the left hand side is observable. On the other hand, the factors $\int_{R^{n-1}}\big(|\nabla\eta|^2 - (t'\cdot\nabla\eta)^2\big)\,dx'$ and $(t'\cdot\omega')$ in the right hand sides, are controllable (except the case $n = 2$), that is, these factors are determined explicitly from the Dirichlet data. Then from (6) we obtain a 2×2 system of equations which can be solved for $\gamma(0)$ and $\frac{\partial\gamma}{\partial x_n}(0)$ simultaneously. When $n = 2$ the factor $\int_{R^{n-1}} \left(|\nabla\eta|^2 - (t'\cdot\nabla\eta)^2\right) dx'$ vanishes. So in this case we are able to reconstruct $\frac{\partial\gamma}{\partial x_n}(0)$ immediately.

Hereafter we assume that $D_{x'}^{\alpha'} D_{x_n}^{\alpha_n}\gamma$ is continuous around $x = 0$ for any multi-index (α', α_n) such that $|\alpha'| + \alpha_n \leq 3, \alpha_n \leq 1$. Also we take $\eta(x')$ to be any function in $C_0^3(R^{n-1})$ compactly supported in a neighborhood of $x' = 0$ and put

$$\phi_N(x') = e^{\sqrt{-1}\,Nx'\cdot t'}\eta(x'), \qquad \psi_N(x') = e^{\sqrt{-1}\,\frac{N}{2}x'\cdot t'}\eta(x'). \tag{7}$$

Theorem 4 *(Reconstruction in a Weak Form).*

(i) $\displaystyle\lim_{N\to\infty} N^{-1} < \Lambda_\gamma\phi_N,\ \overline{\phi_N} >= \int_{R^{n-1}} \gamma(x',0)\ \eta^2(x')\ dx'$

(ii) $\displaystyle\lim_{N\to\infty}\Big[4 < \Lambda_\gamma\psi_N,\ \overline{\psi_N} > -2 < \Lambda_\gamma\phi_N,\ \overline{\phi_N} >\Big] = \int_{R^{n-1}} \frac{\partial\gamma}{\partial x_n}(x',0)\ \eta^2(x')\ dx'.$

Theorem 5 *(Reconstruction of Fourier transform of conductivity). Let $\omega' \in R^{n-1}$. Then*

(i) $\displaystyle\lim_{N\to\infty} N^{-1} < \Lambda_\gamma(\phi_N\, e^{\sqrt{-1}x'\cdot\omega'}),\ \overline{\phi_N} >= \int_{R^{n-1}} \gamma(x',0)\ \eta^2(x')\, e^{\sqrt{-1}x'\cdot\omega'} dx'$

(ii)
$$\begin{aligned}&\lim_{N\to\infty}\Big[4 < \Lambda_\gamma(\psi_N\, e^{\sqrt{-1}x'\cdot\omega'}),\ \overline{\psi_N} > -2 < \Lambda_\gamma(\phi_N\, e^{\sqrt{-1}x'\cdot\omega'}),\ \overline{\phi_N} >\Big]\\ &= \int_{R^{n-1}} \frac{\partial\gamma}{\partial x_n}(x',0)\ \eta^2(x')\, e^{\sqrt{-1}x'\cdot\omega'} dx'\\ &\quad +(t'\cdot\omega')\int_{R^{n-1}} \gamma(x',0)\ \eta^2(x')\, e^{\sqrt{-1}x'\cdot\omega'} dx'.\end{aligned} \tag{8}$$

In (8), for given $\omega' \in R^{n-1}$ we may take $t' \in R^{n-1}$ so that $t'\cdot\omega' = 0$ ($n > 2$). Then we get the Fourier transform of the normal derivative of γ cut off around $x' = 0$.

Outline of Proof : We briefly sketch the idea for the proof of Theorem 4. This idea can be applied also to the proof of Theorem 5. Full details are given by Nakamura and Tanuma[6]. From the definition (2) we can write

$$N^{-1} < \Lambda_\gamma\phi_N,\ \overline{\phi_N} >= N^{-1}\int_\Omega \gamma(x)\ \nabla u_N\cdot\nabla\overline{\Phi_N}\ dx,$$

where $u_N \in H^1(\Omega)$ satisfies

$$\nabla_x \cdot (\gamma \nabla u_N) = 0 \quad \text{in} \quad \Omega, \qquad u_N|_{\partial\Omega} = \phi_N, \tag{9}$$

and Φ_N is an $H^1(\Omega)$ extension of ϕ_N of the form

$$\Phi_N(x) = e^{\sqrt{-1}\,Nx'\cdot t'} e^{-Nx_n} \eta(x').$$

Note that this Φ_N is the first term of an asymptotic solution to (9), because the leading term of $\nabla\Phi_N$ for large N becomes

$$\nabla\Phi_N \sim N \begin{pmatrix} \sqrt{-1}\,t' \\ -1 \end{pmatrix} e^{\sqrt{-1}\,Nx'\cdot t'} e^{-Nx_n} \eta(x') \tag{10}$$

and $\begin{pmatrix} \sqrt{-1}\,t' \\ -1 \end{pmatrix} \cdot \begin{pmatrix} \sqrt{-1}\,t' \\ -1 \end{pmatrix} = 0$. Then it is proved that

$$N^{-1} < \Lambda_\gamma \phi_N,\ \overline{\phi_N} > \sim\ N^{-1} \int_\Omega \gamma(x)\ \nabla\Phi_N \cdot \nabla\overline{\Phi_N}\ dx,$$

and from (10) we get

$$\sim \int_\Omega \gamma(x)\ 2Ne^{-2Nx_n} \eta^2(x')\ dx$$

for large N. Now in this integrant, the sequence

$$\left\{2Ne^{-2Nx_n}\right\}_{N=1}^{\infty} \tag{11}$$

converges to $\delta_{x_n \geq 0}$, " the delta function on the half line $x_n \geq 0$ ", as $N \to +\infty$ in the sense that

$$\int_0^\infty 2Ne^{-2Nx_n} \alpha(x_n)\ dx_n = \int_0^\infty 2e^{-2y_n} \alpha(\frac{y_n}{N})\ dy_n \longrightarrow \alpha(0) \tag{12}$$

as $N \to +\infty$ for any $\alpha(x_n) \in C^0([0,\infty))$. Therefore we get

$$\int_\Omega \gamma(x)\ 2Ne^{-2Nx_n} \eta^2(x')\ dx \longrightarrow \int_{R^{n-1}} \gamma(x',0)\ \eta^2(x')\ dx' \quad (N \longrightarrow +\infty),$$

which proves (i).

We have used a sequence converging to $\delta_{x_n \geq 0}$, which enable us to extract the value of $\gamma(x)$ at $x_n = 0$. So, for (ii), we first propose taking a sequence which is obtained by differentiating each term of the sequence (11):

$$\left\{\frac{d}{dx_n} 2Ne^{-2Nx_n}\right\}_{N=1}^{\infty}.$$

This sequence converges to " the derivative of $\delta_{x_n \geq 0}$ ", and using this we may expect to extract the x_n-derivative of $\gamma(x)$ at $x_n = 0$. [a] However, we get from the integration by parts

$$\int_0^\infty -\Big(\frac{d}{dx_n} 2Ne^{-2Nx_n}\Big)\alpha(x_n)\, dx_n = 2N\alpha(0) + \int_0^\infty 2Ne^{-2Nx_n}\alpha'(x_n)\, dx_n.$$

Although the second term tends to $\alpha'(0)$ as $N \to +\infty$, we must have the first term which goes to infinity. This is because $\delta_{x_n \geq 0}$ is not a usual delta function defined on the whole line $-\infty < x_n < \infty$. This implies that we should first take a sequence converging to $\delta_{x_n \geq 0}$, each term of which vanishes at $x_n = 0$ and then make a new sequence by differentiating each term of that sequence. Like the proof (12) for $2Ne^{-2Nx_n} \to \delta_{x_n \geq 0}$ $(N \to +\infty)$, we easily see that

$$2Ne^{-Nx_n} \longrightarrow 2\delta_{x_n \geq 0} \quad (N \longrightarrow +\infty).$$

Since $2Ne^{-2Nx_n} - 2Ne^{-Nx_n}$ vanishes at $x_n = 0$, the sequence

$$\Big\{\frac{d}{dx_n}(2Ne^{-2Nx_n} - 2Ne^{-Nx_n})\Big\}_{N=1}^{\infty} = \Big\{2N^2e^{-Nx_n} - 4N^2e^{-2Nx_n}\Big\}_{N=1}^{\infty}$$

is a desired one. In fact,

$$\begin{aligned}&\int_0^\infty \frac{d}{dx_n}(2Ne^{-2Nx_n} - 2Ne^{-Nx_n}) \cdot \alpha(x_n)\, dx_n \\ &= \int_0^\infty (2Ne^{-Nx_n} - 2Ne^{-2Nx_n})\alpha'(x_n)\, dx_n \longrightarrow \alpha'(0) \quad (N \longrightarrow +\infty).\end{aligned}$$

Therefore

$$\begin{aligned}&\int_\Omega \gamma(x)\Big(2N^2e^{-Nx_n} - 4N^2e^{-2Nx_n}\Big)\eta^2(x')\, dx \\ &\qquad \longrightarrow \int_{R^{n-1}} \frac{\partial\gamma}{\partial x_n}(x',0)\, \eta^2(x')\, dx' \quad (N \longrightarrow +\infty).\end{aligned} \tag{13}$$

Thus, if we choose

$$\Psi_N(x) = e^{\sqrt{-1}\,\frac{N}{2}x'\cdot t'}e^{-\frac{N}{2}x_n}\eta(x'),$$

the first term of an asymptotic solution to

$$\nabla_x \cdot (\gamma\nabla v_N) = 0 \quad \text{in} \quad \Omega, \qquad v_N|_{\partial\Omega} = \psi_N, \tag{14}$$

[a] Recall that as a linear functional the first derivative of the delta function maps the test function to the minus of its first derivative at the origin.

then we get for large N

$$
\begin{aligned}
&4 < \Lambda_\gamma \psi_N,\ \overline{\psi_N} > -2 < \Lambda_\gamma \phi_N,\ \overline{\phi_N} > \\
&= \int_\Omega \gamma(x)\Big(4\nabla v_N \cdot \nabla\overline{\Psi_N} - 2\nabla u_N \cdot \nabla\overline{\Phi_N}\Big)dx \\
&\sim \int_\Omega \gamma(x)\Big(4\nabla \Psi_N \cdot \nabla\overline{\Psi_N} - 2\nabla \Phi_N \cdot \nabla\overline{\Phi_N}\Big)dx. \qquad (15)
\end{aligned}
$$

Since

$$
\nabla\Psi_N \sim \frac{N}{2}\begin{pmatrix}\sqrt{-1}\,t' \\ -1\end{pmatrix} e^{\sqrt{-1}\,\frac{N}{2}x'\cdot t'} e^{-\frac{N}{2}x_n}\eta(x'),
$$

we see that the dominant term of the last integral (15) for large N is given by the left hand side of (13). [b]

3 Estimate of Convergence

When we assume higher order regularity on γ around $x = 0$, we can get the estimates for the convergences in the formulas of the previous section. As examples, we give the estimates for the formula in Theorem 3 and the formula (ii) of Theorem 4. The proof is given by Nakamura and Tanuma[6].

Theorem 6 *(i) Suppose that $D_{x'}^{\alpha'} D_{x_n}^{\alpha_n}\gamma$ is continuous around $x = 0$ for any multi-index (α', α_n) such that $|\alpha'| + 2\alpha_n \le 4$. Letting $\eta(x') \in C_0^6(R^{n-1})$ satisfy (3), we take $\phi_N(x')$ and $\psi_N(x')$ in (4) and (5) respectively. Then there exists a constant C which depends on the values $\partial_{x'}^{\alpha'}\partial_{x_n}^{\alpha_n}\gamma$ ($|\alpha'| + 2\alpha_n \le 4$) in a neighborhood of $x = 0$ such that*

$$
\begin{aligned}
&\Bigg| N^{\frac{n-1}{2}}\Big[4 < \Lambda_\gamma \psi_N,\ \overline{\psi_N} > -2 < \Lambda_\gamma \phi_N,\ \overline{\phi_N} >\Big] \\
&-\frac{\partial}{\partial x_n}\gamma(0) - 3\,\gamma(0)\int_{R^{n-1}} \left(|\nabla_{x'}\eta|^2 - (t'\cdot\nabla_{x'}\eta)^2\right)dx'\Bigg| \\
&\le C\,N^{-\frac{1}{2}}.
\end{aligned}
$$

[b]More precisely, Φ_N and Ψ_N should be the summations up to the second terms of the asymptotic solutions to (9) and (14) respectively. However, it can be proved that these second terms do not have any effect on the leading term of the integral (15) (see Nakamura and Tanuma [6]).

(ii) Suppose that $D_{x'}^{\alpha'} D_{x_n}^{\alpha_n} \gamma$ is continuous around $x = 0$ for any multi-index (α', α_n) such that $|\alpha'| + \alpha_n \leq 4, \alpha_n \leq 2$. Letting $\eta(x')$ be any function in $C_0^4(R^{n-1})$, we take $\phi_N(x')$ and $\psi_N(x')$ in (7). Then there exists a constant C which depends on the values $\partial_{x'}^{\alpha'} \partial_{x_n}^{\alpha_n} \gamma$ $(|\alpha'| + \alpha_n \leq 4, \alpha_n \leq 2)$ in a neighborhood of $x = 0$ such that

$$\left| \left[4 < \Lambda_\gamma \psi_N, \ \overline{\psi_N} > -2 < \Lambda_\gamma \phi_N, \ \overline{\phi_N} > \right] - \int_{R^{n-1}} \frac{\partial \gamma}{\partial x_n}(x', 0) \ \eta^2(x') \ dx' \right|$$
$$\leq C N^{-1}.$$

Acknowledgments

The first author is partly supported by Grant-in-Aid for Scientific Research (B) (No. 14340038), Society for the Promotion of Science, Japan. The second author is partly supported by Grant-in-Aid for Scientific Research (C) (No. 13640115), Society for the Promotion of Science, Japan.

References

1. R. M. Brown, *Recovering the conductivity at the boundary from the Dirichlet to Neumann map: a pointwise result*, J. Inverse and Ill-posed Prob. **9(6)**, 567–574 (2001).
2. H. Kang and K. Yun, *Boundary determination of conductivities and Riemannian metrics via local Dirichlet-to-Neumann operator*, preprint.
3. A. I. Nachman, *Global uniqueness for a two dimensional inverse boundary value problem*, Ann. of Math. **142**, 71–96(1995).
4. G. Nakamura and K. Tanuma, *Local determination of conductivity at the boundary from Dirichlet to Neumann map*, Inverse Problems **17**, 405–419 (2001).
5. G. Nakamura and K. Tanuma, *Direct determination of the derivatives of conductivity at the boundary from the localized Dirichlet to Neumann map*, Comm. Korean Math. Soc. **16**, 415–425(2001).
6. G. Nakamura and K. Tanuma, *Reconstruction of conductivity and its normal derivative at the boundary from the localized Dirichlet to Neumann map* ,preprint.
7. G. Nakamura and K. Tanuma, *Reconstruction of elastic tensor of anisotropic elasticity at the boundary from the localized Dirichlet to Neumann map* , preprint.

8. R. L. Robertson, *Boundary identifiability of residual stress via the Dirichlet to Neumann map*, Inverse Problems **13**, 1107–1119 (1997).
9. J. Sylvester and G. Uhlmann, *Inverse boundary value problem at the boundary-continuous dependence*, Comm. Pure Appl. Math. **61**, 197–219 (1988).

HOCHSTADT-LIEBERMAN TYPE THEOREM FOR A NON-SYMMETRIC SYSTEM OF FIRST-ORDER ORDINARY DIFFERENTIAL OPERATORS

IGOR TROOSHIN
Institute for Problems of Precision Mechanics and Control
Russian Academy of Sciences, Saratov, Russia
E-mail:lisu@cocoa.plala.or.jp

MASAHIRO YAMAMOTO
Department of Mathematical Sciences, The University of Tokyo
3-8-1 Komaba, Meguro, Tokyo 153 Japan
E-mail:myama@ms.u-tokyo.ac.jp

We consider an eigenvalue problem for a nonsymmetric first order differential operator $Au(x) = \begin{pmatrix} 0 & 1 \\ 1 & 0 \end{pmatrix} \frac{du}{dx}(x) + Q(x)u(x)$, $0 < x < 1$, where Q is a 2×2 matrix whose components are of C^1 class on $[0,1]$. Assuming that $Q(x)$ is known in the half interval of $(0,1)$, we prove the uniqueness in an inverse eigenvalue problem of determining $Q(x)$ from the spectra.

1 Introduction and the Main Result

We consider a non-symmetric first-order differential operator $A_{Q,j,J}$ in $\{L^2(0,1)\}^2$:

$$(A_{Q,j,J}u)(x) = B\frac{du}{dx}(x) + Q(x)u(x), \qquad u \in D(A_{Q,j,J}),$$

where $B = \begin{pmatrix} 0 & 1 \\ 1 & 0 \end{pmatrix}$, $u(x) = \begin{pmatrix} u_1(x) \\ u_2(x) \end{pmatrix}$, $Q(x) = \begin{pmatrix} q_{11}(x) & q_{12}(x) \\ q_{21}(x) & q_{22}(x) \end{pmatrix}$ and

$$D(A_{Q,j,J}) = \left\{ u = \begin{pmatrix} u_1 \\ u_2 \end{pmatrix} \in \{H^1(0,1)\}^2; \right.$$
$$\left. u_2(0) + ju_1(0) = 0, \quad u_2(1) + Ju_1(1) = 0 \right\}.$$

Similarly we can define operators A_{Q,j,J^*}, $A_{P,h,H}$, A_{P,h,H^*}, etc.

We assume that $P = (p_{k\ell})_{1\le k,\ell\le 2}$, $Q = (q_{k\ell})_{1\le k,\ell\le 2} \in \{C^1[0,1]\}^4$, where $p_{k\ell}$, $q_{k\ell}$ are complex-valued, $j, J, \in R \cup \{\infty\} \setminus \{-1, 1\}$. Henceforth we set $i = \sqrt{-1}$.

Throughout this paper, $L^2(0,1)$ and $H^1(0,1)$ are the Lebesgue space and the Sobolev space of complex-valued functions respectively, and $\{H^1(0,1)\}^2$, $\{L^2(0,1)\}^2$ denote the product spaces.

Furthermore we adopt the following: Let $\alpha, \beta \in R$. If $J = \infty$, then the equality $\alpha + \beta J = 0$ means $\beta = 0$. Then we need not distinguish the cases $H = \infty$, $J = \infty$, etc. from the cases $H \neq \infty, J \neq \infty$, etc.

We also set

$$\frac{1+\kappa}{1-\kappa} = \frac{1-\kappa}{1+\kappa} = -1 \qquad \text{if } \kappa = \infty.$$

It was proved in Trooshin and Yamamoto[10] that the spectrum $\sigma(A_{Q,j,J})$ of the operator $A_{Q,j,J}$ consists entirely of eigenvalues, which are simple, except for a finite number of eigenvalues whose algebraic multiplicities are finite and the geometric multiplicities are one.

Let us put $tr\, P = p_{11} + p_{22}$ and

$$\begin{aligned} r_1(x) = &(p_{11}(x) + p_{12}(x) - p_{21}(x) - p_{22}(x) \\ &- q_{11}(x) + q_{12}(x) - q_{21}(x) + q_{22}(x)) \\ &+ (p_{11}(x) - p_{12}(x) + p_{21}(x) - p_{22}(x) \\ &- q_{11}(x) - q_{12}(x) + q_{21}(x) + q_{22}(x)) \\ &\times \exp\left(\int_0^x (p_{11}(s) + p_{22}(s) - q_{11}(s) - q_{22}(s))ds\right) \end{aligned} \tag{1}$$

and

$$\begin{aligned} r_2(x) = &(p_{11}(x) + p_{12}(x) - p_{21}(x) - p_{22}(x) \\ &+ q_{11}(x) - q_{12}(x) + q_{21}(x) - q_{22}(x)) \\ &+ (-p_{11}(x) + p_{12}(x) - p_{21}(x) + p_{22}(x) \\ &- q_{11}(x) - q_{12}(x) + q_{21}(x) + q_{22}(x)) \\ &\times \exp\left(\int_0^x (p_{11}(s) + p_{22}(s) - q_{11}(s) - q_{22}(s))ds\right) \end{aligned} \tag{2}$$

for $0 \leq x \leq 1$. In this paper, we prove the following theorem:

Theorem 1 *Let $P, Q \in \{C^1[0,1]\}^4$, $tr\, P$, $tr\, Q$ be real-valued, $j \in R \setminus \{-1, 1\}$, $H \in R \cup \{\infty\} \setminus \{-1, 1\}$. Then*

$$\left\{ \begin{array}{c} Q(x) = P(x),\ 0 \leq x \leq \frac{1}{2} \\ \sigma(A_{Q,j,J}) = \sigma(A_{P,j,H}), \\ \textit{the algebraic multiplicity of } \lambda \textit{ is equal} \\ \textit{for any } \lambda \in \sigma(A_{Q,j,J}) \end{array} \right. , \tag{3}$$

implies

$$r_1(x) = r_2(x), \quad 0 \le x \le 1 \tag{4}$$

$$\log \frac{(1-J)(1+H)}{(1+J)(1-H)} = \int_0^1 (p_{11}(s) + p_{22}(s) - q_{11}(s) - q_{22}(s))ds. \tag{5}$$

Here and henceforth log means the principal value of the logarithm.

Remark 1 *In the case of $Q,\ P \in \{L^2(0,1)\}^4$, we can prove the same result as Theorem 1 and we omit details for conciseness.*

The following uniqueness follows from our theorem:

Corollary 1 *Let $P, Q \in \{C^1[0,1]\}^4$, $tr\,P, tr\,Q$ be real-valued, $j \in R\backslash\{-1,1\}$, $H \in R \cup \{\infty\} \setminus \{-1,1\}$. Moreover we assume either*

(i) $p_{11} = q_{11}$, $p_{12} = q_{12}$ *on* $[0,1]$,

(ii) $p_{21} = q_{21}$, $p_{22} = q_{22}$ *on* $[0,1]$,

(iii) $p_{11} = q_{11}$, $p_{21} = q_{21}$ *on* $[0,1]$,

(iv) $p_{12} = q_{12}$, $p_{22} = q_{22}$ *on* $[0,1]$.

Then Eq. (3) implies that the rest two components are equal on $[0,1]$.

The corollary is an analogue of the Hochstadt-Lieberman[3] theorem in the Sturm-liouville case. See also Malamud[4], Mochizuki *et al* [5], Ramm[7], Rio *et al* [8] and Sakhnovitch[9].

In a general case, we can characterize the coefficient matrices by two spectra:

Theorem 2 *(Trooshin and Yamamoto[10,11]). Let $P, Q \in \{L^2(0,1)\}^4$, $tr\,P, tr\,Q$ be real-valued, $j, h \in R \setminus \{-1,1\}$, $H, H^*, J, J^* \in R \cup \{\infty\} \setminus \{-1,1\}$ and $J \ne J^*$. Then*

$$\begin{cases} \sigma(A_{Q,j,J}) = \sigma(A_{P,h,H}) \quad \text{and} \quad \sigma(A_{Q,j,J^*}) = \sigma(A_{P,h,H^*}), \\ \text{the algebraic multiplicities of the eigenvalues are equal} \\ \text{for all } \lambda \in \sigma(A_{Q,j,J}) \text{ and for all } \mu \in \sigma(A_{Q,j,J^*}) \end{cases}$$

if and only if following relations hold:

$$\frac{1-h}{1-j}(p_{11}(x) + p_{12}(x) - p_{21}(x) - p_{22}(x)$$

$$-q_{11}(x) + q_{12}(x) - q_{21}(x) + q_{22}(x))$$

$$+\frac{1+h}{1+j}(p_{11}(x) - p_{12}(x) + p_{21}(x) - p_{22}(x)$$

$$-q_{11}(x) - q_{12}(x) + q_{21}(x) + q_{22}(x))$$

$$\times \exp\left(\int_0^x (p_{11}(s) + p_{22}(s) - q_{11}(s) - q_{22}(s))ds\right) = 0, \quad 0 \le x \le 1.$$

$$\frac{1-h}{1-j}(p_{11}(x) + p_{12}(x) - p_{21}(x) - p_{22}(x)$$

$$+q_{11}(x) - q_{12}(x) + q_{21}(x) - q_{22}(x))$$

$$+\frac{1+h}{1+j}(-p_{11}(x) + p_{12}(x) - p_{21}(x) + p_{22}(x)$$

$$-q_{11}(x) - q_{12}(x) + q_{21}(x) + q_{22}(x))$$

$$\times \exp\left(\int_0^x (p_{11}(s) + p_{22}(s) - q_{11}(s) - q_{22}(s))ds\right) = 0. \quad 0 \le x \le 1.$$

$$\log \frac{(1+j)(1-J)(1-h)(1+H)}{(1-j)(1+J)(1+h)(1-H)}$$

$$= \log \frac{(1+j)(1-J^*)(1-h)(1+H^*)}{(1-j)(1+J^*)(1+h)(1-H^*)}$$

$$= \int_0^1 (p_{11}(s) + p_{22}(s) - q_{11}(s) - q_{22}(s))ds.$$

It follows from this theorem that in the case of the same boundary conditions on one end, the two spectra uniquely determine at most two components of P, Q (see (i)–(iv) in Corollary). It is an analogue of the Borg[1] theorem in the Sturm-liouville case.

We will conclude this section with fundamental properties of spectrum which are essential for the proof of our main result.

Theorem 3 *(Trooshin and Yamamoto[10]).*

(i) There exist a natural number N *and* $\Sigma_1, \Sigma_2 \subset \sigma(A_{Q,j,J})$: *the spectrum such that* $\sigma(A_{Q,j,J}) = \Sigma_1 \cup \Sigma_2$, $\Sigma_1 \cap \Sigma_2 = \emptyset$ *and the following properties hold:*

(1) Σ_1 *consists of* $2N - 1$ *eigenvalues including the algebraic multiplicities.*

(2) Σ_2 *consists of eigenvalues with algebraic multiplicity 1. The geometric multiplicity of all eigenvalues is 1.*

Moreover with a suitable numbering $\{\lambda_n\}_{n\in Z}$ *of* $\sigma(A_{Q,j,J})$, *we have an asymptotic behavior*

$$\lambda_n = \frac{1}{2}\log\frac{(1+j)(1-J)}{(1-j)(1+J)} + \frac{1}{2}\int_0^1 (q_{11}(s) + q_{22}(s))ds + n\pi i + O(\frac{1}{|n|}),$$

as $|n| \to \infty$.

(ii) The set of all the root vectors $\{\phi_n\}_{n\in Z}$ *of* $A_{Q,j,J}$ *is a Riesz basis in* $\{L^2(0,1)\}^2$.

Here a Riesz basis means a basis equivalent to an orthonormal basis (e.g., Gohberg and Krein[2]), and we call $u \neq 0$ a *root vector* of an operator A for λ if $(A-\lambda)^m u = 0$ for some $m \in N$. Moreover $\{\varphi_n\}_{n\in Z}$ is a *Riesz basis* in $\{L^2(0,1)\}^2$ if and only if each $u \in \{L^2(0,1)\}^2$ has a unique expansion

$$u = \sum_{n=-\infty}^{\infty} c_n\varphi_n$$

with $c_n \in C$, $n \in Z$ and

$$M^{-1}\sum_{n=-\infty}^{\infty} |c_n|^2 \leq \|u\|^2_{\{L^2(0,1)\}^2} \leq M\sum_{n=-\infty}^{\infty} |c_n|^2,$$

where a constant $M > 0$ is independent of u.

2 Proof of the Main Theorem

Assume (3). Then we have to prove (4) – (5). We set

$$\sigma(A_{P,j,H}) = \sigma(A_{Q,j,J}) = \{\lambda_n\}_{n\in Z}.$$

Then it follows from the asymptotic behavior of the eigenvalues in theorem 3 that

$$\frac{1}{2}\log\frac{(1+j)(1-H)}{(1-j)(1+H)} + \frac{1}{2}\int_0^1 (p_{11}(s) + p_{22}(s))ds + n\pi i + o(1)$$

$$= \frac{1}{2}\log\frac{(1+j)(1-J)}{(1-j)(1+J)} + \frac{1}{2}\int_0^1 (q_{11}(s)+q_{22}(s))ds + n\pi i + o(1)$$

as $|n| \longrightarrow \infty$. Since

$$\int_0^1 (p_{11}(s)+p_{22}(s))ds, \quad \int_0^1 (q_{11}(s)+q_{22}(s))ds \in R$$

and

$$\frac{(1+j)(1-H)}{(1-j)(1+H)}, \quad \frac{(1+j)(1-J)}{(1-j)(1+J)} \in R,$$

which implies (5).

Next we will prove (4).

In Trooshin and Yamamoto[10] the following lemma was proved:

Lemma 1 *For* $\lambda \in C$, *if* $\psi = \psi(\cdot, \lambda) \in \{C^1[0,1]\}^2$ *satisfies*

$$B\frac{d\psi}{dx}(x) + Q(x)\psi(x) = \lambda\psi(x), \quad 0 < x < 1$$

and

$$\psi(0) = \begin{pmatrix} 1 \\ -j \end{pmatrix},$$

then $\phi = \phi(\cdot, \lambda) \in \{C^1[0,1]\}^2$ *defined by*

$$\phi(x,\lambda) = R(x)\psi(x,\lambda) + \int_0^x K(x,y)\psi(y,\lambda)dy, \qquad 0 \le x \le 1,$$

satisfies

$$B\frac{d\phi}{dx}(x) + P(x)\phi(x) = \lambda\phi(x), \quad 0 < x < 1$$

and

$$\phi(0) = \begin{pmatrix} 1 \\ -j \end{pmatrix}.$$

Here we put

$$R(x) = \begin{pmatrix} a(x) & b(x) \\ b(x) & a(x) \end{pmatrix}, \quad 0 \le x \le 1,$$

$$a(x) = \exp(-d_1(x))\cosh d_2(x), \; b(x) = -\exp(-d_1(x))\sinh d_2(x),$$

$$d_1(x) = \frac{1}{2}\int_0^x (p_{12}(y) + p_{21}(y) - q_{12}(y) - q_{21}(y))dy, \quad 0 \le x \le 1,$$

$$d_2(x) = \frac{1}{2}\int_0^x (p_{11}(y) + p_{22}(y) - q_{11}(y) - q_{22}(y))dy, \quad 0 \le x \le 1,$$

and $K(x,y)$ is the solution of the following problem:

$$\begin{cases} B\frac{\partial K}{\partial x}(x,y) + P(x)K(x,y) - K(x,y)Q(y) = -\frac{\partial K}{\partial y}(x,y)B, & (x,y) \in \Omega, \\ K_{12}(x,0) = jK_{11}(x,0), \quad K_{22}(x,0) = jK_{21}(x,0), & 0 \le x \le 1, \\ K_{12}(x,x) - K_{21}(x,x) = \frac{1}{4}\exp(-d_1(x) - d_2(x))r_1(x) & 0 \le x \le 1, \\ K_{11}(x,x) - K_{22}(x,x) = \frac{1}{4}\exp(-d_1(x) - d_2(x))r_2(x), & 0 \le x \le 1, \end{cases}$$

where r_1, r_2 are given by (1)–(2), and

$$\Omega = \{(x,y); 0 < y < x < 1\}.$$

The equality $Q(x) = P(x)$, $0 \le x \le \frac{1}{2}$ implies that $r_1(x) = r_2(x) = 0$, $0 \le x \le \frac{1}{2}$, and, consequently, $K_{12}(x,x) = K_{21}(x,x)$, $K_{11}(x,x) = K_{22}(x,x)$, $0 \le x \le \frac{1}{2}$.

Let us now consider the following problem

$$\begin{cases} B\frac{\partial K}{\partial x}(x,y) + P(x)K(x,y) - K(x,y)Q(y) = -\frac{\partial K}{\partial y}(x,y)B, & (x,y) \in \Omega_1, \\ K_{12}(x,0) = jK_{11}(x,0), \quad K_{22}(x,0) = jK_{21}(x,0), & 0 \le x \le 1, \\ K_{12}(x,x) = K_{21}(x,x), \quad K_{11}(x,x) = K_{22}(x,x), & 0 \le x \le \frac{1}{2}, \end{cases}$$

where Ω_1 is the domain

$$\Omega_1 = \{0 < y < x < 1, x + y < 1\}.$$

We can show by a usual method by characteristics (e.g., Petrovsky[6]), that this problem has a trivial solution $K(x,y) = 0$, $(x,y) \in \overline{\Omega_1}$. In particular,

$$K(x, 1-x) = 0, \quad \frac{1}{2} \le x \le 1.$$

We define the root space $L(A_{Q,j,J}, \lambda_0)$ by

$$L(A_{Q,j,J}, \lambda_0) = \{u \in D(A_{Q,j,J}); (A_{Q,j,J} - \lambda_0)^m u = 0 \text{ for some } m \in N.\}$$

Henceforth we set

$$(Mu)(x) = R(x)u(x) + \int_0^x K(x,y)u(y)dy, \qquad 0 \le x \le 1,$$

where K and R are defined by Lemma 1 .

We have proved in Trooshin and Yamamoto[10] :

Lemma 2 *We assume that* $\lambda_0 \in \sigma(A_{P,j,H}) \cap \sigma(A_{Q,j,J})$ *and that* $\dim L(A_{P,j,H}, \lambda_0) = \dim L(A_{Q,j,J}, \lambda_0) = m$. *Then we can choose a basis* $\{\psi_k\}_{1\le k\le m}$ *in* $L(A_{Q,j,J}, \lambda_0)$ *such that* $\{M\psi_k\}_{1\le k\le m}$ *is a basis in* $L(A_{P,j,H}, \lambda_0)$.

By Theorem 3 in Section 1, we have

$$\sigma(A_{Q,j,J}) = \Sigma_1 \cup \Sigma_2 \equiv \{\mu_n\}_{n\in N} \cup \{\kappa_\ell\}_{1\le\ell\le m},$$

where the algebraic multiplicities of μ_n and κ_ℓ are respectively 1 and $\chi_\ell \ge 2$. Let ψ_n, $n \in Z$ be an eigenvector of $A_{Q,j,J}$ for μ_n satisfying $\psi_n(0) = \begin{pmatrix} 1 \\ -j \end{pmatrix}$, and let $\{\psi_{\ell k}\}_{1\le k\le\chi_\ell}$, $1 \le \ell \le m$ be a basis of $L(A_{Q,j,J}, \kappa_\ell)$ which is chosen for κ_ℓ in Lemma 2. By $\sigma(A_{P,j,H}) = \sigma(A_{Q,j,J})$ and Lemma 2, we see that the function $M\psi_n$ is an eigenvector of $A_{P,j,H}$ for μ_n, and $\{M\psi_{\ell k}\}_{1\le k\le\chi_\ell}$ is a basis in $L(A_{P,j,H}, \kappa_\ell)$ for $1 \le \ell \le m$. For simplicity, we renumber $\{\psi_n\}_{n\in Z} \cup (\cup_{1\le\ell\le m}\{\psi_{\ell k}\}_{1\le k\le\chi_\ell})$ as $\{\psi_n\}_{n\in Z}$, and we set $\phi_n = M\psi_n$, $n \in Z$, $\phi_n = \begin{pmatrix} \phi_n^{(1)} \\ \phi_n^{(2)} \end{pmatrix}$, $\psi_n = \begin{pmatrix} \psi_n^{(1)} \\ \psi_n^{(2)} \end{pmatrix}$.

Therefore

$$\phi_n^{(2)}(1) + H\phi_n^{(1)}(1) = 0, \qquad n \in Z.$$

Here we assume that $H, J, \ne \infty$. (In other cases, the proof is similar.) Since $\phi_n = M\psi_n$, $n \in Z$, we obtain

$$\psi_n^{(1)}(1)\{(H-J)a(1) + (1-JH)b(1)\} + K_n^{(2)} + HK_n^{(1)} = 0, \quad n \in Z. \tag{6}$$

Here we put

$$\begin{cases} K_n^{(1)} = \int_0^1 \{K_{11}(1,y)\psi_n^{(1)}(y) + K_{12}(1,y)\psi_n^{(2)}(y)\}dy, \\ K_n^{(2)} = \int_0^1 \{K_{21}(1,y)\ psi_n^{(1)}(y) + K_{22}(1,y)\psi_n^{(2)}(y)\}dy. \end{cases}$$

We can directly verify

$$(H-J)a(1) + (1-JH)b(1) = 0. \tag{7}$$

Hence (6) and (7) imply

$$K_n^{(2)} + HK_n^{(1)} = 0, \qquad n \in Z.$$

Since the system $\{\psi_n\}_{n\in Z}$ is complete in $\{L^2(0,1)\}^2$ by theorem 3 , these equalities yield

$$K_{21}(1,y) + HK_{11}(1,y) = 0, \quad 0 \le y \le 1$$

and

$$K_{22}(1,y) + HK_{12}(1,y) = 0, \quad 0 \le y \le 1.$$

Let us now consider the following problem

$$\begin{cases} B\frac{\partial K}{\partial x}(x,y) + P(x)K(x,y) - K(x,y)Q(y) = -\frac{\partial K}{\partial y}(x,y)B, & (x,y) \in \Omega_2, \\ K_{12}(x,0) = jK_{11}(x,0), \quad K_{22}(x,0) = jK_{21}(x,0), & 0 \le x \le 1 \\ K(x,1-x) = 0, \quad \frac{1}{2} \le x \le 1, \\ K_{21}(1,y) + HK_{11}(1,y) = 0, \quad 0 \le y \le 1 \\ K_{22}(1,y) + HK_{12}(1,y) = 0, \quad 0 \le y \le 1, \end{cases}$$

where

$$\Omega_2 = \{0 < y < x < 1, x + y > 1\}.$$

We can show by a usual method by characteristics that this problem has a trivial solution $K(x,y) = 0, (x,y) \in \overline{\Omega_2}$. In particular, $K(x,x) = 0$, $\frac{1}{2} \le x \le 1$, which implies $r_1(x) = r_2(x) = 0$, $0 \le x \le 1$ and completes the proof.

Acknowledgments

This paper is partially supported by Sanwa Systems Development Co., Ltd. (Tokyo, Japan).

References

1. G. Borg, *Eine Umkehrung der Sturm-Liouvilleschen Eigenwertfrage*, Acta Math. **78**, 1–96(1946).
2. I.C. Gohberg and M.G. Krein, *Introduction to the Theory of Linear Nonselfadjoint Operators* (American Mathematical Society, Providence, Rhode Island, 1969).
3. H. Hochstadt and B. Lieberman, *An Inverse Sturm-Liouville Problem with Mixed Given Data*, SIAM J. Appl. Math. **34**, 676–680(1978).
4. M.M. Malamud, *Uniqueness Questions in Inverse Problems for Systems of Differential Equations on a Finite Interval*, Trans. Moscow Math. Soc. **60**, 173–224(1999).
5. K. Mochizuki and I. Trooshin, *Inverse Problem for Interior Spectral Data of Sturm-Liouville Operator*, J. Inv. Ill-Posed Problems **9**, 425–433(2001).
6. I.G. Petrovsky, Lectures on Partial Differential Equations (Interscience, New York, 1954).

7. A.G. Ramm, *Property C for ODE and Applications to Inverse Problems*, in *Operator Theory and Applications. (Winnipeg, MB, 1998)*; (Fields Inst. Commun. **25**, 15-75, AMS, Providence, RI, (2000)).
8. R. del Rio, F.Gesztesy, and B.Simon, *Inverse Spectral Analysis with Partial Information on the Potential, III. updating boundary conditions*, Internat. Math. Res. Notices (IMRN) **15**, 751–758(1997).
9. L. Sakhnovitch, *Half-inverse Problems on the Finite Interval*, Inverse Problems **17**, 527–532(2001).
10. I. Trooshin and M. Yamamoto, *Riesz Basis of Root Vectors of a Nonsymmetric System of First-order Ordinary Differential Operators and Application to Inverse Problems*, to appear in Appl. Anal..
11. I. Trooshin and M. Yamamoto, *Spectral Properties of Non-symmetric System of Ordinary Differential Operators*, in (UTMS 2001-11, University of Tokyo, 2001).

SOLVING THE MKDV HIERARCHY WITH INTEGRAL TYPE OF SOURCE BY INVERSE SCATTERING TRANSFORMATION

YUNBO ZENG* AND SHUO YE

Department of Mathematical Science, Tsinghua University, Beijing, 100084, P. R. China

**E-mail: yzeng@math.tsinghua.edu.cn*

The mKdV hierarchy with integral type of source (mKdVHWS), which consist of the reduced AKNS eigenvalue problem with $r = q$ and the mKdV hierarchy with extra term of the integration of square eigenfunction, is investigated. We propose a method to find the explicit evolution equation for eigenfunction of the auxiliary linear problems of the mKdVHWS. Then we determine the evolution equations of scattering data corresponding to the mKdVHWS and solve the equation in the mKdVHWS by inverse scattering transformation.

1 Introduction

The nonlinear Schrödinger equation with integral type of source (NLSEWS) is relevant to some problems of plasma physics and solid state physics [1]. The NLSEWS in some case was studied by so called $\bar{\partial}$-method in Leon and Latifi[1]. Later it was shown in Mel'nikov[2] that the NLSEWS can be integrated by the inverse scattering method for the Dirac operator. The key point of the application of the inverse scattering method to integration of the NLSEWS in Mel'nikov[2] is the use of the determining relations playing the same role as different operator representations of the Lax type of nonlinear evolution equations integrable by various modifications of this method. Just using the determining relations Mel'nikov obtained the evolution equations for all the scattering data of the Dirac operator corresponding to NLSEWS. Similar method was used to investigate the KdV equation with integral type of source (KdVWS) in Mel'nikov[3]. The reason for use of the determining relations in Mel'nikov[2,3] is that the evolution equation of eigenfunction for eigenvalue problem corresponding to the NLSEWS and KdVWS was not found. In fact, establish of these determining relations and derivation of the evolution equations for all scattering data in Mel'nikov[2,3] are quite complicated and required some skill.

In present paper we investigate the new mKdV hierarchy with integral type of sources (mKdVHWS), which consist of the reduced AKNS eigenvalue problem with $r = q$ and the mKdV hierarchy with extra term of the integration of square eigenfunction. We first present a method to construct the zero-curvature representation for mKdVHWS by finding the explicit evolution equation for eigenfunction of the auxiliary linear problem for mKdVHWS. Then we present a way to determine the evolution equation for the scattering data corresponding to the mKdVHWS , which implies that the mKdVHWS can be integrated by the inverse scattering method. Comparing with the method by using determining relation in Mel'nikov[2,3], the method proposed in this paper for determining the evolution equation of the scattering data is quite natural and simple. This general method can be applied to other (1+1)-dimensional soliton equations with integral type of source.

2 The mKdV Hierarchy with Integral Type of Source

Consider the reduced AKNS eigenvalue problem for $r = q$ (see Ablowitz and Segur[4])

$$\begin{pmatrix} \phi_1 \\ \phi_2 \end{pmatrix}_x = U \begin{pmatrix} \phi_1 \\ \phi_2 \end{pmatrix}, \quad U = \begin{pmatrix} -\lambda & q \\ q & \lambda \end{pmatrix}. \tag{1}$$

The adjoint representation of (1) leads (see Newell[5])

$$V_x = [U, V] = UV - VU. \tag{2}$$

Set

$$V = \sum_{i=0}^{\infty} \begin{pmatrix} a_i & b_i \\ c_i & -a_i \end{pmatrix} \lambda^{-i} \tag{3}$$

Eq. (2) yields

$$a_0 = -1, b_0 = c_0 = a_1 = 0, b_1 = c_1 = q$$

$$a_2 = \frac{1}{2}q^2, b_2 = -c_2 = -\frac{1}{2}q_x, \cdots,$$

and in general

$$b_{2m+1} = c_{2m+1} = Lb_{2m-1}, \quad b_{2m} = -c_{2m} = -\frac{1}{2}Db_{2m-1},$$

$$a_{2m+1} = 0, \quad a_{2m} = 2D^{-1}qb_{2m}. \tag{4}$$

where

$$L = \frac{1}{4}D^2 - qD^{-1}qD, \quad D = \frac{\partial}{\partial x}, \quad DD^{-1} = D^{-1}D = 1.$$

Set

$$V^{(2n+1)} = \sum_{i=0}^{2n+1} \begin{pmatrix} a_i & b_i \\ c_i & -a_i \end{pmatrix} \lambda^{2n+1-i} \tag{5}$$

and take

$$\begin{pmatrix} \phi_1 \\ \phi_2 \end{pmatrix}_{t_{2n+1}} = V^{(2n+1)} \begin{pmatrix} \phi_1 \\ \phi_2 \end{pmatrix}. \tag{6}$$

The compatibility conditions of Eqs. (1) and (6) give rise to the mKdV hierarchy[4]

$$q_{t_{2n+1}} = -2b_{2n+2} = Db_{2n+1} = D\frac{\delta H_{2n+1}}{\delta q}, \quad n = 0, 1, \cdots, \tag{7}$$

where

$$H_{2n+1} = \frac{2a_{2n+2}}{2n+1}.$$

Using (1), we have

$$\frac{\delta\lambda}{\delta q} = \phi_1^2 - \phi_2^2, \qquad L(\phi_1^2 - \phi_2^2) = \lambda^2(\phi_1^2 - \phi_2^2). \tag{8}$$

As proposed in Mel'nikov[2,3] and Zeng *et al* [7,8], the mKdV hierarchy with integral type of source is defined by

$$q_{t_{2n+1}} = D[b_{2n+1} + \int_{-\infty}^{\infty} C(t,\zeta)(\phi_1^2(x,t,\zeta) - \phi_2^2(x,t,\zeta))d\zeta] \tag{9a}$$

$$\phi_{1,x} = -i\zeta\phi_1 + q\phi_2, \quad \phi_{2,x} = q\phi_1 + i\zeta\phi_2 \qquad \zeta \in (-\infty, \infty) \tag{9b}$$

we assume $q(x, t_{2n+1})$ tends rather quickly to zero as $x \to \pm\infty$, and

$$\phi_1(x,t,\zeta) \sim aexp(-i\zeta x), \qquad \phi_2(x,t,\zeta) \sim bexp(i\zeta x), \quad x \to -\infty \tag{10}$$

where $C = C(t, \zeta)$, a=$a(t, \zeta)$ and b=$b(t, \zeta)$ are complex functions of $t \geq 0$ and $\zeta \in (-\infty, \infty)$. Moreover we assume that the functions C, a and b are chosen so that the right-hand side of equation (9) determines the function absolutely integrable over x along the whole real axis. One can easily verify that the requirement will certainly be satisfied if the function E and Γ of the form as argued in Mel'nikov[2]

$$E = |C(t,\zeta)|[|a(t,\zeta)| + |b(t,\zeta)|]^2, \Gamma = |\frac{\partial}{\partial\zeta}[C(t,\zeta)a^2(t,\zeta)]| + |\frac{\partial}{\partial\zeta}[C(t,\zeta)b^2(t,\zeta)]|$$

at any $t \geq 0$ satisfy the condition

$$\int_{-\infty}^{\infty} [E(t,\zeta) + \Gamma(t,\zeta) + \Gamma^2(t,\zeta)]d\zeta < \infty$$

3 The Lax Representation

Following the method in Zeng *et al* [6,7,8], we first consider

$$D[b_{2n+1} + \int_{-\infty}^{\infty} C(t,\zeta)(\phi_1^2(x,t,\zeta) - \phi_2^2(x,t,\zeta))d\zeta] = 0 \tag{11a}$$

$$\phi_{1,x} = -i\zeta\phi_1 + q\phi_2, \quad \phi_{2,x} = q\phi_1 + i\zeta\phi_2 \qquad \zeta \in (-\infty,\infty) \tag{11b}$$

According to (4), (8) and (11), we may define

$$\tilde{a}_i = a_i, \tilde{b}_i = b_i, \tilde{c}_i = c_i, \qquad i = 0, 1, \cdots, 2n,$$

$$\tilde{b}_{2n+2m+1} = \tilde{c}_{2n+2m+1} = -\int_{-\infty}^{\infty} (i\zeta)^{2m} C(t,\zeta)[\phi_1^2(x,t,\zeta) - \phi_2^2(x,t,\zeta)]d\zeta,$$

$$\tilde{b}_{2n+2m+2} = -\int_{-\infty}^{\infty} (i\zeta)^{2m+1} C(t,\zeta)[\phi_1^2(x,t,\zeta) + \phi_2^2(x,t,\zeta)]d\zeta,$$

$$\tilde{c}_{2n+2m+2} = \int_{-\infty}^{\infty} (i\zeta)^{2m+1} C(t,\zeta)[\phi_1^2(x,t,\zeta) + \phi_2^2(x,t,\zeta)]d\zeta,$$

$$\tilde{a}_{2n+2m+2} = -2\int_{-\infty}^{\infty} (i\zeta)^{2m+1} C(t,\zeta)\phi_1(x,t,\zeta)\phi_2(x,t,\zeta)d\zeta,$$

$$\tilde{a}_{2n+2m+1} = 0, \qquad m = 0, 1, \cdots,$$

Then

$$N^{(2n+1)} = \begin{pmatrix} A^{(2n+1)} & B^{(2n+1)} \\ C^{(2n+1)} & D^{(2n+1)} \end{pmatrix}$$

where θ is some constant and

$$A^{(2n+1)} = \sum_{k=0}^{2n} a_k \lambda^{2n+1-k} + \theta + \int_{-\infty}^{\infty} \frac{2(i\zeta)(i\eta)C(t,\eta)\phi_1(x,t,\eta)\phi_2(x,t,\eta)}{(i\zeta)^2 - (i\eta)^2} d\eta,$$

$$B^{(2n+1)} = \sum_{k=0}^{2n} b_k \lambda^{2n+1-k} + \int_{-\infty}^{\infty} \frac{i\zeta(i\zeta - i\eta)C(t,\eta)\phi_2^2(x,t,\eta) - i\zeta(i\zeta + i\eta)C(t,\eta)\phi_1^2(x,t,\eta)}{(i\zeta)^2 - (i\eta)^2} d\eta,$$

$$C^{(2n+1)} = \sum_{k=0}^{2n} c_k \lambda^{2n+1-k} + \int_{-\infty}^{\infty} \frac{i\zeta(i\zeta + i\eta)C(t,\eta)\phi_2^2(x,t,\eta) - i\zeta(i\zeta - i\eta)C(t,\eta)\phi_1^2(x,t,\eta)}{(i\zeta)^2 - (i\eta)^2} d\eta,$$

$$D^{(2n+1)} = -\sum_{k=0}^{2n} a_k \lambda^{2n+1-k} + \theta - \int_{-\infty}^{\infty} \frac{2(i\zeta)(i\eta)C(t,\eta)\phi_1(x,t,\eta)\phi_2(x,t,\eta)}{(i\zeta)^2 - (i\eta)^2} d\eta,$$

also satisfies the adjoint representation (2), i.e.

$$N_x^{(2n+1)} = [U, N^{(2n+1)}], \tag{12}$$

which, in fact, gives rise to the Lax representation of (11). Since (11) is the stationary equation of (9), it is easy to find that the zero-curvature representation for the mKdV hierarchy with integral type of source (9) is given by

$$U_{t_{2n+1}} - N_x^{(2n+1)} + [U, N^{(2n+1)}] = 0, \tag{13}$$

with the auxiliary linear problems

$$\begin{pmatrix} \psi_1 \\ \psi_2 \end{pmatrix}_x = \begin{pmatrix} -\lambda & q \\ q & \lambda \end{pmatrix} \begin{pmatrix} \psi_1 \\ \psi_2 \end{pmatrix} = \begin{pmatrix} -i\zeta & q \\ q & i\zeta \end{pmatrix} \begin{pmatrix} \psi_1 \\ \psi_2 \end{pmatrix}, \tag{14a}$$

where $\lambda = i\zeta$ and

$$\psi_{1,t_{2n+1}}(x, t_{2n+1}, \zeta) = (A^{(2n+1)} + \theta)\psi_1 + B^{(2n+1)}\psi_2,$$

$$\psi_{2,t_{2n+1}}(x, t_{2n+1}, \zeta) = C^{(2n+1)}\psi_1 + (-A^{(2n+1)} + \theta)\psi_2 \tag{14b}$$

In this way we find the explicit evolution equations of eigenfunction ψ. Indeed, this kind of evolution equation of eigenfunction was not obtained in Mel'nikov[2,3].

4 Evolution Equation for the Reflection Coefficients

We define the eigenfunctions $f^-(x,\zeta) = (f_1^-(x,\zeta), f_2^-(x,\zeta))^T$, $\bar{f}^-(x,\zeta) = (\bar{f}_1^-(x,\zeta), \bar{f}_2^-(x,\zeta))^T$, $f^+(x,\zeta) = (f_1^+(x,\zeta), f_2^+(x,\zeta))^T$ and $\bar{f}^+(x,\zeta) = (\bar{f}_1^+(x,\zeta), \bar{f}_2^+(x,\zeta))^T$ for the equation (14a), and the following asymptotics are fulfilled at any $\zeta \in (-\infty, \infty)$

$$f^-(x,\zeta) \sim \begin{pmatrix} 1 \\ 0 \end{pmatrix} e^{-i\zeta x}, \quad \bar{f}^-(x,\zeta) \sim \begin{pmatrix} 0 \\ -1 \end{pmatrix} e^{i\zeta x}, as \;\; x \to -\infty \tag{15a}$$

$$f^+(x,\zeta) \sim \begin{pmatrix} 0 \\ 1 \end{pmatrix} e^{i\zeta x}, \quad \bar{f}^+(x,\zeta) \sim \begin{pmatrix} 1 \\ 0 \end{pmatrix} e^{-i\zeta x}, as \;\; x \to +\infty \tag{15b}$$

As is known, the functions $f^-(x,\zeta)$ and $f^+(x,\zeta)$ admit an analytical continuation in the parameter ζ into the upper half-plane $\mathrm{Im}\zeta > 0$, and the functions $\bar{f}^-(x,\zeta)$ and $\bar{f}^+(x,\zeta)$ admit an analytical continuation in the parameter ζ into the lower half-plane $\mathrm{Im}\zeta < 0$. It is easily seen that at any real $\zeta \in (-\infty, \infty)$ the pair of functions $f^-(x,\zeta)$ and $\bar{f}^-(x,\zeta)$ forms a fundamental system of solutions to (14a). Hence, we may define

$$f^+(x,\zeta) = S_{12}(\zeta)\bar{f}^-(x,\zeta) + S_{22}(\zeta)f^-(x,\zeta) \tag{16a}$$

$$\bar{f}^+(x,\zeta) = S_{11}(\zeta)\bar{f}^-(x,\zeta) + S_{21}(\zeta)f^-(x,\zeta) \tag{16b}$$

where the quantities $S_{11} = S_{11}(\zeta)$, $S_{12} = S_{12}(\zeta)$, $S_{21} = S_{21}(\zeta)$ and $S_{22} = S_{22}(\zeta)$ are independent of x. Taking account of (15) and (16) we get that at any $\zeta \in (-\infty, \infty)$ the equality

$$S_{11}(\zeta)S_{22}(\zeta) - S_{12}(\zeta)S_{21}(\zeta) = 1. \tag{17}$$

Under the assumption that $q(x,t)$ vanishes rapidly as $|x| \to \infty$, we have

$$a_0 = -1, \;\; b_0 = c_0 = 0, \;\; lim_{|x|\to\infty} a_j = lim_{|x|\to\infty} b_j = lim_{|x|\to\infty} c_j = 0,$$

$$j = 1, 2, \cdots, 2n.$$

We denote the parameter θ in (14b) corresponding to $f^+(x,\zeta)$ by θ^+ and $\bar{f}^+(x,\zeta)$ by $\bar{\theta}^+$, respectively. Substituting $f^+(x,\zeta)$, $\bar{f}^+(x,\zeta)$ into (14b) respectively, we have

$$\frac{\partial f_1^+(x,\zeta)}{\partial t_{2n+1}} = \{\sum_{k=0}^{2n} a_k (i\zeta)^{2n+1-k} + \theta^+ - \oint_{-\infty}^{\infty} (\frac{\zeta}{\eta-\zeta} + \frac{\zeta}{\eta+\zeta}) H(\eta) d\eta$$
$$-\pi(i\zeta)H(\zeta) + \pi(i\zeta)H(-\zeta)\} f_1^+(x,\zeta)$$
$$+\{\sum_{k=0}^{2n} (b_k (i\zeta)^{2n+1-k} + \oint_{-\infty}^{\infty} \frac{\zeta}{\eta-\zeta} H_1(\eta) d\eta + \oint_{-\infty}^{\infty} \frac{\zeta}{\eta+\zeta} H_2(\eta) d\eta$$
$$+\pi(i\zeta)H_1(\zeta) - \pi(i\zeta)H_2(-\zeta)\} f_2^+(x,t,\zeta), \tag{18a}$$

$$\frac{\partial f_2^+(x,\zeta)}{\partial t_{2n+1}} = \{\sum_{k=0}^{2n} c_k (i\zeta)^{2n+1-k} - \oint_{-\infty}^{\infty} \frac{\zeta}{\eta-\zeta} H_2(\eta) d\eta - \oint_{-\infty}^{\infty} \frac{\zeta}{\eta+\zeta} H_1(\eta) d\eta$$
$$-\pi(i\zeta)H_2(\zeta) + \pi(i\zeta)H_1(-\zeta)\} f_1^+(x,\zeta)$$
$$+\{\sum_{k=0}^{2n} (-a_k (i\zeta)^{2n+1-k} - \oint_{-\infty}^{\infty} (\frac{i\zeta}{i\zeta-i\eta} - \frac{i\zeta}{i\zeta+i\eta}) H(\eta) d\eta + \theta^+$$
$$+\pi(i\zeta)H(\zeta) - \pi(i\zeta)H(-\zeta)\} f_2^+(x,\zeta), \tag{18b}$$

$$\frac{\partial \bar{f}_1^+(x,\zeta)}{\partial t_{2n+1}} = \{\sum_{k=0}^{2n} a_k (i\zeta)^{2n+1-k} + \bar{\theta}^+ - \oint_{-\infty}^{\infty} (\frac{\zeta}{\eta-\zeta} + \frac{\zeta}{\eta+\zeta}) H(\eta) d\eta$$
$$+\pi(i\zeta)H(\zeta) - \pi(i\zeta)H(-\zeta)\} \bar{f}_1^+(x,\zeta)$$
$$+\{\sum_{k=0}^{2n} (b_k (i\zeta)^{2n+1-k} + \oint_{-\infty}^{\infty} \frac{\zeta}{\eta-\zeta} H_1(\eta) d\eta + \oint_{-\infty}^{\infty} \frac{\zeta}{\eta+\zeta} H_2(\eta) d\eta$$
$$-\pi(i\zeta)H_1(\zeta) + \pi(i\zeta)H_2(-\zeta)\} \bar{f}_2^+(x,\zeta), \tag{18c}$$

$$\frac{\partial \bar{f}_2^+(x,\zeta)}{\partial t_{2n+1}} = \{\sum_{k=0}^{2n} c_k (i\zeta)^{2n+1-k} - \oint_{-\infty}^{\infty} \frac{\zeta}{\eta-\zeta} H_2(\eta) d\eta - \oint_{-\infty}^{\infty} \frac{\zeta}{\eta+\zeta} H_1(\eta) d\eta$$
$$+\pi(i\zeta)H_2(\zeta) - \pi(i\zeta)H_1(-\zeta)\} \bar{f}_1^+(x,\zeta)$$
$$+\{\sum_{k=0}^{2n} (-a_k (i\zeta)^{2n+1-k} - \oint_{-\infty}^{\infty} (\frac{i\zeta}{i\zeta-i\eta} - \frac{i\zeta}{i\zeta+i\eta}) H(\eta) d\eta + \bar{\theta}^+$$
$$-\pi(i\zeta)H(\zeta) + \pi(i\zeta)H(-\zeta)\} \bar{f}_2^+(x,\zeta), \tag{18d}$$

where the integral $\oint$ is taken as the principal value, and the quantities $\theta^+, \bar{\theta}^+$ will be determined in the next section,

$$H(\eta) = C(t,\eta)\phi_1(x,t,\eta)\phi_2(x,t,\eta),$$

$$H_1(\eta) = C(t,\eta)\phi_1^2(x,t,\eta), \quad H_2(\eta) = C(t,\eta)\phi_2^2(x,t,\eta). \tag{19}$$

As $x \to -\infty$, we find that the following asymptotics are valid:

$$\oint_{-\infty}^{\infty} \frac{\zeta}{\eta-\zeta} H_1(\eta)d\eta \sim \pi(i\zeta)C(t,\zeta)a^2(\zeta,t)e^{-2i\zeta x},$$

$$\oint_{-\infty}^{\infty} \frac{\zeta}{\eta+\zeta} H_1(\eta)d\eta \sim \pi(i\zeta)C(t,-\zeta)a^2(-\zeta,t)e^{2i\zeta x},$$

$$\oint_{-\infty}^{\infty} \frac{\zeta}{\eta-\zeta} H_2(\eta)d\eta \sim -\pi(i\zeta)C(t,\zeta)b^2(\zeta,t)e^{2i\zeta x}, \tag{20}$$

$$\oint_{-\infty}^{\infty} \frac{\zeta}{\eta+\zeta} H_2(\eta)d\eta \sim -\pi(i\zeta)C(t,-\zeta)b^2(-\zeta,t)e^{-2i\zeta x},$$

Substituting (16) into (18) and using (20), as $x \to -\infty$, we have

$$\begin{aligned}\frac{\partial S_{22}(\zeta)}{\partial t_{2n+1}} &= \{-(i\zeta)^{2n+1} + \theta^+ - \oint_{-\infty}^{\infty}(\frac{\zeta}{\eta-\zeta} + \frac{\zeta}{\eta+\zeta})h(\eta)d\eta \\ &\quad -\pi(i\zeta)h(\zeta) + \pi(i\zeta)h(-\zeta)\}S_{22}(\zeta) \\ &\quad -\{2\pi(i\zeta)h_1(\zeta) - 2\pi(i\zeta)h_2(-\zeta)\}S_{12}(\zeta),\end{aligned}$$

$$\begin{aligned}\frac{\partial S_{12}(\zeta)}{\partial t_{2n+1}} &= \{(i\zeta)^{2n+1} + \theta^+ + \oint_{-\infty}^{\infty}(\frac{\zeta}{\eta-\zeta} + \frac{\zeta}{\eta+\zeta})h(\eta)d\eta \\ &\quad +\pi(i\zeta)h(\zeta) - \pi(i\zeta)h(-\zeta)\}S_{12}(\zeta),\end{aligned} \tag{21}$$

$$\begin{aligned}\frac{\partial S_{21}(\zeta)}{\partial t_{2n+1}} &= \{-(i\zeta)^{2n+1} + \bar{\theta}^+ - \oint_{-\infty}^{\infty}(\frac{\zeta}{\eta-\zeta} + \frac{\zeta}{\eta+\zeta})h(\eta)d\eta \\ &\quad +\pi(i\zeta)h(\zeta) - \pi(i\zeta)h(-\zeta)\}S_{21}(\zeta),\end{aligned}$$

$$\begin{aligned}\frac{\partial S_{11}(\zeta)}{\partial t_{2n+1}} &= \{(i\zeta)^{2n+1} + \bar{\theta}^+ + \oint_{-\infty}^{\infty}(\frac{\zeta}{\eta-\zeta} + \frac{\zeta}{\eta+\zeta})h(\eta)d\eta \\ &\quad -\pi(i\zeta)h(\zeta) + \pi(i\zeta)h(-\zeta)\}S_{11}(\zeta) \\ &\quad -\{2\pi(i\zeta)h_2(\zeta) - 2\pi(i\zeta)h_1(-\zeta)\}S_{21}(\zeta),\end{aligned}$$

where

$$h(\eta) = C(t,\eta)a(\eta,t)b(\eta,t)$$

$$h_1(\eta) = C(t,\eta)a^2(\eta,t), \quad h_2(\eta) = C(t,\eta)b^2(\eta,t)$$

One can easily see that if $C = 0$ or $a = b = 0$ then the resultant system (21) coincides with those equations which appear in the case of the mKdV hierarchy without a source. Also one can verify that system (21) is consistent with equality (17). Using (21), we find that the reflection coefficients

$$R_1(\zeta,t) = \frac{S_{11}(\zeta)}{S_{21}(\zeta)}, \quad R_2(\zeta,t) = \frac{S_{22}(\zeta)}{S_{12}(\zeta)} \tag{22}$$

satisfies the equation

$$\begin{aligned}\frac{\partial R_1(\zeta)}{\partial t_{2n+1}} &= 2\{(i\zeta)^{2n+1} + \oint_{-\infty}^{\infty}(\frac{\zeta}{\eta-\zeta}+\frac{\zeta}{\eta+\zeta})h(\eta)d\eta \\ &\quad -\pi(i\zeta)h(\zeta)+\pi(i\zeta)h(-\zeta)\}R_1(\zeta) \\ &\quad -\{2\pi(i\zeta)h_2(\zeta)-2\pi(i\zeta)h_1(-\zeta)\},\end{aligned} \tag{23a}$$

$$\begin{aligned}\frac{\partial R_2(\zeta)}{\partial t_{2n+1}} &= 2\{-(i\zeta)^{2n+1} - \oint_{-\infty}^{\infty}(\frac{\zeta}{\eta-\zeta}+\frac{\zeta}{\eta+\zeta})h(\eta)d\eta \\ &\quad -\pi(i\zeta)h(\zeta)+\pi(i\zeta)h(-\zeta)\}R_2(\zeta) \\ &\quad -\{2\pi(i\zeta)h_1(\zeta)-2\pi(i\zeta)h_2(-\zeta)\}.\end{aligned} \tag{23b}$$

Then, it follows from (23) that the evolution of the reflection coefficients R_1, R_2 are influenced by the integral type of source which is integration of the square eigenfunctions belonging to the continuous spectrum of the spectral problem (1). For the case $r = q$, there is no discrete eigenvalue for the spectral problem (1) if the potential $q = q(x,t)$ tends rather quickly to zero as $|x| \to \infty$. The evolution equations for the reflection coefficients are presented by (23) which implies that the mKdV hierarchy with integral type of source can be solved by the inverse scattering method.

References

1. J.Leon and A.Latifi, *Solution of an initial-boundary value problem for coupled nonlinear waves*, J. Phys. **A 23**, 1385–1403 (1990).
2. V. K.Mel'nikov, *Integration of the nonlinear Schrödinger equation with a source.* Inverse Probl **8**, 133–147 (1992).

3. V. K. Mel'nikov, *Integration of the Korteweg-de Vries equaion with a source*, Inverse Problem **6**, 233–246 (1990).
4. M. J. Ablowitz and H. Segur,*Solitons and the Inverse Scattering Transform* (Philadephia: SIAM, 1981).
5. A.C.Newell, *Solitons in Mathematics and Physics* (Philadephia: SIAM, 1985).
6. Yunbo Zeng and Yishen Li, *The deduction of the Lax representation for constrained flows from the adjoint representation*, J. Phys. A: Math. Gen. **26**, L273–L278 (1993).
7. Yunbo Zeng , *New factorization of the Kaup-Newell hierarchy*, Physica **D 73**, 171–188 (1994).
8. Yunbo Zeng and Yishen Li , *The Lax representation and Darboux transformation for constrained flows of AKNS hierarchy*, Acta Mathematica Sinica, New Series **12**, 217–224 (1996).

Section III

Numerical Methods

NUMERICAL DIFFERENTIATION ON THE NONUNIFORM GRID AND ITS ERROR ESTIMATE

JIN CHENG
Department of Mathematics,Fudan University
Shanghai 200433, China
e.mail: jcheng@fudan.edu.cn

XIANZHENG JIA
Institute of Mathematics, Fudan University
Shanghai 200433, China

YANBO WANG
Department of Mathematics, Fudan University
Shanghai 200433, China
e.mail: yanbo@public.sta.net.cn

Numerical differentiation on the nonuniform grid arises from many practical problems. It is a typical ill-posed problem. In this paper, we will discuss the numerical differentiation problem by the Tikhonov regularization method. The constructive algorithm and its error estimate are given.

1 Introduction

Derivative is a basic and important conception in calculus which can be used to describe the properties of the functions. It is defined as the limitation of the difference of a function. This means that, if one wants to calculate the derivative at one point, the values of the function at infinite points are needed. While for the practical problems, it became difficult to define the suitable derivatives of a function if we only know the values of the function at finite points. One of the possible methods is to use a difference operator to approximate the derivative. The difficulty of this way is that the step size should be chose suitably. Otherwise the errors in the data may be enlarged. The reason is that differentiation is an ill-posed problem (Tikhonov and Arsenin[8]).

Recently, based on the conditional stability estimate for the ill-posed problem, Cheng and Yamamoto proposed a regularization parameter choice method for the Tikhonov regularization which is a most effective method for the ill-posed problem. In this paper, we will discuss the numerical differentiation problem by the Tikhonov regularization method. The numerical differentiation problem is transformed to a optimization problem which we can construct the solution stably. The parameter choice strategy is based on the results in Cheng and Yamamoto[1].

This paper is organized as

- Section 2: The formulation of the problem and main results
- Section 3: Proofs of the results
- Section 4: Numerical Examples
- Section 5: Conclusions.

2 The formulation of the problem and main results

Define the spaces

$$L^2(0,1) = \{g \,|\, (\int_0^1 g^2(x)dx)^{1/2} < \infty\},$$

$$H^k(0,1) = \{g \,|\, g \in L^2(0,1),\, g^{(k)} \in L^2(0,1)\},$$

The corresponding norms are defined as

$$\|g\|_{L^2(0,1)} = (\int_0^1 |g(x)|^2 dx)^{1/2},$$

$$\|g\|_{H^k(0,1)} = \left(\|g\|^2_{L^2(0,1)} + \|g^{(k)}\|^2_{L^2(0,1)}\right)^{\frac{1}{2}}$$

where $^{(k)}$ is the k-th order derivative with respect to x.

Let $y = y(x) \in H^2(0,1)$ be a function on [0,1], $\triangle = 0 = x_0 < x_1 < \ldots < x_n = 1$ is a grid on the interval [0,1] and δ is a given small constant.

We set

$$h_i = x_{i+1} - x_i, \qquad i = 0,1,2,\cdots,n-1 \qquad h = \max_{0 \le i \le n-1} h_i.$$

The problem we want to discuss is that:
Given the approximation values $\widetilde{y}_j$, $j = 1,2,\cdots,n$ which satisfy

$$|\tilde{y}_i - y(x_i)| \le \delta, \qquad i = 1,2,\cdots,n$$

we want to find a function $f_*(x) \in H^2(0,1)$ such that $f_*^{'}$ is an approximation of the function $y^{'}(x)$. Here $'$ denotes the first order derivative with respect to x.

Moreover, we assume that the data on the end points are exact, i.e.

$$\widetilde{y}_0 = y(0), \quad \widetilde{y}_n = y(1).$$

Remark 1 *This assumption is not essential. We can remove this assumption by some techniques (see Hanke and Scherzer[5]).*

We define the Tikhonov regularization functional as follows:

$$\Phi(f) = \sum_{i=1}^{n-1} \frac{h_{i-1} + h_i}{2} (f(x_i) - \widetilde{y}_i)^2 + \alpha \left\| f'' \right\|^2_{L^2(0,1)} \quad (1)$$

where α is the regularized parameter.

Instead of the original numerical differentiation problem, we will consider the optimization problem and use the derivative f_*' of the minimizer f_* of the functional (1) to approximate y'.

We consider the following two problems:

Problem 1 Find the minimizer f_* of the functional (1), i.e. $f_* \in H^2(0,1)$ satisfies $f_*(0) = y(0)$, $f_*(1) = y(1)$ and, for any $f \in H^2(0,1)$ with $f(0) = y(0)$, $f(1) = y(1)$, it holds

$$\Phi(f_*) \leq \Phi(f). \quad (2)$$

Problem 2 If the minimizer of the functional (1) exists, find a way to select the regularized parameter α so that $f_*'(x)$ is an approximation of $y'(x)$.

Remark 2 *When the division $\triangle$ is a uniform grid, the Tihkonov regularized functional becomes:*

$$\Phi(f) = \frac{1}{n-1} \sum_{j=1}^{n-1} (\widetilde{y}_j - f(x_j))^2 + \alpha \|f''\|^2_{L^2(0,1)}.$$

This has been discussed in (Hanke and Scherzer[5], Wang, Jia and Cheng[9]).

Remark 3 *Concerning on the second problem, there are many possible choices of the regularized parameter α (Engl, Hanke and Neubauer[2]). One of the effective ways is called the discrepancy principle in which the α can be obtained by solving the following nonlinear equation:*

$$\sum_{i=1}^{n-1} \frac{h_{i-1} + h_i}{2} (f_*(x_i) - \widetilde{y}_i)^2 = \delta^2.$$

The reader can find the related results in Engl, Hanke and Neubauer[2], Groetsch[3], Hanke and Scherzer[5].

Recently, based on the conditional stability estimate for the ill-posed problems, Cheng and Yamamoto proposed a simple method to choose parameter

α(Cheng and Yamamoto[1]). This strategy has been shown that it works for a large class of ill-posed problems.

Since the differentiation has some conditional stability, i.e., for $f \in H^2(0,1)$ and $f(0)=0$, $f(1)=0$, it holds that

$$\left\| f' \right\|^2_{L^2(0,1)} \le \|f\|_{L^2(0,1)} \left\| f'' \right\|_{L^2(0,1)}.$$

Therefore, we will try apply the method in Cheng and Yamamoto[1] to our problem.

Our main results are as follows:

Theorem 1 *There is a unique solution f_* of Problem 1 and f_* can be constructed by the following way:*

1. *f_* is a natural cubic spline function, i.e. f_* is a polynomial of the third order in every subinterval (x_i, x_{i+1}), $i = 0, 1, \cdots, n-1$, and $f_*''(0) = f_*''(1) = 0$.*
2. *f_* is a twice continuous differentiable function, i.e.*

$$f_*^{(p)}(x_i-) = f_*^{(p)}(x_i+), \quad i = 1, 2, \ldots, n-1, \quad p = 0, 1, 2.$$

   where $f_*(x_i+) = \lim_{x \to x_i+} f(x)$ and $f_*(x_i-) = \lim_{x \to x_i-} f(x)$.
3. *The third order derivative $f_*^{(3)}$ satisfies*

$$f_*'''(x_i-) - f_*'''(x_i-) = \frac{1}{\delta^2} \frac{h_{i-1} + h_i}{2} (\widetilde{y}_i - f_*(x_i)), \qquad i = 1, 2, \ldots, n-1$$

   at the nodes x_i, $i = 1, 2, \cdots, n-1$.

Theorem 2 *Suppose that f_* is the minimizer of the functional (1).Let $\alpha = \delta^2$. If $y \in H^2(0,1)$, then we have*

$$\|f_*' - y'\|_{L^2(0,1)} \le (2h + 4\sqrt{\delta} + \frac{h}{\pi}) \parallel y'' \parallel_{L^2(0,1)} + h + 2\sqrt{\delta}. \tag{3}$$

3 Proofs of Main Results

3.1 Preliminaries

First, we state some results about the spline functions:

Definition 1 *We call a $h(x)$ the natural spline on interval $[0,1]$ if the function $h(x) \in C^2(0,1)$ satisfies the following conditions:*

1. $h''(0) = h''(1) = 0$;

2. $h(x)$ *is a cubic polynomial on interval* $[x_i, x_{i+1}]$, $i = 0, 1, \cdots, n-1$.

Lemma 1 *Suppose that* s *is a natural cubic spline on* $[0,1]$ *and interpolates at the points* $(x_i, y(x_i))$, $i = 0, 1, \ldots, n$, *then* s'' *is the best approximation of* y'' *in* $L^2(0,1)$ *space, i.e.*

$$\left\|s'' - y''\right\|^2_{L^2(0,1)} + \left\|s''\right\|^2_{L^2(0,1)} = \left\|y''\right\|^2_{L^2(0,1)}.$$

The proof can be found in Hammerlin and Hoffmann[4].

Lemma 2 *Suppose that* s *is a natural cubic spline on* $[0,1]$ *and interpolates at the points* $(x_i, y(x_i))$, $i = 0, 1, \ldots, n$, *then we have*

$$\left\|s' - y'\right\|_{L^2(0,1)} \le \frac{h}{\pi}\left\|y''\right\|_{L^2(0,1)}.$$

Proof: From the result in Strang and Fix[7], we can see that

$$\left\|s' - y'\right\|_{L^2(x_i,x_{i+1})} \le \frac{x_{i+1} - x_i}{\pi}\left\|y''\right\|_{L^2(x_i,x_{i+1})}.$$

Since

$$h = \max_{0 \le i \le n-1} h_i,$$

we can obtain

$$\left\|s' - y'\right\|_{L^2(0,1)} \le \frac{h}{\pi}\left\|y''\right\|_{L^2(0,1)}.$$

The proof is complete.

Lemma 3 *Let* g *be a differentiable function on interval* $[0,1]$ *and* χ *be a piecewise constant spline which is defined as:*

$$\chi(x) = \chi_i = \frac{1}{x_i - x_{i-1}}\int_{x_{i-1}}^{x_i} g(x)dx, \qquad i = 1, 2, \cdots, n$$

for $x \in [x_{i-1}, x_i)$, *then we have*

$$\|g - \chi\|_{L^2(0,1)} \le h\left\|g'\right\|_{L^2(0,1)}.$$

Proof: See Schumaker[6].

3.2 Proof of Theorem 1

We will prove Theorem 1 in two steps:

Step 1: First we will prove that the natural cubic spline f_* which is constructed by the algorithm in Theorem 1 is the unique minimizer of the functional (1). Our method is similar to the method which is used in Hanke and Scherzer[5].

For any function $g \in H^2(0,1)$ with $g(0) = g(1) = 0$, it is easy to verify that:

$$\int_0^1 g''(x)f_*''(x)dx = g'(1)f_*(1) - g'(0)f_*''(0) - \int_0^1 g'(x)f_*'''(x)dx$$
$$= -\sum_{i=1}^{n} g'(x)f_*'''(x)dx.$$

Since $f_*(x)$ is a natural cubic spline, then f_*''' is piecewise constant function. By the boundary conditions of g, we have

$$\int_0^1 g''(x)f_*''(x)dx = -\sum_{i=1}^{n} f_*'''(x)\,|_{(x_{i-1},x_i)} g(x)\big|_{x=x_{i-1}}^{x_i}$$
$$= \sum_{i=1}^{n-1} g(x_i)(f_*'''(x_i+) - f_*'''(x_i-))$$
$$= \sum_{i=1}^{n-1} g(x_i)\frac{1}{\delta^2}\frac{h_{i-1}+h_i}{2}(\widetilde{y}_i - f_*(x_i)). \qquad (4)$$

Let $f \in H^2(0,1)$ with $f(0) = \widetilde{y}_0$, $f(1) = \widetilde{y}_n$. It can calculated directly that

$$\Phi(f) - \Phi(f_*) = \sum_{i=1}^{n-1} \frac{h_{i-1}+h_i}{2}(f(x_i) - f_*(x_i))(f(x_i) + f_*(x_i) - 2\widetilde{y}_i)$$
$$+\delta^2 \left\| f'' - f_*'' \right\|_{L^2(0,1)}^2 + 2\delta^2 \int_0^1 \left(f''(x) - f_*''(x)\right) f_*''(x)dx.$$

We denote $g(x) = f(x) - f_*(x)$. By (1), we have

$$\Phi(f) - \Phi(f_*) = \sum_{i=1}^{n-1} \frac{h_{i-1}+h_i}{2}(f(x_i) - f_*(x_i))(f(x_i) + f_*(x_i) - 2\widetilde{y}_i)$$
$$+\delta^2 \parallel f'' - f_*'' \parallel_{L^2(0,1)}^2$$

$$+2\sum_{i=1}^{n-1}(f(x_i)-f_*(x_i))\frac{h_{i-1}+h_i}{2}(\widetilde{y}_i-f_*(x_i))$$
$$=\sum_{i=1}^{n-1}\frac{h_{i-1}+h_i}{2}(f(x_i)-f_*(x_i))^2+\delta^2\left\|f^{''}-f_*^{''}\right\|^2_{L^2(0,1)}$$
$$\geq 0. \tag{5}$$

This means that $\Phi(f_*)\leq\Phi(f)$, i.e. f_* is the minimizer of the functional $\Phi(f)$.

Step 2: We will prove that f_* is the unique minimizer of the functional $\Phi(f)$. Assume that there are another minimizer f_1 of the functional $\Phi(f)$. Then, from (5), we have that $\left\|f_1^{''}-f_*^{''}\right\|_{L^2(0,1)}=0$. Therefore, $f_1(x)-f_*(x)$ must be a linear function. Since $f_1(0)=f_*(0)$ and $f_1(1)=f_*(1)$, we obtain that $f=f_*$.

The proof is completed.

3.3 Proof of Theorem 2

First we prove a Lemma which will be used in our proof.

Lemma 4 *Suppose that f_* is the unique minimizer of the functional $\Phi(f)$. Then it holds that*

$$\left\|f_*^{''}\right\|^2_{L^2(0,1)}\leq 1+\left\|y^{''}\right\|^2_{L^2(0,1)}. \tag{6}$$

Proof: Since f_* is the minimum solution of functional $\Phi(f)$, we have

$$\delta^2\left\|f_*^{''}\right\|^2_{L^2(0,1)}\leq\Phi(f_*)\leq\Phi(y)$$
$$=\sum_{i=1}^{n-1}\frac{h_{i-1}+h_i}{2}(y(x_i)-\widetilde{y}_i)^2+\delta^2\left\|y^{''}\right\|^2_{L^2(0,1)}$$
$$\leq\delta^2+\delta^2\left\|y^{''}\right\|^2_{L^2(0,1)}$$
$$\leq\delta^2(1+\left\|y^{''}\right\|^2_{L^2(0,1)}).$$

Hence

$$\left\|f_*^{''}\right\|^2_{L^2(0,1)}\leq 1+\left\|y^{''}\right\|^2_{L^2(0,1)}.$$

The proof is completed.

Now we will prove Theorem 2.

Let s be a natural cubic spline which interpolates the exact data $\{y(x_i)\}$, i.e. $s(x_i) = y(x_i)$, $i = 0, 1, \cdots, n$.

We denote

$$e(x) = f_*(x) - s(x)$$

where $f_*(x)$ is the minimizer of the functional $\Phi(f)$.

We consider a piecewise constant spline $\chi(x) \in L^2(0,1)$ which is defined as

$$\chi(x) = \chi_i = \frac{1}{h_i}\int_{x_i}^{x_{i+1}} e'(x)dx, \qquad x \in [x_i, x_{i+1}), \quad i = 0, 1, \cdots, n-1.$$

Then, we have that

$$\begin{aligned}
\|e'\|^2_{L^2(0,1)} &= \int_0^1 (e' - \chi)e'dx + \int_0^1 e'\chi dx \\
&= \int_0^1 (e' - \chi)e'dx + \sum_{i=0}^{n-1} \chi_i \int_{x_i}^{x_{i+1}} e'dx \\
&= \int_0^1 (e' - \chi)e'dx + \sum_{i=0}^{n-1} \chi_i(e(x_{i+1}) - e(x_i)) \\
&= \int_0^1 (e' - \chi)e'dx + \sum_{i=1}^{n-1} e(x_i)(\chi_{i-1} - \chi_i) \\
&\quad + e(1)\chi_{n-1} - e(0)\chi_0
\end{aligned}$$

By $e(0) = e(1) = 0$, we obtain

$$\begin{aligned}
\|e'\|^2_{L^2(0,1)} &= \int_0^1 (e' - \chi)e'dx + \sum_{i=1}^{n-1} e(x_i)(\chi_{i-1} - \chi_i) \\
&=: I_1 + I_2
\end{aligned}$$

Next we will estimate I_1 and I_2 respectively.

By the Cauchy-Schwartz inequality and Lemma 3, it is easy to verify that

$$I_1 \le \|e'\|_{L^2(0,1)} \cdot \left\|e' - \chi'\right\|_{L^2(0,1)} \le h\,\|e'\|_{L^2(0,1)}\,\|e''\|_{L^2(0,1)}$$

By Lemma 4, we obtain

$$\|f''_*\|_{L^2(0,1)} \le \sqrt{1 + \| y'' \|^2_{L^2(0,1)}} \le 1 + \|y''\|_{L^2(0,1)}$$

and

$$\begin{aligned}\|e''\|_{L^2(0,1)} &= \left\|f_*''\right\|_{L^2(0,1)} + \|s''\|_{L^2(0,1)} \\ &\leq 1 + \|y''\|_{L^2(0,1)} + \|s''\|_{L^2(0,1)} \\ &\leq 1 + 2\|y''\|_{L^2(0,1)} .\end{aligned}$$

Hence

$$I_1 \leq h\left(\|e'\|_{L^2(0,1)} + 2\|e'\|_{L^2(0,1)} \cdot \|y''\|_{L^2(0,1)}\right) \tag{7}$$

For I_2, we have

$$\begin{aligned}I_2^2 &= \left\{\sum_{i=1}^{n-1}\left(\frac{h_{i-1}+h_i}{2}\right)^{\frac{1}{2}} e(x_i)\left(\frac{2}{h_{i-1}+h_i}\right)^{\frac{1}{2}}(\chi_{i-1}-\chi_i)\right\}^2 \\ &\leq \sum_{i=1}^{n-1}\frac{h_{i-1}+h_i}{2}e^2(x_i)\sum_{i=1}^{n-1}\frac{2}{h_{i-1}+h_i}(\chi_{i-1}-\chi_i)^2.\end{aligned}$$

By the definition of $\chi(x)$, we obtain estimate:

$$\begin{aligned}|\chi_{i-1}-\chi_i| &= \left|\frac{1}{h_{i-1}}\int_{x_{i-1}}^{x_i} e'dx - \frac{1}{h_i}\int_{x_i}^{x_{i+1}} e'dx\right| \\ &= \left|\frac{x_i - x_{i-1}}{h_{i-1}}\int_0^1 e'(h_{i-1}\tau + x_{i-1})d\tau - \frac{x_{i+1}-x_i}{h_i}\int_0^1 e'(h_i\tau + x_i)d\tau\right| \\ &= \left|\int_0^1 (e'(h_{i-1}\tau + x_{i-1}) - e'(h_i\tau + x_i))d\tau\right| \\ &= \left|\int_0^1\int_{h_i\tau + x_i}^{h_{i-1}\tau + x_{i-1}} e''(x)dxd\tau\right| \\ &\leq \left|\int_0^1\int_{x_{i-1}}^{x_{i+1}} \mid e''(x)\mid dxd\tau\right| \\ &\leq (x_{i+1}-x_{i-1}) \parallel e'' \parallel_{L^2(0,1)} \\ &= (h_{i-1}+h_i) \parallel e'' \parallel_{L^2(0,1)}\end{aligned}$$

Since f_* is the minimizer of the functional $\Phi(f)$, we have

$$\sum_{i=1}^{n-1}\frac{h_{i-1}+h_i}{2}e^2(x_i) = \sum_{i=1}^{n-1}\frac{h_{i-1}+h_i}{2}\left(f_*(x_i) - y(x_i)\right)^2$$

$$\leq 2\sum_{i=1}^{n-1} \frac{h_{i-1}+h_i}{2} \left\{(f_*(x_i)-\widetilde{y}_i)^2 + (\widetilde{y}_i - y(x_i))^2\right\}$$
$$\leq 2\sum_{i=1}^{n-1} \frac{h_{i-1}+h_i}{2} (\widetilde{y}_i - y(x_i))^2 + 2\Phi(y)$$
$$\leq 2\delta^2 + 2\delta^2(1+\|y''\|^2_{L^2(0,1)})$$
$$\leq 2\delta^2(2+\|y''\|^2_{L^2(0,1)}).$$

Hence

$$I_2^2 \leq \sum_{i=1}^{n-1} \frac{h_{i-1}+h_i}{2} e^2(x_i) \sum_{i=1}^{n-1} 2(h_{i-1}+h_i)\,\|e''\|^2_{L^2(0,1)}$$
$$\leq 2\delta^2(2+\|y''\|^2_{L^2(0,1)}) \sum_{i=1}^{n-1} 2(h_{i-1}+h_i)\,\|e''\|^2_{L^2(0,1)}$$
$$\leq 8\delta^2(2+\|y''\|^2_{L^2(0,1)})\,\|e''\|^2_{L^2(0,1)}\,.$$

Therefore, we have

$$I_2 \leq \sqrt{8}\sqrt{2+\|y''\|^2_{L^2(0,1)}}\,\|e''\|_{L^2(0,1)}\,\delta$$
$$\leq \sqrt{8}(\sqrt{2}+\|y''\|_{L^2(0,1)})(1+2\,\|y''\|_{L^2(0,1)})\delta \tag{8}$$

Thus, by the estimations for I_1 and I_2, we obtain

$$\|e'\|^2_{L^2(0,1)} \leq h\,\|e'\|_{L^2(0,1)}\,(1+2\,\|y''\|_{L^2(0,1)})$$
$$+\sqrt{8}\delta(\sqrt{2}+\|y''\|_{L^2(0,1)})(1+2\,\|y''\|_{L^2(0,1)}).$$

This yields

$$\|e'\|_{L^2(0,1)} \leq 2\sqrt{\delta}(1+2\,\|y''\|_{L^2(0,1)}) + h(1+2\,\|y''\|_{L^2(0,1)})$$
$$\leq (2h+4\sqrt{\delta})\,\|y''\|_{L^2(0,1)} + h + 2\sqrt{\delta}.$$

By Lemma 2, we can obtain

$$\|f_*' - y'\|_{L^2(0,1)} \leq \|e'\|_{L^2(0,1)} + \|s'-y'\|_{L^2(0,1)}$$
$$\leq (2h+4\sqrt{\delta}+\frac{h}{\pi})\,\|y''\|_{L^2(0,1)} + h + 2\sqrt{\delta}.$$

The proof is completed.

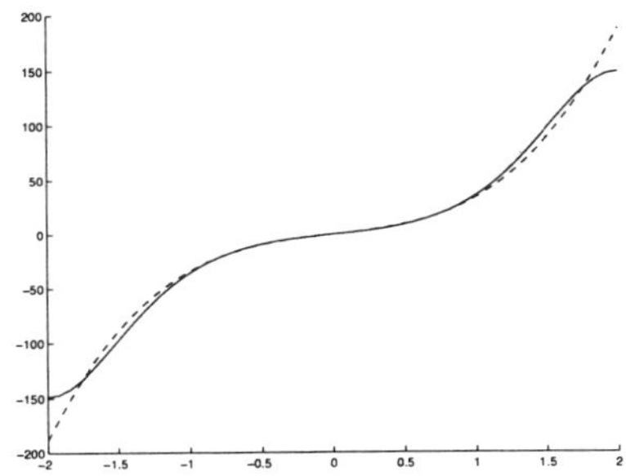

Figure 1. $y' = 20x^3 + 14x$

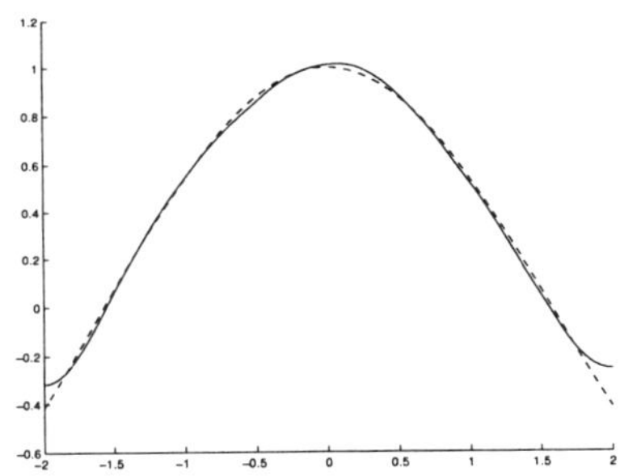

Figure 2. $y' = cos(x)$

4 Numerical Examples

All the examples in our paper are run on SGI 540 NT Workstation(256M memory,double CPU) with MATLAB.

Here,we choose two examples about the numerical differentiation of smooth functions. The regularized parameter is chosen as $\alpha = \delta^2$.

The first function we chose is $y(x) = 5x^4 + 7x^2 + 1$. We take $n = 50, \delta = 0.05, h_{max} = 0.0926$. The numerical result is shown in Figure 1. The broken line is $y'(x)$ and the real line is $f'_*(x)$. We can see that $y'(x)$ and $f'_*(x)$ are of nearly superposition.

The second function we chose is $y(x) = sin(x)$. We take $n = 100, \delta = 0.05, h_{max} = 0.0683$. The numerical result is shown in Figure 2. The broken line is $y'(x)$ and the real line is $f'_*(x)$.

5 Conclusions

A numerical differentiation problem arises from many branches of sciences and engineering. In this paper, based on Tikhonov regularization, we propose a simple way for the numerical differentiation problem. The numerical results show that our method is simple and effective. We have some other applications of our method. We will report our results elsewhere.

Acknowledgments

This research is partly supported by NSF of China (No. 19971016) and Nonlinear Science Laboratory in Fudan University, China.

References

1. J. Cheng and M. Yamamoto, *One new strategy for a priori choice of regularizing parameters in Tikhonov's regularization,* Inverse Problems **16** , L31–L38(2000).
2. H. W. Engl, M. Hanke and A. Neubauer, *Regularization of Inverse Problems* (Kluwer Academic Publishers Group, Dordrecht, 1996).
3. C. W. Groetsch, *The Theory of Tikhonov Regularization for Fredholm Equations of the First Kind. Research Notes in Mathematics,* 105. Pitman (Advanced Publishing Program, Boston, Mass.-London, 1984).
4. G.Hammerlin and K.-H.Hoffmann, *Numerical Mathematics* (Springer-Verlag,New York,1991).
5. M. Hanke and O. Scherzer, *Inverse problems light: numerical differentiation* Amer. Math. Monthly **108(6)**, 512–521 (2001).
6. L.L.Schumaker, *Spline Functions: Basic Theory* (Wiley,New York,1981).
7. G.Strang and G.J.Fix, *An Analysis of the Finite Element Method* (Prentice-Hall,Englewood Cliffs,N.J.,1973).
8. A.N. Tikhonov and V.Y. Arsenin, *Solutions of Ill-posed Problems* (Winston and Sons, Washington, 1977).
9. Y. B. Wang, X. Z. Jia and J. Cheng, *A Stable Numerical Differentiation Method and its application,* Preprint (2002).

WAVELET METHODS FOR THE SIDEWAYS HEAT EQUATION

CHU-LI FU, CHUN-YU QIU AND YOU-BIN ZHU

Department of Mathematics, Lanzhou University, Lanzhou, 730000, P. R. China

E-mail: fuchuli@lzu.edu.cn

We consider the inverse heat conduction problem for the one-dimensional case

$$\begin{array}{ll} u_t = u_{xx}, & x \geq 0,\ t \geq 0, \\ u(x,0) = 0, & x \geq 0, \\ u(1,t) = g(t), & t \geq 0. \end{array}$$

It is well-known that the problem is ill-posed in the sense that the solution (if it exists) does not depend continuously on the data. The Meyer wavelets are applied by T. Regińska earliest at 1995 to formulate a regularized solution, but unfortunately it convergent to exact one not in all interval $(0,1]$ when data error tends to zero. The present paper gives out another wavelet regularization method for the problem. Our results not only make up the defect of Regińska's result but also improve the convergence near $x = 0$ which is a difficult problem left over by some papers.

1 Introduction

In many industrial applications one wishes to determine the temperature on the surface of a body, where the surface itself is inaccessible for measurements[1]. In such cases one is restricted to internal measurements, and from these one wants to compute the surface temperature.

In one-dimensional case, assuming that the body is large, this situation can be modelled as following problem for the heat equation in quarter plane: Determine the temperature $u(x,t)$ for $0 \leq x \leq 1$ from temperature measurements $g(\cdot) := u(1,\cdot)$, such that it satisfies

$$\begin{cases} u_{xx} = u_t, & x \geq 0, t \geq 0, \\ u(x,0) = 0, & x \geq 0, \\ u(1,t) = g(t), & t \geq 0,\ u|_{x\to\infty} \text{ bounded}. \end{cases} \tag{1}$$

As we consider the problem in $L^2(R)$ with respect to variable t, we extend $u(x,\cdot)$, $g(\cdot)$, $f(\cdot) = u(0,\cdot)$ and other functions appearing in the paper be zero for $t < 0$. The notations $\|\cdot\|$, $(\cdot,\cdot)$ denote the L^2-norm and scalar product in $L^2(R)$ respectively and $\hat{h}(\xi) = \frac{1}{\sqrt{2\pi}} \int_{-\infty}^{\infty} e^{-i\xi t} h(t)dt$ be the Fourier transform of $h(t)$. As a solution of the problem (1) we understand a function $u(x,t)$ satisfying (1) in the classical sense; for every fixed $x \in [0, \infty)$, the function $u(x,\cdot)$ belongs to $L^2(R)$. Furthermore, for the uniqueness of the solution, we require that $||u(x,\cdot)||$ be bounded.

Of course, since g is assumed to be measured data, there will be measurement errors, and we would actually have as a data some function $g_m \in L^2(R)$, for which

$$||g_m - g|| = ||g_m - u(1,\cdot)|| \leq \varepsilon, \tag{2}$$

where g and g_m represent exact and measured data at $x = 1$ respectively. We also impose a priori bound on the solution of problem (1) at line $x = 0$:

$$||f|| = ||u(0,\cdot)|| \leq M. \tag{3}$$

This problem has been discussed by many authors, (see e.g. Carasso[2] and the references therein). But wavelet technique or wavelet regularization method for approximating solution of (1) on interval $(0,1)$ has been used earliest by T. Regińska [3]. Unfortunately, there exist defects in the proof of main conclusion in Regińska[3] (see Fu *et al* [4]). Although a revised result is given in Fu *et al* [4], the convergence $||u(x,\cdot) - u_{\varepsilon j}(x,\cdot)|| \to 0$ when $\varepsilon \to 0$ only holds in the interval $(e^*, 1)$ but not in $(0,1)$, where $u_{\varepsilon j}(x,t)$ is the regularized solution and e^* is a small constant, $0.037513 < e^* < 0.037514$. The aim of this paper is to present a new approach for solving this problem, which will make up the weakness in Regińska[3] and improve the convergence of regularization solution near $x = 0$, which is a difficult problem left over by some papers (Carasso[2] and Eldén[5]).

Now we are ready to examine the sideways heat equation in frequency space. Let $u(x,t)$ be the solution of problem (1). Then the Fourier transform $\hat{u}(x,\xi)$ of $u(x,t)$ about variable t satisfies

$$\begin{cases} i\xi\hat{u}(x,\xi) = \hat{u}_{xx}(x,\xi), \ \xi \in R, \\ \hat{u}(1,\xi) = \hat{g}(\xi), \\ \hat{u}|_{x\to\infty} \text{ bounded.} \end{cases} \tag{4}$$

It is easy to know [2,3,4,5] the solution of (4) is

$$\hat{u}(x,\xi) = e^{\sqrt{i\xi}(1-x)}\hat{g}(\xi), \tag{5}$$

where $\sqrt{i\xi}$ is the principal square root of $i\xi$, and

$$\hat{f}(\xi) = e^{\sqrt{i\xi}}\hat{g}(\xi), \tag{6}$$

$$|e^{\sqrt{i\xi}(1-x)}|^2 = e^{(1-x)\sqrt{2|\xi|}}, \quad x \geq 0. \tag{7}$$

The solution of problem (1) can be expressed

$$u(x,t) = \frac{1}{\sqrt{2\pi}} \int_{-\infty}^{\infty} e^{\sqrt{i\xi}(1-x)} \hat{g}(\xi) e^{i\xi t} d\xi. \tag{8}$$

Since the principal value of $\sqrt{i\xi}$ has a positive real part and the Fourier transform $\hat{u}(x,\cdot)$ of solution $u(x,\cdot)$ is in $L^2(R)$, so from (5) we know that $\hat{g}(\xi)$, which is the Fourier transform of exact data function, must decay rapidly as $\xi \to \infty$, small errors in high frequency components can blow up and completely destroy the solution for $0 \le x < 1$. As measured data $g_m(t)$, its Fourier transform $\hat{g}_m(\xi)$ is merely in $L^2(R)$. The wavelets regularization method described in the next section will effectively attenuate or eliminate the high frequencies in $\hat{g}_m$.

2 Wavelets Regularization and Error Estimate

Let φ be the Meyer scaling function defined by its Fourier transform [6]

$$\hat{\varphi}(\xi) = \begin{cases} (2\pi)^{-\frac{1}{2}}, & |\xi| \le \frac{2}{3}\pi, \\ (2\pi)^{-\frac{1}{2}} \cos[\frac{\pi}{2}\nu(\frac{3}{4\pi}|\xi| - 1)], & \frac{2}{3}\pi \le |\xi| \le \frac{4}{3}\pi, \\ 0, & \text{otherwise}, \end{cases} \tag{9}$$

where ν is a C^k function $(0 \le k \le \infty)$ with

$$\nu(x) = \begin{cases} 0, \; x \le 0, \\ 1, \; x \ge 1, \end{cases}$$

and $\nu(x) + \nu(1-x) = 1$. Then $\hat{\varphi}$ is a C^k function and the corresponding wavelet ψ is given by its Fourier transform

$$\hat{\psi}(\xi) = \begin{cases} (2\pi)^{-\frac{1}{2}} e^{i\xi/2} \sin[\frac{\pi}{2}\nu(\frac{3}{2\pi}|\xi| - 1)], & \frac{2}{3}\pi \le |\xi| \le \frac{4}{3}\pi, \\ (2\pi)^{-\frac{1}{2}} e^{i\xi/2} \cos[\frac{\pi}{2}\nu(\frac{3}{4\pi}|\xi| - 1)], & \frac{4}{3}\pi \le |\xi| \le \frac{8}{3}\pi, \\ 0, & \text{otherwise}, \end{cases} \tag{10}$$

and the supports of $\hat{\varphi}$ and $\hat{\psi}$ are respectively

$$\operatorname{Supp} \hat{\varphi} = [\,-\tfrac{4}{3}\pi,\ \tfrac{4}{3}\pi\,],$$

$$\operatorname{Supp} \hat{\psi} = [\,-\tfrac{8}{3}\pi,\ -\tfrac{2}{3}\pi\,] \cup [\,\tfrac{2}{3}\pi,\ \tfrac{8}{3}\pi\,].$$

From Daubechies[6], we see that the functions

$$\psi_{jk}(t) := 2^{\frac{j}{2}}\psi(2^j t - k), \qquad j,k \in Z,$$

constitute an orthonormal basis of the $L^2(R)$. It is easy to see

$$\hat{\psi}_{jk}(\xi) = 2^{-\frac{j}{2}} e^{-ik2^{-j}\xi}\hat{\psi}(2^{-j}\xi),$$

and

$$\text{Supp}\,\hat{\psi}_{jk} = [\,-\frac{8}{3}\pi 2^j,\, -\frac{2}{3}\pi 2^j\,] \cup [\,\frac{2}{3}\pi 2^j,\, \frac{8}{3}\pi 2^j\,], \quad k \in Z. \tag{11}$$

The multiresolution analysis (MRA) $\{V_j\}_{j\in Z}$ of Meyer wavelet generated by

$$V_j = \overline{\{\varphi_{jk} :\ k \in Z\}}, \quad \varphi_{jk}(t) = 2^{\frac{j}{2}}\varphi(2^j t - k), \quad j,k \in Z,$$

and

$$\text{Supp}\,\hat{\varphi}_{jk} = [\,-\frac{4}{3}\pi 2^j,\, \frac{4}{3}\pi 2^j\,], \quad k \in Z. \tag{12}$$

The orthogonal projection on the space V_J is given by

$$P_J f := \sum_{k\in Z}(f, \varphi_{Jk})\varphi_{Jk}, \quad f \in L^2(R),$$

while

$$Q_J f := \sum_{k\in Z}(f, \psi_{Jk})\psi_{Jk}, \quad f \in L^2(R),$$

denotes the orthogonal projection on the wavelet space W_J with $V_{J+1} = V_J \oplus W_J$. It is easy to see by (12), that

$$\widehat{P_J f}(\xi) = 0 \quad \text{for} \quad |\xi| \geq \frac{4}{3}\pi 2^J, \tag{13}$$

and for $j > J$ it follows from (11)

$$\widehat{Q_j f}(\xi) = 0 \quad \text{for} \quad |\xi| < \frac{4}{3}\pi 2^J. \tag{14}$$

Since

$$(I - P_J)f = \sum_{j\geq J} Q_j f,$$

and from (14) we have

$$((I - P_J)f)\hat{}\,(\xi) = \widehat{Q_J f}(\xi), \quad \text{for} \quad |\xi| < \frac{4}{3}\pi 2^J. \tag{15}$$

The following inequality for differential operators $D^k, k \in N$ is known [7].

Lemma 1 *Let $\{V_j\}_{j\in Z}$ be Meyer's MRA and $D^k = \frac{d^k}{dt^k}$ be differential operators. Suppose $J \in N$, then for all $f \in V_J$ we have*

$$\|D^k f\| \le C2^{(J-1)k}\|f\|, \quad k \in N, \tag{16}$$

where C is a positive constant.

Defining the operator $T_x : g \longmapsto u(x,t)$, i.e. $u(x,t) = T_x g$, by (5):

$$\hat{u}(x,\xi) = \widehat{T_x g}(\xi) = e^{\sqrt{i\xi}(1-x)}\hat{g}(\xi), \quad 0 \le x < 1. \tag{17}$$

Then we have

Lemma 2 *Let $\{V_j\}_{j\in Z}$ be Meyer's MRA and suppose $J \in N$. Then for $\forall\, g \in V_J$ we have*

$$\|T_x g\| \le C\exp(2^{(J-1)/2}(1-x))\|g\|, \quad 0 \le x < 1. \tag{18}$$

where C is a constant.

Proof: For convenience we will denote different constants appearing in the proof by same C. It is easy to show that

$$e^{(1-x)\sqrt{|\xi|/2}} \le 2\sqrt{2}|\cosh\sqrt{i\xi}(1-x)|. \tag{19}$$

From (17) ,(16), (19), (7) and Hölder inequality we know that for $\forall\, g \in V_J$ holds

$$\begin{aligned}
||T_x g|| &= [\int_{-\infty}^{\infty} |e^{\sqrt{i\xi}(1-x)}\hat{g}(\xi)|^2 d\xi]^{\frac{1}{2}} \\
&= [\int_{-\infty}^{\infty} |e^{(1-x)\sqrt{|\xi|/2}}\hat{g}(\xi)|^2 d\xi]^{\frac{1}{2}} \\
&\leq C[\int_{-\infty}^{\infty} |\cosh(\sqrt{i\xi}(1-x))\hat{g}(\xi)|^2 d\xi]^{\frac{1}{2}} \\
&= C[\int_{-\infty}^{\infty} |\sum_{k=0}^{\infty} \frac{(1-x)^{2k}(i\xi)^k}{(2k)!}\hat{g}(\xi)|^2 d\xi]^{\frac{1}{2}} \\
&\leq C\sum_{k=0}^{\infty} \frac{(1-x)^{2k}}{(2k)!}[\int_{-\infty}^{\infty} |(i\xi)^k \hat{g}(\xi)|^2 d\xi]^{\frac{1}{2}} \\
&= C\sum_{k=0}^{\infty} \frac{(1-x)^{2k}}{(2k)!}||D^k g|| \\
&\leq C\sum_{k=0}^{\infty} \frac{(1-x)^{2k}}{(2k)!}2^{(J-1)k}||g|| \\
&= C\cosh(2^{(J-1)/2}(1-x))||g|| \\
&\leq C\exp\{2^{(J-1)/2}(1-x)\}||g||.
\end{aligned}$$

Let $T_{x,J} := T_x P_J$, we can show that it approximates T_x in a stable way for an appropriate choice of $J \in N$ depending on ε in (2). In fact we have

Theorem 1 *For every fixed $J \in N$, the problem (1) with data $g \in V_J$ is well-posed. Suppose conditions (2) (3) hold, then the problem of calculating $T_{x,J}g_m$ is stable. Furthermore, with*

$$J^* := [\log_2(2(\ln \frac{M}{\varepsilon})^2)], \tag{20}$$

where $[a]$ with square bracket denotes the largest integer less than or equal to $a \in R$, we have

$$||T_x g - T_{x,J^*} g_m|| \leq C_1 M^{1-x} \varepsilon^x, \quad 0 \leq x < 1, \tag{21}$$

where C_1 is a constant independent of ε and M.

Proof:

$$||T_x g - T_{x,J} g_m|| \leq ||T_x g - T_{x,J} g|| + ||T_{x,J} g - T_{x,J} g_m||. \tag{22}$$

From lemma 2 we can see that the second term of the right-hand side of (22) satisfies

$$\begin{aligned}\|T_{x,J}g - T_{x,J}g_m\| &= \|T_x P_J(g-g_m)\| \\ &\le C\exp\{2^{(J-1)/2}(1-x)\}\|P_J(g-g_m)\| \\ &\le C\exp\{2^{(J-1)/2}(1-x)\}\varepsilon. \end{aligned} \tag{23}$$

For the first term of the right-hand side of (22), by (13) we can rewrite it as

$$\begin{aligned}\|T_x g - T_{x,J}g\| &= \|T_x(I-P_J)g\| \\ &= \Big(\int_{-\infty}^{\infty} |e^{\sqrt{i\xi}(1-x)}((I-P_J)g)\hat{}(\xi)|^2 d\xi\Big)^{\frac{1}{2}} \\ &\le \Big(\int_{|\xi|\ge\frac{4}{3}\pi 2^J} |e^{\sqrt{i\xi}(1-x)}\hat{g}(\xi)|^2 d\xi\Big)^{\frac{1}{2}} \\ &\quad + \Big(\int_{|\xi|<\frac{4}{3}\pi 2^J} |e^{\sqrt{i\xi}(1-x)}((I-P_J)g)\hat{}(\xi)|^2 d\xi\Big)^{\frac{1}{2}} \\ &:= \mathrm{I}_1 + \mathrm{I}_2. \end{aligned} \tag{24}$$

Note that (6), we know

$$\begin{aligned}\mathrm{I}_1 &= \Big(\int_{|\xi|\ge\frac{4}{3}\pi 2^J} |e^{\sqrt{i\xi}(1-x)}\hat{g}(\xi)|^2 \mathrm{d}\xi\Big)^{\frac{1}{2}} \\ &= \Big(\int_{|\xi|\ge\frac{4}{3}\pi 2^J} |e^{-x\sqrt{i\xi}}\hat{f}(\xi)|d\xi\Big)^{\frac{1}{2}} \\ &\le e^{-x\sqrt{\frac{2}{3}\pi 2^J}}\|f\| \le \exp\{-x2^{J/2}\}\|f\|. \end{aligned} \tag{25}$$

By (15), lemma 2 and note that $Q_J g \in W_J \subset V_{J+1}$, it is easy to see that I_2 satisfies

$$\begin{aligned}\mathrm{I}_2 &= \Big(\int_{|\xi|<\frac{4}{3}\pi 2^J} |e^{\sqrt{i\xi}(1-x)}\widehat{Q_J g}(\xi)|^2 d\xi\Big)^{\frac{1}{2}} \\ &\le \|T_x Q_J g\| \le C\exp(2^{J/2}(1-x))\|Q_J g\|. \end{aligned}$$

Let χ_J be the characteristic function of interval $[-\frac{2}{3}\pi 2^J, \frac{2}{3}\pi 2^J]$. We introduce an operator M_J defined by $\widehat{M_J g} = (1-\chi_J)\hat{g}$. From (11) we know

$$\begin{aligned} Q_J g &= \sum_{k\in Z}(g,\psi_{Jk})\psi_{Jk} = \sum_{k\in Z}(\hat{g},\hat{\psi}_{Jk})\psi_{Jk} \\ &= \sum_{k\in Z}((1-\chi_J)\hat{g},\hat{\psi}_{Jk})\psi_{Jk} = \sum_{k\in Z}(M_J g,\psi_{Jk})\psi_{Jk} = Q_J M_J g. \end{aligned}$$

So by using (6) we have

$$\begin{aligned} \|Q_J g\| &= \|Q_J M_J g\| \le \|M_J g\| = \|\widehat{M_J g}\| \\ &= (\int_{|\xi|\ge\frac{2}{3}\pi 2^J} |\hat{g}(\xi)|^2 d\xi)^{\frac{1}{2}} = (\int_{|\xi|\ge\frac{2}{3}\pi 2^J} |e^{-\sqrt{i\xi}}\hat{f}(\xi)|^2 d\xi)^{\frac{1}{2}} \\ &\le \exp\{-2^{J/2}\}\|f\|. \end{aligned}$$

Therefore

$$\mathrm{I}_2 \le C\exp\{-x2^{J/2}\}\|f\|. \tag{26}$$

Combining (25), (26) with (24) we get

$$\|T_x(I-P_J)g\| \le (C+1)\exp\{-x2^{J/2}\}\|f\|.$$

Together with (23) we obtain

$$\|T_x g - T_{x,J}g_m\| \le C\exp\{2^{(J-1)/2}(1-x)\}\varepsilon + (C+1)\exp\{-x2^{J/2}\}\|f\|. \tag{27}$$

where C is the constant appearing in (18) of lemma 2.

Choosing $J^* := [\log_2(2(\ln\frac{M}{\varepsilon})^2)]$, where the meaning of $[a]$ is the same as in narration of the theorem. We get

$$\begin{aligned} &\exp\{2^{(J^*-1)/2}(1-x)\}\varepsilon \le \exp\{(1-x)\ln\frac{M}{\varepsilon}\}\varepsilon = M^{1-x}\varepsilon^x, \\ &\exp\{-x2^{J^*/2}\} \le \exp\{-x(\ln\frac{M}{\varepsilon})\}M = M^{1-x}\varepsilon^x. \end{aligned}$$

Combining these with (27) we obtain

$$\|T_x g - T_{x,J^*}g_m\| \le (2C+1)M^{1-x}\varepsilon^x = C_1 M^{1-x}\varepsilon^x, \quad 0<x<1.$$

The proof is completed.

Remark 1 *Note that the conclusion of Carasso*[2] *we know that the result is at least close to optimal, therefore we can not expect to find a numerical method for approximating solution of (1) that satisfy a better estimate in L^2-sense except for more delicate choice of coefficient C_1. This suggests that wavelets must be useful for solving the considered ill-posed problem.*

Remark 2 *The result completely overcome the defect left over by Regińska*[3].

Remark 3 *As we see from the estimate (21) (and the conclusion obtained by other authors* [2,5]*, when $x \to 0^+$ the accuracy of the regularized solution become progressively lower. At $x = 0$ they merely imply that the error is bounded by $C_1 M$. These results can not be improved in L^2-sense. But by using Sobolev space $H^s(R)$ we can obtain a better estimate near $x = 0$. We still suppose that the function $g(t)$ and $g_m(t)$ are exact and measured data respectively with*

$$\|g - g_m\|_{H^r} \leq \varepsilon \quad \textit{for some} \quad r \leq 0. \tag{28}$$

Since g_m belongs, in general, to $L^2(R)$, r should not be positive. We also assume

$$f(t) := u(0,t) \in H^s(R) \quad \textit{for} \quad s > r, \tag{29}$$

and

$$\|f\|_{H^s} \leq M. \tag{30}$$

Analogues to theorem 1 we can establish that

Theorem 2 *Suppose the condition (28)-(30) hold, let us select*

$$J^{**} := [\log_2(2(\ln(\frac{M}{\varepsilon}(\ln(\frac{M}{\varepsilon})^{-2(s-r)}))^2)],$$

where the sense of bracket $[a]$ is the same with the theorem 1, then we have estimate

$$\begin{aligned}\|T_x g - T_{x,J^{**}} g_m\|_{H^r} \leq (C + (C+1)(\tfrac{\ln \frac{M}{\varepsilon}}{\ln \frac{M}{\varepsilon} + \ln(\ln \frac{M}{\varepsilon})^{-2(s-r)}})^{2(s-r)}) \\ \cdot M^{1-x} \varepsilon^x (\ln \tfrac{M}{\varepsilon})^{-2(s-r)(1-x)}\end{aligned} \tag{31}$$

where C is a constant.

Note that $s > r$, so at $x = 0$ the estimate (31) becomes

$$\begin{aligned}\|T_0 g - T_{0,J^{**}} g_m\|_{H^r} &= \|f - T_{0,J^{**}} g_m\|_{H^r} \\ &\leq (C + (C+1)(\frac{\ln \frac{M}{\varepsilon}}{\ln \frac{M}{\varepsilon} + \ln(\ln \frac{M}{\varepsilon})^{-2(s-r)}})^{2(s-r)}) \\ &\quad \cdot M(\ln \frac{M}{\varepsilon})^{-2(s-r)} \to 0, \text{ as } \varepsilon \to 0^+.\end{aligned}$$

So the conclusion of theorem 2 is really an improvement of theorem 1 near $x = 0$.

Acknowledgments

The project is supported by the Natural Science Foundation of Gansu province (ZS021-A25-001-Z) and the National Natural Science Foundation of China (No.49875024).

References

1. J. V. Beck, B. Blackwell and S. R. Clair, *Inverse Heat Conduction: Ill-Posed Problems* (Wiley, NewYork, 1985).
2. A. Carasso, *Determining surface temperatures from interior observations,* SIAM J. Appl. Math. **42(3)**, 558–574 (1982).
3. T. Regińska, *Sideways heat equation and wavelets,* J. Comput. Appl. Math. **63**, 209–214 (1995).
4. C. L. Fu, C. Y. Qiu and Y. B. Zhu, *A note on "Sideways heat equation and wavelets" and constant* e^*, Comp & Math with Appl. **43(8/9)**, 1125–1134 (2002).
5. L.Eldén, F.Berntsson and T.Regińska, *Wavelet and Fourier methods for solving the sideways heat equation,* SIAM J. Sci. Comp. **21(16)**, 2187–2205 (2000).
6. I.Daubechies, *Ten Lectures on Wavelets* (SIAM,Philadelphia, 1992).
7. Dinh Nho Háo, A. Schneider and H-J Reinhardt, *Regularization of a non-characteristic Cauchy problem for a parabolic equation,* Inverse Problems **11**, 1247–1263 (1995).

DIRECT SIMULATION OF AN INTEGRAL EQUATION OF THE FIRST KIND

H. IMAI

Department of Applied Physics and Mathematics, Faculty of Engineering, University of Tokushima, Tokushima 770-8506, Japan
E-mail: imai@pm.tokushima-u.ac.jp

T. TAKEUCHI

Department of Applied Physics and Mathematics, Faculty of Engineering, University of Tokushima, Tokushima 770-8506, Japan
E-mail: takeuchi@pm.tokushima-u.ac.jp

Direct numerical simulation to an integral equation of the first kind is carried out by using IPNS(Infinite-Precision Numerical Simulation). Numerical results are very satisfactory in accuracy. Moreover, they also show some interesting facts. These numerical results show IPNS facilitates numerical analysis for such inverse problems.

1 Introduction

Inverse problems are very difficult to be solved. Numerical simulation is inevitable in practical analysis. Many mathematical problems are analyzed by using direct numerical simulation. However, it has been a taboo to inverse problems due to easy corruption by strong oscillation. For avoidance of this oscillation some additional methods are usually used together. In such methods original inverse problems are often transformed into modified problems, then solved. Here we should remark this modification is important from the practical view point, however it is not preferable from the view point of analysis. These additional methods are the regularization, the method of least squares and AI. Restriction of the dimension of the solution space is very popular in concrete analysis, and it is a sort of the regularization. AI facilitates the development of the efficient solver of the problem and the implementation of experiences to the solver. Unfortunately these additional methods are not absolute. This is because in numerical simulation the rounding error spoils their theoretical usefulness.

As for direct numerical simulation of inverse problems some new approaches were carried out. Multiple-precision arithmetic is a keyword. It removes the effect of the rounding error to strong oscillation. Multiple-precision arithmetic was applied to the following integral equation of the first kind.

Problem 1 *Find $u(y)$ such that*

$$\int_0^1 e^{xy}u(y)dy = f(x), \quad f(x) = x^{-2}(x-1)e^x + x^{-2}, \quad f(0) = 1. \tag{1}$$

The exact solution for Problem 1 is $u(y) = y$. Direct numerical simulation was carried out. Numerical results in multiple precision were satisfactory comparing with those in double precision[2]. On the other hand, IPNS(Infinite-Precision Numerical Simulation)[5] was applied to several inverse problems governed by PDF systems[3,4,6,7]. It was also applied to Problem 1 [4]. Numerical results were very satisfactory, however numerical investigation was not in detail. In the paper more numerical investigations are carried out.

2 Application of IPNS

2.1 Infinite-Precision Numerical Simulation

Numerical errors originate from the truncation error in the discretization and the rounding error. Realization of highly accurate numerical simulation needs arbitrary reduction of both errors. For such numerical simulation we proposed a simple method called IPNS(Infinite-Precision Numerical Simulation)[5]. IPNS consists of the arbitrary order approximation and multiple-precision arithmetic. The former is used for the arbitrary reduction of truncation errors in the discretization. The last is used for the arbitrary reduction of rounding errors. For the arbitrary order approximation spectral methods are very useful[1]. Especially, the spectral collocation method is most useful. Its application is same as FDM, so it is easily applicable to nonlinear problems, even to free boundary problems. In the spectral collocation method, the order of approximation can be controlled by the number of collocation points.The multiple-precision arithmetic is now easily available. A lot of FORTRAN subroutines about it are already prepared. Some libraries are free and distributed on the net, e.g. `http://www.lmu.edu/acad/personal/faculty/dmsmith2/FMLIB.html`[8]. IPNS has been applied to many problems and ultimately high accuracy has been seen in numerical results.

2.2 The Way of Application

IPNS was applied to Problem 1 as follows[4]. The problem should be transformed to be defined in the interval $[-1, 1]$. So, we consider the following problem.

Problem 2 *Find $u(y)$ such that*

$$\int_{-1}^{1} e^{xy}u(y)dy = f(x), \quad f(x) = \frac{(2x-1)e^x + e^{-1}}{2x^2}, \quad f(0) = 1. \tag{2}$$

The exact solution for Problem 2 is

$$u(y) = \frac{y+1}{2}. \tag{3}$$

Remark 1 *This problem is derived from Problem 1 as follows. From the transformation $y = \frac{t+1}{2}$, Problem 1 becomes*

$$\int_{-1}^{1} e^{\frac{tx}{2}x} u\left(\frac{t+1}{2}\right) dt = 2e^{-\frac{1}{2}x} f(x).$$

Taking $\frac{x}{2} = \xi$ and $u\left(\frac{t+1}{2}\right) = v(t)$, then

$$\int_{-1}^{1} e^{\xi t} v(t)dt = 2e^{-\xi} f(2\xi) = \frac{2(2\xi-1)e^{\xi} + 2e^{-\xi}}{4\xi^2} = \frac{(2\xi-1)e^{\xi} + e^{-\xi}}{2\xi^2}.$$

Then it is easy to see Problem 2 is derived from this. Such transformation is necessary for application of Chebyshev polynomials, however it is not restriction.

Remark 2 *Problem 2 is a special case of the following general case:*

$$\int_{-1}^{1} k(x,y)u(y)dy = f(x). \tag{4}$$

To Problem 2, the spectral collocation method with Chebyshev polynomials is applied to the integrand as follows[4]:

$$e^{xy}u(y) \cong \sum_{k=0}^{N} \tilde{g}_k(x) T_k(y). \tag{5}$$

From the inversion formula

$$\tilde{g}_k(x) \cong \frac{2}{Nc_k} \sum_{j=0}^{N} \frac{1}{c_j} e^{xy_j} u_j T_k(y_j), \quad k = 0, 1, \cdots, N, \tag{6}$$

where

$$y_j = \cos\frac{j\pi}{N}, \quad j = 0,\ 1, \cdots,\ N, \tag{7}$$

$$u_j\ :\ \text{the computed value of } u(y_j), \qquad j = 0,\ 1, \cdots,\ N, \tag{8}$$

$$c_j = \begin{cases} 1, & j = 1,\ 2, \cdots,\ N-1, \\ 2, & j = 0,\ N. \end{cases} \tag{9}$$

$\{y_j\}$ are called C-G-L(Chebyshev-Gauss-Lobatto) points[1]. Thus

$$e^{xy}u(y) \cong \sum_{k=0}^{N} \frac{2}{Nc_k} \sum_{j=0}^{N} \frac{1}{c_j} e^{xy_j} u_j T_k(y_j) T_k(y). \tag{10}$$

Then

$$\int_{-1}^{1} e^{xy}u(y)dy \tag{11}$$

$$\cong \int_{-1}^{1} \left\{ \sum_{k=0}^{N} \frac{2}{Nc_k} \sum_{j=0}^{N} \frac{1}{c_j} e^{xy_j} u_j T_k(y_j) T_k(y) \right\} dy \tag{12}$$

$$= \frac{2}{N} \sum_{k=0}^{N} \sum_{j=0}^{N} \frac{1}{c_k c_j} e^{xy_j} u_j T_k(y_j) \int_{-1}^{1} T_k(y)dy \tag{13}$$

$$= \frac{2}{N} \sum_{\substack{k=0 \\ k\neq 1}}^{N} \sum_{j=0}^{N} \frac{1}{c_k c_j} e^{xy_j} u_j \cos\frac{jk\pi}{N} \cdot \frac{1+(-1)^k}{1-k^2}. \tag{14}$$

We choose proper points $\{x_l\}$, $l = 0,\ 1, \cdots,\ N$ on which equation is satisfied. Set $f_l = f(x_l)$, then we have the following linear system:

$$\sum_{j=0}^{N} a_{lj} u_j = \frac{N}{2} f_l, \quad l = 0,\ 1, \cdots,\ N \tag{15}$$

where

$$a_{lj} = \sum_{\substack{k=0 \\ k\neq 1}}^{N} \frac{1}{c_k c_j} e^{x_l y_j} \cos\frac{jk\pi}{N} \cdot \frac{1+(-1)^k}{1-k^2}. \tag{16}$$

After solving this linear system, $u(y)$ is reconstructed as follows :

$$u(y) \cong \sum_{k=0}^{N}\sum_{j=0}^{N} \frac{2}{Nc_k c_j} u_j T_k(y_j) T_k(y). \tag{17}$$

Remark 3 *The same discretization can be carried out to the general case in Remark 2.*

Remark 4 *Eqs. (5) and (6) are not inversion formulae for $u(y)$ except for the case $x = 0$. This means this reconstruction is not obvious. In the general case in Remark 2 this exceptional case is that $k(x, y)$ is independent of y.*

Remark 5 *Our discretization is done not to $u(y)$ but to $e^{xy}u(y)$. This is because it is applicable to the general case in Remark 2 without numerical integration which generates the additional truncation error. However, we are not sure that our discretization is best.*

The above linear system (15) is very ill-conditioned, so numerical computations must be carried out in multiple-precision. This is IPNS.

3 Numerical Results

Figure 1 shows errors for Problem 2 with C-G-L points : $x_l = \cos\dfrac{l\pi}{N}$, $l = 0, 1, \cdots, N$. Here,

$$\text{error} = \max_{0\leqq j\leqq N} |u_N(y_j) - u(y_j)|, \quad y_j = \cos\frac{j\pi}{N}, \quad j = 0, 1, \cdots, N. \tag{18}$$

$u(y)$ is the exact solution, and $u_N(y)$ is the right-hand side of Eq. (17). Here we should remark $u_N(y_j) = u_j$. If the rounding error is not small enough, the error defined by Eq. (18) grows explosively before obtaining good results. This shows the linear system (15) is very ill-conditioned. At the same time, if the rounding error is small enough, the error reduces successively. In Figure 1(d) the regression line by the method of least squares is log (error) =

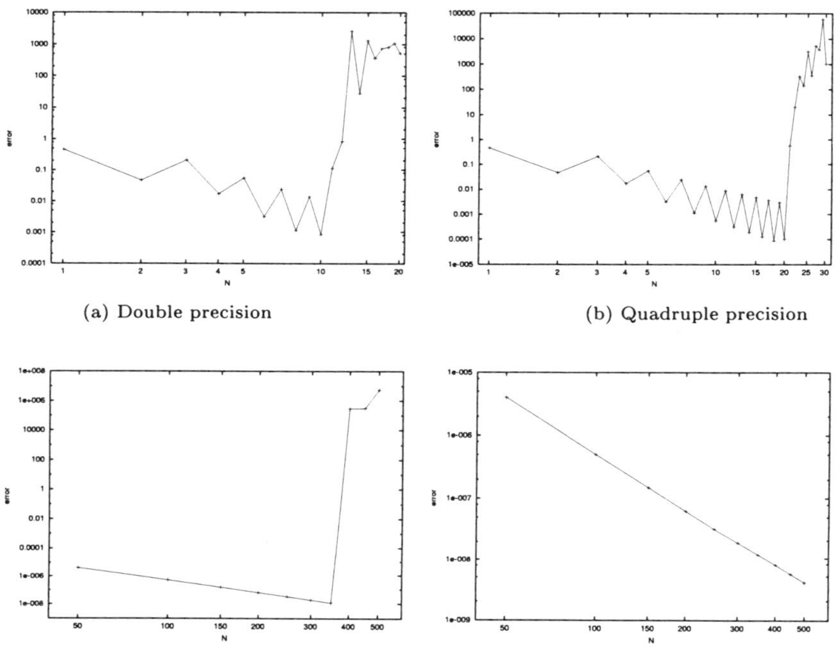

(a) Double precision

(b) Quadruple precision

(c) 1000 digits

(d) 2000 digits

Figure 1. Behavior of maximum errors for Problem 2.

$-3.00 * \log N - 0.686$ with the correlation coefficient $\rho = -1.00$. This means IPNS works well.

Figure 2 shows error dependence on the choice of $\{x_l\}$ for Problem 2 (400 digits). Error by using C-G-L points : $x_l = \cos\dfrac{l\pi}{N}, \quad l = 0, 1, \cdots, N$ is compared with error by using equally spaced points in $[-1, 1]$: $x_l = -1 + \dfrac{2l}{N}, \quad l = 0, 1, \cdots, N$. There is almost no difference in the behavior of error reduction.

Figure 3 shows error dependence on the interval where $\{x_l\}$ are distributed. Equally spaced points in $[-10^{-m}, \; 10^{-m}]$: $x_l = 10^{-m}\left(-1 + \dfrac{2l}{N}\right), \quad l = 0, 1, \cdots, N$ are used for obtaining the linear sys-

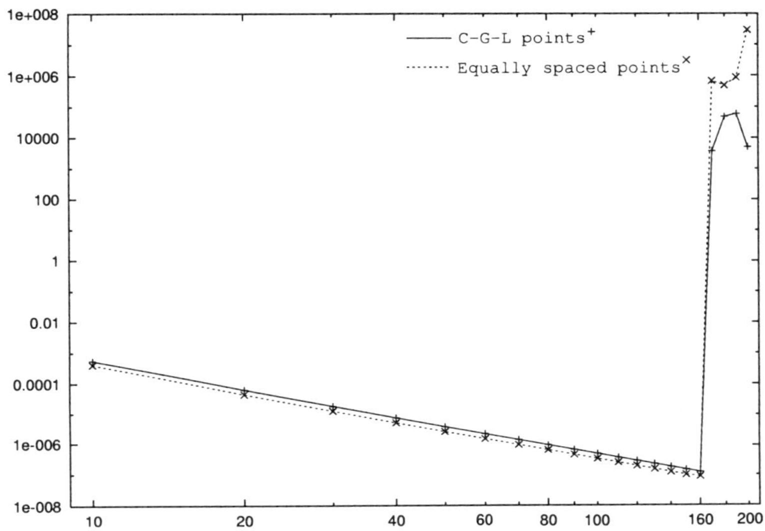

Figure 2. Error dependence on the choice of $\{x_l\}$ for Problem 2(400 digits).

tem (15). Behavior of errors is quite different whether N is odd or not. For even N spectral accuracy is seen.

Remark 6 *Remark 4 may mean the appearance of spectral accuracy seen in Figure 3.*

4 Conclusion

Direct numerical simulation to an integral equation of the first kind is carried out by using IPNS. Numerical results are very satisfactory in accuracy. Moreover, they also show some interesting facts. IPNS sometimes needs long CPU time and huge memory space because it involves multiple-precision arithmetic. However, numerical computation with several hundreds digits is already practical. IPNS facilitates numerical analysis for inverse problems. Direct simulation to inverse problems is not a taboo now.

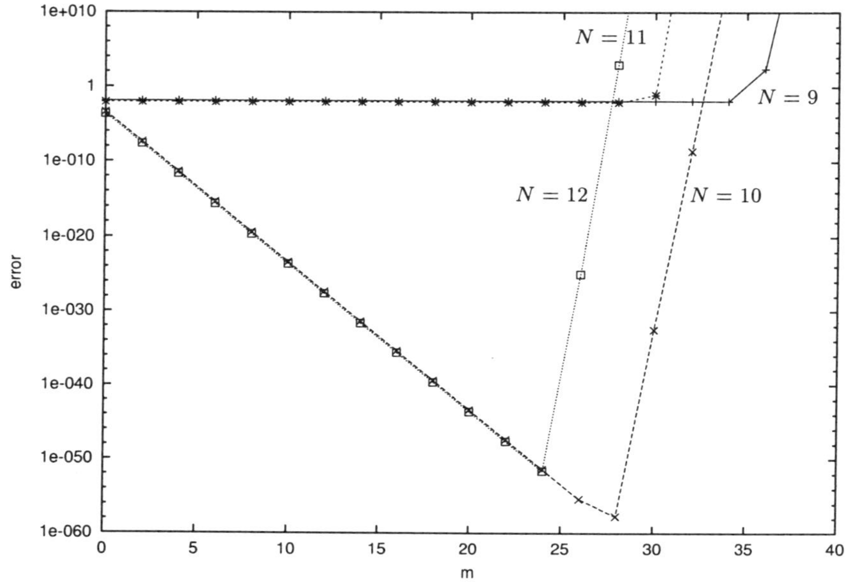

Figure 3. Error dependence on the interval including $\{x_i\}$(400 digits).

Acknowledgments

This work is partially supported by Grants-in-Aids for Scientific Research (No. 13640119), from the Japan Society of Promotion of Science.

References

1. C. Canuto et al., *Spectral Methods in Fluid Dynamics* (Springer, New York, 1998).
2. H. Fujiwara and Y. Iso, *Numerical Challenge to Ill-posed Problems by Fast Multiple-Precision System*, in *Proc. the 50th Japan National Congress on Theoretical and Applied Mechanics* , 419 (2001).
3. H. Imai and T. Takeuchi, *Application of the Infinite-Precision Numerical Simulation to an Inverse Problem*, NIFS-PROC **40**, 38–47(1999).
4. H. Imai and T. Takeuchi, *Some Advanced Applications of the Spectral Collocation Method,* GAKUTO Int. Ser. Math. Sci. Appl. **17**, 323(2001).

INVERSE PROBLEM OF RECONSTRUCTING THE PARABOLIC EQUATION'S INITIAL VALUE AND THE HEAT RADIATIVE COEFFICIENT *

YONGJI TAN

Department of Mathematics, fudan university, Shanghai, China
E-mail: yjtan@fudan.edu.cn

CHUNXIA JIA

Department of Mathematics, Shanghai Normal University, Shanghai, China
E-mail: math12@sohu.com

In this paper, the numerical method to reconstruct the radiative coefficient and initial condition simultaneously by measuring the domain temperature at a fixed time and the temperature of a subdomain all the time is studied. By least-square technique this inverse problem can be formulated into a variational problem and discretized into nonlinear programming problem with the cost function depending on the numerical solution of the corresponding direct problem of heat equation. We obtain the numerical solution of the direct problem by finite difference method and radial basis function (RBF)method respectively and derive the gradient formula for cost function, then implement the numerical reconstruction by quasi-Newton technique. In the case of the measuring data with noise, we use regularization method. Numerical results show that this method is available.

1 Introduction

Consider the following initial-boundary problem with the radiative coefficient:

$$\frac{\partial u}{\partial t} = \Delta u + p(x)u, \text{ in } \Omega \times (0,T) \tag{1}$$

$$u(x,0) = \mu(x), \text{ in } \Omega \tag{2}$$

$$u(x,t) = \eta(x,t), \text{ in } \partial\Omega \times (0,T) \tag{3}$$

where $u = u(x,t)$ is an unknown temperature function, $p(x)$ is radiative coefficient, the physical domain Ω is an open bounded domain in $R^d(d = 1,2,3)$, with a piecewise smooth boundary $\partial\Omega$.

In this paper, we mainly investigate the numerical method for reconstructing the initial temperature distribution $\mu(x)$ and the heat radiative coefficient $p(x)$ in $(1) - -(2)$. It is well known that given only the measurement of temperature at a fixed time $\tau(> 0)$, the reconstruction of the initial temperature

*PROJECT 10171020 SUPPORTED BY NSFC.

is highly ill-posed, let alone the case that we intend to recover both initial temperature and radiative coefficient here. Having some extra observation of the temperature, say in a small subregion of the physical domain ω along the time direction, it is possible to reconstruct $\mu(x)$and $p(x)$. M. Yamamoto etc. proved that the problem is conditionly stability, by formulating it into a variational problem, and using finite element discretization and gradient method, they achieved the numerical reconstruction.

Let $\psi_\tau(x)$ be the measurement of $u(x,\tau)$, $\psi(x,t)$ be the measurement of $u(x,t)$ in $\omega \times (0,\tau)$, the problem of reconstructing $p(x)$ and $\mu(x)$ can be formulated into that to find $(p(x), \mu(x), u(x,t))$ which satisfies (1), (2), (3) and the following equations:

$$u(x,\tau) = \psi_\tau(x), \qquad x \in \Omega \tag{4}$$
$$u(x,t) = \psi(x,t),\ (x,t) \in \omega \times (0,\tau) \tag{5}$$

By least-square method, the problem can be formulated into minimizing the following cost function with constrained conditions(1)–(3)

$$Q(p(x), \mu(x)) = \int_\Omega (u(x,\tau) - \psi_\tau(x))^2 dx + \int_0^\tau \int_\omega (u(x,t) - \psi(x,t))^2 dxdt \tag{6}$$

In this paper, for given discrete $p(x)$ and $\mu(x)$, we will solve boundary value problem (1)–(3) by finite difference method and RBF method respectively and therefore obtain the discrete value of $Q(p,\mu)$. After deriving the gradient formula for $Q(p,\mu)$, we implement the reconstruction of p and μ by quasi-Newton technique.

In the case of the existing measuring errors, we use regularization method with both regularization terms

$$\alpha||\mu - \mu_g||^2_{L^2} + \beta||p - p_q||^2_{L^2}$$

and

$$\alpha||\mu - \mu_g||^2_{L^2} + \beta||p||^2_{H^1}$$

where the constant α and β are regularization parameters, μ_q and p_q are guesses of μ and p respectively.

2 Minimizing the cost function by quasi-Newton method

Consider approximate values $p_i = p(x_i)$ and $\mu_i = \mu(x_i)$ of $p(x)$ and $\mu(x)$ at discrete point sets $\{x_1, x_2, \cdots, x_N\}$ in Ω. Given p_i, $\mu_i(i = 1, 2, \cdots, N)$, we can numerically obtain the temperature value $u(x_i, t_j)$ at (x_i, t_j), and then find the approximate value of the functional $Q(p, \mu)$, therefore $Q(p, \mu)$ can be approximately expressed as $Q(p_1, p_2, \cdots, p_N, \mu_1, \mu_2, \cdots, \mu_N)$. Let

$$\rho = (p_1, p_2, \cdots, p_N, \mu_1, \mu_2, \cdots, \mu_N)'$$

the cost function can be written as

$$Q = Q(\rho) \tag{7}$$

In quasi-Newton method an iteration sequence $\rho^0, \rho^1, \cdots$ is designed to approximate the minimum of $Q(\rho)$. The descending direction after kth iteration is:

$$d^k = -[\nabla^2 Q(\rho^k)]^{-1} \nabla Q(\rho^k) \tag{8}$$

where $\nabla^2 Q(\rho^k)$ is Hesse matrix $Q(\rho)$ evaluated at $\rho = \rho^k$. We use H^k to approximate $\nabla^2 Q(\rho^k)$, which is obtained by updating BFGS formula:

$$H^k = H^{k-1} + \frac{(s_{k-1} - H^{k-1} y_{k-1})(s_{k-1} - H^{k-1} y_{k-1})'}{(s_{k-1} - H^{k-1} y_{k-1})' y_{k-1}} \tag{9}$$

where

$$s_k = \rho^{k+1} - \rho^k, \; y_k = \nabla Q(\rho^{k+1}) - \nabla Q(\rho^k)$$

Therefore, the main process of quasi-Newton method is as follows:

1. initialization: select proper initial point $\rho^0 \in R^N$, let $H^0 = I$, $k = 0$, calculate $Q(\rho^0)$ and $\nabla Q(\rho^0)$;

2. compute the fastest descending direction: $d^k = -H^k \nabla Q(\rho^k)$;

3. one dimensional search: solving one dimensional optimization problem

   $$\min_{t>0} Q(\rho^k + t d^k)$$

   to obtain $t = t^k$, let $\rho^{k+1} = \rho^k + t^k d^k$;

4. updating matrix H^k: using BFGS formula to update H^k, get H^{k+1};

5. compute $\nabla Q(\rho^{k+1})$, if $\|\nabla Q(\rho^{k+1})\| < \varepsilon$ (a given small value), stop; otherwise, goto 2

From above, we know the main job is to calculate $Q(\rho^k)$ and $\nabla Q(\rho^k)$, especially the calculation of $\nabla Q(\rho^k)$ is troublesome. If the finite element method is used to solve the direct problem, sensitivity coefficient method and adjoint state method are often used to calculate the gradient, where many linear algebraic equation systems should be solved, nevertheless, if we use difference method or RBF, it is possible to obtain the explicit expression of the gradient.

3 the calculation of cost function and its gradient

3.1 RBF method

For a given function $g(x)$, we take

$$g_j(x) = g(\|x - x_j\|) \quad j = 1, \cdots, N \tag{10}$$

as radial basis, where $x_1, \cdots, x_N$ are given grids. Let

$$u(x,t) = \sum_{j=1}^{N} \alpha_j(t) g_j(x) \tag{11}$$

Denoting $u(x_i, n\Delta t)$ by u_i^n, $\frac{\partial u}{\partial t}(x, n\Delta t)$ can be discretized into $\frac{u_i^{n+1} - u_i^n}{\Delta t}$, where Δt is time increment . (1)–(3)can be written as:

$$u_i^{n+1} = \Delta t \sum_{j=1}^{N} \alpha_j^n \Delta g_j(x_i) + (1 + \Delta t p_i) \sum_{j=1}^{N} \alpha_j^n \Delta g_j(x_i) \tag{12}$$

$$\sum_{j=1}^{N} \alpha_j^n g_j(x_i) = u_i^n \tag{13}$$

$$\sum_{j=1}^{N} \alpha_j^n g_j(x_i) = \eta(x_i, n\Delta t) \qquad x_i \in \partial\Omega \tag{14}$$

$$\sum_{j=1}^{N} \alpha_j^n g_j(x_i) = \psi(x_i, n\Delta t) \qquad x_i \in \omega \tag{15}$$

By (13)−−(15), we can determine α_j^n, and furthermore by (12) we can deduce u_i^{n+1}. The matric form of (13) − −(15) can be written as

$$\tilde{A}\alpha^n = R^n \tag{16}$$

where

$$\alpha^n = (\alpha_1^n, \alpha_2^n, \cdots, \alpha_N^n)$$

the matrix form of (12) is

$$U^{n+1} = \Delta t \cdot B \cdot \alpha^n + G \cdot \tilde{A} \cdot \alpha^n \tag{17}$$

where:

$$U^n = (u_1^n, u_2^n, \ldots u_N^n)'$$

u_j^n denotes the value of u at the jth grid point after n iterations. $\tilde{A}$ and B are matrices depend only on the grids coordinate, G is a diagonal matrix. It is easy to see that we can obtain the derivative of U^{n+1} with respect to the parameter and obtain $\nabla Q(\rho)$.

3.2 Finite difference method

For convenience, we discuss the one dimensional case and assume that $\Omega = (0,1)$. In this case, boundary condition (3) can be written as

$$u(0,t) = \eta_1(t), \; u(1,t) = \eta_2(t)$$

and the difference scheme of (1) is

$$\frac{u_j^{n+1} - u_j^n}{h} = \frac{u_{j+1}^n - 2u_j^n + u_{j-1}^n}{k^2} + pu_j^n$$

therefore we have

$$u_j^{n+1} = Au_{j-1}^n + Bu_j^n + Cu_{j+1}^n \tag{18}$$

where $A = \frac{h}{k^2}$, $B = 1 + hp - \frac{2h}{k^2}$, $C = \frac{h}{k^2}$.

Suppose that the number of internal point is N, by above formulas, we get

$$\begin{pmatrix} u_1^{n+1} \\ u_2^{n+1} \\ \vdots \\ u_{N-1}^{n+1} \\ u_N^{n+1} \end{pmatrix} = \begin{pmatrix} B & C & 0 & \cdots & 0 & 0 \\ A & B & C & \cdots & 0 & 0 \\ \cdot & \cdot & \cdot & \cdot & \cdot & \cdot \\ \cdot & \cdot & \cdot & \cdot & \cdot & \cdot \\ \cdot & \cdot & \cdot & \cdot & \cdot & \cdot \\ 0 & 0 & 0 & \cdots & B & C \\ 0 & 0 & 0 & \cdots & A & B \end{pmatrix} \begin{pmatrix} u_1^n \\ u_2^n \\ \vdots \\ u_{N-1}^n \\ u_N^n \end{pmatrix} + \begin{pmatrix} Au_0^n \\ 0 \\ \vdots \\ 0 \\ Cu_{N+1}^n \end{pmatrix}$$

or

$$U^{n+1} = D(p)U^n + R^n \tag{19}$$

where:

$$D(p) = \begin{pmatrix} B & C & 0 & \cdots & 0 & 0 \\ A & B & C & \cdots & 0 & 0 \\ \cdot & \cdot & \cdot & \cdot & \cdot & \cdot \\ \cdot & \cdot & \cdot & \cdot & \cdot & \cdot \\ \cdot & \cdot & \cdot & \cdot & \cdot & \cdot \\ 0 & 0 & 0 & \cdots & B & C \\ 0 & 0 & 0 & \cdots & A & B \end{pmatrix} \qquad R^n = \begin{pmatrix} Au_0^n \\ 0 \\ \vdots \\ 0 \\ Cu_{N+1}^n \end{pmatrix} \qquad U^{n+1} = \begin{pmatrix} u_1^{n+1} \\ u_2^{n+1} \\ \vdots \\ u_{N-1}^{n+1} \\ u_N^{n+1} \end{pmatrix}$$

By (19), we have

$$U^s = D^s(p)U^0 + D^{s-1}(p)R^1 + \cdots + D(p)R^{s-1} + R^s \tag{20}$$

Since $\frac{\partial U^s}{\partial D(p)}$ and $\frac{\partial D(p)}{\partial p_j}$ can be obtained easily by the expressions of U^s and $D(p)$, the gradient of Q is not difficult to calculate.

4 numerical result

In this section we show some numerical results. For simplicity, we only consider some one dimensional cases with $\Omega = (0,1)$, $\omega = (0.4, 0.6)$.

4.1 Assume that p is a constant

Let $\eta_1(t) = exp(4t)$, $\quad \eta_2(t) = exp(1+4t)$, and the exact solution $u = exp(x+4t)$. From the solution's expression, we know:

$$\psi(x,t) = exp(x+4t), \qquad \psi_\tau(x) = exp(x+4\tau)$$

The results of using finite difference method and RBF method are given by Table 1 and Table 2 respectively.

Table 1.

	exact value	initial value	numerical value	relative error
p	3	10.5	3.0189	0.0063
μ_2	1.1052	10	1.1049	0.0003
μ_3	1.2214	10	1.2214	0
μ_4	1.3499	10	1.3495	0.0003
μ_8	2.0138	14	2.0127	0.0005
μ_9	2.2255	14	2.2264	0.0004
μ_{10}	2.4596	14	2.4583	0.0005

Table 2.

	exact value	initial value	numerical value	relative error
p	3	10	3.0710	0.0237
μ_2	1.1052	-10	1.0991	0.0055
μ_3	1.2214	-13	1.1755	0.0376
μ_4	1.3499	-9.6	1.3334	0.0122
μ_8	2.0138	-7.3	1.9581	0.0277
μ_9	2.2255	-17.2	2.2872	0.0277
μ_{10}	2.4596	-21.8	2.3419	0.0479

4.2 Assume that $p(x)$ is a function

Now we investigate the case when p is a function and the measurement with error. Let $\eta_1(t) = exp(6t)$, $\eta_2(t) = exp(1+6t)$ and the exact solution $u(x,t) = exp(x^2+6t)$, we have

$$\psi(x,t) = exp(x^2+6t)$$

$$\psi_\tau(x) = exp(x^2+6\tau)$$

We put a random error of 1% on the measurement, the regularization is necessary since the illposedness of the inverse problem.

By taking q_g and μ_g as the estimations of the q and μ with error of 30%, we do the reconstruction.

The results of regularized (Figure 3 - 8) and unregularized (Figure 1, 2) are plotted respectively, where O represents the result obtained by RBF and $*$ represents the result obtained by finite difference method.

Figure 3 - Figure 8 show the results obtained by use of different regularization terms where regularization terms are $\alpha||q - q_g||^2 + \beta||\mu - \mu_g||^2$, $\alpha||q - q_g||^2 + \beta||\mu^{'}(x)||^2$, and $\alpha||q - q_g||^2 + \beta\sum|\mu^{'}(x_{i+1}) - \mu^{'}(x_i)|^2$ respectively.

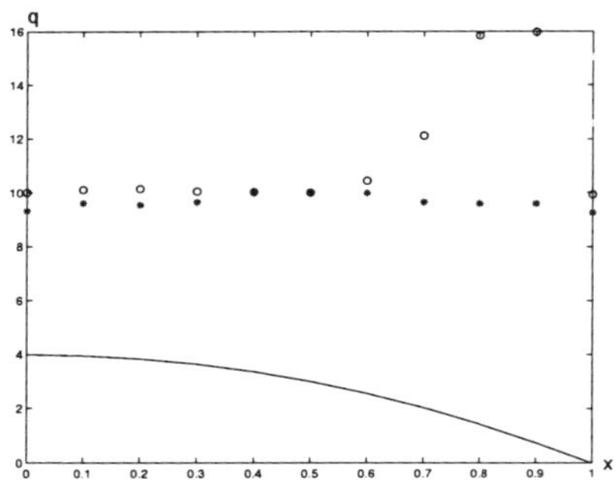

Figure 1: q

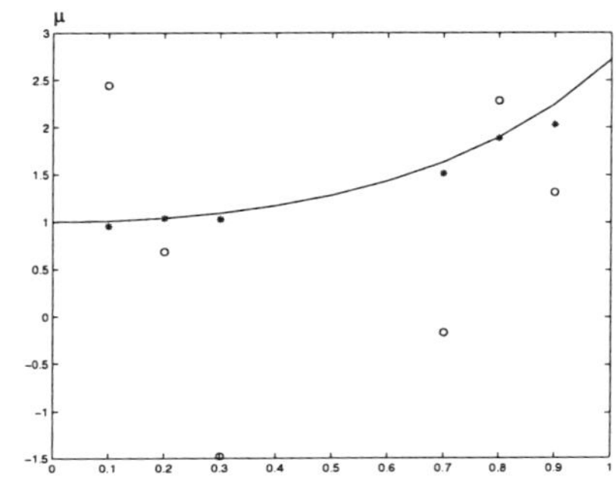

Figure 2: μ

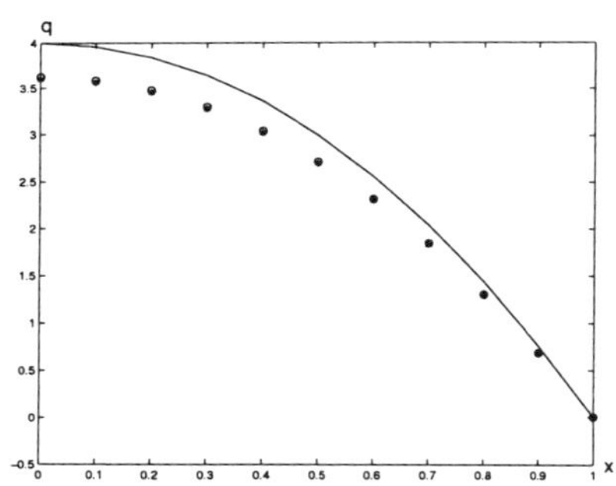

Figure 3: q

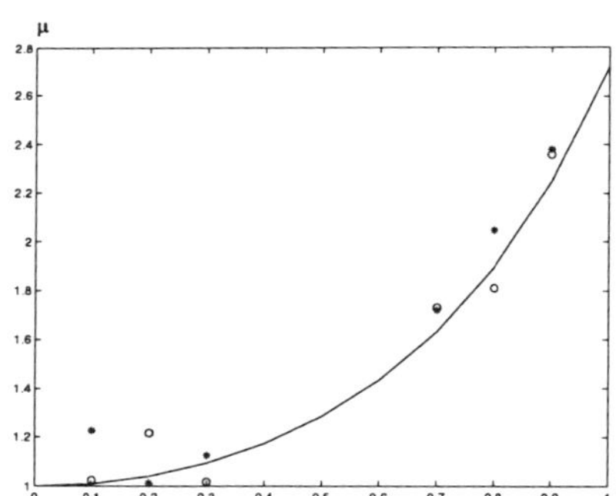

Figure 4: μ

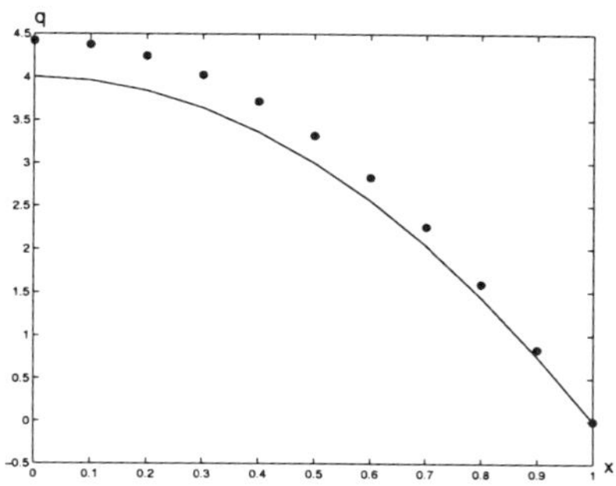

Figure 5: q

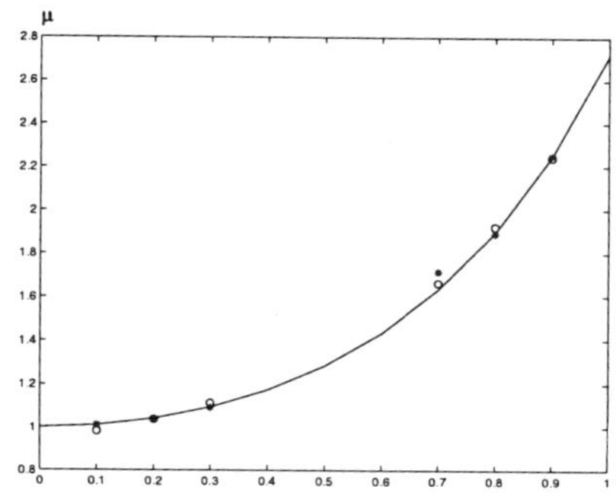

Figure 6: μ

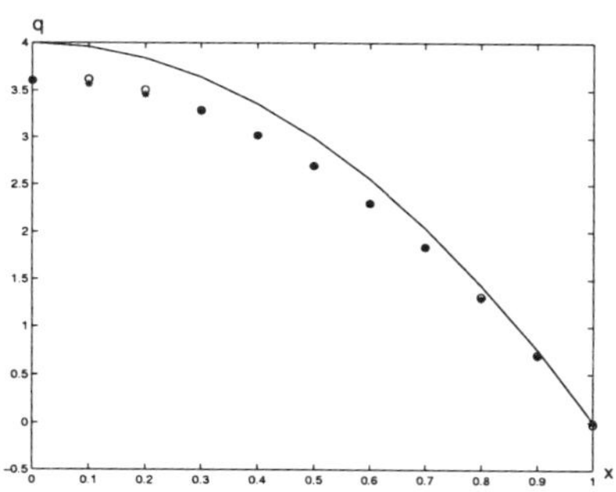

Figure 7: q

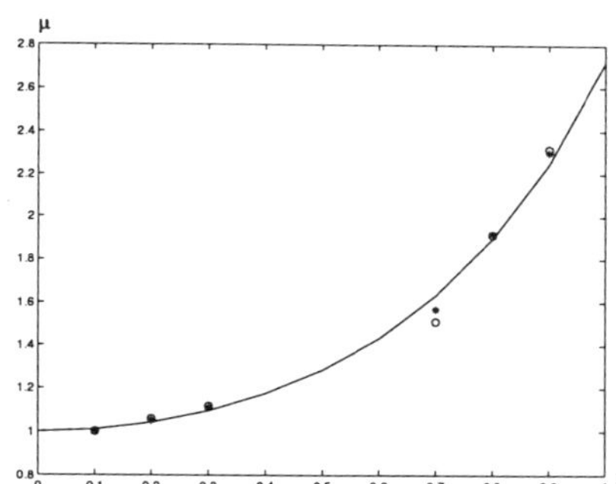

Figure 8: μ

5 Conclusion

Our research shows that to solve the optimization problem from inverse problem based on solving direct problems by FDM and RBF can get the explicit expression of gradient for the cost function and obtain satisfactory results. However, there are still some questions to be studied, such as the existence, uniqueness and stabilities etc.

References

1. Y.C.Hon & Zongmin Wu "A Numerical Computation for Inverse Boundary Determination Problem"
2. Masahiro YAMAMOTO and Jun Zou "Simultaneous reconstruction of the initial temperature and heat radiative coefficient.
3. M.S.Pilant,W.Rundell.An inverse problem for a nonlinear parabolic equation. Commun Partial Differ Equations.1986,11,445–457

NUMERICAL RECONSTRUCTION OF PIECEWISE CONSTANT POTENTIAL FOR ONE DIMENSIONAL HELMHOLTZ EQUATION

FUMING MA AND FANGFANG SUN

School of Mathematics, Jilin University, Changchun, 130012, P. R. China

E-mail: mfm@mail.jlu.edu.cn

In this paper, we study the numerical method of reconstructing potential of one dimensional Helmholtz equation for given impedance function. First, the properties of impedance function with piecewise constant potential for one dimensional Helmholtz Equation was given. Then, numerical method for reconstructing potential was discussed.

1 Problem

Let us consider one dimensional Helmholtz equation as follows:

$$\phi''(x,k) + k^2(1+q(x))\phi(x,k) = 0, \tag{1}$$

where $\phi(x,k)$ is a complex function, k wave number and potential $q(x)$ real function. Setting

$$n(x) = \sqrt{1+q(x)},$$

we assume that

$$0 < n_0 \leq n(x) \leq n_1.$$

For any complex number k,we are concerned with the solutions $\phi_+(x,k)$ and $\phi_-(x,k)$ of equation (1) which are of the form

$$\phi_+(x,k) = \phi_{inc+}(x,k) + \phi_{scat+}(x,k),$$

$$\phi_-(x,k) = \phi_{inc-}(x,k) + \phi_{scat-}(x,k),$$

where

$$\phi_{inc+}(x,k) = e^{ikx}, \quad \phi_{inc-}(x,k) = e^{-ikx},$$

and $\phi_{scat+}(x,k)$, $\phi_{scat-}(x,k)$ satisfy with the outgoing radiation boundary condition:

$$\phi'_{scat\pm}(0,k) + ik\phi_{scat\pm}(0,k) = 0 \tag{2}$$

and

$$\phi'_{scat\pm}(1,k) - ik\phi_{scat\pm}(1,k) = 0. \tag{3}$$

Here and in sequel of this paper, we denote by $f'(x,k)$ the partial derivative $\frac{\partial f}{\partial x}$ for any function $f(x,k)$.

Let

$$C^+ = \{k \in C | \mathrm{Im}(k) \geq 0\}.$$

For any k and $\phi_+(x,k)$,define the impedance functions $p_+(x,k)$ and $p_-(x,k)$ as follows:

$$p_+(x,k) = \frac{\phi'_+(x,k)}{ik\phi_+(x,k)}, \tag{4}$$

$$p_-(x,k) = \frac{\phi'_-(x,k)}{ik\phi_-(x,k)}, \tag{5}$$

Our problem is: for given impedance function $p_+(0,k)$ for all k, reconstruct potential $q(x)$, $x \in [0,1]$.

In the case $q(x) \in C_0^m[0,1]$ and $m > 2$, Chen and Rokhlin (see Chen and Rokhlin[2]) proved that impedance functions $p_+(x,k)$ and $p_-(x,k)$ are defined well for all $x \in R$ and $k \in C^+$, and they satisfy with the following Riccati equations:

$$p'_+(x,k) = -ik(p_+^2(x,k) - (1+q(x))), \tag{6}$$

$$p'_-(x,k) = ik(p_-^2(x,k) - (1+q(x))). \tag{7}$$

In Chen and Rokhlin[2], equations (6) and (7) are used to numerically reconstruct $q(x)$. The numerical results show that, for sufficiently smooth $q(x)$,the method in Chen and Rokhlin[2] works very well. But in some cases of problems, potential $q(x)$ is discontinues. In this paper we want to extend the method in Chen and Rokhlin[2] to the numerical reconstruction of discontinues $q(x)$.

2 Discussion on Impedance Functions

Let $f(x)$ be the function on $[0,1]$. For any division of $[0,1]$

$$\Delta : 0 = x_0 < x_1 < \cdots < x_n = 1,$$

define

$$V(\Delta) = \sum_{i=1}^{n} |f(x_i) - f(x_{i-1})|.$$

We denote by $TV(f)$ total variation of function defined on $[0,1]$, i.e.,

$$TV(f) = \sup_{\Delta}\{V(\Delta)\}.$$

In this section, we will rewrite impedance function $p_+(x,k)$ as $p(x)$ for reason of simplification.

Consider Riccati equation

$$p'(x) = ik(p^2(x) - (1 + q(x))). \tag{8}$$

we have that

Theorem 1 *Assume that $p(x)$ be the solution of equation (8) satisfying with condition $p(0) = p_0$, $p_0 > 0$, $q(x) \in C[0,1]$, $q(x) > -1$ and for $x \notin [0,1]$, $q(x) = 0$. Then for $TV(\ln(1+q(x))) < +\infty$, $p(x)$ is defined well for $x \in [0,1]$ and*

$$\sup_{x\in[0,1]} \{|p(x)|, |p'(x)|\} \le M < +\infty. \tag{9}$$

proof: Let $\eta(x) = 1 + q(x)$. It is easy to prove that there exist piecewise constant functions $\{q_n(x)\}$, $x \in [0,1]$, $n = 0, 1, \cdots$, such that

$$q_n(x) \to q(x), \quad \text{in} \quad L^1[0,1], \quad n \to \infty,$$

and $TV(\ln(1+q_n(x))) \le \tilde{M} < +\infty$. Denote by $p_n'(x)$ the solutions of equation (8) with $q(x) = q_n(x)$ and $p(0) = p_0$. By using of the method in J.Sylvester[1], we can prove that there exists constant $M > 0$, such that

$$\sup_{x\in[0,1]} \{|p_n(x)|, |p_n'(x)|\} \le M.$$

Finally,from Arzelà-Ascoli theorem and the uniqueness of solution of initial value problem for ordinary differential equations,we can get the estimate (9).

By use of the above theorem, we can prove the following results for impedance functions $p_+(x,k)$ and $p_-(x,k)$:

Theorem 2 *Assume that $q(x)$ be continue on $(-\infty, +\infty)$, $q(x) > -1$, and for $x \notin [0,1]$, $q(x) = 0$. In addition, assume that $TV(\ln(1 + q(x))) < +\infty$. Then for all $k \in (0, +\infty)$ and $x \in [0,1]$, impedance functions $p_+(x,k)$ and $p_-(x,k)$ are defined well, and satisfy with equation (6) and (7).*

Furthermore, we have that

Theorem 3 *Assume that $q(x)$ is piecewise constant function, $q(x) > -1$,and for $x \notin [0,1]$, $q(x) = 0$. Then impedance function $p_+(x,k)$ is defined well for all $k \in (0,+\infty)$ and $x \in [0,1]$. Furthermore, $p_+(x,k)$ is a continue on x and $p_+(x,k)$ satisfies equation (6) at point x if x is not discontinue point of $q(x)$.*

3 Numerical Reconstruction of Potential $q(x)$

In this section, we consider the numerical method to reconstruct potential $q(x)$. Our numerical method is based on the following idea:

To find $q(x)$ by solving the system which consists of Riccati equation

$$p'_+(x,k) = -ik(p_+^2(x,k) - (1+q(x))), \quad x \in [0,1] \tag{10}$$

and

$$\sqrt{1+q(x)} = \lim_{a\to+\infty} \frac{1}{a}\int_0^a Rep_+(x,k)dk, \tag{11}$$

which is from Chen and Rokhlin[2] (Trace theorem),with initial conditions

$$p_+(0,k) = p_0(k), \quad \forall k > 0$$

and

$$q(0) = 0.$$

By this way, we can get the approximation $q_h(x)$ to $q(x)$. After this, we optimize the functional

$$f(q_h) = \int_0^a [p_0(k) - p_0(q_h,k)]^2 dk \tag{12}$$

to get the better approximation of $q(x)$, where $p_0(q_h,k)$ is the impedance function of problem (1)–(3) defined by (4) for $q(x) = q_h(x)$ and can be obtained by solving (1)–(3) numerically.

In the numerical implementation of the above method, we set $p_1(x,k) = \text{Re}p_+(x,k)$, $p_2(x,k) = \text{Im}p_+(x,k)$, so that equation (11) can be reformulated as the system

$$p'_1(x,k) = 2p_1(x,k)p_2(x,k), \tag{13}$$

$$p'_2(x,k) = -k[p_1^2(x,k) - p_2^2(x,k) - (1+q(x))]. \tag{14}$$

For solving numerically this ordinary differential system,we use the following difference scheme

$$p_1^{(j+1)} = \frac{p_1^{(j)}}{1-2khp_2^{(j)}}, \tag{15}$$

$$p_2^{(j+1)} = \frac{p_2^{(j)} - khp_1^{(j+1)}p_1^{(j+1)} + kh(1+q^{(j)})}{1 - khp_2^{(j)}}, \tag{16}$$

where h is difference step-size and $x_j = jh, p_l^{(j)} = p_l(x_j, k)$, $j = 0, 1, \cdots, M, l = 1, 2$. This is an explicit and stable scheme. For computing $q(x)$ numerically from equation (11), we can choose a large enough a, and substitute equation (11) by

$$\sqrt{1+q(x)} = \frac{1}{a}\int_0^a Rep_+(x,k)dk, \tag{17}$$

then,for $l = 1, 2, \cdots, M-1$, get

$$q(x_l) = \sqrt{\frac{2\hat{h}}{a}[\sum_{j=1}^{N-1} p_1(x_l, k_j) + \frac{1}{2}(p_1(x_l, 0) + p_1(x_l, a))]} - 1, \tag{18}$$

where $\hat{h} = a/N$, $k_j = j\hat{h}, j = 0, 1, \cdots, N$, by use of trapezoid formula for integrating (17).

By the above method, we did some numerical experiments for piecewise constant function $q(x)$, numerical results of reconstruction for $q(x)$ are satisfying.

Acknowledgments

This work is partly supported by Special Funds for Major State Basic Research Projects in China (G1999032802) and National Nature Science Foundation of China (Foundation item:10076006).

References

1. J.Sylvester, *A convergent layer stripping algorithm for radially symmetric impedance tomography problem,* Comm. in PDE **17**, 1955–1994(1992).
2. Y.Chen and V.Rokhlin, *On the inverse scattering problem for the Helmholtz equation in one dimension*, Inverse problems **8**, 365–391(1992).

ALGEBRAIC SOLUTION FOR THE INVERSE SOURCE PROBLEM OF THE POISSON EQUATION

T. NARA AND S. ANDO

The University of Tokyo, 7-3-1, Hongo, Bunkyo, Tokyo, 113-0033, JAPAN
E-mail: nara@nii.ac.jp

In this paper, a non-iterative, algebraic method for an inverse source problem of the three-dimensional Poisson equation is proposed. The method is based on the multipole expansion of the potential by point sources. Via the multipole expansion coefficients of the sectoral harmonics and the particular tesseral harmonics, the relations between the source parameters and the surface integral of the boundary data are derived. These relations are reduced into an algebraic equation of N-th degree for N source positions projected onto the xy-plane. The number of the sources N is obtained by the property of the leading principal minors of the Hankel matrix composed of the multipole expansion coefficients of the sectoral harmonics. Stability of our algorithm is analyzed, and a numerical simulation is shown.

1 Introduction

The inverse source problem of the Poisson equation has many important applications in science and engineering, such as estimation of current sources inside the brain from the electric potential or the magnetic field measured on the head surface. So far, many numerical algorithms for estimation of several spatially localized sources have been proposed. Though the most basic method is the iterative algorithm which minimizes the error between the boundary data and the solutions of the direct problem[1], the direct estimation of the point source parameters by the boundary data has been studied as well for acceleration of algorithms or calculation of the initial values for the iterative algorithms. Ohe *et al* [2] proposed the method to estimate the positions of the point sources in the unit circle in two-dimensional space. Assuming that the source strength is known, they derived the equation of N-th degree whose solutions are the positions of the sources. They also showed an algorithm for the estimation of the number of sources[3]. Badia *et al* [4] proposed an explicit algorithm to estimate the source positions as eigen values of a matrix composed of the surface integral of the Cauchy data weighted by harmonic functions. The positions in three-dimensional space, the source strength, and the number of sources with an assumed upper bound can be estimated, though the estimation of the number is unstable as they themselves mentioned.

In this paper, we will derive a relation between the source parameters and the surface integral of the Cauchy data via multipole expansion in Sec.

2. The multipole expansion coefficients of the sectoral harmonics and the particular tesseral harmonics yield the relations between the source positions, strength, and the surface integral of the Cauchy data. It is shown that the relations of the sectoral harmonics are equivalent to the equations used by Badia *et al.*[4]. In Sec. 3, we propose another direct algebraic solution: The N source positions projected onto the xy-plane can be represented as the N solutions of the equation of N-th degree whose coefficients can be expressed by the multipole expansion coefficients of the sectoral harmonics. The source strength and the z-coordinates are also expressed by the projected positions and the multipole expansion coefficients of the sectoral and tesseral harmonics. N is obtained by the leading principal minors of the Hankel matrix composed of the multipole expansion coefficients of the sectoral harmonics. In Sec. 4, the stability of our algorithm is analyzed. A numerical simulation is shown in Sec. 5.

2 Relation Between the Source Parameters and the Boundary Data Via Multipole Expansion Coefficients

Let us consider the three dimensional Poisson equation

$$\Delta V = -f \tag{1}$$

in a bounded domain $G \in \mathbb{R}^3$, where the source term f is assumed to be the point sources:

$$f = \sum_{k=1}^{N} q_k \delta(r - r_k, \theta - \theta_k, \phi - \phi_k), \qquad q_k \neq 0 \ \ (k = 1, 2, ..., N). \tag{2}$$

Our inverse source problem is to estimate the source strength q_k, the source positions r_k, θ_k, ϕ_k, and the number of sources N, from the Cauchy data

$$V|_{\partial G}, \qquad \frac{\partial V}{\partial \nu}|_{\partial G}, \tag{3}$$

where ν is the unit outward normal vector to ∂G.

Let V' be the potential that would exist if the source f were in an infinite medium. Then, it is known[5] that in a multipole expansion of V', at a point outside a sphere which contains G, expressed as

$$V'(r, \theta, \phi) = \frac{1}{4\pi} \sum_{n=0}^{\infty} \sum_{m=0}^{n} (2 - \delta_{m0}) \frac{(n-m)!}{(n+m)!} \cdot \left(\frac{1}{r}\right)^{n+1} P_n^m(\cos\theta)(a_{nm} \cos m\phi + b_{nm} \sin m\phi), \tag{4}$$

the expansion coefficients (multipole coefficients) can be represented by both the surface integral of the boundary data and the source term f as

$$a_{nm} + ib_{nm} = \int_{\partial G} \left(V \frac{\partial}{\partial \nu} \left(r^n P_n^m(\cos\theta) e^{im\phi} \right) - \frac{\partial V}{\partial \nu} r^n P_n^m(\cos\theta) e^{im\phi} \right) dS$$
$$= \int_G f\, r^n\, P_n^m(\cos\theta)\, e^{im\phi}\, dv. \tag{5}$$

When f is the point sources in Eq. (2), Eq. (5) is reduced to the basic relation between the surface integral of the boundary data and the source parameters

$$a_{nm} + ib_{nm} = \int_{\partial G} \left(V \frac{\partial}{\partial \nu} \left(r^n P_n^m(\cos\theta) e^{im\phi} \right) - \frac{\partial V}{\partial \nu} r^n P_n^m(\cos\theta) e^{im\phi} \right) dS$$
$$= \sum_{k=1}^{N} q_k r_k^n P_n^m(\cos\theta_k) e^{im\phi_k} \tag{6}$$

for $n \geq m \geq 0$. Here, we use the sectoral harmonics component; $\alpha_m \equiv a_{mm} + ib_{mm}$, and the tesseral harmonics at $n = m+1$; $\beta_m \equiv a_{m+1,m} + ib_{m+1,m}$, for the estimation of the source parameters. Let

$$\zeta \equiv x + iy, \qquad \zeta_k \equiv x_k + iy_k, \tag{7}$$

then by substituting

$$r^m P_m^m(\cos\theta) e^{im\phi} = (2m-1)!!\zeta^m, \; r^{m+1} P_{m+1}^m(\cos\theta) e^{im\phi} = (2m+1)!!\zeta^m z, \tag{8}$$

into the Eq. (6), we obtain

$$\alpha_m = \int_{\partial G} \left(V \frac{\partial \zeta^m}{\partial \nu} - \frac{\partial V}{\partial \nu} \zeta^m \right) dS = \sum_{k=1}^{N} q_k \zeta_k^m \tag{9}$$

$$\beta_m = \int_{\partial G} \left(V \frac{\partial \zeta^m z}{\partial \nu} - \frac{\partial V}{\partial \nu} \zeta^m z \right) dS = \sum_{k=1}^{N} q_k \zeta_k^m z_k \tag{10}$$

The multipole expansion coefficient of $n = m$ in Eq. (9) is equivalent to the equation used by Badia *et al.*[4]. Badia *et al.* also mentioned the use of the polynomial $zQ(x, y)$, where Q is a harmonic polynomial. The multipole expansion coefficient of $n = m+1$ in Eq. (10) corresponds to this polynomial, though they used the other $Q(x, y)$ in the numerical simulation[6]. In the next section, we derive another explicit representation of the source parameters. We propose a stable algorithm for the estimation of the number of sources N.

3 Direct Representation of the Source Parameters

3.1 Explicit Expression of Positions and Strength

It is remarkable that the following linear relation between α_i, α_{i-1}, ..., α_{i-N} holds for $i \geq N$:

$$\begin{aligned}\alpha_i &= \sum_{k=1}^{N} q_k \zeta_k^i \\ &= (\zeta_1 + ... + \zeta_N) \sum_{k=1}^{N} q_k \zeta_k^{i-1} - (\zeta_1\zeta_2 + ... + \zeta_{N-1}\zeta_N) \sum_{k=1}^{N} q_k \zeta_k^{i-2} \\ &\quad + ... + (-1)^{N-1} \zeta_1 ... \zeta_N \sum_{k=1}^{N} q_k \zeta_k^{i-N} \\ &= K_1 \alpha_{i-1} + K_2 \alpha_{i-2} + ... + K_N \alpha_{i-N},\end{aligned} \tag{11}$$

where K_j is the j-th elementary symmetric polynomial of $\zeta_1, \zeta_2, ..., \zeta_N$:

$$K_j = (-1)^{j-1} \sum_{k_1 < k_2 < ... < k_j} \zeta_{k_1} \zeta_{k_2} \cdots \zeta_{k_j}. \tag{12}$$

Thus, by Eq. (11) for $i = N, N+1, ..., 2N-1$, we obtain

$$\begin{pmatrix} \alpha_N \\ \vdots \\ \alpha_{2N-1} \end{pmatrix} = \begin{pmatrix} \alpha_0 & \alpha_1 & \ldots & \alpha_{N-1} \\ \alpha_1 & \alpha_2 & \ldots & \alpha_N \\ \vdots & \vdots & \vdots & \vdots \\ \alpha_{N-1} & \alpha_{N-2} & \ldots & \alpha_{2N-2} \end{pmatrix} \begin{pmatrix} K_N \\ \vdots \\ K_1 \end{pmatrix}. \tag{13}$$

The estimation of the number N of sources is shown in the next section. Now, let the $m \times m$ Hankel matrix composed of α_0 to α_{2m-2} denote

$$H_m \equiv \begin{pmatrix} \alpha_0 & \ldots & \alpha_{m-2} & \alpha_{m-1} \\ \alpha_1 & \ldots & \alpha_{m-1} & \alpha_m \\ \vdots & \vdots & \vdots & \vdots \\ \alpha_{m-1} & \ldots & \alpha_{2m-3} & \alpha_{2m-2} \end{pmatrix}, \tag{14}$$

then the following lemma holds.

Lemma 1

$$\det H_m = \sum_{1 \leq l_1 < ... < l_m \leq N} \prod_{k=1}^{m} q_{l_k} \prod_{l_i > l_j} (\zeta_{l_i} - \zeta_{l_j})^2 \tag{15}$$

Proof: Let $\boldsymbol{w}_{m,k} \equiv (1\ \zeta_k\ \zeta_k^2\ ...\ \zeta_k^{m-1})^T$, ($T$: transposition). Then

$$H_m = (\sum_{k_1=1}^{N} q_{k_1}\zeta_{k_1}^0 \boldsymbol{w}_{m,k_1} \quad \sum_{k_2=1}^{N} q_{k_2}\zeta_{k_2}^1 \boldsymbol{w}_{m,k_2} \ \cdots \ \sum_{k_m=1}^{N} q_{k_m}\zeta_{k_m}^{m-1} \boldsymbol{w}_{m,k_m}). \tag{16}$$

Thus,

$$\begin{aligned}
\det H_m &= \sum_{k_1=1}^{N}\sum_{k_2=1}^{N}\cdots\sum_{k_m=1}^{N} q_{k_1}\cdots q_{k_m}\zeta_{k_1}^0\cdots\zeta_{k_m}^{m-1}\det(\boldsymbol{w}_{m,k_1}\ \ ...\ \ \boldsymbol{w}_{m,k_m}) \\
&= \sum_{\substack{k_1,\ldots,k_m \\ k_i\neq k_j}} q_{k_1}\cdots q_{k_m}\zeta_{k_1}^0\cdots\zeta_{k_m}^{m-1}\det(\boldsymbol{w}_{m,k_1}\ ...\ \boldsymbol{w}_{m,k_m}) \\
&= \sum_{1\le l_1<l_2<\cdots<l_m\le N}\ \sum_{\substack{\{k_1,\ldots,k_m\} \\ =\{l_1,\ldots,l_m\}}} q_{k_1}\cdots q_{k_m}\zeta_{k_1}^0\cdots\zeta_{k_m}^{m-1}\det(\boldsymbol{w}_{m,k_1}\ ...\ \boldsymbol{w}_{m,k_m}) \\
&= \sum_{1\le l_1<l_2<\cdots<l_m\le N}\ \prod_{k=1}^{m} q_{l_k}\prod_{l_i>l_j}(\zeta_{l_i}-\zeta_{l_j})\ \times \\
&\qquad \sum_{\substack{\{k_1,\ldots,k_m\} \\ =\{l_1,\ldots,l_m\}}} \mathrm{sgn}\begin{pmatrix} k_1 & k_2 & \ldots & k_m \\ l_1 & l_2 & \ldots & l_m\end{pmatrix}\zeta_{k_1}^0\cdots\zeta_{k_m}^{m-1} \\
&= \sum_{1\le l_1<l_2<\cdots<l_m\le N}\ \prod_{k=1}^{m} q_{l_k}\prod_{l_i>l_j}(\zeta_{l_i}-\zeta_{l_j})^2 \qquad \text{Q.E.D.}
\end{aligned} \tag{17}$$

Corolloary 1

$$\det H_m = \begin{cases} \displaystyle\prod_{k=1}^{N} q_k \prod_{i>j}(\zeta_i-\zeta_j)^2 & (m=N) \\ 0 & (m\ge N+1)\end{cases} \tag{18}$$

Assuming that all the projected positions, ζ_k ($k = 1, 2, ..., N$), are different on the xy-plane, $\det H_N \neq 0$ by Corollary 2. Therefore, Eq. (13) can be solved for $K_1, K_2, ..., K_N$ as

$$\begin{pmatrix} K_N \\ \vdots \\ K_1 \end{pmatrix} = H_N^{-1} \begin{pmatrix} \alpha_N \\ \vdots \\ \alpha_{2N-1} \end{pmatrix} = \begin{pmatrix} \alpha_0 & ... & \alpha_{N-2} & \alpha_{N-1} \\ \alpha_1 & ... & \alpha_{N-1} & \alpha_N \\ \vdots & \vdots & \vdots & \vdots \\ \alpha_{N-1} & ... & \alpha_{2N-3} & \alpha_{2N-2} \end{pmatrix}^{-1} \begin{pmatrix} \alpha_N \\ \vdots \\ \alpha_{2N-1} \end{pmatrix}. \tag{19}$$

Since $K_1, ..., K_N$ are defined as the elementary symmetric polynomials as in Eq. (12), $\zeta_1, \zeta_2, ..., \zeta_N$ are the solutions of the equation of N-th degree

$$\zeta^N - K_1\zeta^{N-1} - K_2\zeta^{N-2} - ... - K_N = 0, \tag{20}$$

where $K_1, K_2, ..., K_N$ are given by Eq. (19). In this way, the algebraic equation of N-th degree whose N solutions are the source positions projected onto the xy-plane was derived.

To obtain the source strength and the z-coordinates of the sources, the simultaneous equations of Eq. (9) and Eq. (10) for $m = 0, 1, ..., 2N - 1$

$$\begin{pmatrix} \alpha_0 \\ \alpha_1 \\ \vdots \\ \alpha_{2N-1} \end{pmatrix} = \begin{pmatrix} 1 & 1 & ... & 1 \\ \zeta_1 & \zeta_2 & ... & \zeta_N \\ \vdots & \vdots & \ddots & \vdots \\ \zeta_1^{2N-1} & \zeta_2^{2N-1} & ... & \zeta_N^{2N-1} \end{pmatrix} \begin{pmatrix} q_1 \\ q_2 \\ \vdots \\ q_N \end{pmatrix}, \tag{21}$$

$$\begin{pmatrix} \beta_0 \\ \beta_1 \\ \vdots \\ \beta_{2N-1} \end{pmatrix} = \begin{pmatrix} 1 & 1 & ... & 1 \\ \zeta_1 & \zeta_2 & ... & \zeta_N \\ \vdots & \vdots & \ddots & \vdots \\ \zeta_1^{2N-1} & \zeta_2^{2N-1} & ... & \zeta_N^{2N-1} \end{pmatrix} \begin{pmatrix} q_1 z_1 \\ q_2 z_2 \\ \vdots \\ q_N z_N \end{pmatrix} \tag{22}$$

are used. Since the matrix composed of $\zeta_1, \zeta_2, ..., \zeta_N$ in Eq. (21) and Eq. (22) is the Vandermonde matrix, we obtain

$$\begin{pmatrix} q_1 \\ \vdots \\ q_N \end{pmatrix} = \begin{pmatrix} 1 & ... & 1 \\ \vdots & \ddots & \vdots \\ \zeta_1^{N-1} & ... & \zeta_N^{N-1} \end{pmatrix}^{-1} \begin{pmatrix} \alpha_0 \\ \vdots \\ \alpha_{N-1} \end{pmatrix}, \tag{23}$$

$$\begin{pmatrix} z_1 \\ \vdots \\ z_N \end{pmatrix} = \begin{pmatrix} \frac{1}{q_1} & & 0 \\ & \ddots & \\ 0 & & \frac{1}{q_N} \end{pmatrix} \begin{pmatrix} 1 & ... & 1 \\ \vdots & \ddots & \vdots \\ \zeta_1^{N-1} & ... & \zeta_N^{N-1} \end{pmatrix}^{-1} \begin{pmatrix} \beta_0 \\ \vdots \\ \beta_{N-1} \end{pmatrix}. \tag{24}$$

Our algorithm is summarized by the following theorem.

Theorem 1 *Assuming that the projected source positions $\zeta_1, \zeta_2, ..., \zeta_N$ are different on the xy-plane, they are the solutions of Eq. (20) of N-th degree, where the coefficients of the equation are expressed by Eq. (19). The source strength and the z-coordinates of the source positions are represented by Eq. (23) and Eq. (24), respectively, in terms of the surface integral of the boundary data α_m, β_m defined in Eq. (9) and Eq. (10).*

Especially when $N \leq 4$, we obtain the completely explicit expression of the projected positions in terms of α_m.

3.2 Estimation of the Number N

By the corollary 1, we propose to estimate the number of sources by the following procedure:

1. Compute the leading principal minor of H_m for $m = 1, 2, ...$
2. k which satisfies
$$\det H_k \neq 0, \;\; \det H_{k+1} = 0 \tag{25}$$
is a candidate for the number of sources.
3. Confirm
$$\det H_{k+1} = \det H_{k+2} = ... = \det H_M = 0 \tag{26}$$

If we know N is bounded by a known integer M as Badia *et al.* assumed, we confirm Eq. (26) until the upper bound M. However, in practical engineering applications where data is perturbed by noise, M is determined rather by the accuracy of the measurement of the data. The stability of our algorithm is mentioned in Sec. 4.

4 Stability

Since the elementary symmetric polynomials of the source positions projected onto the xy-plane are obtained by Eq. (19), we should calculate the condition number of H_N to analyze the stability of our algorithm for estimating the positions. Now, the condition number of H_N can be evaluated as

$$\text{cond}(H_k) < \frac{\max_k\{|q_k|\}}{\min_k\{|q_k|\}} \frac{N^2}{\delta^{2(N-1)}} \left(\frac{1-\zeta^N}{1-\zeta} \left((1+\zeta)^N - \zeta^N\right) \right)^2 \tag{27}$$

where $\zeta \equiv \max_k |\zeta_k|$ and $\delta \equiv \min |\zeta_i - \zeta_j|$. The proof of Eq. (27) is as follows: First,

$$
\begin{aligned}
||H_N||_1 = \sum_{j=0}^{N-1}\sum_{i=0}^{N-1} |\alpha_{i+j}| \quad &= \quad \sum_{j=0}^{N-1}\sum_{i=0}^{N-1} \left| \sum_{k=1}^{N} q_k \zeta_k^{i+j} \right| \\
\leq \max_k\{|q_k|\} N \sum_{j=0}^{N-1}\sum_{i=0}^{N-1} \zeta^{i+j} \quad &\leq \quad \max_k\{|q_k|\} N \left(\frac{1-\zeta^N}{1-\zeta} \right)^2 .
\end{aligned} \tag{28}
$$

Second, to calculate $||H_N^{-1}||_1$, let $\tilde{H}_N$ be the adjugate matrix of H_N and $\boldsymbol{v}_{i,k} \equiv (1\ \zeta_k\ \dots\ \zeta_k^{i-2}\ \zeta_k^{i}\ \dots\ \zeta_k^{N-1})^T$. Then, as in the derivation of Eq. (17),

$$
\left(\tilde{H}_N\right)_{ij} = \sum_{1 \leq l_1 < l_2 < \dots < l_{N-1} \leq N} \prod_{k=1}^{N-1} q_{l_k} \prod_{l_i > l_j} (\zeta_{l_i} - \zeta_{l_j})^2 K_{N-j} K_{N-i}, \tag{29}
$$

where K_{N-j}, K_{N-i} is the $N-j, N-i$ th symmetric polynomial of ζ_k. Thus,

$$
\begin{aligned}
\left| \left(H_N^{-1}\right)_{ij} \right| &= \left| \frac{\left(\tilde{H}_N\right)_{ij}}{\det H_N} \right| \\
&= \left| K_{N-j} K_{N-i} \sum_{l=1}^{N} \frac{1}{q_l \prod_{l \neq i} (\zeta_l - \zeta_i)^2} \right| \\
&\leq \frac{N |K_{N-j}| |K_{N-i}|}{\min_k \{|q_k|\} \cdot \delta^{2(N-1)}}.
\end{aligned} \tag{30}
$$

Therefore,

$$
\begin{aligned}
||H_N^{-1}||_1 &\leq \frac{N}{\min_k \{|q_k|\} \cdot \delta^{2(N-1)}} \left(\sum_{i=1}^{N} \binom{N}{i} \zeta^{N-i} \right)^2 \\
&= \frac{N \left((1+\zeta)^N - \zeta^N \right)^2}{\min_k \{|q_k|\} \cdot \delta^{2(N-1)}}
\end{aligned} \tag{31}
$$

By Eq. (28) and Eq. (31), we obtain Eq. (27). The proof is completed.

By Eq. (27), if we can obtain the accuracy of α_k ($k = 0, 1, \dots, 2N$) according to dispersion of the source strength and closeness of the source positions, the relative error of $K_1, K_2, \dots, K_N$ is guaranteed to be small, which results in the small relative error of the projected source positions.

In the estimation of the number of the sources, we should consider 1) the evaluation of the ratio $|\det H_{k+1}|/|\det H_k|$ in Eq. (25) and 2) the upper limitation M until which the leading principal minors are confirmed to be zero in Eq. (26). As for 1), $|\det H_{k+1}|/|\det H_k|$ becomes zero theoretically by Corollary 1, but it does not exactly equal zero when data is perturbed by noise. However,

$$\frac{|\det H_{N+1}|}{|\det H_N|} = |\det \left(\begin{array}{c|c} H_N & \boldsymbol{v} \\ \hline \boldsymbol{v}^T & \alpha_{2N} \end{array} \right) / \det H_N| = |\alpha_{2N} - \boldsymbol{v}^T H_N^{-1} \boldsymbol{v}| \quad (32)$$

where $\boldsymbol{v}^T = (\alpha_N \ \alpha_{N+1} \ ...\alpha_{2N-1})$. Since the condition number of H_k is evaluated in Eq. (27), we find that $|\det H_{N+1}|/|\det H_N|$ becomes sufficiently close to zero if $|\delta\alpha_{2N}|/|\alpha_{2N}|$ is small according to the dispersion of the source positions and the strength. As for 2), our algorithm uses α_0 to α_{2N-1}. Thus, M such that

$$|\delta\alpha_{2M}/\alpha_{2M}| \simeq 1. \quad (33)$$

is the upper limitation to be estimated. M increases as the accuracy of the data increases.

5 Numerical Simulation

Numerical simulation in the case $N = 3$ is shown. The source positions P_1, P_2, P_3 in the Cartesian coordinates and the source strength q_1, q_2, q_3 are set as shown in Table 1. The region G is assumed to be a unit sphere. The data is measured at the intersection points of 20 meridians and 15 parallels on ∂G. The standard deviation of the noise is 1% of the maximum value of the potential on ∂G.

Figure 1 Left shows the relative error of α_m calculated from the data with noise. When $2M = 10$, $|\delta\alpha_{2M}/\alpha_{2M}| \simeq 1$. Thus, the upper limitation of the number of sources that can be estimated under the noise level is around $M = 5$. To determine the number of sources, the leading principal minor of H_m are calculated for $m = 2, 3, \ldots$. (Note that in the present situation $\sum_{k=1}^{3} q_k = 0$, $\det H_1 = \alpha_0 = 0$.) The results are shown in Figure 1 Right. $|\det H_4|$ decreases suddenly to zero, and $|\det H_M| \simeq 0$ till $M = 5$, which enables us to estimate the number of the sources $N = 3$. The estimated positions and strength are shown in Table 1.

6 Conclusion

In this paper, we proposed an algorithm for the reconstruction of the positions, the strength, and the number of point sources in the three-dimensional Poisson equation from the boundary data. The N source positions projected onto the xy-plane are expressed by the solutions of the equation of N-th degree. The number of sources, N, is estimated by the property of the leading principal minors of the Hankel matrix composed of the multipole expansion coefficients of the sectoral harmonics.

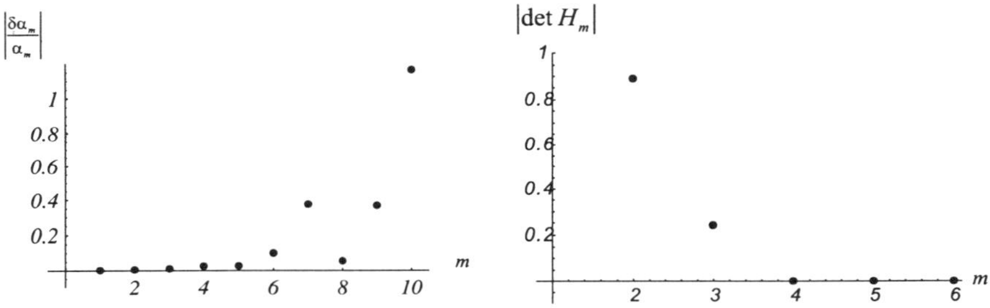

Figure 1. Left: The relative error of $|\delta\alpha_m/\alpha_m|$ Right: $|\det H_m|$

Table 1. The real and estimated sources.

	P_1	P_2	P_3
Real source positions (x, y)	$(0.5, 0.3)$	$(-0.4, 0.6)$	$(-0.5, -0.3)$
Estimated by theoretical data	$(0.499, 0.300)$	$(-0.403, 0.596)$	$(-0.501, -0.297)$
Estimated by noisy data	$(0.494, 0.293)$	$(-0.395, 0.571)$	$(-0.515, -0.290)$
	z_1	z_2	z_3
Real source positions (z)	0.5	-0.4	0.1
Estimated by theoretical data	$0.496 + 0.002i$	$-0.397 - 0.002i$	$0.105 + 0.000i$
Estimated by noisy data	$0.483 + 0.002i$	$-0.374 - 0.005i$	$-0.119 + 0.002i$
	q_1	q_2	q_3
Real source strength	1	-0.6	-0.4
Estimated by theoretical data	$1.003 - 0.001i$	$-0.602 + 0.006i$	$-0.400 - 0.004i$
Estimated by noisy data	$1.004 - 0.005i$	$-0.590 + 0.032i$	$-0.413 - 0.027i$

References

1. K.Ohnaka and K.Uosaki, Int. J. Control **49**, 119(1989).

2. T.Ohe and K.Ohnaka,*A precise estimation method for locations in an inverse logarithmic potential problem for point mass models*, Appl. Math. Modelling **18(8)**, 446–452(1994).
3. T.Ohe and K.Ohnaka, *An estimation method for the number of point masses in an inverse logarithmic potential problem using discrete Fourier transform*, Appl. Math. Modelling **19(7)**, 429–436(1995).
4. A.El.Badia and T.Ha-Duong, *An inverse source problem in potential analysis*, Inverse Problems **16(3)**, 651–663(2000).
5. D.B.Geselowitz, *Multipole Representation for an Equivalent Cardiac Generator*, Proc. IRE **48**, 75 (1960).
6. M.Chafik, A.El.Badia and T.Ha-Duong, in *Inverse Problems in Engineering Mechanics II*, ed. M.Tanaka and G.S.Dulikravich (Elsevier Science Publishers, 2000).

IDENTIFYING PARAMETERS OF LINEAR STOCHASTIC DIFFERENTIAL EQUATIONS FROM INCOMPLETE NOISY MEASUREMENTS

I. V. SEMOUSHIN

School of Mathematics and Mechanics, Ulyanovsk State University, 42 Leo Tolstoy Street, 432970 Ulyanovsk, Russia

E-mail: isem@ulsu.ru

The problem taken as a whole is stated in the paper as follows: Given a system of linear stochastic differential equations with some unknown parameters and unknown stochastic inputs, determine the estimates of these parameters from the incomplete noisy measurements of the solution to the equation. The purpose of the paper is to present a possible numerical solution to this inverse mathematical problem.

1 Problem Statement

Let a linear stochastic differential equation be given in the form

$$dx(t^c)/dt^c = F_c(t^c)x(t^c) + \Psi_c(t^c)a + \Xi_c(t^c)b + \Sigma_c(t^c)s(t^c) \tag{1}$$

with the state $x(t^c)$ ($x \in \mathbf{R}^{n_x}$), two constant parameters a ($a \in \mathbf{R}^{n_a}$) and b ($b \in \mathbf{R}^{n_b}$), and a stochastic process $s(t^c)$ ($s \in \mathbf{R}^{n_s}$) generated by

$$ds(t^c) = \Gamma_c s(t^c)dt^c + dW(t^c) \tag{2}$$

from a vector-valued Wiener process $W(t^c)$ of some positive definite diffusion Q_c ($Q_c > 0$) for all t^c, matrices Γ_c and Q_c are constant. In Eq. (1) and Eq. (2), index c stands for 'continuous' showing that matrices F_c, Ψ_c, Ξ_c, Σ_c, Γ_c and Q_c belong to the continuous-time system described by these equations. System matrices F_c, Ψ_c, Ξ_c and Σ_c may depend of continuous time t^c and in such a case they are composed of piecewise continuous functions of t^c.

Eq. (1) is propagated forward from the initial state $x(t_0^c)$. For any particular case, $x(t_0^c)$ assumes a specific value, which may not be known precisely a priory, and so, $x(t_0^c)$ is modelled as a random vector that is normally distributed with the mean $\hat{x}(t_0^c)$ and covariance P_0:

$$\begin{gathered} E\{x(t_0^c)\} = \hat{x}(t_0^c) \\ E\{[x(t_0^c) - \hat{x}(t_0^c)][x(t_0^c) - \hat{x}(t_0^c)]^T\} = P_0 \end{gathered} \tag{3}$$

Assume the following *uncertainty conditions*: In the above relations unknown are statistics $\hat{x}(t_0^c)$ and P_0, vectors a and b, and matrices Ξ_c, Σ_c, Γ_c and Q_c. Only matrices $F_c(t^c)$ and $\Psi_c(t^c)$ are assumed to be known for all t^c.

Although both parameters a and b are assumed to be unknown, only one of them, namely a, is to be estimated from incomplete noisy measurements of solution $x(t^c)$. Measurements are represented by a vector $z \in \mathbf{R}^m$ available at discrete points $t_1^c, t_2^c, \dots t_N^c$ equally spaced in time t^c so that $t_{t+1}^c - t_t^c = \tau$, and are modelled by the relation

$$z(t_t^c) = H(t_t^c)x(t_t^c) + v(t_t^c) \tag{4}$$

with $H(t_t^c)$ a known m-by-n measurement matrix and $v(t_i^c)$ an m-vector discrete-time noise with the known statistics (for all t_t^c):

$$\begin{aligned} E\{v(t_t^c)\} &= 0 \\ E\{v(t_i^c)v(t_j^c)^T\} &= \begin{cases} R(t_i^c) > 0 & t_i^c = t_j^c \\ 0 & t_i^c \neq t_j^c \end{cases} \end{aligned} \tag{5}$$

(For simplicity we take $t_0^c = 0$, then $t_t^c = t\tau$.) Measurements Eq. (4) are called incomplete because $\forall t_i^c : \text{rank} H(t_i^c) < n$. One sample of measurement sequence $\{z(t_t^c, \omega)\}$, where ω is a point of a fundamental sample space Ω, provides measurement history $\{z(t_t^c, \omega_j) = z_t,\ t = 1, 2, \dots N\}$ formed by the measurement numbers that become available at time t_t^c for some $\omega_j \in \Omega$.

Problem: Given the sample of measurement sequence, it is necessary to identify parameter a with a prescribed accuracy under the uncertainty conditions.

2 Discrete-time Model

Let the sample period τ be short compared to the system's natural transients, then a first order approximation to the standard discrete-time model of system Eq. (1), Eq. (2) can be used [1]. So we have the model

$$\begin{aligned} x_{t+1} &= \Phi_t x_t + \Psi_t a + \Xi_t b + \Sigma_t c_t \\ c_{t+1} &= \Lambda c_t + w_t \end{aligned} \tag{6}$$

which is completely defined in discrete time $t = 0, 1, \dots N$ by the relations

$$\begin{aligned} \Phi_t &= I + \tau F_c(t\tau), & \Psi_t &= \tau \Psi_c(t\tau) \\ \Xi_t &= \tau \Xi_c(t\tau), & \Sigma_t &= \tau \Sigma_c(t\tau) Q^{1/2} \\ \Lambda &= Q^{-1/2}(I + \tau \Gamma_c) Q^{1/2}, & Q &= \tau Q_c \\ c_t &= Q^{-1/2} s(t\tau), & w_t &= Q^{-1/2}[W(t_{t+1}^c) - W(t_t^c)] \end{aligned} \tag{7}$$

where random vectors w_t are taken from the standard independent Gaussian sequence, i.e.

$$w_t \sim \mathcal{N}(0, I), \qquad E\{w_t w_{t'}^T\} = \begin{cases} I & t = t' \\ 0 & t \neq t' \end{cases} \tag{8}$$

and, of characteristics (7), only Φ_t and Ψ_t are known.

Thus, linear stochastic difference equation Eq. (6) motivated by discrete-time measurements Eq. (4) has been built. Here we used a square root $Q^{1/2}$ which can be find from Cholesky decomposition of matrix Q.

Next, we apply Cholesky decomposition (of LDL^T or UDU^T type this time) to matrix $R(t_i^c)$ in Eq. (5), in order to replace Eq. (4) by the discrete-time measurement model

$$z_t = H_t x_t + v_t \tag{9}$$

with some known matrix H_t and noise characteristics

$$v_t \sim \mathcal{N}(0, R_t), \quad E\{v_t v_{t'}^T\} = \begin{cases} R_t = \text{diag}(r_t^{(1)}, \dots r_t^{(m)}) & t = t' \\ 0 & t \neq t' \end{cases} \tag{10}$$

By Eqs. (6) and (9) together with (7) and (8) and (10), the discrete-time model is fully determined in the form characteristic for the Kalman filtering theory. Summands $\Xi_t b$ and $\Sigma_t c_t$ in Eq. (6) represent the systematic (non-random) component and, correspondingly, random component of model uncertainty, so the theory can not be used directly.

3 Adaptive Filter-Identifier, *AFI*

Let us introduce an augmented vector $y_t^T = [x_t^T \mid a_t^T]$ with $a_t \equiv a$ for all t and voluntarily consider the sum $\Xi_t b + \Sigma_t c_t$ in Eq. (6) as a result of passing the noise w_t through an artificial matrix Σ_t^a thought of as a predesigned matrix in the interests of the appropriate filter-identifier building. Equating Σ_t^a to Σ_t of Eq. (6) implies that c_t is artificially replaced by w_t. Sometimes it looks quite natural, and we do so in Section 5. Hence, as a basis for the Kalman filter construction, we use the following equations:

$$y_{t+1} = \Phi_t^a y_t + \Gamma_t^a w_t, \quad y \in \mathbf{R}^{n_F}, \quad n_F = n_x + n_a \tag{11}$$

with

$$\Phi_t^a = \begin{bmatrix} \Phi_t & \Psi_t \\ 0 & I \end{bmatrix}, \quad \Gamma_t^a = \begin{bmatrix} \Sigma_t^a \\ 0 \end{bmatrix} \tag{12}$$

To estimate vector y_t, we have a lot of numerically stable versions of the Kalman algorithm to choose from Bierman [2].

Having chosen one of versions, for example, Potter's mechanization, we have to do the second step: to provide a means for adaptivity of the algorithm, in other words, to accommodate the algorithm to uncertainty in the real data model Eq. (6) and Eq. (9). After wide range comparative study of many approaches to adaptive filtering (e.g. Mehra [3]), we choose the method of fictitious noise introduced into the algorithm (cf. Kaufman *et al* [4]). With this modification, write down Potter's algorithm in two consequent steps.

(i) Time propagation ($t = 0, 1, \ldots$):

$$\hat{y}_{t+1}^{-} = \Phi_t^a \hat{y}_t^{+}, \quad P_{t+1}^{-} = \Phi_t^a P_t^{+} \Phi_t^{aT} + \Gamma_t^a \Gamma_t^{aT} + G q_t G^T$$

where $G = \text{diag}\ \{g_i\}$ is a pre-selected matrix and q_t is a covariance of a fictitious noise introduced into filter at time t. To ensure numerical stability, square root mechanization $P^{-} = S^{-}(S^{-})^T$, $P^{+} = S^{+}(S^{+})^T$ is used:

$$\begin{bmatrix} (S_{t+1}^{-})^T \\ 0 \end{bmatrix} = \mathrm{T} \begin{bmatrix} (S_t^{+})^T \Phi_t^{aT} \\ B_t^T \end{bmatrix}, \qquad B_t = \text{diag}\ \{b_i\}_t$$

where S^{-} and S^{+} are the low triangular matrices and T denotes the modified Gram-Schmidt triangularization. The diagonal entries of B_t at time t may be chosen in different ways. The simplest way would be:

$$\forall i, t: \quad \{b_i\}_t = \left[\, g_i \sqrt{q_t} \,\right]$$

(ii) Scalar measurement update ($t = 1, 2, \ldots$).

1. Set initial values:

$$\hat{y}_t^{(0)} = \hat{y}_t^{-}, \quad \hat{S}_t^{(0)} = \hat{S}_t^{-}, \quad \delta_t^{(0)} = 0$$

2. For $k = 1, 2, \ldots, m$ where m is a dimension of the measurement vector z_t, $h^{(k)}$ is the k-th row of matrix H_t and $z_t^{(k)}$ is the k-th element of z_t, compute:

$$\begin{aligned}
f_t^{(k)} &= S_t^{(k-1)} (h^{(k)})^T \\
\alpha_t^{(k)} &= 1/[(f_t^{(k)})^T f_t^{(k)} + r_t^{(k)}] \\
\gamma_t^{(k)} &= 1/[1 + \sqrt{\alpha_t^{(k)} r_t^{(k)}}] \\
K_t^{(k)} &= S_t^{(k-1)} f_t^{(k)} \alpha_t^{(k)} \\
S_t^{(k)} &= S_t^{(k-1)} - \gamma_t^{(k)} K_t^{(k)} (f_t^{(k)})^T \\
\nu_t^{(k)} &= z_t^{(k)} - (h^{(k)})^T y_t^{(k-1)} \\
\hat{y}_t^{(k)} &= \hat{y}_t^{(k-1)} + K_t^{(k)} \nu_t^{(k)} \\
\delta_t^{(k)} &= \delta_t^{(k-1)} + (\nu_t^{(k)})^2 \alpha_t^{(k)}
\end{aligned}$$

3. Obtain results of Step (ii):

$$\delta_t = (1/m)\delta_t^{(m)} - 1, \qquad \hat{y}_t^{+} = \hat{y}_t^{(m)},\ S_t^{+} = S_t^{(m)}$$

As a measure for filter optimality, δ_t was introduced in Semoushin[5]. The new what we make now is an adaptive mechanism to determine value q_t at Step (i).

Adaptive filter mechanism:
Compute two values

$$\bar{\delta}_t = (1/t)\sum_{j=1}^{t}\alpha^{t-j}\delta_j, \qquad s_t = \sqrt{m/(2t)}\sum_{j=1}^{t}\alpha^{t-j}\delta_j$$

where $\alpha \in (0,1)$, and afterwards obtain $\sqrt{q_t}$ according to one of the following 15 formulae (NAD stands for 'Number of adaptation formula').

$$NAD = 1: \quad \sqrt{q_t} = \gamma|\bar{\delta}_\tau|$$

$$NAD = 2: \quad \sqrt{q_t} = \gamma|\bar{\delta}_t|$$

$$NAD = 3: \quad \sqrt{q_t} = \begin{cases} \gamma|\bar{\delta}_\tau| & \text{if } |s_\tau| \geq \eta \\ \sqrt{q_{t-1}} & \text{otherwise} \end{cases}$$

$$NAD = 4: \quad \sqrt{q_t} = \begin{cases} \gamma|\bar{\delta}_t| & \text{if } |s_\tau| \geq \eta \\ \sqrt{q_{t-1}} & \text{otherwise} \end{cases}$$

$$NAD = 5: \quad \sqrt{q_t} = \gamma|s_\tau|$$

$$NAD = 6: \quad \sqrt{q_t} = \gamma|s_t|$$

$$NAD = 7: \quad \sqrt{q_t} = \alpha\sqrt{q_\tau} + \gamma(|\bar{\delta}_\tau| - |\bar{\delta}_{\tau-T}|)$$

$$NAD = 8: \quad \sqrt{q_t} = \alpha\sqrt{q_\tau} + \gamma|\bar{\delta}_\tau| \text{ sign } (|\bar{\delta}_\tau| - |\bar{\delta}_{\tau-T}|)$$

$$NAD = 9: \quad \sqrt{q_t} = \alpha\sqrt{q_{t-1}} + \gamma(|\bar{\delta}_t| - |\bar{\delta}_{t-1}|)$$

$$NAD = 10: \quad \sqrt{q_t} = \alpha\sqrt{q_\tau} + \gamma|\bar{\delta}_\tau| \text{ sign } (|s_\tau| - |s_{\tau-T}|)$$

$$NAD = 11: \quad \sqrt{q_t} = \alpha\sqrt{q_{t-1}} + \gamma|\bar{\delta}_t| \text{ sign } (|s_t| - |s_{t-1}|)$$

$$NAD = 12: \quad \sqrt{q_t} = \alpha\sqrt{q_\tau} + \gamma(|s_\tau| - |s_{\tau-T}|)$$

$$NAD = 13: \quad \sqrt{q_t} = \alpha\sqrt{q_{t-1}} + \gamma|\bar{\delta}_t| \text{ sign } (|\bar{\delta}_t| - |\bar{\delta}_{t-1}|)$$

$$NAD = 14: \quad \sqrt{q_t} = \alpha\sqrt{q_{t-1}} + \gamma(|s_t| - |s_{t-1}|)$$

$$NAD = 15: \quad \sqrt{q_t} = \sqrt{q_{t-1}} \quad (\text{ no adaptation})$$

Some parameters of these formulae have to be chosen experimentally. Let us demonstrate such a choice by executing a wide range of computational experiments with a concrete application problem. As a result, we have chosen

$$\alpha = 0.99,\ \gamma = 0.1,\quad \tau = t - (t \mod T),\quad T = 500,\ \text{and}\quad \eta = 1$$

4 Application Problem

4.1 Extended Inertial Navigation System Error Model, EINSEM

This model includes 15 constant values being factors of separate error sources, and 15 state variables: 9 error state variables and 6 random inputs modelled as first order Gauss-Markov processes (see for example [6]). Notice that notations x, y and z stand here for axes of a gyro-stabled platform (GSP).

The fifteen constant values are classified into 5 groups of 3 each, taken along the axes as follows.

1. n_{Gx}, n_{Gy}, n_{Gz} – the gyro constant drift rate.
2. K_{Ax}, K_{Ay}, K_{Az} – non-linearity factors of accelerometer scaling coefficients.
3. K_{Gx}, K_{Gy}, K_{Gz} – the gyro characteristic first-order non-linearity factors due to non-symmetric center of mass position.
4. l_{Gx}, l_{Gy}, l_{Gz} – the gyro characteristic second-order non-linearity factors due to non-equal gimbal rigidity along the GSP axes.
5. $K_{DMx}, K_{DMy}, K_{DMz}$ – the actuator (gyro motor) characteristic non-linearity factors along the axes.

These values are referred to as parameters and introduced here so that design engineers could estimate any offending component contributing to the total INS error. The nine error state variables are described by the following stochastic differential equations (prime $'$ means derivative).

Errors in indicated position

$$\Delta\varphi' = \Delta\nu_x/r, \quad \Delta\lambda = \Delta\nu_y/r, \quad \Delta h' = \Delta\nu_z$$

Errors in indicated velocity

$$\begin{aligned}\Delta\nu_x' &= -f_z'\beta + f_y\delta + m_{Ax} + f_x K_{Ax}\\ \Delta\nu_y' &= f_z'\alpha - f_x\delta + m_{Ay} + f_y K_{Ay}\\ \Delta\nu_z' &= -f_y\alpha + f_x\beta + m_{Az} + f_z' K_{Az}\end{aligned}$$

The two angular errors (α, β) in the indicated vertical (α being the angular deflection of the vertical in the east/west direction and β being the angular deflection of the vertical in the north/south direction)

$$\begin{aligned}\alpha' &= -r^{-1}\Delta\nu_y + \omega_z\beta - \omega_y\delta + m_{Gx} + n_{Gx}\\ &\quad + f_x K_{Gx} + f_x f_z' l_{Gx} + (\omega_x - \omega_{x0})K_{DMx}\\ \beta' &= r^{-1}\Delta\nu_x - \omega_z\alpha + \omega_x\delta + m_{Gy} + n_{Gy}\\ &\quad + f_y K_{Gy} + f_y f_z' l_{Gy} + (\omega_y - \omega_{y0})K_{DMy}\end{aligned}$$

The angular error δ in the indicated azimuth (azimuthal deflection)

$$\delta' = \Delta\varphi\Omega\cos\varphi + \omega_y\alpha - \omega_x\beta + m_{Gz} + n_{Gz} \\ + f'_z K_{Gz} + f_y f'_z l_{Gz} + (\omega_z - \omega_{z0})K_{DMz}$$

In the above equations, r is the Earth's large half-axis; Ω is the Earth's angular velocity; g is the gravity acceleration; f_x, f_y, f_z are the projections of the vehicle acceleration on the platform axes and $f'_z = f_z - g$; $\omega_x, \omega_y, \omega_z$ are the projections of Ω on the platform axes; $\omega_{x0}, \omega_{y0}, \omega_{z0}$ are the initial values of $\omega_x, \omega_y, \omega_z$ (at time $t = 0$); and φ is the latitude.

The six random inputs with variances σ_i^2 and correlation intervals γ_i^{-1} are assumed to be mutually independent and modelled by the equations

$$m'_i + \gamma_i m_i = \sigma_i\sqrt{2\gamma_i}w_i; \quad i = Ax, Ay, Az, Gx, Gy, Gz$$

where w_i are mutually independent standard white Gaussian noises.

4.2 Real Data Mathematical Model, RDMM

For conducting simulated tests, we have designed the $RDDM$ including:

1. An INS Error Model. The $INSEM$ may be of desired (pre-selected) composition/dimension as compared to $EINSEM$ (Section 4.1). We have an easy possibility to formulate the $INSEM$ on the basis of $EINSEM$ by including or excluding a selection of parameters and/or variables at our own will.

2. A Kinematical INS Model that generates f_x, f_y, f_z and $\omega_x, \omega_y, \omega_z$ along with φ for the $INSEM$.

3. A Vehicle Motion Model. The VMM generates the geographical components of vehicle velocity for the $KINSM$.

4.3 Filter INS Error Model, FINSEM

Separately to $RDMM$ (Section 4.2), which places at our disposal all the model values of Sections 1 and 2, we construct the Filter INS Error Model corresponding to Eq. (11) and Eq. (12).

5 Computational Experiments

We present here two tasks solved by computational experiments with $RDMM$ and AFI (based on $FINSEM$):

1. Determining the 'best' adaptive filter mechanism of proposed in Section 3.

2. Wide range testing the so determined 'best' mechanism.

Task 1. From $EINSEM$, we select the following values for Sections 1 to 3:

$$x = \begin{bmatrix} \Delta\varphi \\ \Delta\nu_x \\ \beta \end{bmatrix}, \quad a = \begin{bmatrix} n_{Gy} \\ K_{Ax} \\ K_{Gy} \end{bmatrix}, \quad b = \begin{bmatrix} l_{Gy} \\ K_{DMy} \end{bmatrix}, \quad s = \begin{bmatrix} m_{Ax} \\ m_{Gy} \end{bmatrix}, \quad z = \Delta\varphi$$

$$F_c = \begin{bmatrix} 0 & r^{-1} & 0 \\ 0 & 0 & -f_z' \\ 0 & r^{-1} & 0 \end{bmatrix}, \Psi_c = \begin{bmatrix} 0 & 0 & 0 \\ 0 & f_x & 0 \\ 1 & 0 & f_y \end{bmatrix}, \quad \Xi_c = \begin{bmatrix} 0 & 0 \\ 0 & 0 \\ f_y f_z' & \omega_y - \omega_{y0} \end{bmatrix}$$

$$\Sigma_c = \begin{bmatrix} 0 & 0 \\ 1 & 0 \\ 0 & 1 \end{bmatrix}, \quad \Gamma_c = \begin{bmatrix} -\gamma_{Ax} & 0 \\ 0 & -\gamma_{Gy} \end{bmatrix}, Q_c = \begin{bmatrix} 2\gamma_{Ax}\sigma_{Ax}^2 & 0 \\ 0 & 2\gamma_{Gy}\sigma_{Gy}^2 \end{bmatrix}$$

After this, all the necessary values for Section 3 are easily found.

Task 2. From $EINSEM$, we select the following values for Sections 1 to 3:

$$\begin{aligned} x &= (\Delta\varphi, \Delta\lambda, \Delta h, \Delta\nu_x, \Delta\nu_y, \Delta\nu_z, \alpha, \beta, \delta)^T \\ a &= (n_4, n_5, n_6, K_1, K_2, K_3, K_4, K_5, K_6)^T \\ b &= (l_4, l_5, l_6, K_7, K_8, K_9)^T \\ c &= (c_i)^T, i = 1, \ldots, 6; \; c_i = m_i/(\sigma_i\sqrt{2\gamma_i\tau}) \end{aligned}$$

Numerical indexing here and symbolical indexing of the same values in Section 4.1 are equivalent to each other, i.e. $1 \equiv Ax$, $2 \equiv Ay$, $3 \equiv Az$, $4 \equiv Gx$, $5 \equiv Gy$, $6 \equiv Gz$, $7 \equiv DMx$, $8 \equiv DMy$, and $9 \equiv DMz$. Instead of writing down values for Section 1, we show now the values obtained for Sections 2 and 3:

$$\Phi_t = \begin{bmatrix} I & \Phi_{12} & 0 \\ 0 & I & \Phi_{23} \\ \Phi_{31} & \Phi_{32} & \Phi_{33} \end{bmatrix}, \Phi_{12} = \begin{bmatrix} \phi_1 & 0 & 0 \\ 0 & \phi_1 & 0 \\ 0 & 0 & \tau \end{bmatrix}, \quad \Phi_{23} = \begin{bmatrix} 0 & -\phi_4 & \phi_3 \\ \phi_4 & 0 & -\phi_2 \\ -\phi_3 & \phi_2 & 0 \end{bmatrix}$$

$$\Phi_{32} = \begin{bmatrix} 0 & -\phi_1 & 0 \\ \phi_1 & 0 & 0 \\ 0 & 0 & 0 \end{bmatrix}, \quad \Phi_{33} = \begin{bmatrix} 1 & \phi_8 & -\phi_7 \\ -\phi_8 & 1 & \phi_6 \\ \phi_7 & -\phi_6 & 1 \end{bmatrix}, \Psi_t = \begin{bmatrix} 0 & 0 & 0 \\ 0 & B & 0 \\ A & 0 & B \end{bmatrix}$$

$\phi_1 = \tau/r$, $\phi_2 = \tau f_x$, $\phi_3 = \tau f_y$, $\phi_4 = \tau f_z'$, $\phi_5 = \tau\Omega\cos\phi_t$, $\phi_6 = \tau\omega_x$, $\phi_7 = \tau\omega_y$, $\phi_8 = \tau\omega_z$, in Φ_{31} is only one non-zero element ϕ_5. Matrices A, B, C, D and Λ

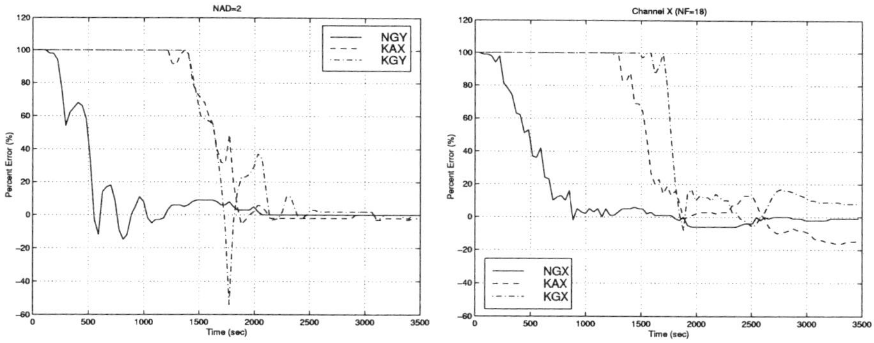

Figure 1. Left: identifying a by $NAD = 2$ in Task 1. Right: in Task 2, channel X.

are diagonal:

$$\begin{aligned}
A &= \text{diag}\ \{\tau, \tau, \tau\} \\
B &= \text{diag}\ \{\tau f_x, \tau f_y, \tau f_z'\} \\
C &= \text{diag}\ \{\tau f_x f_z', \tau f_y f_z', \tau f_y f_z'\} \\
D &= \text{diag}\ \{\tau(\omega_x - \omega_{x0}), \tau(\omega_y - \omega_{y0}), \tau(\omega_z - \omega_{z0})\} \\
\Lambda &= \text{diag}\ \{(1 - \tau\gamma_i)\}, \quad i = 1, \ldots, 6
\end{aligned}$$

and 0 stands for 3×3 zero matrix, and I for the unit matrix. Matrix $\Sigma_t = [\sigma_{ij}]$, $j = 1, \ldots 6$, and matrix $H = [h_{ij}]$, $i = 1, 2, 3$, are defined by their entries:

$$\sigma_{ij} = \begin{cases} \tau\sigma_j\sqrt{2\gamma_j\tau} & i = j + 3 \\ 0 & i \neq j + 3 \end{cases} \qquad h_{ij} = \begin{cases} 1 & i = j \\ 0 & i \neq j \end{cases}$$

Selected experimental results are presented in Figs 1 and 2. The figure plots show the percentage errors in parameter estimates so that we can see when 10%-corridor of accuracy has been reached.

6 Conclusions

All the works conducted in this paper allows to draw the following key conclusions:

1. The most suitable way to adaptively estimate unknown parameters of linear stochastic differential equations from incomplete noisy measurements is the combination of the Extended Model Approach and the Covariance Matching Approach, the latter using the fictitious noise of covariance q.

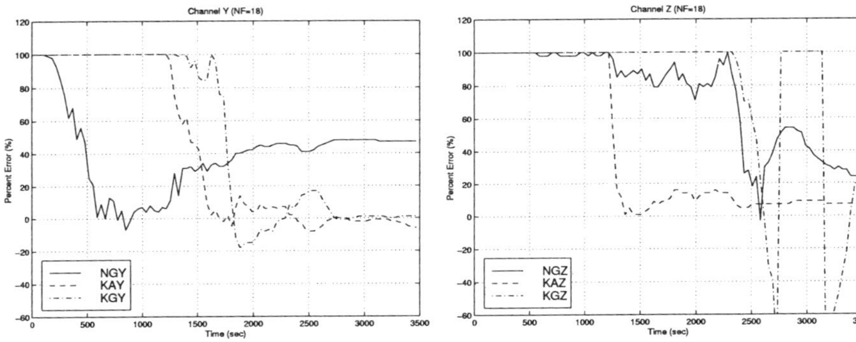

Figure 2. Left: identifying a by $NAD = 2$ in Task 2, channel Y. Right: in channel Z.

2. The most efficient way to tune the fictitious noise RMS value $\sqrt{q}$ to optimality for the extended filter-estimator is defined by the two formulae:

   (a) $\bar{\delta}_t = \alpha\bar{\delta}_{t-1} + (\delta_t - \alpha\bar{\delta}_{t-1}), \quad \alpha \approx 0.98$

   (b) $\sqrt{q_t} = \gamma|\bar{\delta}_t|, \quad \gamma \approx 0.1$

3. In the inertial navigation application, the vehicle trajectory has proven to have a profound impact in identification as it changes parameter observability conditions (in Task 2 we used three 180° turns in heading and four ±60° turns in pitch after take-off).

References

1. P. Maybeck, *Stochastic Models, Estimation, and Control* (Acad. Press, New-York, 1978).
2. G. Bierman, *Factirization Methods for Discrete Sequential Estimation* (Acad. Press, New-York, 1977).
3. R.K. Mehra, IEEE Trans. Automat. Contr. **17**, 5 (1972).
4. H. Kaufman and D. Beaulier, IEEE Trans. Automat. Contr. **17**, 5 (1972).
5. I.V. Semoushin, Technicheskaya Cybernetika, The USSR Academy of Sciences **1**, 6 (1979).
6. C. Broxmeyer, *Inertial Navigation Systems* (McGraw-Hill Book Co., New York, 1956).

A MESHLESS SCHEME FOR SOLVING INVERSE PROBLEMS OF LAPLACE EQUATION

Y.C.HON

Department of Mathematics, City University of Hong Kong,
E-mail: maychon@cityu.edu.hk

T.WEI

Department of Mathematics, City University of Hong Kong,
Department of Mathematics, Lanzhou University, Lanzhou, 730000, P. R. China
E-mail: tingwei@lzu.edu.cn

In this paper, we present a meshless numerical method to solve inverse problems for Laplace equation which are the descriptions of a steady-state heat conduction problem. The temperature and heat flux on unspecified boundary can be determined simultaneously. The basic idea of our proposed method is to approximate the solution of problem by a linear combination of fundamental solution of Laplace operator. The numerical results of several examples involving smooth or non-smooth geometries show that the proposed method is efficient and accurate.

Key words: Inverse problem for Laplace equation, Meshless method.

1 Introduction

We consider a multidimensional steady-state heat conduction problem. Let Ω be a bounded and simply connected domain in $\mathcal{R}^d$, $d = 2,3$ with Lipschitzian boundary. Suppose that Γ_1 and Γ_2 are two open parts of boundary $\partial\Omega$ and $\Gamma_1 \bigcup \Gamma_2 \neq \partial\Omega$, where Γ_2 can be empty set.

Find a temperature distribution $u \in C^2(\Omega) \cap C^1(\overline{\Omega})$ that satisfies,

$$\triangle u = 0, \qquad x \in \Omega, \tag{1}$$

$$u|_{\Gamma_1} = \varphi, \tag{2}$$

$$\frac{\partial u}{\partial n}|_{\Gamma_2} = \psi, \tag{3}$$

$$u(x_j) = h_j, \qquad j = 1,2,\cdots,m, \tag{4}$$

where $\triangle$ is the d-dimensional Laplace operator; φ and ψ are respectively the temperature and heat flux data on boundary Γ_1 and Γ_2; $\frac{\partial u}{\partial n}$ is the outward normal derivative of u at Γ_2 and $\{x_j\}|_{j=1}^m$ is a set of measurement locations in the interior of Ω.

Denote $I_M = \{x_1, x_2, \cdots, x_m\}$ and consider the following two special cases.

Problem 1 $\Gamma_1 = \Gamma_2$ *and* $I_M = \emptyset$. *In this case problem (1)–(3) is called a Cauchy problem for Laplace equation which arises in many applications such as non-destructive testing* [1] *, electro-cardiology* [8] *and steady-state heat conduction* [2,5]. *Many numerical computational methods have been researched for past fifty years* [10,15,5,3].

Problem 2 $\Gamma_2 = \emptyset, \Gamma_1 \neq \emptyset$ *and* $I_M \neq \emptyset$. *This problem given by (1),(2) and (4) is one kind of steady-state inverse heat conduction problem*[12]. *From the temperature measurements inside solid body , we need to determine temperature distribution and heat flux on unspecified boundary.*

Note that these problems are severely ill-posed, *i.e.* the solutions do not depend continuously on the boundary data or inside measured data, and small errors in the data can destroy the numerical solution.

In this paper, we only consider these two cases, but the numerical technique can be applied to general cases, for example problem in reference paper[2] . Our proposed meshless method is the application of the method of fundamental solution (*MFS*) and radial basis function (*RBF*) on inverse problems for elliptic equation [16,6,14,4,12,13]. In the last decade, the development in applying fundamental solution with radial function as a truly meshless method for approximating the solutions of PIES has drawn the attention of many researchers in science and engineering. Being meshless, fast convergent and the extensible to high dimension problems make the MFS very attractive in solving problems with complex geometry. More details of the MFS method can be found in the review papers of Fairweather and Karageorghis[7] and Golberg and Chen[9].

2 Meshless Method

Denote by $F(x, x^*)$ the fundamental solution of the Laplace operator $\triangle$:

$$F(x,x^*) = \begin{cases} -\frac{1}{2\pi} ln|x - x^*|, & d = 2, \\ \frac{1}{4\pi|x-x^*|}, & d = 3, \end{cases} \tag{5}$$

where x and x^* are points in $\mathcal{R}^d$ and $|x - x^*|$ denote the distance between the point x and x^*. When the source point x^* is located outside the domain $\overline{\Omega}$, the fundamental solution satisfies Laplace equation exactly in domain Ω.

In the following, we give the fundamental solution method based on collocation. At first we choose collocation points on boundary or inside domain. For the Problem 1, take m points $x_1, x_2, \cdots, x_m$ on Γ_2 and n

points $x_{m+1}, x_{m+2}, \cdots, x_{m+n}$ on Γ_1. In the Problem 2, choose the points $x_1, x_2, \cdots, x_m$ to be measurement locations given by (4) and other points $x_{m+1}, x_{m+2}, \cdots, x_{m+n}$ on Γ_1. For every problem , we need to find $m+n$ source points $x_1^*, x_2^*, \cdots, x_{m+n}^*$ in the exterior of $\overline{\Omega}$. All the collocation points $x_1, x_2, \cdots, x_{m+n}$ are needed to be pairwise distinct points.

Following the idea of RBF's approximation , an approximate solution of Problem 1 and 2 can be expressed in the following linear combination:

$$u^*(x) = \sum_{j=1}^{m+n} \lambda_j F(x, x_j^*), \tag{6}$$

where $\{\lambda_j\}$ are constants to be determined.

By the boundary data or and measurement data inside domain, we can deduce a linear system of equations for problem 1 and problem 2 respectively as follows:

Problem 1

$$\frac{\partial u^*}{\partial n}(x_i) = \psi(x_i), \qquad i = 1, 2, \cdots, m, \tag{7}$$

and

$$u^*(x_i) = \varphi(x_i), \qquad i = m+1, m+2, \cdots, m+n. \tag{8}$$

Problem 2

$$u^*(x_i) = \begin{cases} h_i, & i = 1, 2, \cdots, m, \\ \varphi(x_i), & i = m+1, m+2, \cdots, m+n. \end{cases} \tag{9}$$

In matrix form, the values of undermined coefficients λ_j are found by solving the following system of linear equations

$$A\lambda = b \tag{10}$$

where

$$A = \begin{pmatrix} \frac{\partial F}{\partial n}(x_i, x_j^*) \\ F(x_k, x_j^*) \end{pmatrix} \quad or \quad A = (F(x_l, x_j^*)) \tag{11}$$

and

$$b = \begin{pmatrix} \psi(x_i) \\ \varphi(x_k) \end{pmatrix} \quad or \quad b = \begin{pmatrix} h_i \\ \varphi(x_k) \end{pmatrix} \tag{12}$$

with $i = 1, 2, \cdots, m$, $k = m+1, m+2, \cdots, m+n$ and $l = j = 1, 2, \cdots, m+n$.

Once the system of equations are assembled, they are solved using a Matlab solver.

3 Numerical Experiments

In the situation of measurement data including some random noises, we use man-made noisy data $\widetilde{h}_i = h_i + \sigma rand(i)$ to compute the approximation , where h_i is the exact data and $rand(i)$ is a random number between $[-1, 1]$ and the value of σ indicates the error level.

For showing the accuracy of approximate solution, we choose enough test points in domain $\overline{\Omega}$ and then calculate the Root Mean Square error by the following formula

$$RMS(u) = \sqrt{\frac{1}{N}\sum_{i=1}^{N}(u_i - u_i^*)^2} \tag{13}$$

where N is the number of test points in domain $\overline{\Omega}$, u_i and u_i^* are respectively exact and approximate temperature at these test points.

In this section, we compute three examples for two-dimensional and three-dimensional Problem 1 and Problem 2 in various cases.

3.1 Numerical Tests for Two-dimensional Problem 1 and Problem 2

In the following, we test two examples with exact analytic solutions under four different domains and boundary conditions.

Case 1: Take $\Omega = \{ (x_1, x_2) \,|\, 0 < x_1 < 1, 0 < x_2 < 1\}$ and $\Gamma_1 = \{ (x_1, x_2) \mid x_2 = 0,\ 0 < x_1 < 1\}$, $\Gamma_2 = \Gamma_1$. Collocation points are shown in Figure 1(a).

Case 2: Let $\Omega = \{ (x_1, x_2) \mid x_1^2 + x_2^2 < 1\}$, and $\Gamma_1 = \{ (x_1, x_2) \mid x_1^2 + x_2^2 = 1,\ x_1 > 0,\ x_2 > 0\}$, $\Gamma_2 = \Gamma_1$. Collocation points are given in Figure 1(b).

Case 3: Take $\Omega = \{ (x_1, x_2) \,|\, 0 < x_1 < 1, 0 < x_2 < 1\}$ and $\Gamma_1 = \{ (x_1, x_2) \mid x_2 = 0,\ 0 < x_1 < 1\}$, $\Gamma_2 = \emptyset$. Measurement and collocation points are shown in Figure 2(a).

Case 4: Let $\Omega = \{ (x_1, x_2) \mid x_1^2 + x_2^2 < 1\}$, and $\Gamma_1 = \{ (x_1, x_2) \mid x_1^2 + x_2^2 = 1,\ x_1 > 0,\ x_2 > 0\}$, $\Gamma_2 = \emptyset$. Measurement and collocation points are given in Figure 2(b).

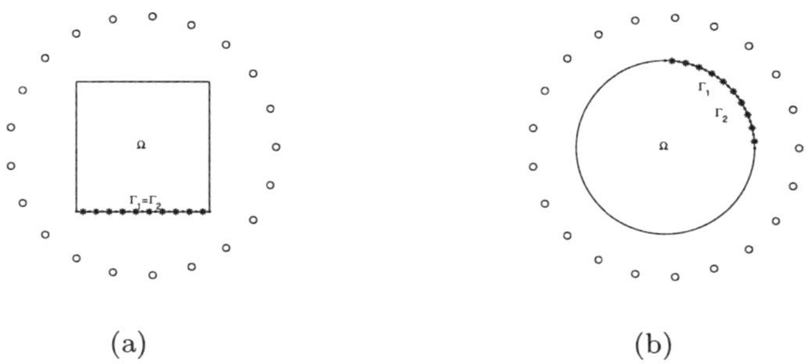

Figure 1. Collocation points on Ω. Dots are collocation points for Dirichlet data , stars represent collocation points for Neumann data and circles are source points.

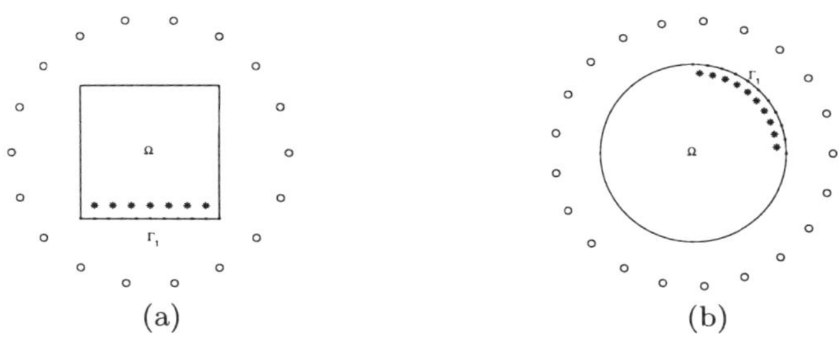

Figure 2. Collocation points on Ω. Dots are collocation points for Dirichlet data , stars represent measurement locations and circles are source points.

Example 1 *The exact solution of (1) is chosen as*

$$u(x_1, x_2) = x_1^3 - 3x_1x_2^2 + e^{2x_2}sin(2x_1) - e^{x_1}cos(x_2) \tag{14}$$

Example 2 *Taking an exact solution of (1) as follow*

$$u(x_1, x_2) = \ln\sqrt{(x_1 + 0.5)^2 + (x_2 + 1.5)^2}. \tag{15}$$

The boundary data φ and ψ can be deduced by simple computation. The numerical results obtained by our method are presented in Table 1 with no noisy data. Table 2 presents RMS error for temperature in domain Ω with noisy Dirichlet data in Problem 1 and noisy measurement data in Problem 2. In our computation, the source points are uniformly distributed on a circle with radius R. And all the collocation points are also chosen uniformly on boundary. Let $m = n - 1$, n, m are the numbers of collocation points. The parameters R and n used in computing will be shown in Tables. The distance from measured point to boundary is chosen as 0.1 in Case 3 and Case 4.

Table 1. The RMS error in domain Ω with no noisy data.

	Example 1.			Example 2.		
	RMS	R	n	RMS	R	n
Case 1	1.9149e-6	3	31	1.2590e-6	3	21
Case 2	3.7165e-4	4	21	6.7530e-4	4	21
Case 3	4.2821e-6	5	21	8.0441e-7	3	21
Case 4	4.5125e-5	5	21	7.6843e-4	5	21

Table 2. The RMS error in domain Ω with noisy data.

	Example 1.				Example 2.			
	RMS	σ	R	n	RMS	σ	R	n
Case 1	0.0248	1e-4	35	31	0.0136	1e-4	60	21
Case 2	0.0147	1e-6	15	21	0.0231	1e-3	80	21
Case 3	0.0126	1e-4	65	31	0.0145	1e-4	55	21
Case 4	0.0498	1e-6	15	31	0.0265	1e-4	80	21

Figure 3(a) indicates the changes of RMS error in term of radius R for Example 1 in case 1. Figure 3(b) give the same description as Figure 3(a) with random noise data($\sigma = 1e - 4$). Our numerical results imply that parameter R plays a role of regularization parameter. As the random noise level increases, available choice for parameter R corresponding to highly accurate approximate solution is decreased .

As an example, in Figure 4 we show the availability of reconstructing the heat flux on unspecified boundary by fundamental solution method.

3.2 Three-dimensional Test Case.

Example 3 *Let* $\Omega = \{ (x_1, x_2, x_3) \mid 0 < x_i < 1,\ i = 1,2,3\}$ *and* $\Gamma_1 = \{ (x_1, x_2, x_3) \mid 0 < x_1 < 1,\ 0 < x_2 < 1,\ x_3 = 0\}$.

Case 5: $\Gamma_2 = \Gamma_1, I_M = \emptyset$.

Case 6: $\Gamma_2 = \emptyset, I_M \subset \{x_3 = h\}$

An exact solution is chosen as

$$u(x_1, x_2, x_3) = x_1^3 x_2 - 3x_1 x_2 x_3^2 + x_2^3 - 3x_2 x_3^2. \tag{16}$$

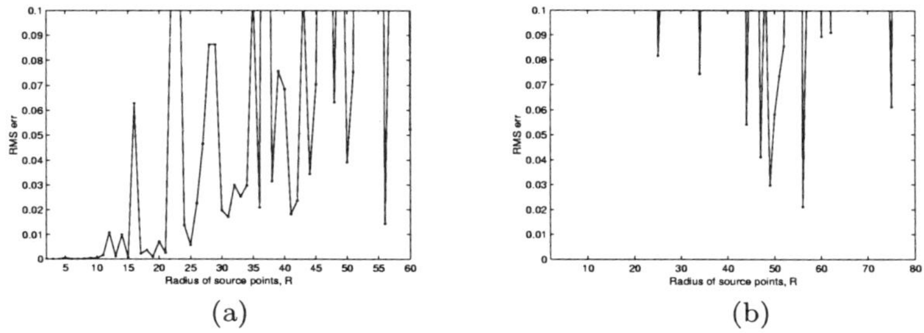

Figure 3. The RMS error for temperature in Ω with respect to parameter R.

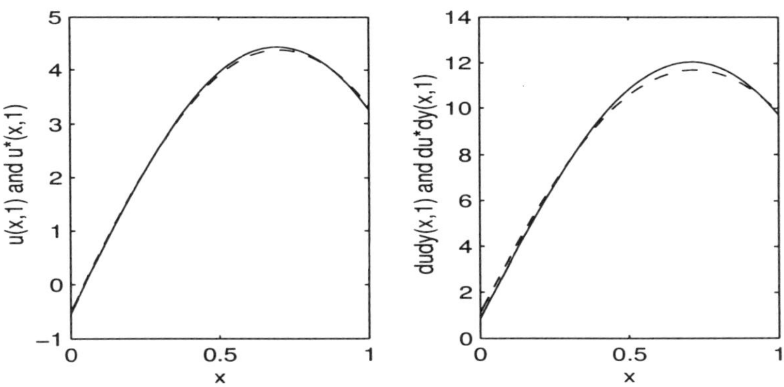

Figure 4. The plots of temperature and heat flux on boundary $x_2 = 1$ for Example 1 in Case 1 with noisy data . $\sigma = 1e-4$, $R = 4, n = 21$.

In our computation, parameter $n = 21 \times 21, m = 20 \times 20$. In case 6, we take $h = 0.1$, h is the distance from measured points to boundary. The difference between exact solution and approximate estimation about temperature and heat flux on surface $x_3 = 1$ have been shown in Figure 5 , Figure 6 for Case 5 and Figure 7 , Figure 8 for case 6.

For the first try, we locate the source points uniformly on a circle outside considered domain. Note that the accuracy of approximate solution changes with respect to the radius of source points and locations of collocation points. So one needs to investigate an optimal method for locating the collocation points and the source points to improve the accuracy of the scheme.

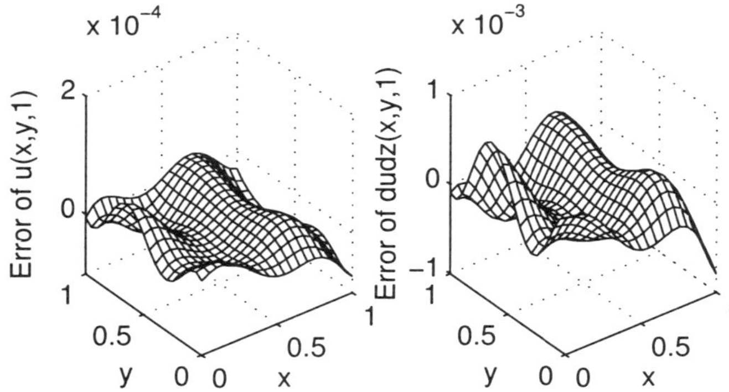

Figure 5. Error of temperature and heat flux on boundary $x_3 = 1$ with no noisy data for Case 5. $R = 4$.

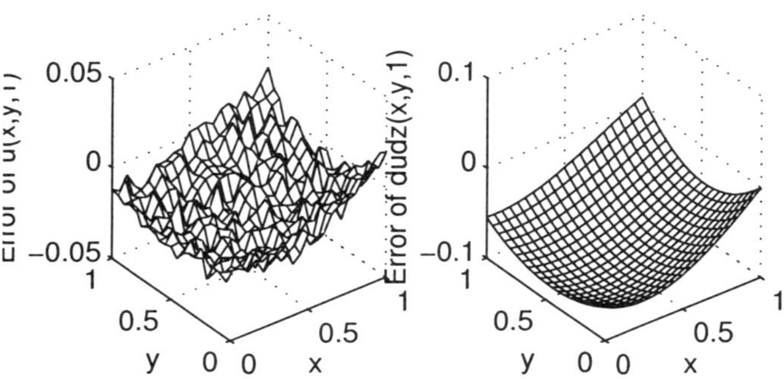

Figure 6. Error of temperature and heat flux on boundary $x_3 = 1$ with noisy data, $\sigma = 1e-5$, $R = 1000$.

4 Conclusion and Future Directions

From the previous section, one can see that the MFS is a powerful mesh-free method for solving inverse problems in nonregular high dimensional geometries. The lack of interior or surface meshing makes the method extremely attractive for complicated boundary condition and inside measurement data. The efficacy of the method has been demonstrated for simply connected domains. But there have been no efforts to show the utility of the method for multiply connected domains. From the numerical experiments , the accuracy

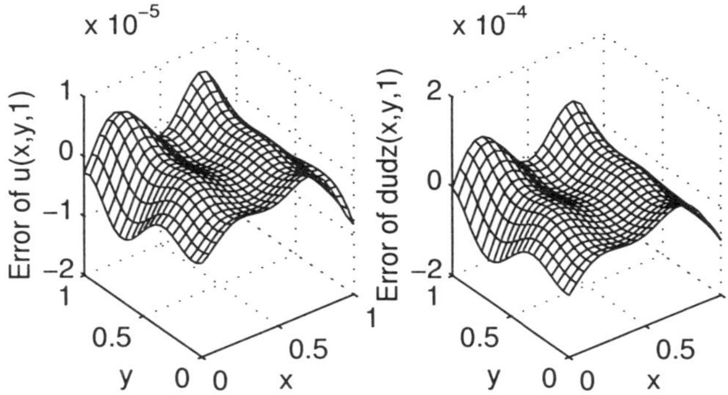

Figure 7. Error of temperature and heat flux on boundary $x_3 = 1$ with no noisy data for Case 6. $R = 4$.

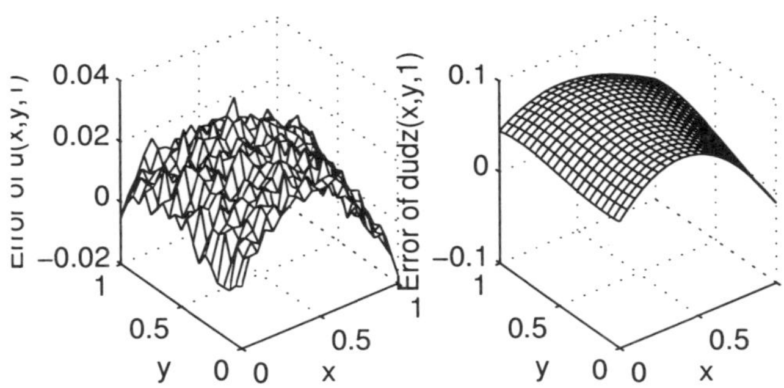

Figure 8. Error of temperature and heat flux on boundary $x_3 = 1$ with noisy data, $\sigma = 1e - 4$, $R = 1000$.

of results would be worse when putting a little large random error into boundary data and measurement data. How to use some regularization method to solve the ill-conditional discrete problem is our further work.

References

1. G. Alessandrini, *Stable determination of a crack from boundary measurements*, Proc. R. Soc. A **123**, 497–516(1993).
2. N.M. AL-Najem, A.M. Osman, M.M. Ei-Refaee and K.M.Khanafer, *Two*

dimensional steady-state inverse heat conduction problems, Int. Comm. Heat Mass Transfer **25**, 541–550(1998).

3. D. D. Ang, N. H. Nghia and N. C. Tam, *Regularized solutions of Cauchy problem for the Laplace equation in an irregular layer: a three dimensional case*, Acta Math. Vietnamica **23** 65–74(1998) .
4. K. Balakrishnan and P. A. Ramachandran, *The method of fundamental solutions for linear diffusion-reaction equations*, Mathematical and Computer Modelling **31**, 221–237 (2000).
5. F. Berntsson and L. Eldén, *Numerical solution of a Cauchy problem for the Laplace equation*, Inverse Problems **17**, 839–853(2001) .
6. A. Bogomonlny, *Fundamental solutions method for elliptic boundary value problems*, SIAM Journal on Numerical Analysis **22**, 644–669 (1985).
7. G. Fairweather and A. Karageorghis, *The method of fundamental solutions for elliptic boundary value problems*, Advances in Computational Mathematics **9**, 69–95(1998).
8. Colli Franzone P and E. Magenes, *On the inverse potential problem of electrocardiology*, Calcolo **16**, 459–538 (1979).
9. M.A. Golberg and C.S. Chen, *The method of fundamental solutions for potential, Helmholtz and diffusion problems*, in *Boundary integral methods-numerical and mathematical aspects, Sounthampton: Computational Mechanic Publications* (ed. M. A. Golberg , 103–176(1998)).
10. D. N. Hào and D. Lesnic, *The Cauchy problem for Laplace's equation via the conjugate gradient method*, IMA J. Appl. Math **65**, 199–217(2000).
11. Y. C. Hon and Z. M. Wu, *A numerical computation for inverse boundary determination problem*, Engineering Analysis with Boundary Elements **24**, 599–606(2000).
12. Y. C. Hon and T. Wei, *A meshless computational method for solving inverse heat conduction problem*, 24th World Conference on Boundary Element Methods, in press.
13. Y. C. Hon and W. Chen, *Boundary knot method for 2D and 3D Helmholtz and convection-diffusion problems with complicated geometry*, International Journal for Numerical Methods in Engineering, in press.
14. M. Katsurada, *The collocation points of the fundamental solution method for the potential problem*, Computers Math. Applic, **31**, 123–137(1996).
15. H. J. Reinhardt, H. Han, and D. N. Hào, *Stability and regularization of a discrete approximation to the Cauchy problem for Laplace's equation*, SIAM J.Numer. Anal. **36**, 890–905 (1999).
16. Y. S. Smyrlis and A. Karageorghis, *Some aspects of the method of fundamental solutions for certain harmonic problems*, Journal of Scientific Computing **16**, 341–371(2001).

STABILIZED SOLUTION AND NUMERICAL SIMULATION FOR A TWO-DIMENSIONAL HAUSDORFF MOMENT PROBLEM

DINGHUA XU AND ZEWEN WANG

Department of Computational Sciences, East China Geological Institute
Fuzhou 344000, Jiangxi Province, P. R. China
E-mail: dhxujc@ecgi.jx.cn

In this paper we consider a two-dimensional Hausdorff moment problem(2-D HMP) to recover an unknown function from a finite number of moments contaminated by noise. It is well known that the 2-D HMP is a severely ill-posed problem. In order to obtain a conditional stability, we transform equivalently the 2-D HMP into two 1-D HMPs. From our derived result on the 1-D HMP by using the integral equation methods, We establish a conditional stability estimate for the 2-D HMP. Based on the conditional stability, we present an algorithm with an error estimate to the reconstruction of the function. Finally we provide some numerical examples to test the theoretical results. The numerical simulation shows the efficiency and sound implementation of the given algorithm.

1 Hausdorff Moment Problems

It is well known that many practical problems such as in Geophysics (eg. Ang *et al*[2], Backus and Gilbert[4], Inglese[9]), medical computerized tomography(eg. Ang *et al*[2], Engl *et al*[8]), nondestructive testing (eg. Engl *et al*[8]), etc., can be formulated into moment problems including linear and nonlinear cases. One of the important moment problems is *Hausdorff moment problems* (HMP): for example, in a one-dimensional HMPs, the 1-D HMP is to recover a function $u(x)$ from moments $\{\mu_k\}_{k=0}^{\infty}$ satisfying the following condition:

$$\int_0^1 x^k u(x)dx = \mu_k, \; k = 0, 1, \cdots, \tag{1}$$

or in a two-dimensional HMPs, an unknown function $u(x, y)$ needs be determined from moments $\{\mu_{ij} : i = 0, 1, \cdots ; j = 0, 1, \cdots\}$ satisfying the following condition:

$$\int_0^1 \int_0^1 x^i y^j u(x, y)dxdy = \mu_{ij}, \; i = 0, 1, \cdots ; j = 0, 1, \cdots. \tag{2}$$

Solutions of the HMPs that belong to sufficiently nice function spaces, such as L^p, are unique, see Rudin[11] for instance. The Hausdorff moment problem is severely ill-posed in the sense of Hadamard.

To the practical viewpoint, the function $u(x)$ or $u(x,y)$ has to be recovered from only a finite number of moments $\{\mu_k : k = 0,1,\cdots,N\}$ or $\{\mu_{ij} : i = 0,1,\cdots,N_1; j = 0,1,\cdots,N_2\}$, N,N_1 and N_2 are fixed natural numbers. This case shows that the solution is of no uniqueness and no stability. A satisfactory algorithm for the ill-posed problem should involve the full set of data, including information on noise, and a priori information on the solutions. Generally speaking, one cannot meet the above requirements. However, it may be worth noting that we can turn to some stabilized algorithm to solve it. Hence, we have to make an in-depth discussion on the structure of well-posedness, especially the error estimate in some reasonable Sobolev spaces, which coincide with the purpose of practical uses, and establish stabilized algorithms for computation of the solution of the Hausdorf moment problem.

Backus and Gilbert (see Backus and Gilbert[4], Kirsch *et al* [10]) constructed an efficient numerical method, well-known called Backus-Gilbert Method, for geophysical use. Later a variaty of stabilized algorithms, such as regularization method (eg. Ang *et al* [2]) and approximation method (eg. Ang *et al* [1], Askey *et al* [3], Talenti[13]), are derived for solving Hausdorff moment problem theoretically or numerically. For general linear moment problems, we can refer to Inglese[9], Shohat and Tamarkin[12], Wang[14] etc.. But they all only obtained global stability and error estimates. By their methods, one could not derive local estimates. In practical uses the local estimate is essential, for example, the determination of boundaries of inaccessible objects and of internal structures needs the local estimates. Recently the author of this paper gives a novel local stability estimates for the 1-D HMP by the integral equation method and establish a regularization algorithm for solving unknown functions, see Xu *et al* [15]. In this paper we will establish a conditional stability estimate for the 2-D HMP, on which we present an algorithm to show reconstruction of the function and prove an error estimate for the algorithm. The same regularization method can be found in the paper of Cheng and Yamamoto[6].

The paper is organized as follows:

- Section 2 Local Conditional Stability and Tikhonov Regularization Algorithm for the One-Dimensional HMP;
- Section 3 Local Conditional Stability for the Two-Dimensional HMP;
- Section 4 Stabilized Algorithm for the Two-Dimensional HMP;
- Section 5 Numerical Examples;
- Section 6 Some Remarks.

2 Local Conditional Stability and Tikhonov Regularization Algorithm for the One-Dimensional HMP

For convenience in this paper, we rewrite the 1-D HMP as an operator equation. Define an operator equation

$$\mathbf{A}u = \mu,$$

where

$$\mathbf{A}u = ((\mathbf{A}u)_0, (\mathbf{A}u)_1, \cdots, (\mathbf{A}u)_N, \cdots)^T,$$

$$(\mathbf{A}u)_j = \int_0^1 x^j u(x)dx, \; j = 0, 1, \cdots, N, \cdots,$$

$$\mu = (\mu_0, \mu_1, \cdots, \mu_N, \cdots)^T.$$

First we note that the Hausdorff moment problem (1) is equivalent to the integral equation of the first kind (3):

$$\int_0^1 \ln(y-x)u(x)dx = \ln y \cdot \mu_0 - \Sigma_{n=1}^{\infty} \frac{1}{ny^n}\mu_n, \; y \in [2,3]. \tag{3}$$

Direct application of the integral equation method proposed in Bruckner and Cheng [5] gives the following result for the 1-D HMP (1). The proof of the lemma 1 is found in Xu *et al* [15].

Lemma 1 *Let* $\varepsilon = \sqrt{\Sigma_{k=0}^N |\mu_k|^2}$, $C(N) = \sqrt{\frac{4+(N+1)^2}{(N+1)^3 2^{2N+2}}}$, $u_0(x)$ *be the solution of the 1-D HMP (1). Let* $x_0 \in (0,1)$ *is fixed, and* $\eta_1 \equiv dist(x_0, 0) = |x_0 - 0|$. *If there exists a constant* $M_1 > 0$ *such that* $\| u_0 \|_{H^1(0,1)} \leq M_1$, *then we have the following local estimate*

$$|u_0(x_0)| \leq C_1 \frac{1}{|\log \; frac1\sqrt{2} \cdot \varepsilon + C(N)|^{\gamma}}, \tag{4}$$

where $C_1 = C_1(M_1, \eta_1) > 0$ *is a constant which depends only on* M_1 *and* η_1*;* $\gamma \in (0,1)$ *depends only on* η_1, *independent of* u_0 *and* N;$0 < \varepsilon < 1$.

Remark 1 *In the lemma 1, the assumption that* $\| u_0 \|_{H^1(0,1)}$ *is bounded is not strong since we can transform solving Hausdorff moment problem (1) in* $L^2(0,1)$ *into solving the following moment problem in* $H^1(0,1)$

$$\int_0^1 x^k p(x)dx = \frac{\mu_0 - \mu_k}{k}, \ k = 1, 2, \cdots, N,$$

by the transformation

$$p(x) = \int_0^x u(t)dt.$$

If $u(x) \in L^2(0,1)$, then $p(x) \in H^1(0,1)$. We can compute $u(x)$ stably from the $p(x)$ in the sense of the following estimate

$$\| u \|_{L^2(0,1)} \leq \sqrt{M} \| p \|_{L^2(0,1)}^{\frac{1}{2}}$$

if we assume that $\| u' \|_{L^2(0,1)} \leq M$, and $u(1) = 0$. The above inequality can be obtained by means of direct computation, integration by parts and Hölder inequality.

Remark 2 *If the conditions in lemma 1 hold, then for any fixed natural number N, the stability estimate (4) shows that the magnitude of u_0 will decrease by the logarithmic rate as ε decreases. Let*

$$B := \frac{1}{|\log \frac{1}{\sqrt{2}\cdot\varepsilon + C(N)}|^\gamma},$$

Then the upper error B on the solutions decrease steadily as the error on the measurement data ε decreases, but virtually stops decreasing when the error on the data gets small. In other words, improving the accuracy of the data without increasing the number of data need not result in more accurate solutions.

If the conditions in lemma 1 hold, then for any fixed ε, the inequality (4) shows that the magnitude of u_0 can be decreasing as N increases. The upper error B on the solutions decrease steadily as the number of the measurement data N increases up to some limit, but virtually stops decreasing when the number further gets beyond the limit. In other words, increasing the number of data without improving the accuracy need not result in more accurate solutions.

Remark 3 *Under the assumption that $\| u_0 \|_{H^1(0,1)} \leq M_1$, we see that the series $\Sigma_{i=0}^{\infty} |\mu_i|^2$ is convergent. In fact, by the representation of the moments μ_i and integration by parts, we have*

$$\mu_i = \frac{u_0(1)}{i+1} - \frac{1}{i+1} \int_0^1 x^{i+1} u_0'(x)dx, \ i = 0, 1, 2, \cdots,$$

hence

$$|\mu_i|^2 \leq \frac{1}{(i+1)^2}[u_0^2(1) + (\int_0^1 |x^{i+1}u_0'(x)|dx)^2 \\ +2u_0(1)\int_0^1 |x^{i+1}u_0'(x)|dx], \ i = 0, 1, 2, \cdots.$$

By Hölder inequality, we see

$$|\mu_i|^2 \leq \frac{1}{(i+1)^2}[u_0^2(1) + \frac{1}{2i+3}\int_0^1 |u_0'(x)|^2 dx \\ +2u_0(1)\frac{1}{\sqrt{2i+3}}(\int_0^1 |u_0'(x)|^2 dx)^{\frac{1}{2}}], \ i = 0, 1, 2, \cdots.$$

Since $\| u_0 \|_{H^1(0,1)} \leq M_1$, *and the series* $\Sigma_{i=0}^{\infty}\frac{1}{(i+1)^2}$ *is convergent, we know that the series* $\Sigma_{i=0}^{\infty}|\mu_i|^2$ *is convergent. That is to say, if* $\| u_0 \|_{H^1(0,1)} \leq M_1$, *then* $\Sigma_{i=0}^{\infty}|(\mathbf{A}u_0)_i|^2 = \Sigma_{i=0}^{\infty}|\mu_i|^2$ *is convergent,and meanwhile* $\Sigma_{i=0}^{\infty}|\mu_i|^2 \leq C \| u_0 \|_{H^1(0,1)}^2$.

Basing on the above conditional stability, we will next present an algorithm which is efficient and stabilized in computation of solving the 1-D HMP.

For $\delta > 0$ is fixed and $u \in H^1(0,1)$, define a Tikhonov functional

$$G_\alpha(u) = \| \mathbf{A}u - \mu^\delta \|_{l^2}^2 + \alpha \| u \|_{H^1(0,1)}^2 . \tag{5}$$

where α is a positive parameter, $\mu^\delta = (\mu_0^\delta, \mu_1^\delta, \cdots, \mu_N^\delta)^T$, and $\| \mu^\delta - \mu \|_{l^2} < \delta$.

Since $G_\alpha(u) > 0$, there exists $\beta \geq 0$ such that

$$\beta = \inf_{u \in H^1(0,1)} G_\alpha(u).$$

Let u_α^δ satisfy

$$G_\alpha(u_\alpha^\delta) \leq \beta + \delta^2,$$

we call this function u_α^δ a regularized solution of (1) with δ^2, which reflects computational errors in minimizing (5).

Lemma 2 *Suppose the exact solution of the 1-D HMP (1)* $u_0 \in H^1(0,1)$ *and there exist a constant* $M_1 > 0$ *such that* $\| u_0 \|_{H^1(0,1)} \leq M_1$. *Let* $\alpha = \delta^2$. *Then the regularized solution* u_α^δ *pointwise converges to* u_0 *in (0,1), and the following error estimate holds*

$$|u_\alpha^\delta(x_0) - u_0(x_0)| \leq C_2 \frac{1}{|\log \frac{1}{\sqrt{2}\cdot(1+\sqrt{2+M^2})\delta + C(N)}|^\gamma}, \ x_0 \in (0,1), \tag{6}$$

where $C_2 > 0$ *is a constant which only depends on* M_1 *and* x_0; $\gamma \in (0,1)$ *only depends on* x_0.

The proof of the lemma 2 can be found in Xu *et al* [15].

3 Local Conditional Stability for the Two-Dimensional HMP

The two-dimensional HMP can be solved by two class of one-dimensional HMPs accordingly.

First for any fixed the natural number j , we can determine the functions $g_j(x)$ from

$$\int_0^1 x^i g_j(x)dx = \mu_{ij}, i = 0, 1, \cdots, N_1. \tag{7}$$

Second for any fixed variable $x \in [0,1]$, we can further recover the function $u(x,y)$ from

$$\int_0^1 y^j u(x,y)dy = g_j(x), j = 0, 1, \cdots, N_2. \tag{8}$$

Utilizing the lemma 1 twice, we have the following conditional stability of double logarithmic type for the 2-D HMP.

Theorem 1 *Let*

$$\epsilon = \max_{0\le j\le N_2} \sqrt{\Sigma_{i=0}^{N_1}|\mu_{ij}|^2},$$

$$C(N) = \sqrt{\frac{4+(N+1)^2}{(N+1)^3 2^{2N+2}}},$$

$u_0(x,y)$ *be the solution of the 2-D HMP (2). Let* $(x_0,y_0) \in (0,1)\times(0,1)$ *is fixed, and* $\eta_2 = \sqrt{x_0^2+y_0^2}$. *If there exists a constant* $M_2 > 0$ *such that* $\| u_0(x,y) \|_{H^1[(0,1)\times(0,1)]} \le M_2$, *then we have the following local estimate for the 2-D HMP:*

$$|u_0(x_0,y_0)| \le C_3 \frac{1}{|\log[C(N_2)+\sqrt{2}C_1\frac{1}{\sqrt{N_2+1}|\log(C(N_1)+\sqrt{2}\epsilon)|^\gamma}]|^\gamma}, \tag{9}$$

where $C_3 = C_3(M_2,\eta_2) > 0$ *is a constant which depends only on* M_2 *and* η_2; $\gamma \in (0,1)$ *depends only on* η_2, *independent of* u_0 *and* N_1, N_2; C_1 *is given in lemma 1;* $0 < \epsilon < 1$.

The proof of the theorem is evident, so we omit it here.

Remark 4 *The conditional stability of double-logarithmic rate has only been derived for the 2-D HMP. This kind of stability is weaker than single-logarithmic stability, and can be optimized. We are sure that the single-logarithmic stability can be obtained if the similar method, which was proposed in the paper cheng et al* [7], *is adopted.*

4 Stabilized Algorithm for the Two-Dimensional HMP

In order to numerically solve the 2-D HMP, we use the Tikhonov regularization method presented in section 2 twice, that is to say, we do it for two 1-D HMPs (7) and (8) respectively.

On the basis of the conditional stability–Theorem 1 and the regularization method, we can obtain the error estimate.

Theorem 2 *Let*

$$\max_{0\leq j\leq N_2} \sqrt{\Sigma_{i=0}^{N_1}|\mu_{ij}^{\delta} - \mu_{ij}|^2} \leq \delta,$$

$$C(N) = \sqrt{\frac{4+(N+1)^2}{(N+1)^3 2^{2N+2}}},$$

Suppose the exact solution of the 2-D HMP (2) $u_0(x,y) \in H^1[(0,1)\times(0,1)]$, and there exists a constant $M_2 > 0$ such that $\| u_0 \|_{H^1[(0,1)\times(0,1)]} \leq M_2$. Let $\alpha = \delta^2$. Then the regularized solution $u_\alpha^\delta(x,y)$ pointwise converges to $u_0(x,y)$ in $(0,1)\times(0,1)$, and the following error estimate holds

$$|u_\alpha^\delta(x_0,y_0) - u_0(x_0,y_0)| \leq$$

$$C_4 \frac{1}{|\log[C(N_2)+\sqrt{2}(1+\sqrt{2+M_2^2})C_2 \frac{1}{\sqrt{N_2+1}|\log(C(N_1)+\sqrt{2}(1+\sqrt{2+M_2^2})\delta)|^\gamma}]|^\gamma}, \tag{10}$$

where $C_4 = C_4(M_2,\eta_2) > 0$ is a constant which depends only on M_2 and η_2; $\gamma \in (0,1)$ depends only on η_2, independent of u_0 and N_1, N_2.

The proof of the theorem is easily completed by two steps of error estimates for two 1-D HMPs. Here we omit it.

5 Numerical Examples

The following are two illustrative,numerical examples. In these examples, we first compute the moments for exact solutions $u(x,y)$ in (2). If we give a small perturbation for each exact solution $u(x,y)$, then we can calculate its noised moments, the resulting error can be controlled by δ. Thus the parameter α can be given by $\alpha = \delta^2$.

Example 1 *Consider the moment problem*

$$\int_0^1 \int_0^1 x^i y^j u(x,y)dxdy = \frac{1}{(i+3)(j+1)} + \frac{1}{(i+1)(j+3)}, \quad (i = 0, 1, \cdots, N; j = 0, 1, \cdots, M). \tag{11}$$

Its exact solution is $u(x,y) = x^2 + y^2$. The numerical results for approximate solutions $u_\alpha^\delta(x,y)$ are computed and shown for four cases:

(a) $N_1 = N_2 = 10, \delta = 0.01$;

(b) $N_1 = N_2 = 10, \delta = 0.001$;

(c) $N_1 = N_2 = 20, \delta = 0.001$;

(d) $N_1 = N_2 = 30, \delta = 0.001$.

The numerical solution approximates the exact solution very well, see figure 1.

In order to easily observe the efficiency of the presented algorithm, we give some transversal lines $u(x_0, y)$, see figure 2.

Example 2 *Consider the moment problem*

$$\int_0^1 \int_0^1 x^i y^j u(x,y)dxdy = \frac{1}{(i+3)(j+1)} - \frac{1}{(i+1)(j+3)}, \quad (i = 0, 1, \cdots, N; j = 0, 1, \cdots, M). \tag{12}$$

Its exact solution is $u(x,y) = x^2 - y^2$. The numerical results for approximate solutions $u_\alpha^\delta(x,y)$ are computed and shown for three cases:

(a) $N_1 = N_2 = 10, \delta = 0.001$;

(b) $N_1 = N_2 = 20, \delta = 0.001$;

(c) $N_1 = N_2 = 30, \delta = 0.001$.

The numerical solution approximates the exact solution very well, see figure 3.

Similarly we give some transversal lines $u(x_0, y)$ to observe the efficiency of the algorithm, see figure 4.

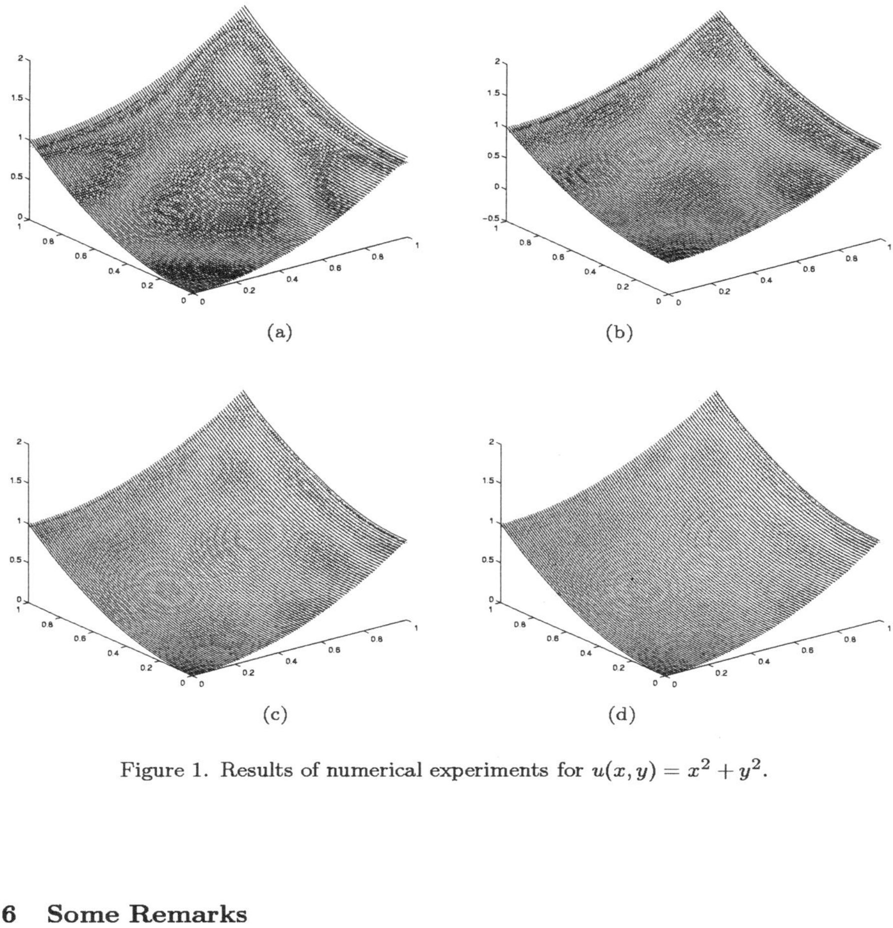

Figure 1. Results of numerical experiments for $u(x, y) = x^2 + y^2$.

6 Some Remarks

Remark 5 *Our results in the paper can be applicable in the numerical treatment for some convolution equations of the first kind and for some inverse problems.*

Remark 6 *Our results in the paper can be used to numerically discuss the analytic continuation for potential functions and t he inversion of Laplace integral transformation.*

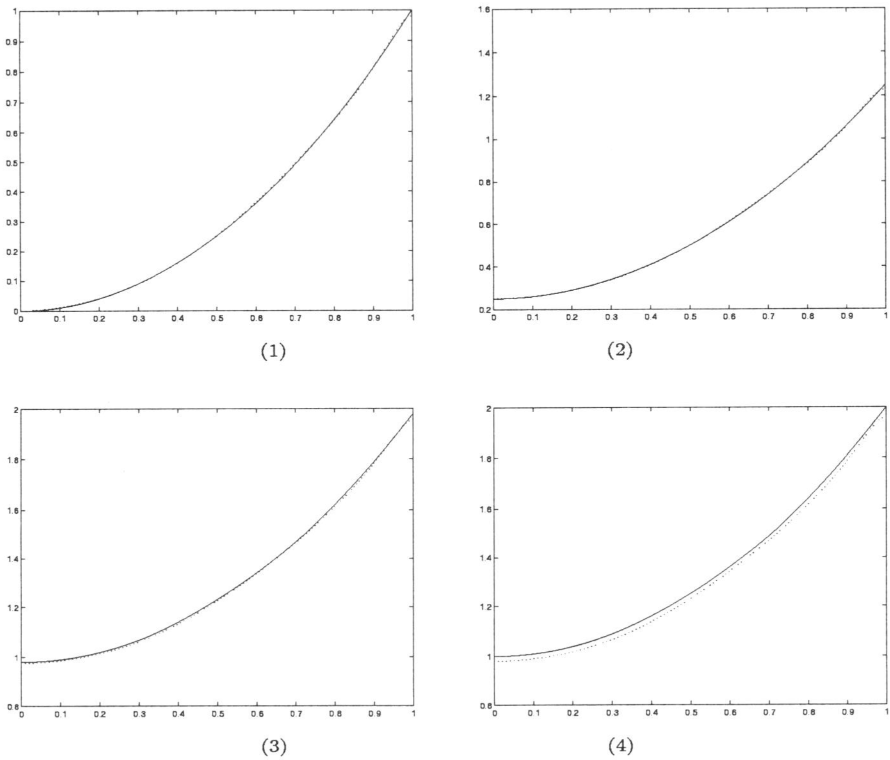

Figure 2. Results of transversal lines $u(x_0, y)$.

Acknowledgments

The authors are supported by the Jiangxi Provincial Natural Scientific Foundation, Shanghai Municipal Natural Scientific Foundation and Scientific Research Program from East China Geological Institute.

References

1. D. D.Ang , R.Gorenflo and D. D.Trong, *A multidimensional Hausdorff moment problem: Regularization by finite moments*, Zeitschrift für Analysis und ihre Anwendungen(Journal for Analysis and its Applications) **18**, 13–25 (1999).

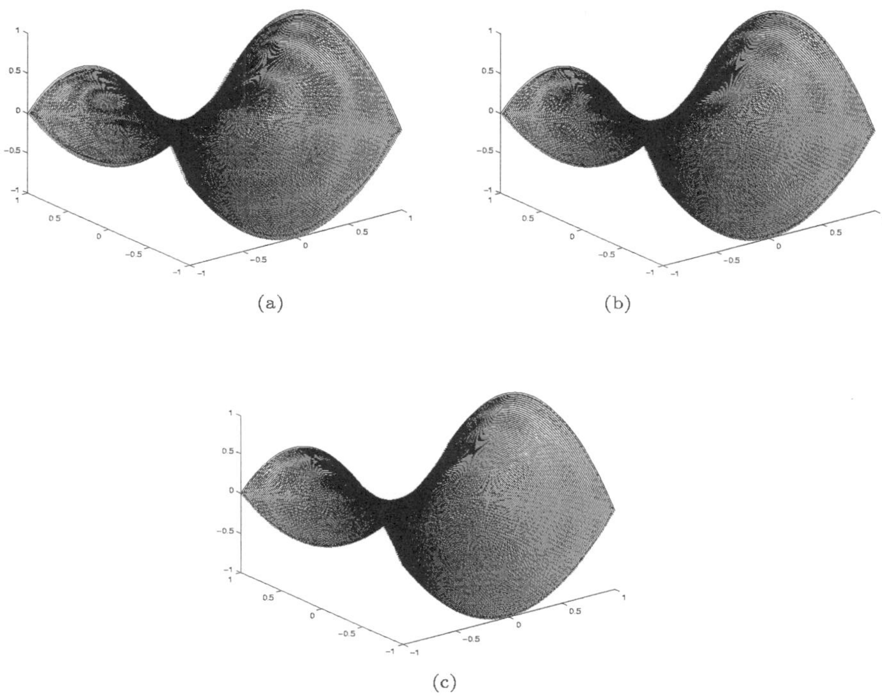

Figure 3. Results of numerical experiments for $u(x,y) = x^2 - y^2$.

2. D.D.Ang, L. K. Vy and R. Gorenflo, *A regularization method for the moment problem*, in *Inverse Problems: Principles and Applications in Geophysics, Technology and Medicine* Math. Research **74** , 37-45 (1993)(Berlin:Akademic Verlag)
3. R. Askey , I. J. Schoenberg and A. Sharma, *Hausdorff moment problem and expansion in Legendre polynomials*, J. Math. Anal. Appl. **86**, 237–245(1983).
4. G. E.Backus and J. F.Gilbert, *The resolving power of gross earth data*, Geophysical Journal of the Royal Astronomical Society **16**, 169–205 (1968).
5. G. Bruckner and J.Cheng, *Tikhonov regularization for an integral equation of the first kind with logarithmic kernel*, J. Inverse and Ill-posed

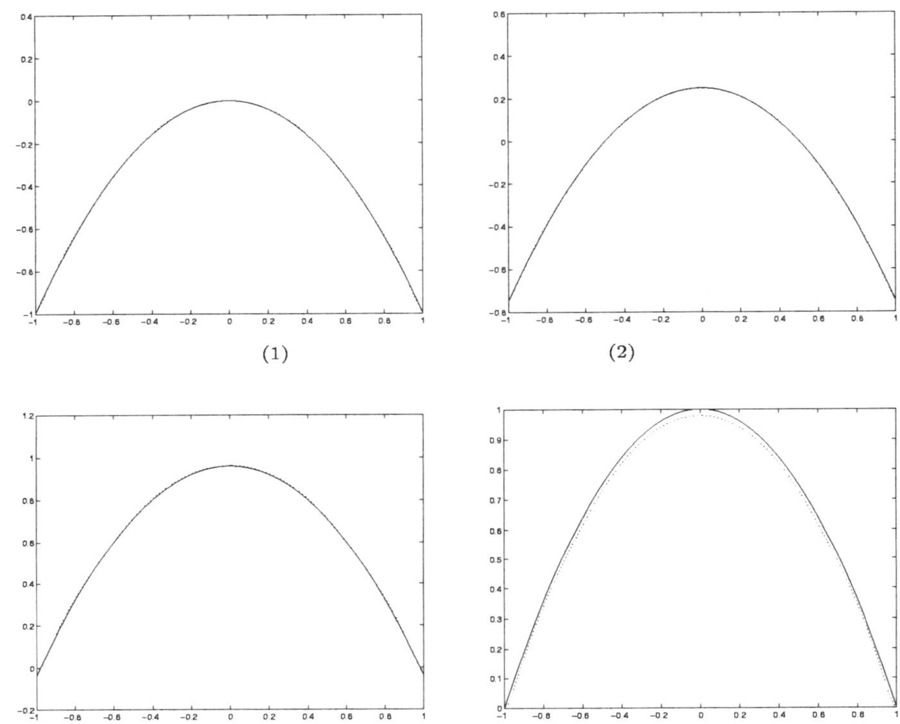

(1) (2) (3) (4)

Figure 4. Results of transversal lines $u(x_0, y)$.

Problems ,–(2000).

6. J.Cheng and M.Yamamoto, *One new strategy for a priori choice of regularizing parameters in Tikhonov regularization*, Inverse Problems **16**, L31–L38 (2000).
7. J. Cheng, D. H. Xu and M.Yamamoto, *An inverse contact problem in the theory of elasticity*, Mathematical Methods in the Applied Sciences **22**, 1001–1015 (1999).
8. H. W.Engl, A. K.Lions and W. Rundell, *Problems in Medical Imaging and Nondetructive Testing* (Springer,NewYork, 1996)
9. G. Inglese, *Recent results in the study of the moment problem*, in *Theory and Practice of Geophysical DataInversion*,ed. A Vogel et al), 73–84 (Braunschweig und Wiesbadan: Vieweg-Verlag, 1992).

10. A. Kirsch, B. Schomburg and G.Berendt, *The Backus-Gilbert method*, Inverse Problems **4**, 771–783(1988).
11. W. Rudin, *Real and Complex Analysis* (McGraw Hill, New York, 1966).
12. J. A. Shohat and J. D. Tamarkin, *The Problem of Moment*, Math. Surveys(Providence, RI: Am. Math. Soc.,1943).
13. G. Talenti, *Recovering a function from a finite number of moments*, Inverse Problems **3**, 501–517 (1987).
14. L. Wang, *A modified method for linear moment problem*, Mathematica Numerica Sinica **21**, 303–308 (1999)(in Chinese).
15. D. H. Xu, S. X.Huang and M. Z.Li, *Local conditional stability and numerical analysis for Hausdorff Moment Problems*, Inverse Problems, submitted.

A NOVEL HYBRID GENETIC ALGORITHM AND ITS APPLICATION TO INVERSE PROBLEMS IN MEMS

Y.G. XU AND G.R. LIU

Center for Advanced Computations in Engineering Science, Singapore-MIT Alliance
Department of Mechanical Engineering, National University of Singapore, 10 Kent Ridge Crescent, Singapore 119260
E-mail: smaxuyg@nus.edu.sg; mpeliugr@nus.edu.sg

H. OHTSUBO

Department of Naval Architecture and Ocean Engineering Faculty of Engineering, The University of Tokyo, Japan

A novel hybrid genetic algorithm is proposed in this paper for solving inverse problems in microelectromechanical systems (MEMS). The new algorithm presents two hybridization operations in order to speed up the convergence process. It takes only 4.1% $\sim$ 4.7% number of function evaluations required by the conventional genetic algorithm to reach global optima for the benchmark functions tested. The new algorithm is then used for solving two inverse problems. One is the identification of flow-pressure characteristic parameters of the valve-less micropumps. The other is the identification of material property parameters and bonding quality of the piezoelectric patches. Numerical simulations have shown the very satisfactory results.

1 Introduction

Hybrid genetic algorithms (GAs) have been known as the effective optimization technique for solving the complicated optimization problems [1,2,3]. As the hybrid algorithms combine the globe explorative power of conventional GAs with the local exploitation behaviors of deterministic optimization methods, they usually outperform the conventional GAs or deterministic optimization methods to be individually used in engineering practice.

In this study, a new hybrid genetic algorithm (called nhGA) is proposed. It presents two hybridization operations. The first one is to use a simple interpolation method to move the best individual produced by the conventional genetic operations to an even better neighboring point in each of generations. The second one is to use a hill-climbing search to move a randomly selected individual to its local optimum. This may be done only when the first hybrid operation fails to improve the best individual consecutively in several generations. Compared with the other hybrid GAs, the nhGA is not only excellent in the convergence performance, but also very simple and easy to be implemented in engineering practice.

As an effective optimization method, the nhGA is used for solving two inverse problems in MEMS. The first one is to identify the dynamic flow-pressure characteristic parameters of the valve-less micropumps. The second one is to identify the material property parameters and bonding quality of the piezoelectric patches. Both of them have demonstrated the excellent performance of the nhGA for inverse problems.

2 Hybrid Genetic Algorithm (nhGA)

2.1 Algorithm Description

Basically, the nhGA proposed in this study is the further development for the hybrid GA called hGA[4]. As the hGA has been discussed in detail in Ref. of Xu *et al* [4], which may be used as a reference to explain the mechanism of nhGA, it is decided herein to only give a brief description for the implementation process of nhGA as follows:

(1) j=0, start up the evolutionary process.

  (a) Select the operation parameters including population size N, crossover possibility p_c, mutation possibility p_m, random seed i_d, control parameter α and β (see Xu *et al* [4]), etc.

  (b) Initialize N individuals, $P(j)=(p_{j1}, p_{j2}, \ldots, p_{jN})$, using a random method. Every individual p_{ji} (i=1,...,N) is a candidate solution.

  (c) Evaluate the fitness values of $P(j)$.

(2) Check the termination condition. If "yes", the evolutionary process ends. Otherwise, $j = j+1$ and proceed to next step.

(3) Carry out the conventional genetic operations in order to generate the offspring, i.e. the next generation of solutions, $C(j)=(c_{j1}, c_{j2}, \ldots, c_{jN})$. These operations to be used include niching[5], selection[1], crossover[1], elitism[5], etc.

(4) Implement the first hybridization operation.

  (a) Construct the move direction $\bar{d}$ of best individual.

$$\bar{d} = (c_j^b - c) \tag{1}$$

$$c = \begin{cases} c_{j-1}^b & c_j^b \neq c_{j-1}^b \\ c_j^s & c_j^b = c_{j-1}^b \end{cases} \tag{2}$$

where c_{j-1}^b is the best individual in $C(j\text{-}1)$ at the $(j-1)$-th generation, c_j^b and c_j^s are the best and second best individuals in $C(j)$ at the j-th generation, respectively.

(b) Generate two new individuals c_1, c_2, and evaluate their fitness values.

$$c_1 = c_j^b + \alpha \bar{d} \tag{3}$$

$$c_2 = c_{j-1}^b + \beta \bar{d} \tag{4}$$

where α and β are control parameters. They are recommended to be within $0.1 \sim 05$ and $0.3 \sim 0.7$, respectively.

(c) Select a better individual c_m,

$$f(c_m) = \max\{f(c_1), f(c_2)\} \quad c_m \in \{c_1, c_2\} \tag{5}$$

$f(.)$ is the fitness function.

(d) Replace the individual c_j^b in $C(j)$ with the individual c_m. This results in an upgraded offspring $C_u(j)$=$(c_{j1}, c_{j2}, \ldots, c_m, \ldots, c_{jN-1})$.

(e) Check if there occurs population convergence in $C_u(j)$. If "yes", implement restarting strategy[4] to generate the new $C(j)$.

(5) Check if the best individual keeps unimproved consecutively in the M generations (M=3~5). If "yes', implement the second hybrid operation as follows.

(a) Randomly select a individual c_{ji} in $C_u(j)$.

(b) Take c_{ji} as an initial point to start the hill-climbing search.

(c) Replace individual c_{ji} with the local optimum c_{jL} obtained by the hill-climbing search.

(6) Go back to step (2).

It is clear from the above description that the newly proposed nhGA, compared with the previous hGA, does not incur any deterioration of population diversity when incorporated with the hybridization operations.

2.2 Performance Tests

Three benchmark functions are used to test the nhGA. Each of benchmark functions has lots of local optima and one or more global optima. Figure 1 shows the search space of function $F1$.

$$\text{F1: } f(x_1, x_2) = \prod_{i=1}^{2} \sin(5.1\pi x_i + 0.5)^{80} e^{-4 \log 2 (x_i - 0.0667)^2 / 0.64}$$

$$\pi = 3.14159, \quad 0 < x_i < 1.0, \quad i = 1, 2$$

$$\text{F2: } f(x_1, x_2, x_3) = \sum_{i=1}^{10} \{e^{-ix_1/10} - e^{-ix_2/10} - [e^{-i/10} - e^{-i}]x_3\}^2$$

$$-5 < x_i < 15, \quad i = 1, 2, 3$$

$$\text{F3: } f(x_1, ..., x_5) = \pi\{10\sin(\pi x_1)^2 + \sum_{i=1}^{4} [(x_i - 1)^2(1 + 10\sin(\pi x_{i+1})^2]\}/5 + (x_5 - 1)^2$$

$$\pi = 3.14159, -10 < x_i < 10, \quad i = 1, \ldots, 5$$

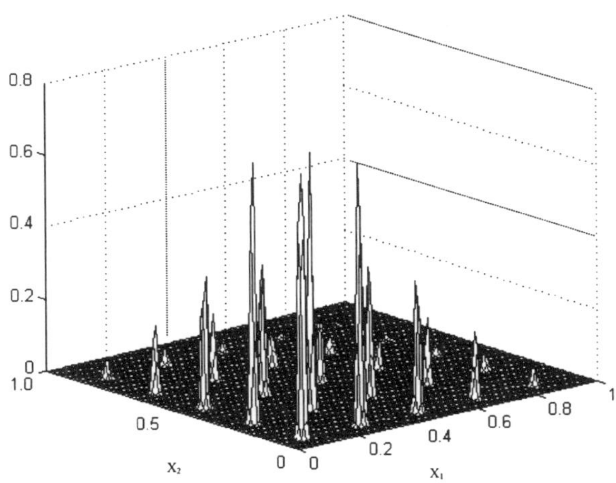

Figure 1. Search space of benchmark function F1.

For each of benchmark functions, the nhGA runs 10 times with the different random seed i_d. The 10 random seeds are -1×10^2, -5×10^2, -1×10^4, -1.5×10^4, -2×10^4, -3×10^4, -3.5×10^4, -4×10^4, -4.5×10^4, -5×10^4, respectively. The other operation parameters are N=5, p_c=0.5, p_m=0.02, α=0.2, $\beta = 0.5$

and M=3. Tournament selection, one child, niching, elitism are chosen to use. Table 1 shows the mean numbers of function evaluations, $\overline{n}$ and $\overline{n}_m$, that are taken to reach the global optima using the nhGA and conventional mGA[5], respectively. It can be found that the nhGA demonstrates a much faster convergence than the conventional mGA.

Table 1. Mean numbers of function evaluations to convergence.

No.	Global Optimum	Func value	$\bar{n}$	$\bar{n}_m$	$\bar{n}/\bar{n}_m$ (%)
F1	(0.0669, 0.0669)	1.0	141	3365	4.2
F2	(1, 10, 1)	0.0	237	5745	4.1
F3	(1, 1, 1, 1, 1, 1)	0.0	6637	139915	4.7

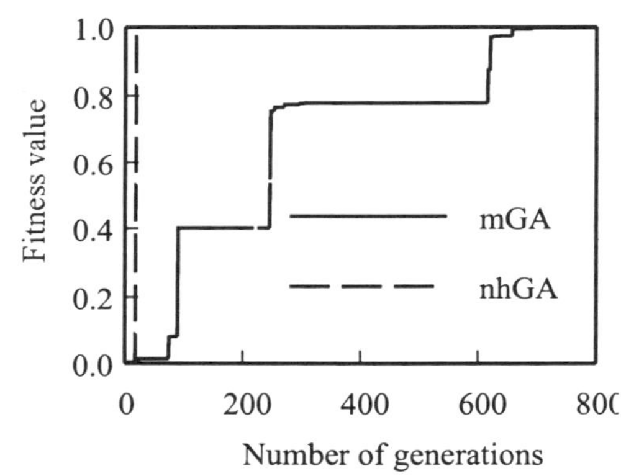

Figure 2. Convergence process in view of generations.

Figure 2 shows the convergence processes of benchmark function $F1$ when using the nhGA against the mGA, from which comparison of the convergence processes between nhGA and mGA can be seen more clearly.

3 Inverse Problem Solving

3.1 Parameter Identification of the Valve-less Micropumps

Figure 3 schematically shows a valve-less micropump. The pressure-loss coefficients, ζ_p and ζ_n, in the flow channels can be optimally solved from the following objective function[6]:

$$\min E(\zeta_p, \zeta_n) = (\sum_{i=1}^{n} |Q_i(\zeta_p, \zeta_n) - Q_i^m|^2)^{\frac{1}{2}} \tag{6}$$

$$\zeta_{pmax} \leq \zeta_p \leq \zeta_{pmin}, \ \zeta_{nmax} \leq \zeta_n \leq \zeta_{nmin}$$

$$i = 1, \ldots, K$$

$Q_i(\zeta_p, \zeta_n)$ is the mean flux calculated from a complicated model[5] using the trial ζ_p and ζ_n, Q_i^m is the measured mean flux at the i-th trial. K is the number of trials.

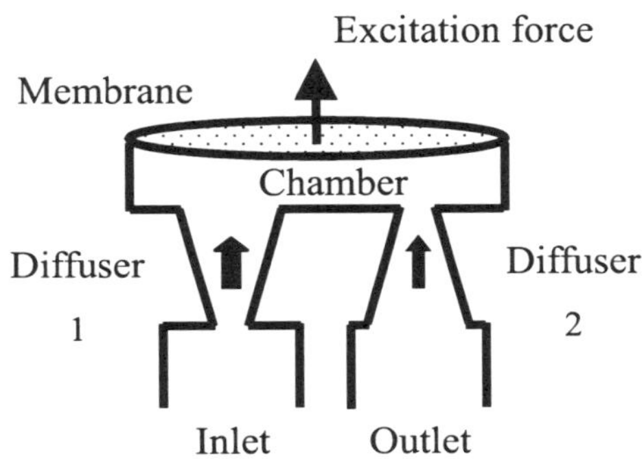

Figure 3. Cross-sectional view of a micropump.

Table 2. Solutions for 3 simulated cases.

	n	ζ_p	ζ_n	$e(\zeta_p)$ (%)	$e(\zeta_n)$ (%)
Case I	790	1.389	0.918	-4.9	-3.4
Case II	767	1.307	0.894	2.1	2.8
Case III	525	1.112	0.443	5.9	5.5

The nhGA is used for solving this problem. Table 2 shows the corresponding solutions for 3 simulated cases. In Table 2, n is the number of function evaluations taken by the nhGA, ζ_p and ζ_n are the solved pressure-loss coefficients, $e(\zeta_p)$ and $e(\zeta_n)$ are the errors with respect to their actual values, respectively. It can be seen that nhGA converges to the satisfactory results very fast. The maximal error of solved ζ_p and ζ_n are only -4.9%, 2.8% and 5.9% for 3 simulated cases, respectively.

3.2 Identification of property Parameters and Bonding Equality of a Piezoelectric Patch

Piezoelectric (PZT) patches have been widely used as actuators and sensors in MEMS. Their property parameters and bonding equalities are usually required to calibrate in order to obtain the accurate analysis results[7]. In this study, we only take account of the dielectric constant $\varepsilon_{31}^{\sigma}$, piezoelectric constant d_{31}, elastic modulus E_{11}^{E} and coefficient ξ which represents the equality of bonding layer[7]. They would be identified using the nhGA. As usually done, an optimization problem is formed as follows to this end.

$$\min MSF = \sum_{i=1}^{N} \left[Re(Y_i) - Re(Y_{mi})\right]^2 \tag{7}$$

where

$$Y = j\omega a(\varepsilon_{11}^{\sigma} - d_{31}^2 E_{11}^E + \frac{d_{31}^2 E_{11}^E Z_a}{Z_a + \xi\ Z_s}\frac{\tan(kl)}{kl}) \tag{8}$$

N is the number of frequency sampling, $\mathrm{Re}(Y_i)$ and $\mathrm{Re}(Y_{mi})$ are the real parts of calculated and measured electric admittance of PZT patch at sampling point i, respectively. Figure 4 shows the effect of the coefficient ξ on the electric admittance for a one-dimension example [7]. The other parameters in Eq. (8) can be found in Ref. Xu and Liu[7].

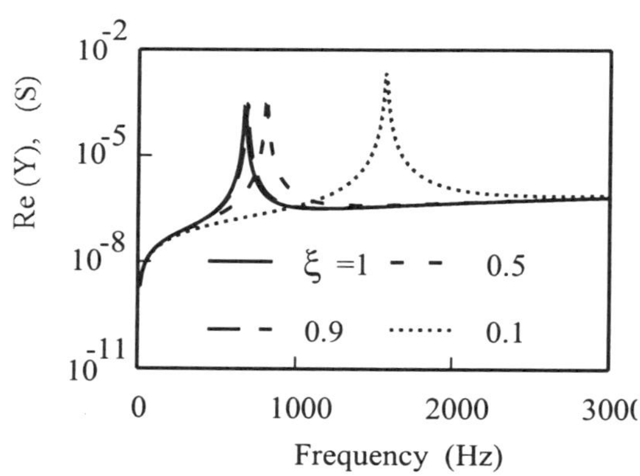

Figure 4. Effect of coefficient ξ on admittance.

We have set 3 simulated cases, in which the 4 parameters to be identified are 85%, 100% and 115% of their nominal values, respectively[7]. With the given parameter values in each case, the electric admittance calculated from Eq. (8) is taken as the measured Y_m. Then, these parameters are allowed to vary within the range of from 50% to 150% off from their nominal values. The nhGA is used to find the optimal solution. It is found out that the maximal errors of identified 4 parameters with respect to their specified values are only 4.3%, 3.7% and 4.8%, respectively. The computation costs are also very low. The maximal number of function evaluations required is 873.

4 Conclusions

In this study, a novel nhGA is proposed and validated using 3 benchmark functions. It is also used to solve two typical inverse problems in MEMS. Numerical examples have demonstrated its effectiveness and efficiency. This provides a new choice for solving complicated optimization problems as well as inverse problems in engineering practice.

References

1. D. E. Goldberg, *Genetic Algorithms in Search, Optimization, and Machine Learning* (Addison-Wesley Publishing Company, USA, 1989).
2. T. Back, U. Hammel and H. P. Schwefel, *Evolutionary computation: comments on the history and current state*, IEEE Trans. Evol. Comput. **1**, 3–17 (1997).
3. M. Gen and R. W. Chen, *Genetic Algorithms and Engineering Design* (John Wiley & Sons, New York, 1997).
4. Y. G. Xu, G. R. Liu and Z. P. Wu, *A novel hybrid genetic algorithm using local optimizer based on heuristic pattern move*, Appl. Artif. Intell. An Int. J. **15**, 601–631 (2001).
5. D. L. Carroll, *Genetic algorithms and optimizing chemical oxygen-iodine lasers*, Developments in Theor. and Appl. Mech. Uni. of Alabama **10**, 411–424 (1996).
6. Y. G. Xu, G. R. Liu, L. S. Pan and N. Y. Ng, *Parameter Identification of Dynamic Flow-Pressure Characteristics in Valve-less Micropumps*, Sensors and Actuators A (2001),submitted.
7. Y. G. Xu and G. R. Liu, *A Modified Electro-Mechanical Impedance Model of Piezoelectric Actuator-Sensor for Debonding Detection of Composite Repair Patch*, J. of Intelli. Materi. Sys. Struc. (2001), Submitted .

Section IV

Solutions to Applied Inverse Problems

RESTORING IMAGES WITH REGULARIZATION IN UNCORRELATED TRANSFORM DOMAIN

YIK-HING FUNG, YUK-HEE CHAN* AND WAN-CHI SIU
Department of Electronic and Information Engineering, Hong Kong Polytechnic University, Hong Kong
E-mail: enyhfung@polyu.edu.hk, enyhchan@polyu.edu.hk, enwcsiu@polyu.edu.hk

SHEUNG-ON CHOY
Department of Applied Computing, Hong Kong Open University, Hong Kong
E-mail: sochoy@ouhk.edu.hk

Conventional spatially adaptive regularized image restoration schemes weight the amount of regularization according to the spatial content of an image. A better performance can be achieved by first separately decorrelating the available information about the signal under analysis into uncorrelated components and then weighting the amount of regularization performed to these components accordingly. An iterative restoration algorithm is proposed accordingly to restore images which are blurred and corrupted with color noise.

1 Introduction

In image restoration, an image degradation process can be generally formulated by $y = Hx + n$, where x and y are the lexicographically ordered original and degraded images, n is a noise vector and H represents a linear degradation operator[1].

Solving this equation directly to get x from the observable y is basically an ill-posed problem[2]. Restoration methods based on regularization theory[3] are widely used instead to get an estimate of x, say $\hat{x}$. The constrained optimal method[4] is the simplest methodology to realize regularization. In this method, an algebraic objective function of $\hat{x}$ is defined based on different constraint sets. The solution is then obtained by minimizing the objective function with respect to $\hat{x}$.

In general, two constraints are used. One of them tries to keep the solution faithful to the information provided by the observed version y, which is usually given as $||y - H\hat{x}||^2 < \epsilon$, while the other one tries to remain the solution faithful to the *a priori* information about the original image, which can be generally given as $||L\hat{x} - L\check{x}||^2 < e$. Here, $\check{x}$ represents our existing knowledge about the solution, which is represented in a form of image, and L is a linear

*CORRESPONDING AUTHOR

operator used to extract features of a given input. The bounds ϵ and e, respectively, tell the relative significance of the constraints to the solution and hence should be used to weight the contribution of the constraints in constructing the objective function[5]. The objective function derived from this idea is given as $J = ||y - H\hat{x}||^2 + \alpha||L(\hat{x} - \check{x})||^2$, where $\alpha = \epsilon/e$.

Obviously, different elements of $y - H\hat{x}$ make different amount of contribution to the error function $||y - H\hat{x}||^2$ in fulfilling the constraint $||y - H\hat{x}||^2 < \epsilon$. Their contribution should be weighted so as to have a good solution $\hat{x}$. A similar case happens when we investigate the elements of $L(\hat{x} - \check{x})$. By taking these factors into account, the objective function should be modified to be

$$J = ||y - H\hat{x}||_R^2 + \alpha||L(\hat{x} - \check{x})||_S^2 \tag{1}$$

Here $|| \bullet ||_R^2$ and $|| \bullet ||_S^2$ denote weighted norms. This generalized formulation describes almost all spatially adaptive regularized restoration methods reported in the literature[6,7,8,9,10]. In general, these methods differ by their ways to evaluate R and S. To reduce the complexity of their realization, both R and S are usually oversimplified to be diagonal matrices in these methods. Accordingly, each element of $y - H\hat{x}$ and $L(\hat{x} - \check{x})$ is weighted separately. This implies the elements are considered to be independent of each other. This is obviously not true as adjacent image pixels in an image are highly correlated. By using such a simplified weighting approach, the weighting effect of different weighting factors may counteract each other and hence not be able to provide a good restoration result effectively.

To solve this problem, $y - H\hat{x}$ and $L(\hat{x} - \check{x})$ are considered as two different signals and separately decomposed into a number of uncorrelated channels by transforms for weighting. By doing this, two advantages can be gained. First, it is easier for one to determine the weighting factor for a particular channel as uncorrelated channels do not interact with each other. Weighting a particular channel will not affect the other channels. The second advantage also comes from the decorrelation property of the image transform. In practical circumstances, one has to estimate the weighting factors from either the distorted image or the *a priori* information, so there must be some estimation errors. The less correlated the channels are, the less sensitive is the restoration result to the estimation errors in the channels.

In this paper, based on the aforementioned idea, the transform theory[11] is used to decorrelate images into uncorrelated transform components and these components are then weighted according to their variances. Simulation results show that the proposed approach can improve the restoration performance as compared with other conventional spatially adaptive approaches[6,7,8,9,10].

Note there are some reported literatures which consider a distorted image as a multichannel signal and restore it in the frequency domain[12]. However, their motivations and implementations are quite different from those of this paper. Generally speaking, decorrelating the signal before weighting is not their basic concern. In these approaches, the discrete Fourier transform (DFT) is typically used to decompose the distorted image into a number of frequency channels for subsequent restoration. Since DFT components are still correlated, these approaches can be considered as simultaneously performing spatial weighting schemes to a number of subband images.

2 Algorithms

Suppose b is a vector of random variables. The value of each random variable is of a certain uncertainty but its statistical characteristics are known or can be estimated. Without lose of generality, it is assumed that $\mathrm{E}[b] = \vec{0}$, where $\mathrm{E}[\bullet]$ is the expectation operator and $\vec{0}$ denotes the zero vector. Note, since its statistical characteristics are known, b can always be zero-meaned. Assume T is the unitary Karhunen-Loève transform (KLT) of b [13]. Then T can completely decorrelate b and $\mathrm{E}[Tbb^tT^t]$ is a diagonal matrix. The ith diagonal element of the matrix, denoted as $\langle \mathrm{E}[Tbb^tT^t]\rangle_{ii}$, is the variance of $[Tb]_i$, where $[Tb]_i$ is the ith element of Tb. In formulation, we have $\langle \mathrm{E}[Tbb^tT^t]\rangle_{ii} = \mathrm{E}[[Tb]_i^2]$. Obviously, $\mathrm{E}[[Tb]_i^2]$ indicates the relative degree of uncertainty of $[Tb]_i$ with respect to the other elements of Tb. This information can hence be used to weight the contribution of each element of Tb to $||Tb||^2$. Specifically, the weighting factor should be proportional to $1/\mathrm{E}[[Tb]_i^2]$. If $\mathrm{E}[bb^t]$ is the *a priori* information we know about b, then $\mathrm{E}[[Tb]_i^2]$ can be easily determined as $\mathrm{E}[[Tb]_i^2] = \langle \mathrm{E}[Tbb^tT^t]\rangle_{ii} = \langle T(\mathrm{E}[bb^t])T^t\rangle_{ii}$.

Based on the idea described, the objective function can be given as

$$J = ||T_1(y - H\hat{x} - M)||_R^2 + \alpha||T_2(L(\hat{x} - \check{x}) - M_f)||_S^2 \tag{2}$$

where T_1 and T_2 are the KLTs for $y - H\hat{x} - M$ and $L(\hat{x} - \check{x}) - M_f$ respectively. Here, $M = \mathrm{E}(y - H\hat{x})$ and $M_f = \mathrm{E}[L(\hat{x} - \check{x})]$. The weighting parameter α should be determined as $\alpha = \epsilon_1/e_1$, where ϵ_1 and e_1 are the bounds of $||T_1(y - H\hat{x} - M)||_R^2$ and $||T_2(L(\hat{x} - \check{x}) - M_f)||_S^2$ respectively. The weighting matrices R and S are diagonal matrices intrinsically and their ith diagonal elements can be determined as

$$r_i = \frac{1}{\langle \mathrm{E}[T_1(y - H\hat{x} - M)(y - H\hat{x} - M)^tT_1^t]\rangle_{ii}} \tag{3}$$

$$s_i = \frac{1}{\langle \mathrm{E}[T_2(L(\hat{x} - \check{x}) - M_f)(L(\hat{x} - \check{x}) - M_f)^tT_2^t]\rangle_{ii}} \tag{4}$$

Note eqns. (2)-(4) provide the general formulations for restoring a degraded image.

Now consider the case when a smoothness constraint is applied. In such a case, we can let $L(\hat{x}-\check{x})$ be $C\hat{x}$, where C is a spatial 2D highpass Laplacian filter represented in matrix form. As $C\hat{x}$ theoretically contains no low frequency component, it is safe to assume $M_f = \mathrm{E}[C\hat{x}] = \vec{0}$ to simplify the analysis. When $y - Hx$ is a stationary zero-mean color noise, it can be modelled with a linear noncausal all-pole signal model such that $F(y - Hx)$ is a zero-mean white noise of variance σ_n^2, where F is, represented in matrix form, the corresponding filter derived based on the signal model of $y - Hx$. In that case, we can assume $M = \mathrm{E}[y - H\hat{x}] = \vec{0}$ and $\mathrm{E}[(F(y - H\hat{x}))(F(y - H\hat{x}))^t] = \sigma_n^2 I$. This implies $T_1 = F$ and $r_i = 1/\sigma_n^2$. Hence, the objective function (2) can be simplified as

$$J = \frac{1}{\sigma_n^2}||F(y - H\hat{x})||^2 + \alpha||T_2 C\hat{x}||_S^2 \tag{5}$$

The minimization of J with respect to $\hat{x}$ results in the normal equation

$$(H^t F^t F H + \alpha\sigma_n^2 C^t T_2^t S T_2 C)\hat{x} = H^t F^t F y \tag{6}$$

In general, $\hat{x}$ cannot be evaluated directly from this equation as it requires the inversion of a huge matrix. An alternative approach is to use a steepest-descent algorithm to approximate $\hat{x}$ iteratively. This approach leads to the following iterative equation:

$$\begin{aligned} &\hat{x}_0 = \beta H^t F^t F y \\ &\hat{x}_{k+1} = \hat{x}_k + \beta(H^t F^t F(y - H\hat{x}_k) - \alpha\sigma_n^2 C^t T_2^t S T_2 C\hat{x}_k) \end{aligned} \tag{7}$$

where $\hat{x}_k$ is the estimate of $\hat{x}$ at the $\mathrm{k^{th}}$ iteration. The iteration converges if β satisfies the condition $0 < \beta < 2/\lambda$, where λ is any eigenvalue of the matrix $H^t F^t F H + \alpha\sigma_n^2 C^t T_2^t S T_2 C$.

The weighting matrix S can be estimated at each iteration based on the available form of the restored image $\hat{x}_k$. By substituting the assumptions mentioned earlier into eqn. (4), we have

$$s_i = \frac{1}{\langle \mathrm{E}[T_2 C\hat{x}_k (T_2 C\hat{x}_k)^t]\rangle_{ii}} \tag{8}$$

Note that $\langle \mathrm{E}[T_2 C\hat{x}_k (T_2 C\hat{x}_k)^t]\rangle_{ii}$ is in fact the variance of the $\mathrm{i^{th}}$ element of $T_2 C\hat{x}_k$. In practice, it is estimated with the ensemble $\Omega = \{T_2(C\hat{x}_k)^{\langle m,n\rangle} : |m|, |n| \leq d\}$, where d is an integer parameter which defines the size of the ensemble and $(C\hat{x}_k)^{\langle m,n\rangle}$ denotes the shift version of $C\hat{x}_k$ obtained by shifting all its elements m steps up and n steps right in the spatial domain. Specifically, its estimated value θ_i is given as

$$\theta_i = \frac{1}{(2d+1)^2} \sum_{m=-d}^{d} \sum_{n=-d}^{d} [T_2((C\hat{x}_k)^{\langle m,n \rangle} - M_c)]_i^2, \tag{9}$$

where

$$M_c = \frac{1}{(2d+1)^2} \sum_{m=-d}^{d} \sum_{n=-d}^{d} (C\hat{x}_k)^{\langle m,n \rangle} \tag{10}$$

In contrast to the approaches which make use of local spatial variance[6,7,8], the proposed approach approximates weighting factors in the transform domain. As we have mentioned in previous section, it would be helpful to obtain a better restoration result because of the lower sensitivity to approximation error in the transform domain.

The value of θ_i could fluctuate violently from i to i, so s_i may not be stable if one directly lets s_i be $1/\theta_i$ with eqn. (8). In order to make s_i stable, the equation $s_i = 1/(1+\kappa\theta_i)$ is used instead to confine s_i in the interval (0,1]. The parameter κ is a tuning parameter that can be adjusted experimentally to make the weighting effect be able to provide a good restoration result from the human visual point of view.

Due to the difficulty encountered in determining a KLT kernel of large size and the huge computational complexity required for realizing a KLT[11], an approximation of the KLT involved in the proposed scheme is performed. In practice, the discrete cosine transform (DCT) is used instead to decorrelate the unstacked image $C\hat{x}$. This is because, according to the image transform theory, an image can typically modelled as a highly correlated 2D Markov-I signal and the DCT is asymptotically equivalent to the KLT in decorrelating such signals of this kind[13]. Other reasons for using the DCT are that there are a number of fast algorithms for its realization and its realization complexity is much lower than that of the KLT.

3 Simulation Studies

Simulations were carried out to evaluate the performance of the proposed restoration scheme on a set of 256-level gray-scale digital images of size 256×256 each. In particular, it was hoped to find out whether weighting decorrelated components is more effective than weighting correlated components in providing a good image restoration performance. To achieve this, a conventional scheme was realized as well for comparison[6]. These two schemes are more or less the same except that the proposed scheme weights the components after decorrelating them while the other one does not. Hereafter, they are, respectively, referred to as non-spatially adaptive weighting (NAW) and spatially adaptive weighting (SAW) schemes.

In the realization of the proposed scheme, the decorrelation transform was approximated with periodic 8×8 two-dimensional DCT transform kernels. Specifically, to decorrelate the unstacked image $C\hat{x}$, it was first partitioned into a number of non-overlapped subimages of size 8×8 and then an 8×8 DCT was performed on each of them. Images are actually not stationary signal. Using block-based transform enables the weighting matrix to adapt to the local characteristics of an image and saves realization effort as compared with using a single 256×256 DCT.

As for the realization of the SAW scheme[6], the solution was obtained by the following iterative equations:

$$\begin{aligned} \hat{x}_0 &= \beta_o H^t y \\ \hat{x}_{k+1} &= \hat{x}_k + \beta_o(H^t(y - H\hat{x}_k) - \alpha_o C^t S' C\hat{x}_k) \end{aligned} \tag{11}$$

Here, S' is a diagonal matrix whose i^{th} diagonal element is given as $s'_i = 1/(1+\kappa\theta'_i)$, where θ'_i is the local spatial variance of the corresponding pixel. In particular, we have $\theta'_i = \frac{1}{(2d+1)^2}\sum_{m=-d}^{d}\sum_{n=-d}^{d}[(\hat{x}_k)^{\langle m,n\rangle} - M_s]_i^2$ and $M_s = \frac{1}{(2d+1)^2}\sum_{m=-d}^{d}\sum_{n=-d}^{d}(\hat{x}_k)^{\langle m,n\rangle}$.

In our simulation, testing images were first blurred and then color noise was added to the blurred images. The color noise was generated by filtering a white noise of variance σ_n^2 with a separable 2D filter composed of 2 identical 1D filters of system function $(-.2z+1.04-.2/z)$. The variance σ_n^2 was adjusted to generate a color noise achieving a particular SNR level, where SNR is defined as SNR=10log(variance of signal/variance of noise). For the NAW scheme, the 1D filter used to make up F was estimated as a 11-tap symmetric filter with a least-squares method. Parameters α_o and α were, respectively, estimated to be $\alpha_o = \sigma_{nc}^2/10\|Cy\|_{S'}^2$ and $\alpha = \|F(y-Hy)\|^2/10\sigma_n^2\|TCy\|_{S}^2$, where σ_{nc}^2 was the variance of the color noise. For both schemes, $\kappa = 0.05$ and $d = 1$ were used. The termination criterion was $\|\hat{x}_k - \hat{x}_{k+1}\|^2/\|\hat{x}_k\|^2 < 0.000005$. Table 1 summaries the objective performance of both schemes in different cases. On average, the SNR improvements achieved by the NAW scheme can be 0.4dB and 1dB, respectively, higher than that of the SAW scheme when handling noisy defocus-blurred images and noisy motion-blurred images.

Figure 2 shows parts of the restoration results of different schemes when defocus blur is involved while Figure 3 shows the case when motion blur is involved. Figure 1 shows the corresponding parts of the original testing images for reference. The SAW results generally contain more high-frequency noise, which implies that the NAW scheme is more effective in removing the color noise.

4 Conclusions

Conventional spatially adaptive regularized image restoration schemes weight the amount of regularization performed to different pixels of the solution according to the spatial content of an image. In this paper, a different approach is presented. In this approach, the signals under analysis are first separately decorrelated into a number of uncorrelated components by making use of the image transform theory and then these components are weighted accordingly. Based on this idea, an effective adaptive iterative restoration algorithm is also proposed for restoring images which are blurred and corrupted with color noise. Simulation results show that weighting decorrelated components provides a better restoration result than weighting highly correlated image pixels.

Acknowledgments

This work was supported by Center for Multimedia Signal Processing, The Hong Kong Polytechnic University, Hong Kong.

References

1. H. C. Andrews and B. R. Hunt, *Digital image restoration* (Prentice-Hall, New Jersey, 1977).
2. A.N.Tikhonov and V. Y. Arsenin, *Solutions of ill-posed problems* (W.H.Winston,Washington, D.C., 1977).
3. N. B. Karayiznnis and A. N. Venetsanopoulos, *Regularization Theory In Image Restoration-The Stabilizing Functional Approach*, IEEE trans. on ASSP **38**, 1155–1179 (1990).
4. M.I. Sezan and A. M. Tekalp, *Survey of recent developments in digital image restoration*, Optical Engineering **29(5)**, 393–404 (1990).
5. G. Demoment, *Image reconstruction and restoration: Overview of common estimation structures and problems*, IEEE trans. on ASSP **37(12)**, 2024–2036 (1989).
6. A.K.Katsaggelos, J.Biemond, R.W.Schafer and R.M.Mersereau, *A regularized iterative image restoration algorithm*, IEEE trans. on Signal Processing **39(4)**, 914–929 (1991).

Table 1. SNR performance of various algorithms in restoring noisy blurred images.

	SNR (dB)					
	defocus blur 5 × 5 pixels			motion blur 1 × 9 pixels		
	input	SAW	NAW	input	SAW	NAW
Noise added: 25dB						
Lenna	17.04	20.05	20.46	15.26	18.49	19.48
Cameraman	15.83	18.93	19.08	15.43	18.14	19.11
Noise added: 20dB						
Lenna	15.77	19.18	19.68	14.37	17.16	18.35
Cameraman	14.86	17.85	18.17	14.53	16.96	17.98
Noise added: 15dB						
Lenna	13.18	18.19	18.80	12.35	15.94	17.22
Cameraman	12.76	17.01	17.39	12.57	15.93	16.96

7. S. N. Efstratiadis and A. K. Katsaggelos, *Adaptive iterative image restoration with reduced computational load,* Optical Engineering **29**, 1458–1468 (1990).
8. R. L.Lagendijk, J. Biemond and D. E. Boekee, *Regularized iterative image restoration with ringing reduction,* IEEE trans. on ASSP **36**, 1874–1888(1988).
9. A.K. Katsaggelos, *Iterative image restoration algorithms,* Optical Engineering **28(7)**, 735–748(1989).
10. S. J. Reeves, *Optimal space-varying regularization in iterative image restoration,* IEEE trans. on image processing **3(3)**, 319–324 (1994).
11. A.K. Jain, *Fundamentals of digital image processing* (Prentice-Hall, Englewood Cliffs, NJ, USA, 1989).
12. M. G. Kang and A. K. Katsaggelos, *Frequency-domain adaptive iterative image restoration and evaluation of the regularization parameter,* Optical Enginering **33(10)**, 3222–3232(1994).
13. K. R. Rao and P. Yip, *Discrete cosine transform: algorithms, advantages and applications* (Academic Press, 1990).

Figure 1. Corresponding portions of testing images for reference

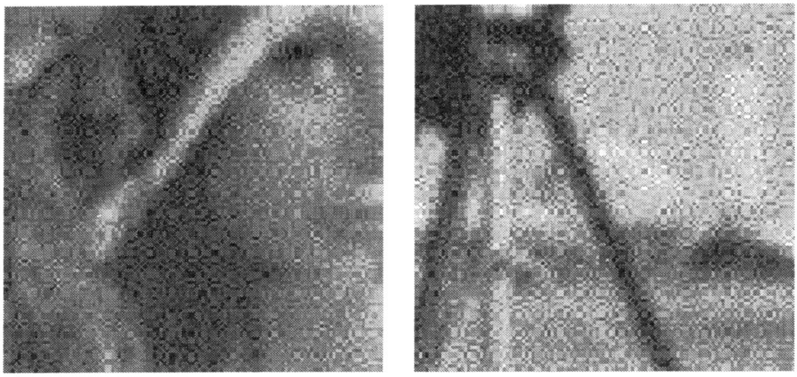

Degraded input

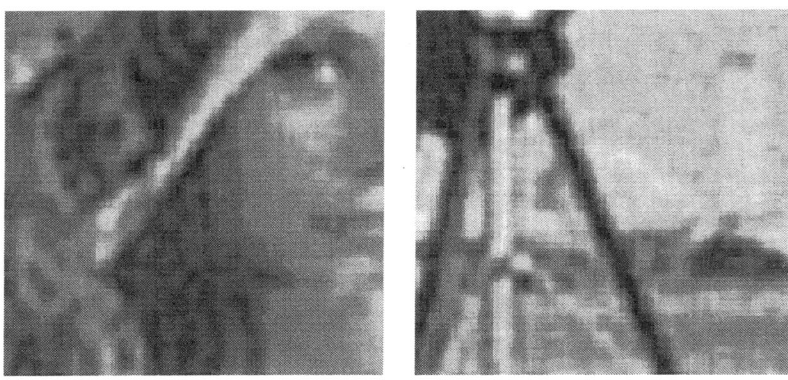

Result of NAW

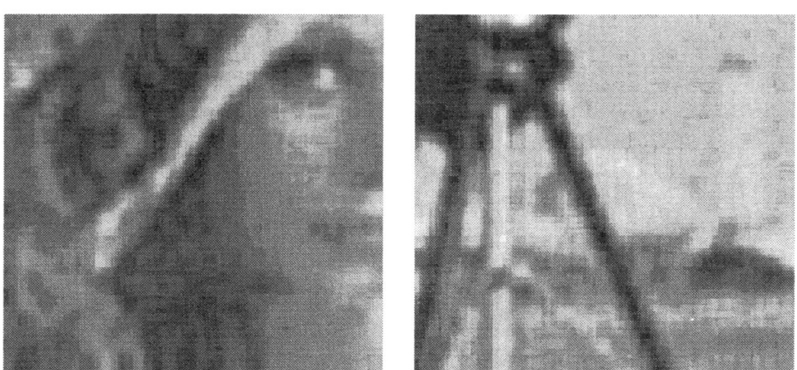

Result of SAW

Figure 2. Comparison of the restoration performance of different approaches (Distortion: defocus blur 5×5 pixels + 15dB color noise)

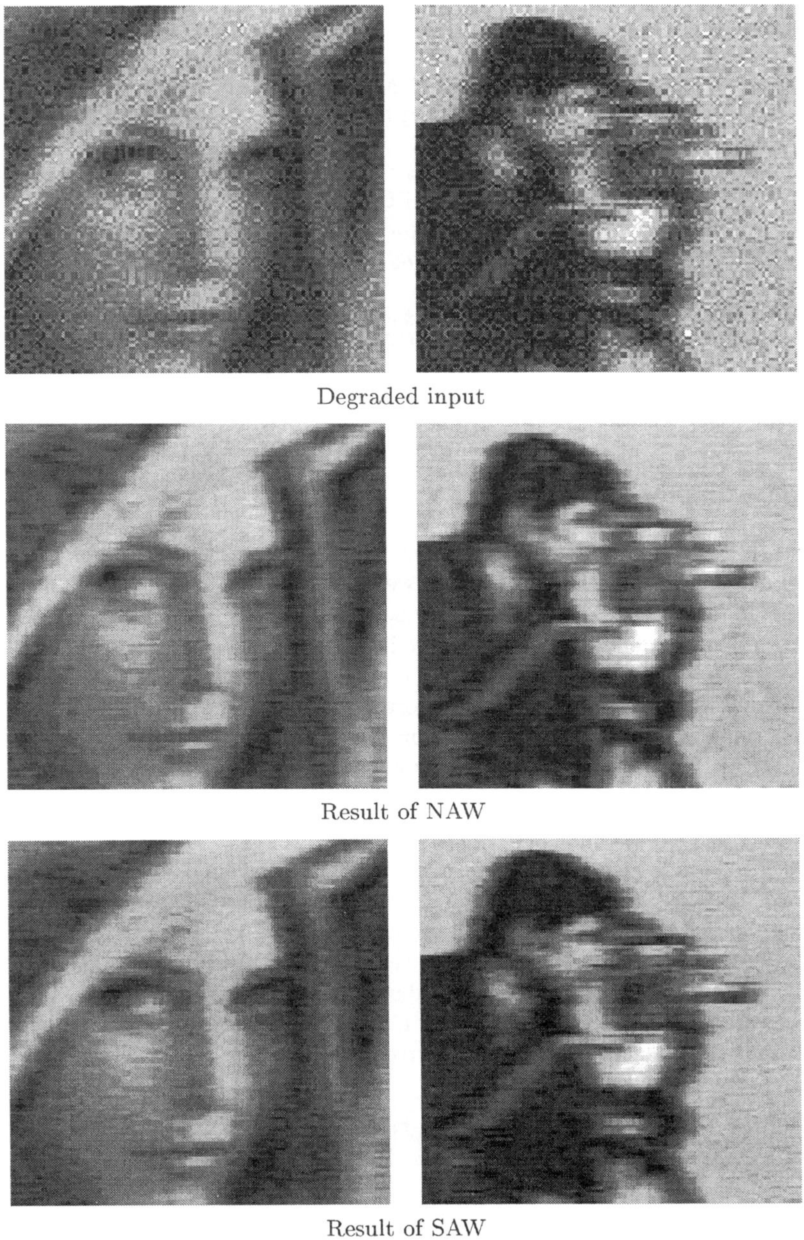

Figure 3. Comparison of the restoration performance of different approaches (Distortion: motion blur 1×9 pixels + 15dB color noise)

SIMULATED ANNEALING METHOD IN ELECTRICAL IMPEDANCE TOMOGRAPHY

Z. GIZA, S. F. FILIPOWICZ, J. SIKORA

Department of Electrical Engineering, Warsaw University of Technology, Koszykowa 75, 00-662 Warsaw, POLAND

This paper presents a new method of the image reconstruction inside the body for Electrical Impedance Tomography (EIT). The advantage of using Simulated Annealing optimization method is the elimination of calculations of the objective function gradient and the opportunity to search the whole range of decision parameters values to find the lowest value of the objective function.

1 Introduction

The theory of Inverse Problems have been intensively developed in recent years. For many technical problems we are unable to collect sufficient measurement data to achieve satisfactory solutions. This is difficult especially in Electrical Impedance Tomography. It is the kind of Inverse Problem, which relies on the identification of material coefficients inside the region under consideration. There is a strict assumption that data can only be collected from the periphery of the object, not from the internal part. This assumption makes the problem more difficult to solve.

2 Inverse Problem Formulation

In order to use Simulated Annealing method for EIT image reconstruction, let us consider the following model (schematically presented in the Figure 1).

Let us assume that vector $\mathbf{u}$ represents the value of electric potentials and vector γ represents conductivity distribution inside the region under consideration [2]. The inverse transformation T^{-1} gives us γ distribution, which minimizes the objective function defined as follows [3]:

$$F = \sum_{j=1}^{p} F_j = \sum_{j=1}^{p} \frac{1}{2}(f_j - v_{0j})^T (f_j - v_{0j}) = \frac{1}{2}\sum_{j=1}^{p}\sum_{i=1}^{n_{de}}(f_{ji} - v_{0ji})^2 \qquad (1)$$

where:

j – projection angle (positions of the energy source),

$\mathbf{f}_j$, $\mathbf{v}_{0j}$ – the vectors of calculated and measured potentials at the boundary for the j-th projection angle;

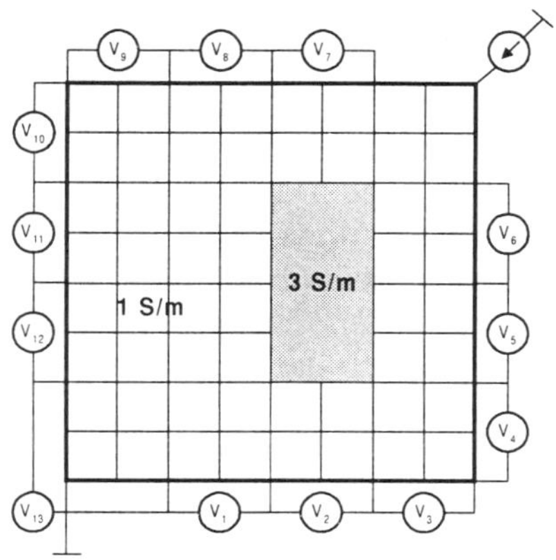

Figure 1. Region consideration with data collecting system

n_{de} – the number of measurements collected according to the protocol presented in Figure 2.

3 The Simulated Annealing Algorithm

The term *annealing* comes from the science of metallurgy. During the melting process metal is heated to a high temperature. The absorbed energy causes the atoms to vibrate. The sudden cooling of the metal captures the microstructure of the atoms in unstable state. The material structure, because of internal tensions, becomes fragile and breakable. But when the cooling process is slow enough, simultaneously with the temperature decrement, the structure of the atoms creates homogenous crystals, which results the material in being resistant to mechanical forces and free from internal tensions. As a result, metal achieves stable ordered structure of crystals [6,7].

The simulated annealing image reconstruction algorithm for EIT can be formulated as follows. The algorithm will iteratively reconstruct an image, that the values of calculated vector $\mathbf{f}_j$ fit best the values of vector $\mathbf{v}_{0j}$ for each j-th projection angle. The vector $\mathbf{f}_j$ represents the current state of the reconstructed image. We assumed that minimization of the difference between the measured voltage data set $\mathbf{v}_{0j}$ and calculated data set $\mathbf{f}_j$ gives us the reconstruction image, which will resemble the sought-after original image. Therefore, the objective function expresses the difference between these both data sets [1].

At the starting point, all values of material coefficients (projecting variables) are set to the same value of background conductivity and the objective function value is calculated.

Then the cooling process starts. During this process all values of projecting variables are slightly disturbed. The objective function value is calculated for each set of the varied values. If the set of projecting variables corresponds closer to the real material coefficient distribution, the values of calculated voltage data $\mathbf{f}$ fits the measured data $\mathbf{v}_0$ more closely, and the value of objective function $\mathbf{F}$ is lower. When this value is lower than the previously calculated value, the corresponding set of projecting variables is chosen to the next step of calculation, if not, we compute the value of probability:

$$p = \exp\left(-\frac{\mathbf{F}_i - \mathbf{F}_{i-1}}{T}\right) \tag{2}$$

and decide, if the objective function value might be used to the further calculations.

In the next steps of calculations, the temperature of the system is decreased and the verifications of the objective function value are performed. An important consideration for the simulated annealing algorithm is the proper choice of initial and final temperatures, and the manner the temperature will be decreased. One of the methods is to multiply the current temperature value by the constant value c. This value can be calculated from the following formula:

$$c = \exp\left(\ln\frac{\frac{T_k}{T_p}}{L-1}\right) \tag{3}$$

where:

T_k – final temperature,
T_p – initial temperature,
L – number of iterations.

The stop criteria of the calculations might be:

1. The value of the objective function is sufficiently low,
2. The number of computing iterations is large enough,
3. The temperature of the annealing is low enough.

After a defined number of iterations, the set of projecting variables should correspond to the minimal value of the objective function (final energy of the system). The value of the objective function should be lower than the value at the starting point. This means that the computed distribution of the material coefficient has become more similar to the original distribution.

4 Numerical Experiments

In order to use the SA algorithm for EIT image reconstruction let us consider the following model of an object. The conductivity of the two-dimensional region was set to 1 [S/m]. The conductivity of the object was 3 [S/m] (Figure 2). The finite element network was generated inside the region. The projecting variables were chosen as the material parameters (conductivity) located in nodes of the discretization network. The energy of the system is equal to the objective function designed as an error between the voltages measured at the boundary of the region and the voltages calculated in current computing iteration.

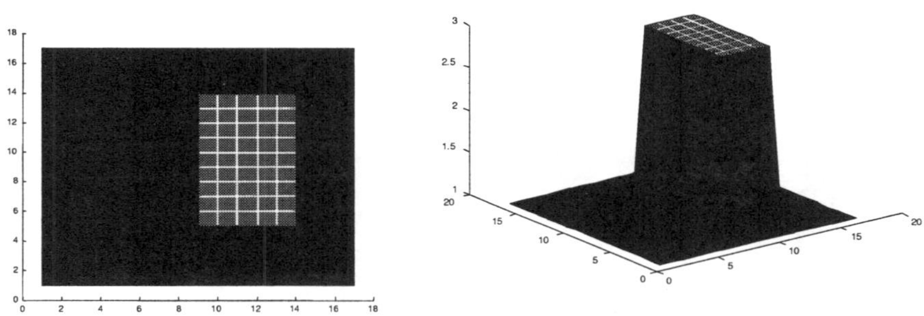

Figure 2. Object type **A**

At the very beginning of the testing of the SA image reconstruction algorithm, the initial temperature and the temperature decrement were assumed randomly using the trial and error method. The results are presented in Figure 3. These results were treated as the starting point of the further researches to discover the limits of the values.

In the next step of the tests, the range of the temperature decrement was sought. The results of this experiment are presented in Figure 4. It can be seen that the value of this coefficient should be taken from the range: 0.001 $\div$ 0.01.

Then the different values of the initial temperature T_p were tested. At the beginning, this value was set to 0.8. The influence of this value on the reconstructed image can be observed in Figure 5, Figure 6 and Figure 7.

It can be seen that if the initial temperature is high (Figure 5 and Figure

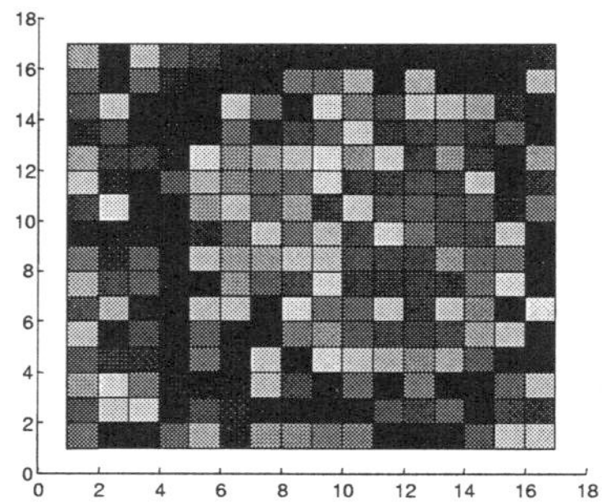

Figure 3. The image of object type **A** for randomly chosen parameters (T_p = 0.8, T_k = 0.072793, E_{min} = 0.006759)

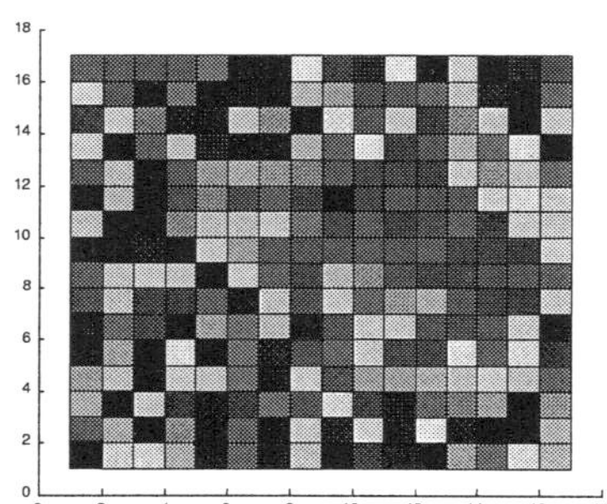

a)

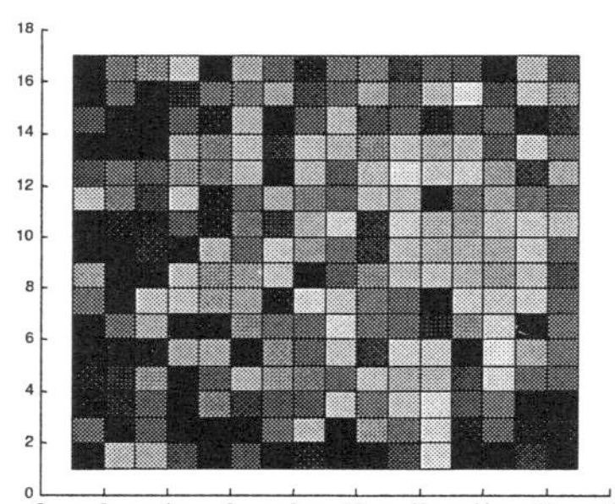

b)

Figure 4. The images of object type **A** for the different values of the temperature decrement coefficient (T_p = 0.8): a) 0.001, T_k = 0.4002, E_{min} = 0.027842, b) 0.1, T_k = 0.0008, E_{min} = 0.517599

6d)), the background conductivity is strongly affected, especially near the boundary. The shape of the object is unclear. The best achieved results are shown in the Figure 7 and Figure 6e).The initial temperatures was ranged: $0.5 \div 0.9$.

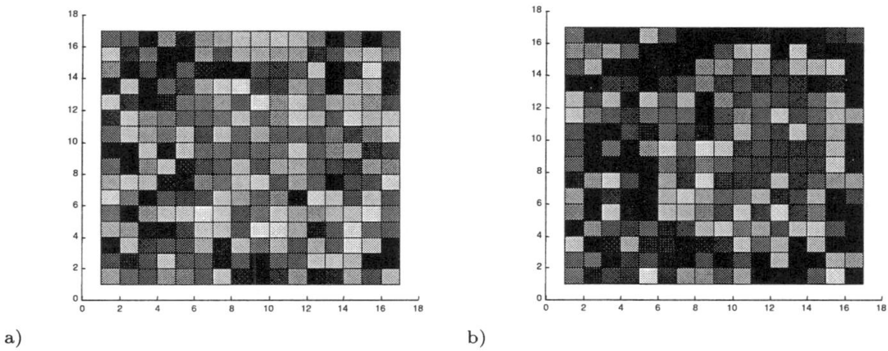

Figure 5. The images of object type A for different values of the initial temperature: a) T_p = 1.00, T_k = 0.090992, E_{min}= 0.007346, b) T_p = 10.0, T_k = 0.909918, E_{min}= 0.091334

4.1 Image Filtering Algorithm

As shown in the previously presented figures, the simple simulated annealing algorithm is unable to reconstruct the clear shape and the precise placement of the object. Therefore, to improve results an additional filtering algorithm was developed and applied to the reconstruction algorithm based on the simulated annealing [5].

The conductivity of each discretization network node is calculated as the average from the conductivity of neighbour-nodes and the node, where the average is calculated (Figure 8).

The following figures present the images of the object type **A** generated using the simulated annealing method extended with image filtering algorithm.

The best results were achieved for the following parameters: T_p = 0.54, L = 2000, and the filtering performed every 250 iterations (Figure 9d)).

Using the parameters obtained from the trial and error method of the theoretical experiments, the SA reconstruction image algorithm was tested on a real object [4]. The object consisted of two elements. The dimensions of each element were 3x10 cm. This object was inserted into a tank filled with saline. Each element was placed parallel to the boundary of the tank 2cm from the edges (Figure 11).

The difficulty of this problem was the correct identification of the object location and the reconstruction of the real dimensions of the two elements. It was important to properly reconstruct the background conductivity between the elements of the object. Additional problems were caused by the proximity of the object elements to the boundary, because the regions are especially sensitive due to the influence of the electrodes.

The results of the experiment are shown in Figure 12 and Figure 13 respectively for the resolution of 16x16 and 32x32 elements of the image.

It can be observed that a better quality image is obtained for the higher spatial image resolution, despite the fact that the number of the projecting variable was equal to 1089. By comparison, in the 16x16 elements resolution, the number of those variables was equal only to 289.

5 Conclusions

The completed researches lead to the following conclusions. The parameters, which most affect the quality of the reconstructed image are: the initial temperature T_p and the number of iterations L.

There is no strict rule to obtain the parameters values of the simulated annealing algorithm. Each experiment requires carrying out the simulations, which will give the set of the best parameters of the algorithm. The trial and error method of seeking the parameters is time consuming.

The schedule of the researches was directed to search for the best parameters in the case of one-element object, but values obtained were successfully applied to the case of the two-elements object.

For the majority of images, the parameters have to be sought again.

The main problem of the SA algorithm appears to be no unambiguous stop criteria. In most cases, the main stop criteria might be the number of the iterations or the final system temperature value. In comparison to deterministic methods, where the value of the gradient is checked throughout the calculations, for the simulated annealing there is no information about achieving the best solution.

It was necessary to implement the filtering algorithm of the image to obtain results with a quality comparable to the classical methods of EIT, where the deterministic techniques of optimization are used.

On the basis of the results, we conclude that the use of stochastic optimization methods in Electrical Impedance Tomography requires at least primary knowledge, such as the number of elements of the object or the range of the material parameters values.

References

1. Z. Giza, *Methods of identification of material coefficient distribution in Electrical Impedance Tomograph,* (Doctorial Dissertation, Warsaw University of Technology, IETiME, Warsaw 2000) (in Polish).
2. T. Kurztkowski, J. Sikora and M. Miosz, *Electrical Impedance Tomography Based on Higher Order Finite Element Approximation of Conductivity,* in *Nonlinear Electromagnetic Systems*, ed. A. J. Moses and A. Basak (IOS Press, 270–273 (1996)).
3. J. Sikora, *Algorytmy numeryczne w tomografii impedancyjnej i wiroprdowej* (Oficyna Wydawnicza Politechniki Warszawskiej, Warszawa 2000).
4. S.F.Filipowicz , Z.Giza , J.Sikora and R.Sikora, *New Methods of Imaging in Electrical Impedance Tomography: A Comparative Study,* in *International Symposium on Electromagnetic Fields in Electrical Engineering* (ISEF'99, Pavia, Italy, 23 - 25.09., 493 – 496 (1999)).
5. R.C.Gonzales and P.Wintz, *Digital Image Processing* (Addison-Wesley Publishing Company, 1987).
6. L.Ingber, *Adaptative simulated annealing (ASA),1 Reseach note*, Caltech, Lester Ingber Research. (1993a).
7. S. Kirkpatrick , C.D. Gelatt and M.P.Vecchi, *Optimization by simulated annealing,* Science **220**, 671– 680 (1983).

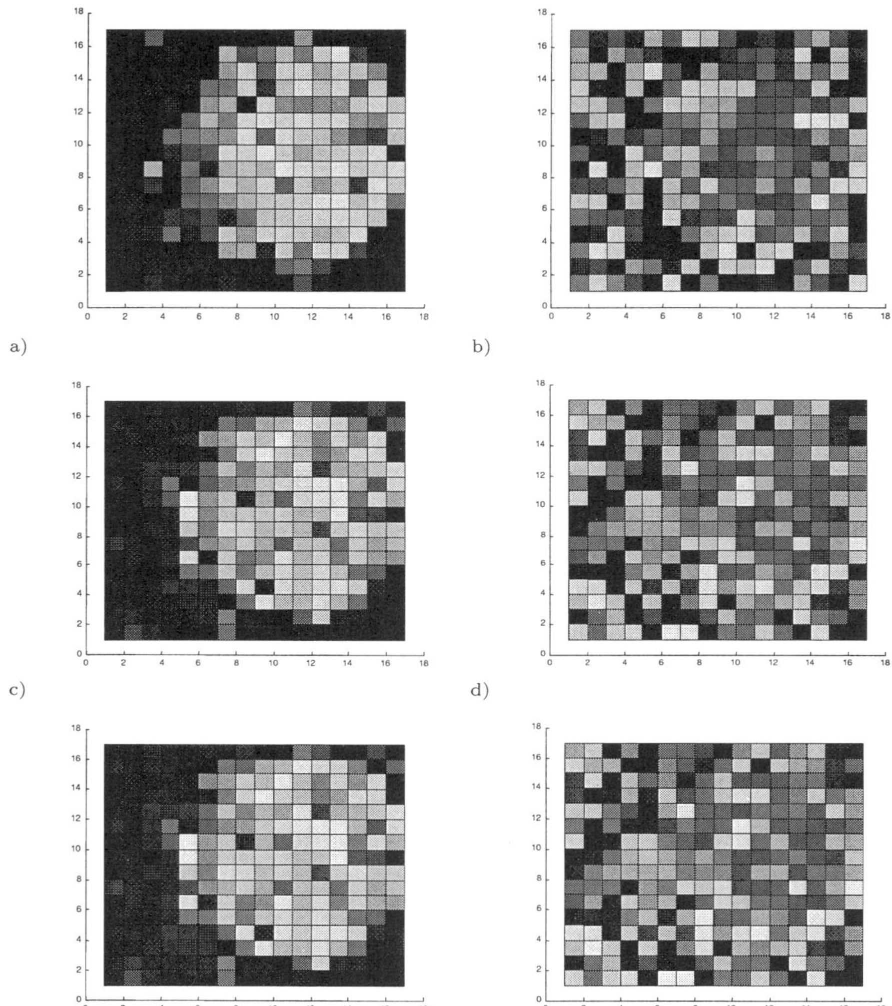

a) b) c) d) e) f)

Figure 6. The images of object type A for different values of the initial temperature: a) $T_p = 0.08$, $T_k = 0.007279$, $E_{min} = 0.307364$, b) $T_p = 2.50$, $T_k = 0.227480$, $E_{min} = 0.017258$, c) $T_p = 0.10$, $T_k = 0.009099$, $E_{min} = 0.232278$, d) $T_p = 4.00$, $T_k = 0.363967$, $E_{min} = 0.024504$, e) $T_p = 0.90$, $T_k = 0.008189$, $E_{min} = 0.253515$, f) $T_p = 0.05$, $T_k = 0.004550$, $E_{min} = 0.486219$

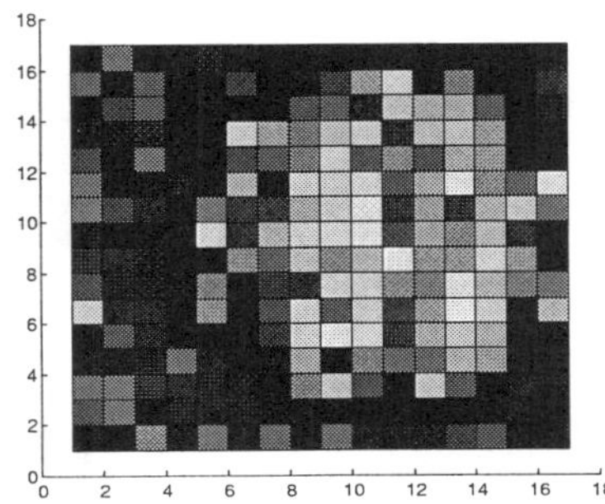

Figure 7. The image of object type A for the following parameters: $T_p = 0.50$, $T_k = 0.045496$, $E_{min} = 0.007508$

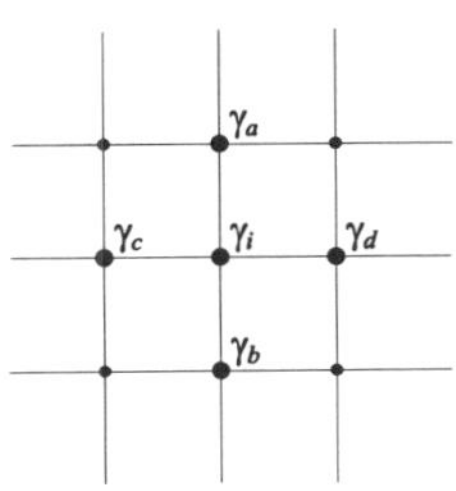

$$\gamma_{i+1} = \frac{\gamma_a + \gamma_b + \gamma_c + \gamma_d + \gamma_i}{5}$$

Figure 8. The algorithm of the image filtering

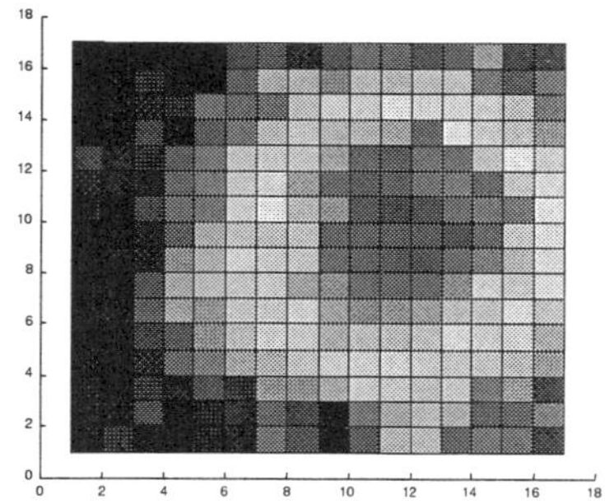

a)

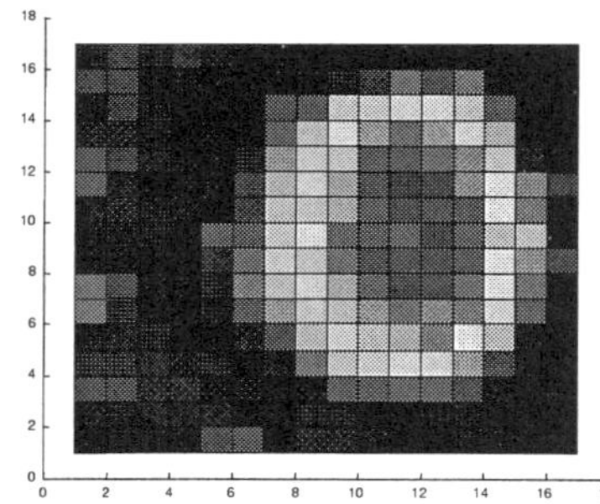

b)

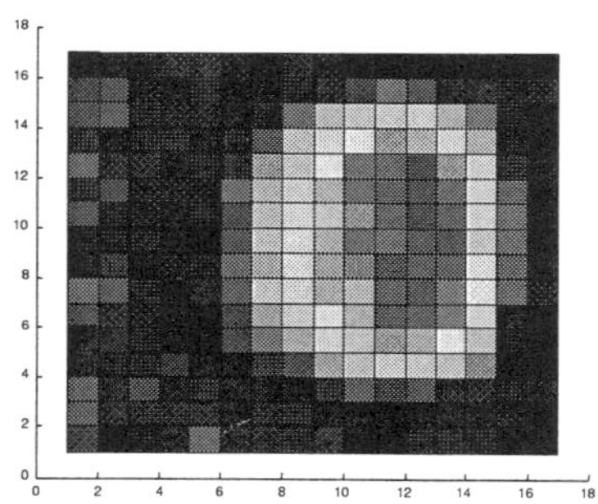

c)

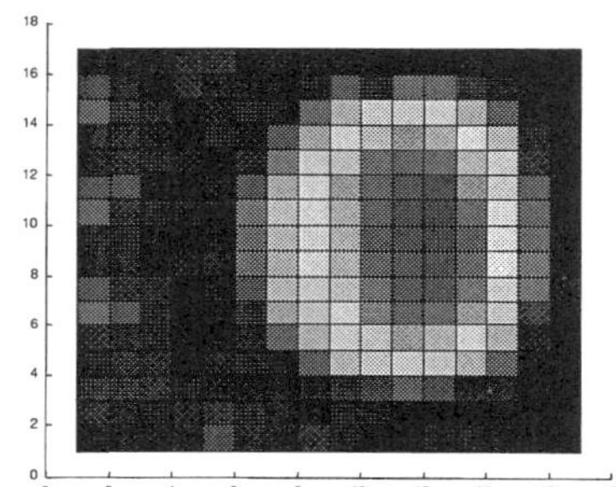

d)

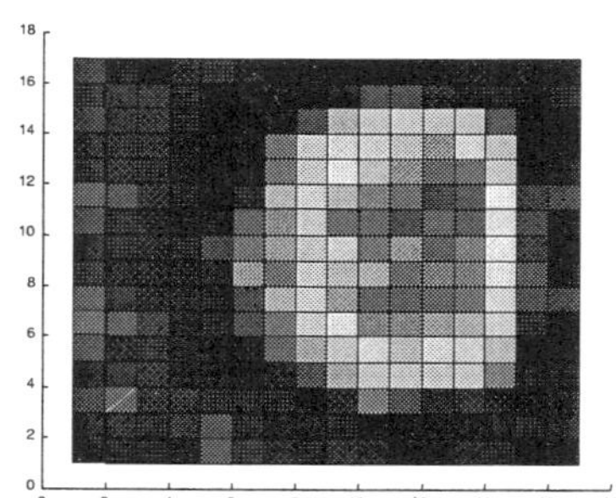

e)

Figure 9. The images of object type A for the SA algorithm extended with the filtering image algorithm (T_p = 0.50, T_k = 0.019057, L = 2000): a) filtering performed every 10 iterations, E_{min} = 0.772703, b) filtering performed every 100 iterations, E_{min} = 0.006235, c) filtering performed every 200 iterations, E_{min} = 0.003889 d) filtering performed every 250 iterations, E_{min} = 0.004397, e) filtering performed every 500 iterations, E_{min} = 0.002760

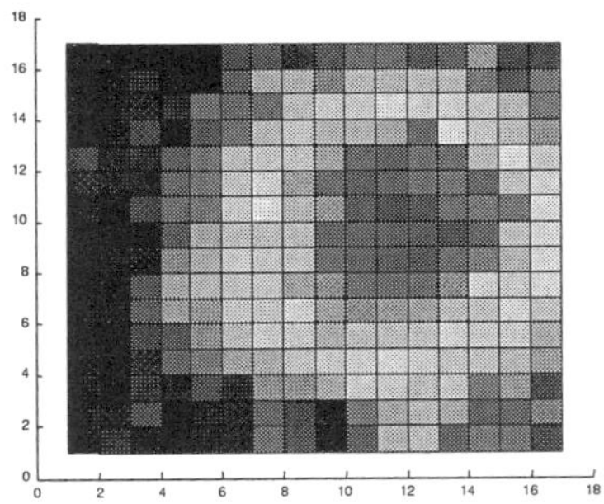

a)

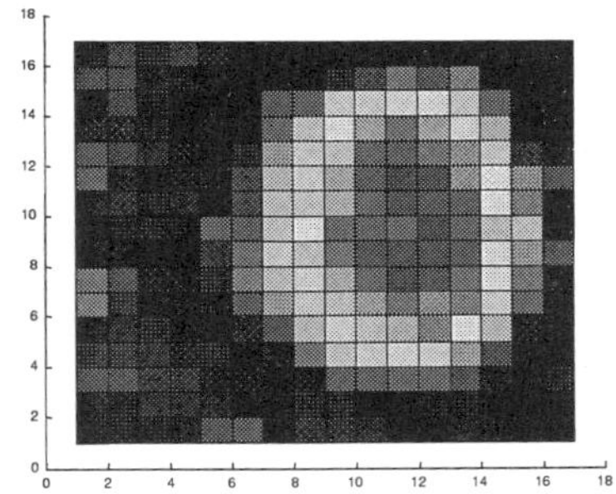

b)

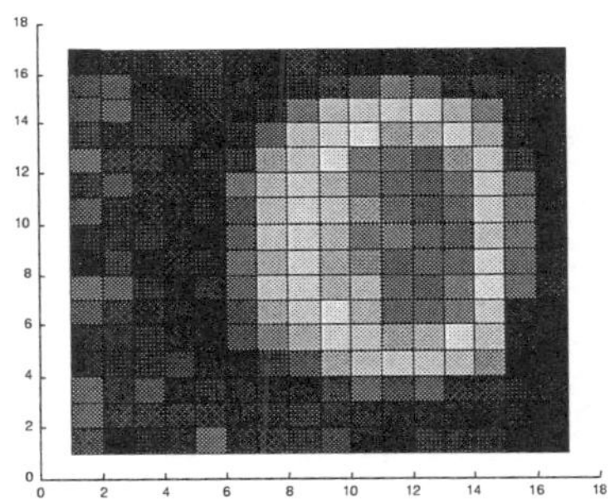

c)

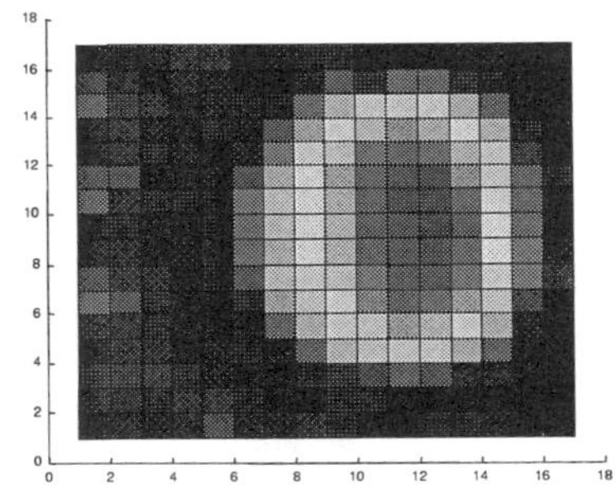

d)

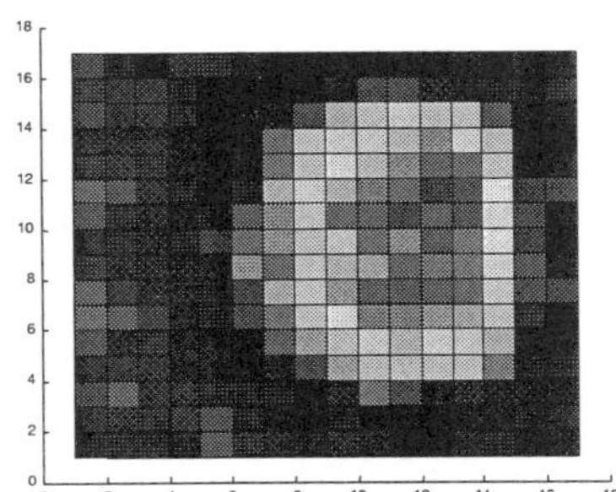

e)

Figure 10. The images of the object type A achieved using the algorithm with the image filtration ($T_p = 0.50$, $T_k = 0.019057$, L = 2000): a) filtering every 10 iterations, $E_{min} = 0.772703$, b) filtering every 100 iterations, $E_{min} = 0.006235$, c) filtering every 200 iterations, $E_{min} = 0.003889$ d) filtering every 250 iterations, $E_{min} = 0.004397$, e) filtering every 500 iterations, $E_{min} = 0.002760$

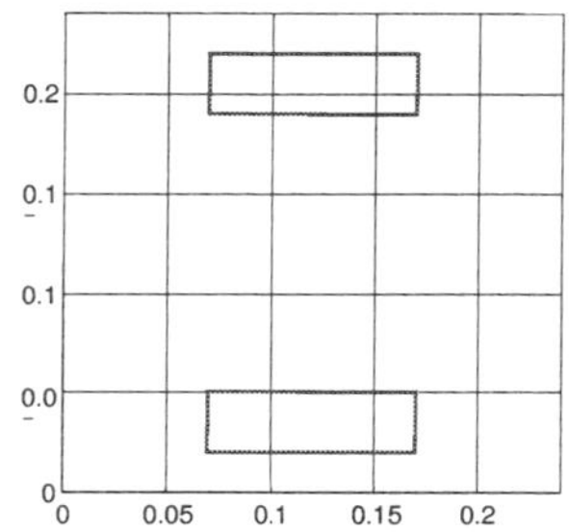

Figure 11. Real object type **R1**

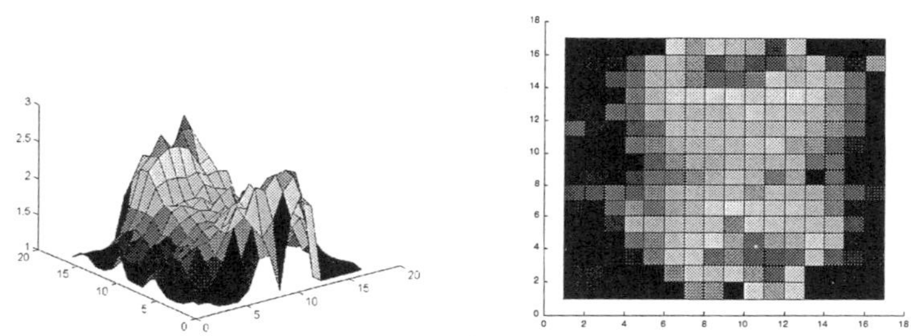

a) b)

Figure 12. The image of object type **R1** for the resolution of 16x16 elements $T_k = 0.019057$, $E_{min} = 0.097026$

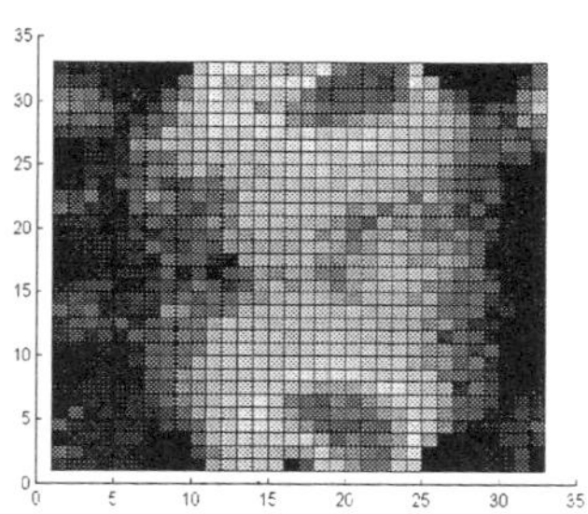

Figure 13. The image of object type **R1** for the resolution of 32x32 elements $T_k = 0.019057$, $E_{min} = 0.087150$

APPLICATION OF TECHNIQUES IN INVERSE PROBLEMS TO VARIATIONAL DATA ASSIMILATION IN METEOROLOGY AND OCEANOGRAPHY

SIXUN HUANG
LMSWE, Nanjing University, 210093,P.R.China
E-mail: huangsx@public1.ptt.js.cn

WEI HAN
P.O. Box 003, Nanjing,211101,P.R.China
E-mail:howiner@263.net

There is an international focus on the developments of data assimilation systems for meteorology and physical oceanography models and there has been considered interests in the "Inverse Problems" of determining poorly known initial boundary conditions and model parameters by incorporating measured data into the numerical model, taking into account both the information about dynamics about the model and the information about the true state which is constrained by a set of measurements. In this paper the data assimilation problem in meteorology and physical oceanography is re-examined using the adjoint methods in combination with regularization ideas in inverse problem, then two sets of numerical experiments are performed to examine whether the proposed approach is capable to reconstruct the accurate initial boundary conditions and model parameters. One set of experiments are using global observations and the other with local observations, the numerical experiments show that variational data assimilation with regularization techniques contribute a lot to the stability and accuracy of the numerical calculation.

1 Introduction of Variational Data Assimilation

The principle of the variational approach in meteorology and physical oceanography is a particular case of the general framework of the optimal control [1] (Lions ,1971). The controls are basically made of the initial boundary conditions and model parameters of the dynamical model. We search for an optimal control which minimizes the misfit between the state of the system and the observations over some time interval. A cost functional J measuring this misfit is user-defined, generally as the sum of weighted squared individual misfits. This cost functional is then minimized by a general optimizer such as a Quasi-Newton.

The dynamical models in meteorology and oceanography can be represented the following nonlinear evolution equation:

$$\begin{cases} \frac{dX}{dt} = F(t, X, \mu) \\ X\mid_{t=0} = U \\ BVC \end{cases} \tag{1}$$

here $X(t)$ is state variable , F is a nonlinear model operator and it is assumed that μ is a poorly known parameter in the model, and there are also errors in initial conditions U and boundary conditions . The performances of numerical forecasts has been improved in the last century, however the accuracy and forecast time limitation is not satisfied. There are four main reasons:

1. In general the initial boundary problems of nonlinear evolution differential equations has only local solution, not the global solution , so it is difficult to expect very long forecast valid time.

2. The errors of initial conditions. The initial conditions are obtained through analyzing the measurements and initialization , the measurements are not error free.

3. The errors of boundary conditions. In the case of limited area model, the boundary conditions are of vital importance to the forecast performance and are very difficult to be prescribed.

4. The errors due to physics parameterization. There are many empirical parameters in the numeric model which are set by experiences. In the light of the above limitations, the variational data assimilation methods are proposed . The system (1) as it is has one unique solution for a given value of μ and initial boundary conditions. Thus if the parameters and initial boundary conditions shall be improved, we need additional information. This can be given by introducing a set of observations of the model variable taken at various locations in time and space.

The particularity of the four dimensional variational data assimilation method is to use the adjoint of the operators involved in the cost function, and in particular the backward adjoint model. This provides an efficient way to compute the gradient of the cost functional [2] (Courtier and Talagrand,1990). From a physical point of view, the approach makes the best use of the physics of the model and easily allows the use of any data that could be represented by the model. It is a physical and versatile approach. This methods have been greatly improved in the last twenty years, nevertheless, there are some limitations. Specific limitations of the basic approach include the need for error estimates, and the limitations imposed on the resolution by the nonlinearities and the limitation is the range of validity of the linear tangent model, and the smoothness of the cost function in general. Another limitation is ill-posedness of the problem , being beset by instabilities and non-uniqueness when identifying parameters distributed in the space time domain, especially when the data is noisy. How to introduce and develop the methods and ideas of inverse problems in mathematical physics to overcome the difficulties in the variational data assimilation is very important and challenging. We have explored in this direction for two years and adopt a simple model to show our efforts and work.

2 Variational Data Assimilation with Global Observation

For illustrational purposed we will use a one-dimension heat-diffusion model for describing the vertical distribution of sea temperature over time as an example. The governing equation is:

$$\frac{\partial T}{\partial t} = \frac{\partial}{\partial z}\left(K\frac{\partial T}{\partial z}\right) - \frac{\upsilon I_0}{\rho_0 C_p} exp\left(-\upsilon z\right) \quad (z,t) \in (0,H) \times (0,\tau), \qquad (2)$$

with the initial conditions $T\,|_{t=0} = U\,(z)$, and boundary conditions at surface $\left[K\frac{\partial T}{\partial z} + \frac{I_0}{\rho_0 C_p}\right]\Big|_{z=0} = \frac{Q(t)}{\rho_0 C_p}$, at bottom $K\frac{\partial T}{\partial z}\,|_{z=H} = 0$. Here $T = T\,(t,z)$ is sea temperature , $K = K\,(t,z)$is vertical eddy diffusion coefficient, ρ_0 is sea water density, C_p is sea water specific heat capacity, υ is light diffusion coefficient, H is the depth of ocean upper layer,I_0 is the transmission component of solar radiation at sea surface, $Q\,(t)$ is net heat flux at sea surface. Based on the theory of partial differential equation [3] , it is known that there exist the unique solution of model (2) if the initial boundary conditions and the model parameters (K, I_0) are known and smooth. Assume υ , ρ_0 and C_p are known constant, the initial boundary conditions $U\,(z)$,$Q\,(t)$ and model parameters $K\,(t,z)$,$I_0\,(t)$ are not known exactly, e.g., they have unknown errors and need to be improved by data assimilation . Now a set of observations of sea temperature $T_{obs}\,(t,z)$ are given. A convenient cost functional formulation J is thus defined as

$$J\left[U,K,Q,I_0\right] = \frac{1}{2}\int_0^{\tau} \left\|T - T_{obs}\right\|^2 dt = \frac{1}{2}\int_0^{\tau}\int_0^{H} \left(T - T_{obs}\right)^2 dz dt, \qquad (3)$$

and the problem becomes: Find the optimal initial boundary conditions $(U\,(z)$,$Q\,(t))$and model parameters $(K\,(t,z)\,,I_0\,(t))$, such that the cost functional is minimum. With the aid of regularization techniques in inverse problem , an additional stable functional which is related to the heat flux and the smoothness of the solution , is introduced to J in order to overcome the ill-posedness and make the calculation stable. Then the improved cost functional is defined as follows:

$$J[U,K,Q,I_0]=\frac{1}{2}\int_0^{\tau}\int_0^{H}(T-T_{obs})^2\,dzdt+\frac{1}{2}\gamma^2\int_0^{\tau}\int_0^{H}K(t,z)\left(\frac{\partial T}{\partial z}\right)^2 dzdt, \tag{4}$$

where $\frac{1}{2}\int_0^{\tau}\int_{-H}^{0}K(t,z)\left(\frac{\partial T}{\partial z}\right)^2 dzdt$ is a stable functional and γ^2 is the regularization parameter.

In order to get the gradient of the cost functional with respect to the control variables , a series of variational calculations was performed, then the following adjoint equation and boundary conditions are obtained:

$$-\frac{\partial P}{\partial t}=\frac{\partial}{\partial z}\left(K\frac{\partial P}{\partial z}\right)-\gamma^2\frac{\partial}{\partial z}\left(K\frac{\partial T}{\partial z}\right)+(T-T_{obs}) \tag{5}$$

with initial conditions $P|_{t=\tau}=0$, boundary conditions at bottom $K\frac{\partial P}{\partial z}|_{z=H}=0$ and at surface$[K\frac{\partial P}{\partial z}-\gamma^2K\frac{\partial T}{\partial z}]|_{z=0}=0$.

The gradients of the cost functional (4) with respect to U, K, Q and I_0 are :

$$\begin{aligned}\nabla_U J=P(0,z),\nabla_K J=-\tfrac{\partial T}{\partial z}\tfrac{\partial P}{\partial z}+\tfrac{1}{2}\gamma^2\left(\tfrac{\partial T}{\partial z}\right)^2,\nabla_Q J=-\tfrac{P(t,0)}{\rho_0C_p},\\ \nabla_{I_0}J=\int_0^{\tau}\tfrac{P(t,0)}{\rho_0C_p}dt-\tfrac{v}{\rho_0C_p}\int_0^{\tau}\int_0^{H}exp(-vz)\cdot P(t,z)\,dtdz.\end{aligned} \tag{6}$$

With these gradients ,the iteration formulas are:

$$\begin{aligned}U^{i+1}=U^i-(\nabla_U J)|_{R^i}\cdot\rho_U^i,\quad K^{i+1}=K^i-(\nabla_K J)|_{R^i}\cdot\rho_K^i,\\ Q^{i+1}=Q^i-(\nabla_Q J)|_{R^i}\cdot\rho_Q^i,\quad I_0^{i+1}=I_0^i-(\nabla_{I_0}J)|_{R^i}\cdot\rho_{I_0}^i,\end{aligned} \tag{7}$$

where ρ_U^i,ρ_K^i, ρ_Q^i and $\rho_{I_0}^i$ are iteration steps for U,K,Q and I_0 respectively. The optimal solution of U,K,Q and I_0 can be get by gradient based iterations, such as conjugate gradient method or Quasi-Newton method. Here we apply the Newton method and the iteration step is adjusted to keep it mono-decreasing during the iteration process.

When the cost functional (4) satisfy the end criterion,

$$J\le\varepsilon, \tag{8}$$

the iteration is ended ,where ε is a given small positive real number .

In order to test the theoretical results above ,twin numerical experiments are performed by numerical method. In the present theoretical framework, the initial condition $U(z)$, the boundary condition $Q(t)$, the eddy diffusion coefficient $K(t,z)$ and the transmission component of solar radiation at sea surface I_0, can be assimilated simultaneously. Nevertheless, we keep our focus on the assimilation of the eddy diffusion coefficient. The ideal model of (2) is given:

$$\frac{\partial T}{\partial t}=\frac{\partial}{\partial z}\left(K(t,z)\frac{\partial T}{\partial z}\right)+f(t,z),(t,z)\in(0,1.0)\times(0,\pi/2) \tag{9}$$

with initial boundary conditions
$T|_{t=0}=U(z)$, $K\frac{\partial T}{\partial z}|_{z=0}=Q(t)$, $K\frac{\partial T}{\partial z}|_{z=\pi/2}=0$. where $f(t,z)=sin(z)[cos(t)-sin(t)]$,$Q(t)=cos(t)$. The true initial conditions are $U(z)=sin(z)$and the true eddy diffusion coefficient is $K(t,z)=1$, and the ideal model (9) has the analytical solution $T(t,z)=sin(z)cos(t)$. We take the true solution $T(t,z)=sin(z)cos(t)$ as the observation data, and add different perturbations to the first guess of initial conditions and eddy diffusion coefficient , then the assimilation process is performed. It is shown that the additional functional plays a important role in the assimilation process especially for the optimization of model parameter and the improved cost functional form (4) is acceptable.

3 Variational Data Assimilation with Local Observation

The forward model is the same as (2), the differences are that the observations are taken only at sea surface , i.e., $T_{obs}(t,0)$ are given, the cost functional are defined:

$$J[U,K]=\frac{1}{2}\int_0^{\tau}(T(t,0)-T_{obs})^2dt+\frac{1}{2}\gamma^2\int_0^{\tau}\int_0^{H}K(t,z)\left(\frac{\partial T}{\partial z}\right)^2dzdt, \tag{10}$$

Through a series of variational calculations, the following adjoint equation and adjoint boundary conditions are obtained:

$$-\frac{\partial P}{\partial t}=\frac{\partial}{\partial z}\left(K\frac{\partial P}{\partial z}\right)-\gamma^2\frac{\partial}{\partial z}\left(K\frac{\partial T}{\partial z}\right), \tag{11}$$

with initial conditions $P|_{t=\tau}=0$,boundary conditions at bottom $K\frac{\partial P}{\partial z}|_{z=H}=0$, and at surface $K\frac{\partial P}{\partial z}|_{z=0}=-(T(0,t)-T_{obs}(t))-\gamma^2\frac{1}{\rho_0 C_p}(Q(t)-I_0)$ and the gradients of the cost functional (10) with respect to $U$and K are obtained:

$$\nabla_U J=P(0,z),\nabla_K J=-\frac{\partial T}{\partial z}\frac{\partial P}{\partial z}+\frac{1}{2}\gamma^2\left(\frac{\partial T}{\partial z}\right)^2. \tag{12}$$

Then two sets numerical experiments are performed.

The first set. The aim is to test the efficiency of the determination of initial conditions . It is designed as: keep K to be true, and a perturbation is added to the initial conditions, $U_0(z) = sin(z) + 0.1 sin(2z)$, $K = 1$, one is calculated with $\gamma = 0$ and the other with $\gamma = 0.001$. The descent of the cost functional are shown in Figure 1(a).

The second set. Keep the initial condition as true, a perturbation is added to the eddy diffusion coefficient, $U_0(z) = sin(z)$, $K = 1 + 0.05(z - H)$,one is calculated with $\gamma = 0$ and the other with $\gamma = 0.001$. The descent of the cost functional are shown in Figure 1(b). From the numerical experiments,

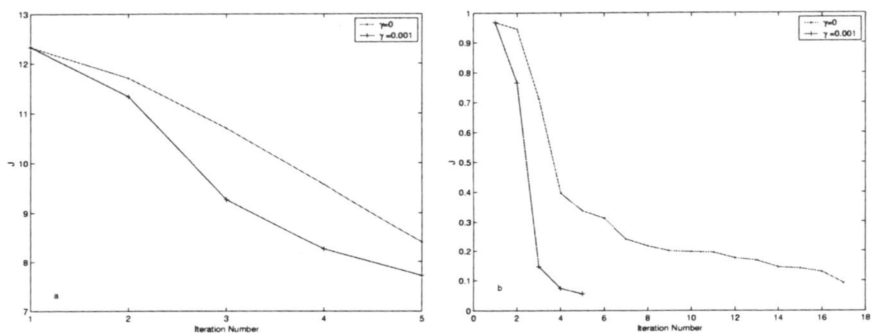

Figure 1. The iteration process of the cost functional with and without regularization. a)for optimizing initial conditions; b) for optimizing the model parameter.

two conclusions can be made:

1. It is ill-posed to determine the initial condition and model parameter which are distributed in space and time with local observations (in the present work which are observations at the boundary) by adjoint method. In the test of numerical experiments , the solution is very sensitive to the first guess and the iteration steps, and the calculation is unstable to some extent without regularization.

2. The introduction of regularization overcame the ill-posedness of the problem to some extent. In the case of local observations, it improved the accuracy and stability of the solution as shown in Figure1, especially for the determination of the model parameter (Figure1.b) in which the descent speed of the cost functional and the accuracy are both improved. However,the ill-posedness of the problem is very complicated, more efforts should be done in future.

Acknowledgments

This research was supported by National Natural Science Foundation of China (No. 40075014 and No. 40175014)

References

1. J.L. Lions, *Optimal Control of Systems Governed by PDE* (Springer-Verlag, New York, 1971).
2. P.Courtier, and O. Talagrand , *Variational assimilation of meteorological observations with the direct and adjoint shallow water equations* , Tellus **42A**, 531–549(1990).
3. A. Friedman,*PDE of Parabolic Type* (Prentice-Hall, Inc. ,1964).

INVERSE RADIUS OF BIOLOGICAL FLOCCULUS IN THE REACTOR OF THE WATER DECONTAMINATION

KE-AN LIU, XI-LIAN WANG, BO HAN, JIA-QI LIU AND HONG-BIN ZHAO

Mathematics Department, Harbin Institute Of Technology, China, P.O.Box:150006

E-mail: keanl@263.net

The biological flocculus in the water disposing reactor can be treated as the spherical cell model. The biological flocculus grows with the time – the volume becomes big, the biological membrane becomes thick, the permeability becomes bad and the interior biophore dies so that to decrease the reactor's decontaminating ability. Properly controlling the volume of biological flocculus can improve the reactor's efficiency. The satisfactory radius of the biological flocculus is obtained by us of the finite-difference method, considering the properly controlling of biological flocculus volume as the mathematical inverse problem of geometrical boundary.

1 Introduction

Commonly in the water decontamination reactor many biological impurity combine together to form the biological flocculus which can be treated a spherical cell model. The biological flocculus grows with the time – the volume becomes big, the biological membrane becomes thick, the permeability becomes bad and the interior biophore dies so that to decrease the reactor's decontaminating ability. So properly controlling the volume of biological flocculus can improve the reactor's efficiency. How to calculate the most appropriate radius of the biological flocculus and to implement the real-time control of the radius is a meaningful topic to upraise the efficiency of the reactor.

2 Mathematical Model of the Biological Flocculus

Given a period of time, we suppose that [1]:

1) Each biological flocculus is approximately a sphere;
2) The density of the biological flocculus doesn't vary with the time and its volume;
3) The whole biological flocculus is homogeneous, and the interior biochemical reaction is merely the function of the local environment. For each biological flocculus cell, we fetch a platelet to build up the mathematical model of the cell material (see Figure 1). Under the stable condition, the substrate equation between r and r+dr is (see Wang[1]):

$$(D\frac{dC}{dr}4\pi r^2)\mid_{r+dr} -(D\frac{dC}{dr}4\pi R)\mid_{r}= r_0 4\pi r^2 dr.$$

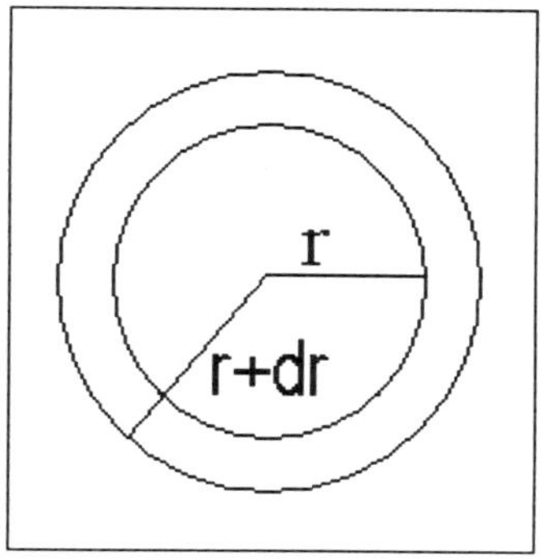

Figure 1. Biological flocculus model.

Here, D is the diffusion coefficient of the substrate in the biological flocculus membrane, C is the substrate density of the solution and r_0 is the substrate decontaminating rate of the biological flocculus unit volume.

Suppose that D is a constant, we divide both side of the equation by $2\pi dr$, and let $dr \to 0$, then

$$D(\frac{d^2C}{dr^2} + \frac{2}{r}\frac{dC}{dr}) = r_0. \tag{1}$$

And the boundary conditions are

$$\frac{dC}{dr}|_{r=0} = 0,$$

$$C|_{r=R} = C_0. \tag{2}$$

Here, R is the radius of the flocculus cell and C_0 is the substrate density.

According to the effective thickness of the biological membrane and the variance range of the substrate density in the biochemical process, there are three cases for the dynamic expression of the reaction:

1) Larger variance range of the substrate density, then the reaction rate equation can be expressed by

$$r_0 = \frac{1}{Y}\frac{\mu_{\max}C}{K_s + C}X,$$

$$r_0(C_0) = \frac{1}{Y}\frac{\mu_{\max}C_0}{K_s + C_0}. \tag{3}$$

In such case, the reaction happens merely inside the biological membrane and the interior organism deceases. Here, X is the density of the micro-organism, Y is the coefficient of the productive rate (the production volume of the micro-organism at a unit volume), K_s is the saturation constant and $\mu_{\max}$ is the maximal growth rate of the micro-organism.

2) Lower substrate density inside the biological flocculus, then the reaction rate can be expressed by the first-order dynamic formula:

$$r_0 = K_1 C.$$

Here, K_1 is the constant of the first-order reaction rate and

$$K_1 = \frac{\mu_{\max}}{Y K_s}.$$

3) Higher substrate density for the whole biological flocculus, then it is the zero-order dynamic reaction. Supposed no restraint of oxygen inside the limited thickness of the biological membrane, because of the higher substrate density, the zero-order dynamic reaction not only exists inside the platelet but can extend to the center of the biological flocculus such that the reaction rate of the whole biological flocculus can be expressed by the zero-order dynamic equation:

$$r_0 = K_2 C.$$

Here, K_2 is the constant of the zero-order reaction rate and

$$K_2 = \frac{\mu_{\max} X}{Y}.$$

The boundary conditions are

$$D\frac{d^2C}{dr^2} = K_2, \frac{dC}{dr}|_{r=0} = 0, C|_{r=R} = C_0.$$

It is the special case for 2) and 3), so we just discuss 1).

The detersive efficiency has positive ratio with the amount of the active organism.Though the surface area increases along with the biological flocculus volume, the interior organism deceases so as to decrease the amount of the liquid inside the reactor and to lower the reactive efficiency. So it is necessary to shatter the biological flocculus at appropriate moment to keep the optimal radius of it.

If the radius of the biological flocculus R is known, the Equations (1), (2) and (3) determine a non-linear boundary problem of ordinary differential equation. But because we hope to control it to increase the efficiency of the reactor, we add a control condition

$$C|_{r=0} = \delta \tag{4}$$

Then the Equations (1), (2), (3) and (4) constitute an inverse problem of solving the boundary value R. Here δ is a very small positive value, meaningly that when the radius R increases to confirm the condition (4), the biological flocculus ought to be shattered.

3 The Solution of the Inverse Problem

For the problem above, to get the numeric solution, we introduce the transform

$$r = ax^2 + bx$$

such that
(1) $r = 0, r = R$ correspond to $x = 1, x = 2$ respectively;
(2) The interval $[0,R]$for r corresponds to the interval $[1,2]$ for x;
Then we have

$$\begin{cases} 0 = a + b \\ R = 4a + 2b \end{cases}$$

Then

$$a = R/2, b = -R/2.$$

According to the derivative rule of the composite function

$$\frac{dc}{dr} = \frac{dc}{dx}\frac{dx}{dr} = \frac{dc}{dx}\frac{1}{2ax + b}$$

$$\frac{d^2c}{dr^2} = \frac{d}{dr}(\frac{dc}{dr}) = \frac{d^2c}{dx^2}\frac{1}{(2ax+b)^2} - \frac{dc}{dx}\frac{2a}{(2ax+b)^3}.$$

We have

$$\frac{d^2c}{dx^2}\frac{1}{(2ax+b)^2} - \frac{dc}{dx}\frac{2a}{(2ax+b)^3} - \frac{K}{D}c = 0.$$

Substitute it with $a = R/2, b = -R/2$,

$$\frac{d^2c}{dx^2} - \frac{dc}{dx}\frac{2}{2x-1} - \frac{K}{D}R^2c(x - \frac{1}{2})^2 = 0.$$

Also transform the boundary condition, we have

$$\frac{dc}{dr}|_{r=0} = 0 \Longrightarrow \frac{dc}{dx}|_{x=1} = 0,$$

$$c|_{r=0} = C^0 \Longrightarrow c|_{x=2} = C^0.$$

Next we divide the interval [1,2] by $1 = x_1 < x_2 < \cdots\cdots < x_{N-1} < x_N = 2$, and substitute the derivative equation with difference quotient

$$\frac{c_{i+1} - 2c_i + c_{i-1}}{h^2} - \frac{c_i - c_{i-1}}{h} \frac{2}{2x_i - 1} - \frac{K}{D} R^2 c_{i-1} (x_i - \frac{1}{2})^2 = 0.$$

To get the iterative expression

$$\begin{aligned} c_i &= 2c_{i-1} - c_{i-2} - \tfrac{2h}{2x_{i-1}-1}(c_{i-1} - c_{i-2}), \\ E_r &= \tfrac{K}{D} R^2 c_{i-1} (x_i - \tfrac{1}{2})^2, \\ c_{i+1} &= c_i + E_r. \end{aligned} \tag{5}$$

Where

$$h = \frac{(2-1)}{N}.$$

Also we substitute the derivative with difference quotient for the boundary and additional conditions:

$$\begin{aligned} &\tfrac{dc}{dr}|_{x=1} = 0 \Longrightarrow \tfrac{c_0 - c_1}{h} = 0 \Longrightarrow c_0 = c_1 \\ &c|_{x=2} = C^0 \Longrightarrow c_N = C^0 \\ &c|_{x=2} < C^0 \Longrightarrow c_N < C^0. \end{aligned} \tag{6}$$

So we get the algorithm to the solution of the inverse problem:

For a given initial value R, let $c_0 = c_1$ =constant, according to the iteration (5), we can compute the value of $c_2, \cdots\cdots, c_{N-1}, \overline{c_N}$. Then we check that if there exists the result of $\overline{c_N} < C^0$. If not, we constitute a functional function $T(R) = ||c_N - \overline{c_N}||^2$, where c_N is the theoretic value, $\overline{c_N}$ is the computed value, and R is unknown.

$$\begin{aligned} T(R) &= (c_N - (c_{N-1} + E_r))^2 \\ &= (c_N - c_{N-1})^2 - 2(c_N - c_{N-1})E_r + E_r^2 \\ &= (c_N - c_{N-1})^2 - 2(c_N - c_{N-1})c_{N-2}\frac{K}{D}h^2 R^2 (x_{N-1} - \frac{1}{2})^2 \\ &\quad + \frac{K^2}{D^2} h^4 R^4 (x_{N-1} - \frac{1}{2})^4 c_{N-2}^2, \end{aligned}$$

Table 1. simulation data one.

Parameter	Value	Unit
D	$1—10\times10^{-10}$	m^2/s
μ_{max}	1.0—2.0	h^{-1}
K_S	0.1—2.0	g/L
Y	0.05	
X	100.0—200.0	mg/L

$T(R)$ is a continuous function of R, and we can compute the value R^* which make $T(R)$ to reach the minimal value, i.e. R^* satisfies that $T'(R^*) = 0$. So we have

$$-4(c_N - c_{N-1})c_{N-2}\frac{K}{D}h^2R(x_{N-1} - \frac{1}{2})^2 + 4\frac{K^2}{D^2}h^4R^3(x_{N-1} - \frac{1}{2})^4c_{N-2}^2 = 0$$

$$R = \sqrt{\frac{2D(c_N - c_{N-1})}{Kh^2(2x_{N-1} - 1)^2}}. \tag{7}$$

Then we substitute such value R to compute $c_2, \cdots\cdots, c_{N-1}, \overline{c_N}$ once more, and repeat the computing procedure again and again until the value satisfies the additional condition.

Algorithm:

1) Input the initial value R^0, c_0, c_1, C^0, N ;

2) Compute $c_2, \cdots\cdots, c_{N-1}, \overline{c_N}$ with iterative expression (5), and store the value of $c_{N-2}, c_{N-1}, \overline{c_N}$;

3) Compare if $\overline{c_N} < C^0$. If so, go to step 4, or else print the value of $\overline{c_N}, R$ and jump to step 5;

4) Compute R with the expression (7), then go to step 2;

5) End.

4 Experimental Simulation

Table 1 and table 2 give some simulation data.

Figure 2 and figure 3 show the correspond simulating computational results for different value of μ_{max}, C_0.

Table 2. simulation data two.

Parameter	Value	Unit
D	2.0×10^{-10}	m^2/s
μ_{max}	1.02	h^{-1}
K_S	0.5	g/L
Y	0.05	
X	150	mg/L

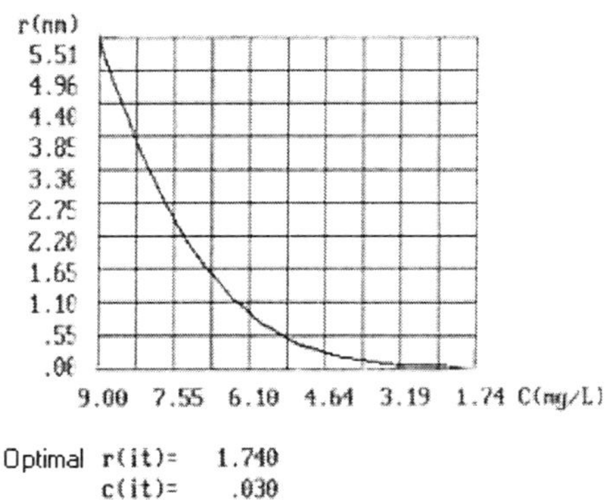

Figure 2. Computational result of table 1.

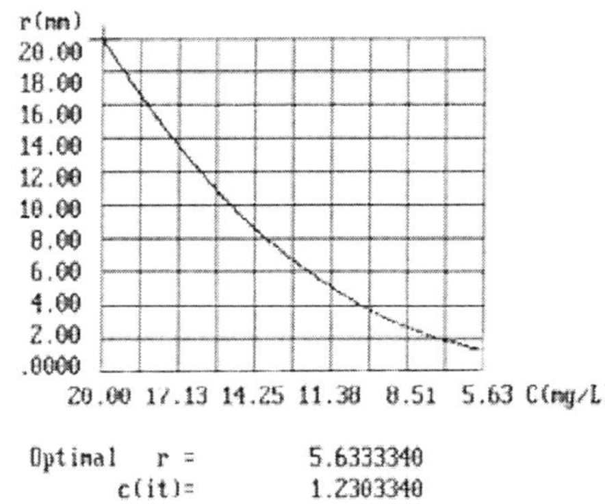

Figure 3. Computational result of table 2.

For example, given the initial reactive rate v=2.034825E-001, the iterative output accuracy eps=3.500000E-002, the maximum iterative number n= 600, the initial liquor density C_0 = 20.000, the growth rate of the micro-organism μ=0.280, the initial radius R=20.00, the saturation density , the density of the micro-organism X =.140, the productive rate Y =.081 and the biological diffusive coefficient D =.800, we can compute the best appropriate radius of the biological flocculus $R(170)$ = 5.6333 (mm) and the substrate density $C(170)$= 1.5423mg/Lafter 170 iteration .

Table 3 gives the computational result for different value of C^0 and μ_{max}.

5 Summary

It has practical significance to control the radius of the biological flocculus in the process of water decontamination. This paper brings forward a mathematical inverse problem model of geometry boundary used to calculate the

Table 3. The best appropriate Radius for different C^0 and μ_{max}.

$\mu_{max}(h^{-1})$ $C^0(mg/L)$	0.66	0.50	0.40	0.33	0.28
20.00	6.24	7.17	8.01	8.82	9.59
10.00	4.77	5.49	6.12	6.75	7.32
4.00	3.48	3.99	4.44	4.92	5.34

best appropriate radius of the biological flocculus, and gives the correspond numeric method for calculating it.

References

1. Nai Zhong Wang, *The Theoretic Foundation of Water Decontamination* (Southwest Communication University Press, Changsha, 246–271(1988)).
2. Jia Qi Liu,*Classification of The Inverse Problem of Mathematical Physics Equations And The Solution to The Improperly Posed Problem*, Applied And Computational Mathematics **4**, 82–96(1983).

CLUSTERING PROBLEMS USING TABU SEARCH TECHNIQUES

MICHAEL K. NG

Department of Mathematics, The University of Hong Kong, Hong Kong
E-mail: mng@maths.hku.hk

Clustering methods partition a set of objects into clusters such that objects in the same cluster are more similar to each other than objects in different clusters according to a dissimilarity measure between objects. Different dissimilarity measures will result in different cluster structures. In this paper, we present a tabu search based clustering algorithm to determine the dissimilarity measure between objects and then to cluster a set of objects based on the computed dissimilarity measure. It is found that the preliminary clustering results produced by the proposed algorithm are high in accuracy.

1 Introduction

Clustering is an inverse problem in data mining. The clustering problem states that partitioning a set of objects into homogeneous clusters if the cluster structure exists in the set of objects. The clustering operation is required in a number of data analysis tasks, such as unsupervised classification and data summation, as well as segmentation of large homogeneous data sets into smaller homogeneous subsets that can be easily managed, separately modelled and analyzed.

Clustering methods partition a set of objects into clusters such that objects in the same cluster are more similar to each other than objects in different clusters according to a dissimilarity measure between objects. Different dissimilarity measures will result in different cluster structures. In this paper, we present a tabu search based clustering algorithm to determine the dissimilarity measure between objects and then to cluster a set of objects based on the computed dissimilarity measure. We first formulate the clustering problem as a mathematical optimization problem:

$$\min_{W,Z} F(W,Z) = \sum_{l=1}^{k} \sum_{i=1}^{n} w_{li}^{\alpha} d(\mathbf{z}_l, \mathbf{x}_i) \tag{1}$$

subject to

$$0 \le w_{li} \le 1, \quad 1 \le l \le k, \ 1 \le i \le n, \tag{2}$$

$$\sum_{l=1}^{k} w_{li} = 1, \quad 1 \le i \le n, \quad \text{and} \quad 0 < \sum_{i=1}^{n} w_{li} < n, \quad 1 \le l \le k, \qquad (3)$$

where n is the number of objects, m is the number of attributes of each object, $k(\le n)$ is a known number of clusters, $\mathcal{X} = \{\mathbf{x}_1, \mathbf{x}_2, ..., \mathbf{x}_n\}$ is a set of n objects with m attributes, $Z = [\mathbf{z}_1, \mathbf{z}_2, ..., \mathbf{z}_k]$ is an m-by-k matrix containing k cluster centers, $W = [w_{li}]$ is an k-by-m fuzzy matrix and $d(\mathbf{z}_l, \mathbf{x}_i)(\ge 0)$ is a certain dissimilarity measure between the cluster center $\mathbf{z}_l$ and the object $\mathbf{x}_i$. In this paper, we assume that the number k and the index α are known in advance. The above optimization problem was first formulated by Dunn [1]. A widely known approach to this problem is the k-means algorithm which was proposed by Ruspini [2] and Bezdek [3].

1.1 *The Dissimilarity Measure*

We assume the set of objects to be clustered is stored in a database table $\boldsymbol{T}$ defined by a set of attributes, $A_1, A_2, ..., A_m$. Each attribute A_j describes a domain of values, denoted by $DOM(A_j)$, associated with a defined semantic and a data type. In this paper, we only consider two general data types, *numeric* and *categorical* and assume other types used in database systems can be mapped to one of the se two types. The domains of attributes associated with these two types are called numeric and categorical respectively. A numeric domain consists of real numbers. A domain $DOM(A_j)$ is defined as categorical if it is finite and unordered, e.g., for any $a, b \in DOM(A_j)$, either $a = b$ or $a \neq b$. An object X in $\boldsymbol{T}$ can be logically represented as a conjunction of attribute-value pairs $[A_1 = y_1] \wedge [A_2 = y_2] \wedge \cdots \wedge [A_m = y_m]$ where $y_j \in$ DOM(A_j) for $1 \le j \le m$. Without ambiguity, we represent X as a vector $[y_1, y_2, \cdots, y_m]$. X is called a categorical object if it has only categorical values. We consider every object has exactly m attribute values. If the value of an attribute A_j is missing, then we denote the attribute value of A_j by ϵ.

In the literature, the Euclidean norm is often used in the clustering algorithm for the numerical data [3]. For the categorical data, the simple matching dissimilarity measure between objects is recently proposed, see Huang and Ng [6]. In this paper, we consider the weighted combined dissimilarity measure between two objects

$$\mathbf{x}_l = [x_{l1}^{(n)}, x_{l2}^{(n)}, \cdots, x_{lm_1}^{(n)} \mid x_{l1}^{(c)}, x_{l2}^{(c)}, \cdots, x_{lm_2}^{(c)}]^T$$

and

$$\mathbf{x}_i = [x_{i1}^{(n)}, x_{i2}^{(n)}, \cdots, x_{im_1}^{(n)} \mid x_{i1}^{(c)}, x_{i2}^{(c)}, \cdots, x_{im_2}^{(c)}]^T$$

$(m = m_1 + m_2)$, is defined as:

$$d(\mathbf{x}_l, \mathbf{x}_i) \equiv \underbrace{\sum_{j=1}^{m_1} \lambda_j^{(n)} |x_{lj}^{(n)} - x_{ij}^{(n)}|^2}_{\text{numerical part}} + \underbrace{\sum_{j=1}^{m_2} \lambda_j^{(c)} \delta(x_{lj}^{(c)}, x_{ij}^{(c)})}_{\text{categorical part}} \tag{4}$$

where

$$\delta(x_{lj}^{(c)}, x_{ij}^{(c)}) = \begin{cases} 0, & \text{if } x_{lj}^{(c)} = x_{ij}^{(c)}, \\ 1, & \text{if } x_{lj}^{(c)} \neq x_{ij}^{(c)}, \end{cases}$$

$$\sum_{j=1}^{m_1} \lambda_j^{(n)} + \sum_{j=1}^{m_2} \lambda_j^{(c)} = 1, \quad \lambda_j^{(n)}, \lambda_j^{(c)} \geq 0, \quad \forall j. \tag{5}$$

Here $\lambda_j^{(n)}$ and $\lambda_j^{(c)}$ are weighting parameters to be determined.

The main aim of this paper is to develop a tabu search based clustering algorithm to determine the weighting parameters $\lambda_j^{(n)}$ and $\lambda_j^{(c)}$ in (1). Based on the computed dissimilarity measure, we expect that we obtain a better clustering result than that using the unweighted dissimilarity measure (i.e., $\lambda_j^{(n)} = \lambda_j^{(c)} = 1$). The outline of the paper is as follows. In Section 2, tabu search based techniques are introduced and the tabu search based clustering algorithm is proposed. In Section 3, the experimental results are presented to illustrate the effectiveness of our new approach. In Section 4, some concluding remarks are given.

2 Tabu Search Based Techniques

Minimization of F in (1) with the constraints in (2) and (3) forms a class of constrained nonlinear optimization problems whose solution is unknown. However, we find that the matrices W and Z are formulated in the following methods, see Huang and Ng [6].

Let Z and the weighting parameters be fixed, i.e., $\mathbf{z}_l$ for $l = 1, 2, ..., k$ are given, we can find W by:

$$w_{li} = \begin{cases} 1, & \text{if } \mathbf{x}_i = \mathbf{z}_l \\ 0, & \text{if } \mathbf{x}_i = \mathbf{z}_h \text{ but h} \neq 1 \\ 1/\sum_{h=1}^{k} \left[\frac{d(\mathbf{z}_l, \mathbf{x}_i)}{d(\mathbf{z}_h, \mathbf{x}_i)}\right]^{\frac{1}{\alpha-1}}, & \text{if } \mathbf{x}_i \neq \mathbf{z}_l \text{ and } \mathbf{x}_i \neq \mathbf{z}_h, 1 \leq h \leq k \end{cases} \tag{6}$$

for $1 \leq l \leq k, 1 \leq i \leq n$.

Let W and the weighting parameters be fixed, we can find Z by the modes and frequency update methods for the categorical and numerical data respectively. Each object is described by m_2 categorical attributes and its jth attribute has n_j categories: $a_j^{(1)}, a_j^{(2)}, ..., a_j^{(n_j)}$ for $1 \leq j \leq m_2$. Let the l-th cluster center be

$$\mathbf{z}_l = [z_{l1}^{(n)}, z_{l2}^{(n)}, ..., z_{lm_1}^{(n)} \mid z_{l1}^{(c)}, z_{l2}^{(c)}, ..., z_{lm_2}^{(c)}]^T.$$

Then $F(W, Z)$ is minimized if and only if

$$z_{lj}^{(n)} = \frac{\sum_{i=1}^{n} w_{li} x_{ij}}{\sum_{i=1}^{n} w_{li}}, \quad 1 \leq l \leq k, \quad 1 \leq j \leq m. \tag{7}$$

and

$$z_{lj}^{(c)} = a_j^{(r)} \quad \text{where} \quad \sum_{i, x_{ij} = a_j^{(r)}} w_{li}^{\alpha} \geq \sum_{i, x_{ij} = a_j^{(t)}} w_{li}^{\alpha}, \quad 1 \leq t \leq n_j. \tag{8}$$

For the weighting parameters, we note that when W and Z are fixed, we solve the following the minimization problem:

$$\min_{\lambda_j^{(n)}, \lambda_j^{(c)}} \sum_{l=1}^{k} \sum_{i=1}^{n} w_{li}^{\alpha} \left(\sum_{j=1}^{m_1} \lambda_j^{(n)} |x_{lj}^{(n)} - x_{ij}^{(n)}|^2 + \sum_{j=1}^{m_2} \lambda_j^{(c)} \delta(x_{lj}^{(c)}, x_{ij}^{(c)}) \right)$$

subject to (5). It is obvious that the above problem is just a linear programming problem which can be solved efficiently.

The usual method towards optimization of F in (1) is to use partial optimization for Z, W and the weighting parameters. In this method we first fix Z and the weighting parameters, and minimize F with respect to W. Then we fix W and the weighting parameters, and minimize F with respect to Z. Then we fix W and Z, and solve the above linear programming problem to determine the weighting parameters. However, the above iterative procedure may only stop at a local optimal solution of the clustering problem [3]. This means that the solution obtained can still be further improved. In the next subsection, tabu search based techniques are incorporated to aim at finding a global solution of the optimization problem (1) and determining the weighting parameters.

Table 1. Tabu search based categorical clustering algorithm.

Tabu Search Based Clustering Algorithm:

Step 1: Initialization
Let Z^u be arbitrary centers and F^u the corresponding objective function value. Let $Z^b = Z^u$ and $F^b = F^u$. Select values for $NTLM$ (tabu list size), P (probability threshold), NH (number of trial solutions), $IMAX$ (the maximum number of iterations for each center), and γ (the iteration reducer). Let $h = 1$, $NTL = 0$ and $r = 1$. Go to Step 2.

Step 2:
Using Z^u, fix all centers and move center $\mathbf{z}_r^u$ by generating NH neighbors $\mathbf{z}_1^t, \mathbf{z}_2^t, ..., \mathbf{z}_{NH}^t$, and evaluate their corresponding objective function values $F_1^t, F_2^t, ..., F_{NH}^t$. Go to Step 3.

Step 3:
(a) Sort F_i^t, $i = 1, ..., NH$ in a nondecreasing order and denote them as $F_{[1]}^t, ..., F_{[NH]}^t$. Clearly $F_{[1]}^t \leq ... \leq F_{[NH]}^t$. Let $e = 1$. If $F_{[1]}^t \geq F^b$, then replace h by $h + 1$. Goto Step 3(b).
(b) If $z_{[e]}$ is not tabu or if it is tabu but $F_{[e]}^t < F^b$, then let $\mathbf{z}_r^u = \mathbf{z}_{[e]}$ and $F^u = F_{[e]}^t$ and go to Step 4. Otherwise generate $u \sim U(0,1)$ where $U(0,1)$ is a uniform density function between 0 and 1. If $F^b < F_{[e]}^t < F^u$ and $u > P$, then let $\mathbf{z}_r^u = \mathbf{z}_{[e]}$ and $F^u = F_{[e]}^t$ and go to Step 4; otherwise, go to Step 3(c).
(c) Check for the next neighbor by letting $e = e + 1$. If $e \leq NH$, go to Step 3(a). Otherwise go to Step 3(d).
(d) If $h > IMAX$, then go to Step 5. Otherwise select a new set of neighbors by go to Step 2.

Step 4:
Insert $\mathbf{z}_r^u$ at the bottom of the tabu list. If $NTL = NTLM$, then delete the top of the tabu list; otherwise let $NTL = NTL + 1$. If $F^b > F^u$, then let $F^b = F^u$ and $Z^b = Z^u$. Go to Step 3 (d).

Step 5:
If $r < k$, then let $r = r + 1$ and reset $h = 1$ and go to Step 2. Otherwise set $IMAX = \gamma(IMAX)$. If $IMAX > 1$, then let $r = 1$ and reset $h = 1$ and go to Step 2; otherwise stop. (Z^b represents the b est centers and F^b is the corresponding best objective function value).

2.1 Tabu Search Based Techniques

Tabu search method [4] is based on procedures designed to cross boundaries of feasibility or local optimality, which are usually treated as barriers, and systematically to impose and release constraints to permit exploration of otherwise forbidden regions. Tabu search is a meta-heuristic that guides a local heuristic search procedure to explore the solution space beyond local optimality.

A fundamental element underlying tabu search is the use of flexible memory. A chief mechanism for exploiting memory in tabu search is to classify a subset of the moves in a neighborhood as forbidden or tabu. The basic elements of tabu search method are defined as follows:

1. *Configuration* is an assignment of values to variables. It is a solution to the optimization problem.

2. *Move* is a specific procedure for getting a trial solution which is feasible to the optimization problem that is related to the current configuration.

3. *Neighborhood* is the set of all neighbors, which are the "adjacent solutions" that can be reached from any current configuration. It may also include neighbors that do not satisfy the given customary feasible conditions.

4. *Candidate subset* is a subset of the neighborhood. It is to be examined instead of the entire neighborhood, especially for large problems where the neighborhood have many elements.

5. *Tabu restrictions* are constraints that prevent the chosen moves to be reversed or repeated. They play a memory role for the search by making the forbidden moves as *tabu*. The tabu moves are stored in a list, called *tabu list*.

6. *Aspiration criteria* are rules that determine when the tabu restrictions can be overridden, thus removing a tabu classification otherwise applied to a move. If a certain move is forbidden by some tabu restrictions then the aspiration criteria, when satisfied, can make this move allowable.

2.2 Clustering Algorithm

Our algorithm in Table 1 is to use tabu search based techniques in order to find a global solution of the clustering problem. In our algorithm, (6) is used to update the partition matrix W. But we do not use (7) and (8) to update the cluster center Z. Similarly, the weighting parameters is determined by using the partition matrix W and Z. Instead Z is generated by the below method and is mapped into a value for each objective function value.

Let Z^t, Z^u, Z^b denote the trial, current and best cluster centers, and F^t, F^u, F^b denote the corresponding trial, current and best objective function values respectively. A number of trial cluster centers Z^t are to be generated through moves from the current cluster centers Z^u. As the algorithm proceeds, the best cluster centers found so far is saved in Z^b. The corresponding objective function values F^t, F^u, F^b are also operated respectively. In Table 1, there are also several parameters. They are described as below:

1. $NTLM$ (tabu list size): It contains the history of the search and represents the maximum number of moves to be stored in the list. The larger (smaller, respectively) the value of $NTLM$, the stronger (less, respectively) the memory of the search and hence the search emphasizes diversification (intensification, respectively).

2. P (probability threshold): It is used to allow moves that are tabu but better than the current solution to be examined because this may lead to a better solution.

3. NH (number of trial solutions): It is the number of trial solutions generated for each center. The larger (smaller, respectively) the value of NH, the more (fewer, respectively) neighbors are examined and hence the search emphasizes diversification (intensification, respectively).

4. $IMAX$ (maximum number of non-improving moves for each center: It decides on how many non-improving moves are allowed for each center before going to the next one. It is observed that when getting close to the solution, the time needed to examine a given center is reduced. Therefore $IMAX$ is determined to be a variable parameter instead of a fix number.

5. γ (reduction factor for $IMAX$): If $IMAX$ non-improving moves are performed, then the next center is considered. When all centers are considered, then $IMAX$ is reduced by a factor γ, where $0 < \gamma < 1$, until it goes below 1, which corresponds to the stopping criteria. The smaller the value of γ, the faster $IMAX$ goes below 1 and hence the fewer passes through the centers the search makes, but this could be at the expense of the solution quality.

One of the most distinctive features of tabu search is the generation of neighborhoods. Since numeric data have naturally ordering, the neighborhood of the center $\mathbf{z}^u$ is defined as follows:

$$N(\mathbf{z}^u) = \{\mathbf{y} = [y_1, y_2, ..., y_m]^T \mid y_i = \mathbf{z}_i^u + \phi d,\ i = 1, 2, ..., m,\ d = 0, -1 \text{ or } +1\}. \quad (9)$$

We note that when $\mathbf{z}^u$ is close to the solution, a small step-size ϕ can be used. The neighbors of $\mathbf{z}^u$ can be generated by picking randomly from $N(\mathbf{z}^u)$.

For categorical values, we use the "distance" concept to make moves from the cluster center. The neighborhood of $\mathbf{z}^u$ is defined as follows:

$$N(\mathbf{z}^u) = \{\mathbf{y} = [y_1, y_2, ..., y_m]^T \mid d_c(\mathbf{y}, \mathbf{z}^u) < d\}, \quad (10)$$

for some positive integers d. In our algorithm, we generate a set of neighbors which are of a certain distance d from the center, i.e., neighbors which have d attributes different from the center. We remark that the distance d can be seen as the number of attributes changed for generating a neighbor, which is the criteria for selecting the neighborhood. These d attributes are randomly chosen among the m given attributes to change their values of categories, where $0 \leq d \leq m$. The greater (smaller, respectively) the value of d, the larger (smaller, respectively) the solution space to be examined and hence the search emphasizes diversification (intensification, respectively).

3 Experimental Results

The tabu search based clustering algorithm is coded in C++ programming language. The heart disease data set is used to test for the algorithm. This data set has 270 records and contains 13 attributes which have been extracted from a larger set of 75 attributes. Each record is characterized by 7 numeric and 6 categorical attributes. The records are classified into 2 classes: absence and presence of heart disease. There are no missing values included in the data set.

We obtain the cluster memberships from the fuzzy matrix W as follows. The record $\mathbf{x}_i$ for $i = 1, 2, ..., n$ is assigned to the l-th cluster if

$$w_{li} = \max_{1 \leq h \leq k} w_{hi}$$

If the maximum is not unique, then $\mathbf{x}_i$ is assigned to the clusters first achieving the maximum. A clustering result is measured by the clustering accuracy r defined as

$$r = \frac{\sum_{l=1}^{k} r_l}{n}$$

Table 2. Clustering results of tabu search based clustering algorithm.

Clustering accuracy	$\lambda^{(n)}$ and $\lambda^{(c)}$ are equal to 1	$\lambda^{(n)}$ and $\lambda^{(c)}$ determined by the algorithm
0.86	0	1
0.85	0	6
0.84	0	17
0.83	14	37
0.82	40	16
0.81	30	10
0.80	10	7
0.79	3	0
0.78	1	1
0.77	0	3
0.76	1	0
0.75	0	1
0.74	0	1
0.73	1	0
Average accuracy	0.814	0.824

where r_l is the number of objects partitioned int o the correct cluster l and n is the total number of objects in the data set.

In the test, we study the case where $\lambda_j^{(n)} = \lambda^{(n)}$, $\lambda_j^{(c)} = \lambda^{(c)}$, $\lambda^{(n)} + \lambda^{(c)} = 1$, the number of clusters is equal to two, and the index $\alpha = 1.1$ as suggested in the paper [5]. Here $\lambda^{(n)}$ and $\lambda^{(c)}$ are the weighting parameters to balance the numeric and categorical parts to avoid favoring either type of attribute. In the clustering process, all numeric attributes in the data set are rescaled to the range of $[0, 1]$ as suggested as in the paper [5]. We partition the heart disease data set into 2 clusters and the initial cluster centers are arbitrarily chosen 2 records from the data set. We also set $\gamma = 0.75$, $P = 0.97$, $IMAX = 100$, $NTLM = 100$, $NH = 100$ and $d = 1$ in the algorithm. Moreover, the algorithm is run 100 times to study the clustering accuracy.

Table 2 shows the clustering results of the tabu search based clustering algorithm using the weighted and unweighted dissimilarity measure. The average clustering accuracy by using the weighted dissimilarity measure is better than that by using the unweighted dissimilarity measure. In the tabu search based algorithm, we find that the weighting parameters are $\lambda^{(n)} = 0.87$ and $\lambda^{(c)} = 0.13$.

4 Concluding Remarks

We have introduced the tabu search based algorithm for clustering a set of objects with weighted dissimilarity measure. The most important result of this work is the procedure that allows the tabu search paradigm to be used for weighted dissimilarity measure in the clustering process. The preliminary results have shown that the tabu search based algorithms are effective in recovering the inherent clustering structures from the data set if such structures exist. In the future work, we plan to test our algorithm for other data sets and extend the algorithm to determine the number k of clusters and the index α in the minimization model.

Acknowledgment

The research was supported in part by HKU CRCG Grant Nos. 10203408, 10203501, 10203907.

References

1. J. C. Dunn, *A fuzzy relative of the ISODATA process and its use in detecting compact well-separated clusters*, J. Cybernet. **3(3)**, 32–57 (1974).
2. E. R. Ruspini, *A new approach to clustering*, Information Control **19**, 22–32 (1969).
3. J. C. Bezedek, *A convergence theorem for the fuzzy ISODATA clustering algorithms*, IEEE Transactions on Pattern Analysis and Machine Intelligence **2**, 1–8 (1980).
4. F. Glover and M. Laguna, *Tabu Search* (Kluwer Academic Publishers, Boston, 1997).
5. Z. Huang, *Extensions to the k-means algorithm for clustering large data sets with categorical values*, Data Mining and Knowledge Discovery **2(3)**, 283–304 (1998).
6. Z. Huang and M. K. Ng, *A fuzzy k-modes algorithm for clustering categorical data*, IEEE Transactions on Fuzzy Systems **7(4)**, 446–452 (1999).

BIOMECHANICALLY CONSTRAINED MULTIFRAME ESTIMATION OF NONRIGID CARDIAC KINEMATICS FROM MEDICAL IMAGE SEQUENCE

HUAFENG LIU AND PENGCHENG SHI

Department of Electrical and Electronic Engineering
Hong Kong University of Science and Technology
Clear Water Bay, Kowloon, Hong Kong
E-mail: {eeliuhf, eeship}@ust.hk

Noninvasive estimation of soft tissue kinematics properties from medical image sequences has many important clinical and physiological implications, such as the diagnosis of heart diseases and the understanding of cardiac mechanics. In this paper, we present a biomechanics based strategy, framed as *a priori* constraints for the ill-posed motion recovery problems, that performs multi-frame estimation of the cardiac motion and deformation parameters. Constructing the heart dynamics system equations from biomechanics principles, we rely on techniques from statistical estimation theory and use a Kalman filter framework to generate smooth estimates of heart kinematics throughout the cardiac cycle. We will demonstrate the application of the strategy to estimate displacements and strains from *in vivo* left ventricular magnetic resonance image sequence, which provides initial displacement measures at the boundaries.

1 Introduction

Ischemic heart diseases often manifest as varying abnormalities of myocardial regional functions, which may be detected through kinematics measurements. Noninvasive assessment of motion parameters throughout the cardiac cycle thus provides an invaluable insight for the diagnosis of the location and extent of ischemic disorders. With the availability of real-time and EEG-gated tomographic images, i.e. x-ray computed tomography (CT), magnetic resonance imaging (MRI), and echocardiography, there have been many image-based efforts on the problems of analyzing the global and local motion of the heart, especially the left ventricle [1].

In a typical method, a relatively sparse set of corresponding feature points, or the so called *landmarks*, is extracted from image sequence first. These salient points can be implanted physical markers [2], crossings of MRI tags [3], or geometrically significant features [4]. Nevertheless, the locations and thus the displacements of these landmarks are still corrupted by noises. The next step of the process is to recover the motion/deformation fields of the entire myocardium from this sparse set of landmark trajectories. It is an ill-posed problem and needs additional constraints to obtain a unique solution.

This can be achieved through mathematically motivated regularization [5], or continuum mechanics based energy minimization [6].

Because of the periodic nature of the heart motion, the importance of multiframe analysis is well recognized yet seldom addressed in a systematic fashion [7]. While most of the current strategies deal with frame-to-frame motion [1], several attempts do try to track the motion over the entire cardiac cycle using explicit temporal modeling. A Kalman filter framework was constructed, assuming elliptic trajectories of cardiac tissue elements, to estimate two-dimensional (2D) left ventricular deformation from MRI phase contrast velocity fields constrained by myocardial contours [8]. A global geometrical evolution analysis followed by regional shape matching was tested on simulated and real endocardial surfaces [9]. And an adaptive filtering scheme with spatial-temporal smoothness and periodicity constraint was proposed to track myocardial contour motion [7].

In this paper, we present a biomechanically constrained framework which performs multiframe estimation of the nonrigid cardiac kinematics from medical image sequence. It grows out of our earlier works on shape-based boundary motion analysis [4] and mechanics-based volumetric kinematics recovery [6], both of which have been carefully validated with animal models. However, instead of tracking frame-to-frame motion, we are proposing a temporal filtering framework to recover the kinematics throughout the cardiac cycle. We differ from previous multiframe efforts [8,7] in two aspects. First, rather than making *ad hoc* mathematical assumptions, we construct the myocardial system dynamics equations from continuum biomechanics point of view, which allows the incorporation of realistic material constraints based on experimental measurement and *a priori* knowledge. Secondly, although we do not have a explicit periodic motion constraints in our modeling of cardiac behavior, we cyclically feed the updated image and image-derived data into our framework until reaching convergence. We will show the experiment results from segmented 2D image data which provides boundary displacement information.

2 Methodology

2.1 Biomechanical Model of the Myocardium

The structure, dynamics, and material of the heart should be modelled in such a way that given image–derived constraints and other measurements or conditions, we would have a realistic yet computationally feasible framework for the recovery of cardiac kinematics. The heart is a nonrigid object that deforms over time and has very complicated biomechanical properties in terms of stress-strain relationships [2]. For computational simplicity, we assume that the myocardium is a linear material, with its stress($[\sigma]$) and strain($[\varepsilon]$) relationship (the constitutive law) obeying the *Hooke's law* [a], and is bounded by endocardial and epicardial contours in 2D:

$$[\sigma] = [D]\,[\varepsilon] \tag{1}$$

Under two dimensional Cartesian coordinate system, assuming the displacement along the x- and y-axis of a point to be $u(x,y)$ and $v(x,y)$ respectively, the strain tensor $[\varepsilon]$ of the point can be expressed as:

$$[\varepsilon] = \begin{bmatrix} \frac{\partial u}{\partial x} \\ \frac{\partial v}{\partial y} \\ \frac{\partial u}{\partial y} + \frac{\partial v}{\partial x} \end{bmatrix} \tag{2}$$

and under plane strain condition, matrix D can be derived to be:

$$[D] = \frac{E}{(1+\nu)(1-2\nu)} \begin{bmatrix} 1-\nu & \nu & 0 \\ \nu & 1-\nu & 0 \\ 0 & 0 & \frac{1-2\nu}{2} \end{bmatrix} \tag{3}$$

Here, E and ν, *Young's modulus* and *Poisson's ratio*, are two material-related constants which have been established experimentally for myocardium in biomechanics literature [10], with value to be around $75,000 Pascal$ and 0.47 respectively. It is clear that under our model, the internal stress caused by the deformation is a function of the displacement vector and some material-specific constants. We are going to use this strain-stress relationship in the finite element representation of the model.

2.2 Finite Element Representation of the Heart

The finite element method is used for the representation of the heart structure and dynamics. A Delaunay triangulated finite element mesh is formed (Fig. 2), bounded by automatically segmented endocardium and epicardium. An isoparametric formulation defined in a natural coordinate system is used, in which the interpolation of the element coordinates and element displacements use the same basis functions. For the tri-nodal linear element, the basis functions are linear functions of the nodal coordinates [12].

The nodal displacement based governing dynamic equation of each element is established under the principle of minimum potential energy. The

[a] Currently, we are in the process of considering and implementing more realistic constitutive models, any of which can be inserted into this framework.

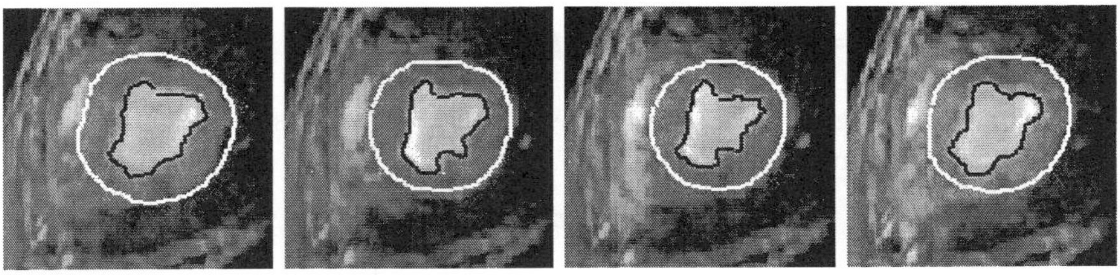

Figure 1. Segmented EEG-gated canine MR image sequence throughout cardiac cycle (sixteen frames in total). From left to right, frames #1, #5, #9, and #13.

equations are assembled together in matrix form as:

$$M\ddot{U} + C\dot{U} + KU = R \tag{4}$$

with M, C and K the mass, damping and stiffness matrices respectively, R the load vector, and U the displacement vector. M is a known function of material density and is assumed temporally constant for incompressible material. K is a function of material constitutive law, and is related to the material-specific Young's modulus and Poisson's ratio which are again assumed constant [b]. C is frequency dependent, and we assume Rayleigh damping with $C = \alpha M + \beta K$.

We want to point out that we also intend to use this framework to enforce certain real physical constraints related to known cardiac pressures. It is also important to note that while the finite element grid provides the basis for approximating a continuous spatial model, the dynamic equations provide the basis of an appropriate temporal model for the matching and predicting of image frames.

2.3 State Space Strategy

The dynamics equation (4) is transformed into a state-space representation of a continuous-time linear system:

$$\dot{x}(t) = A_c x(t) + B_c w(t) \tag{5}$$

where the state vector x, the system matrices A_c and B_c, and the control (input) term w are:

$$x(t) = \begin{bmatrix} U(t) \\ \dot{U}(t) \end{bmatrix}, \quad A_c = \begin{bmatrix} 0 & I \\ -M^{-1}K & -M^{-1}C \end{bmatrix},$$

[b]The material parameters can vary temporally and spatially. In our other work [11], they are treated as random variables with known *a priori* statistics for any given data, and need to be jointly estimated along with kinematics measures.

$$B_c = \begin{bmatrix} 0 & 0 \\ 0 & M^{-1} \end{bmatrix}, \quad w(t) = \begin{bmatrix} 0 \\ R \end{bmatrix}.$$

The observed imaging/imaging-derived data $y(t)$ relates to the state vector through the measurement equation:

$$y(t) = Hx(t) + e(t) \tag{6}$$

where H is the measurement matrix (assumed to be known), and $e(t)$ is the measurement noise which is additive, zero mean, and white ($E[e(t)] = 0, E[e(t)e(s)'] = R_e(t)\delta_{ts}$).

Equations (5) and (6) describe a continuous-time system with discrete-time measurements, or a so-called sampled data system. The input is computed from the system equation, and is piecewise constant over the sampling interval T. Thus, we arrive at the system equations [14]:

$$x((k+1)T) = Ax(kT) + Bw(kT) \tag{7}$$

with $A = e^{A_c T}$ and $B = A_c^{-1}(e^{A_c T} - I)B_c$.

For the general continuous-time system with discrete-time measurements, and including the additive, zero-mean, white process noise v ($E[v(t)] = 0, E[v(t)v(s)'] = Q_v(t)\delta_{ts}$, independent of $e(t)$), the state equation becomes:

$$x(t+1) = Ax(t) + Bw(t) + v(t) \tag{8}$$

2.4 Kalman Filter Estimation

The Kalman filter adopts a form of feedback control in estimation: the filter estimates the process state at some time and then obtains the feedback in the form of (noisy) measurements [13]. Hence, the time update equations of the Kalman filter are responsible for projecting forward (in time) the current state and error covariance estimates to obtain the *a priori* estimates for the next time step, while the measurement update equations are responsible for the feedback – i.e. for incorporating a new measurement into the *a priori* estimate to obtain an improved *a posteriori* estimate. And the final estimation algorithm resembles that of a predictor-corrector algorithm for solving numerical problems.

A recursive procedure is used to perform the state estimation of Equations (6) and (8)(see Kamen *et al* [13]):

1. Initial estimates for state $\hat{x}(t-1)$ and error covariance $P(t-1)$.

2. Time update equations, *the predictions*, for the state

$$\hat{x}^-(t) = A\hat{x}(t-1) + Bw(t) \tag{9}$$

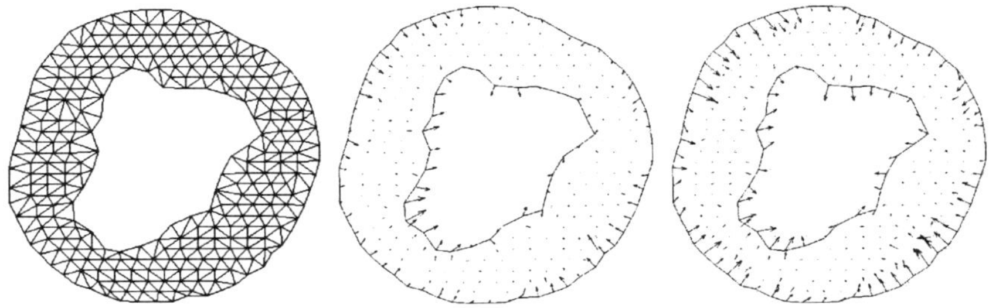

Figure 2. Left: finite element mesh of the myocardium at frame #1 (end of diastole (ED)). Middle: displacement field between frames #1 and #4. Right: displacement field between frames #1 and #8 (end of systole (ES)).

and the error covariance

$$P^{-}(t) = AP(t-1)A^{T} + Q_{v}(t) \tag{10}$$

3. Measurement update equations, *the corrections*, for the Kalman gain

$$L(t) = P^{-}(t)H^{T}(HP^{-}(t)H^{T} + R_{e}(t))^{-1} \tag{11}$$

the state

$$\hat{x}(t) = \hat{x}^{-}(t) + L(t)(y(t) - H\hat{x}^{-}(t)) \tag{12}$$

and the error covariance

$$P(t) = P^{-}(t) - L(t)(HP^{-}(t)H^{T} + R_{e}(t))L^{T}(t) \tag{13}$$

3 Implementation and Experiment

In our current implementation and experiment of the framework, displacement constraints at selected sampling points of myocardial boundaries are used for the recovery of cardiac kinematics. However, other types of imaging/imaging-derived data with acceleration, velocity, and displacement information can be used without fundamental changes to the framework.

3.1 Shape-Based Boundary Displacement

We have proposed a strategy for myocardial boundary motion tracking based on locating and matching differential geometric landmarks [4]. In 2D case, a sparse subset of the contour points are created by choosing shape landmarks

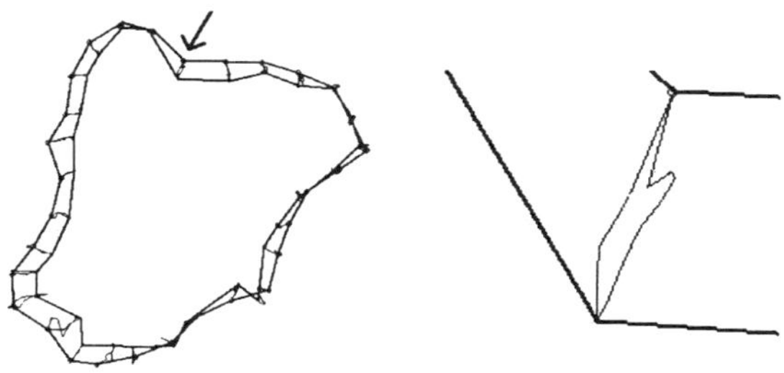

Figure 3. Shape-based boundary displacement constraints at selected endocardial points. Left: the trajectories of the contour points. Right: the blown-up view of one trajectory (arrow pointed point).

that are geometrically significant. Computation of the displacements of these landmarks is carried out using the bending energy matching criterion:

$$e_{bend}(\delta) = \frac{1}{2}EI \int_{C} \left[\kappa_g(\tau+\delta) - \kappa_f(\tau)\right]^2 d\tau \tag{14}$$

where κ_f is the curvature for any given landmark point in the first contour, $\mathcal{C}$ the search region on the second contour, κ_g the curvature of a candidate point within the search region on the second contour, and δ indexes the different candidate points within search region. Among all the candidate points within the search region, the one at $\bar{x}$ which yields the smallest bending energy is chosen as the matched point. This value indicates the *goodness* of the match:

$$m_g(x) = \epsilon_{be}(x, \bar{x}) \tag{15}$$

Meanwhile, the bending energy measures for all other points inside each search region are also recorded as the basis to measure the *uniqueness* of the matching choice. Ideally, the bending energy value of the chosen point should be an outlier (much smaller value) compared to the values of the rest of the points (much larger values). If we denote the mean values of the bending energy measures of all the points inside search window except the chosen point as $\bar{\epsilon}_{be}$ and the standard deviation as σ_{be}, we define the uniqueness measure as:

$$m_u(x) = \frac{\epsilon_{be}(x, \bar{x})}{\bar{\epsilon}_{be} - \sigma_{be}} \tag{16}$$

Obviously for both goodness and unique measures, the smaller the values are the more reliable the match. Combining these two measures together, we

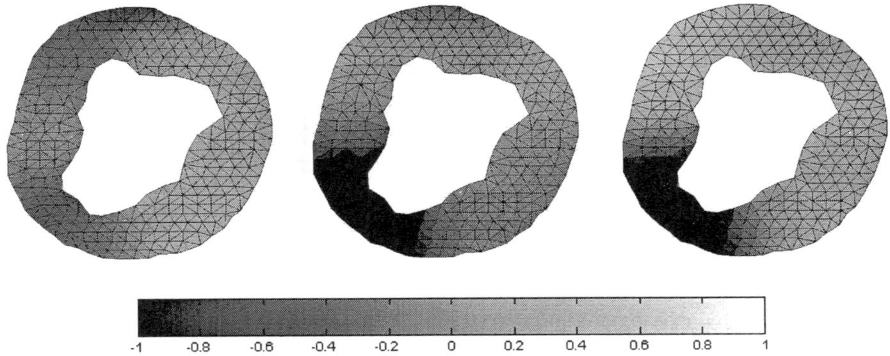

Figure 4. Estimated x-strain (left) y-strain (middle) and shear strain (right) maps between ED and ES.

arrive at one *confidence measure* for the matched point x of point x:

$$c(x) = \frac{1}{k_{1,g} + k_{2,g}m_g(x)} \frac{1}{k_{1,u} + k_{2,u}m_u(x)} \tag{17}$$

where $k_{1,g}, k_{2,g}, k_{1,u}$, and $k_{2,u}$ are scaling constants for normalizing purposes. The confidence measures for all the surface matches are normalized to the range of 0 to 1.

The result of this process for every landmark produces a set of shape-based, best-matched motion vectors for each pair of contours, and each vector has an associated confidence measure. Figure 3 shows the trajectories of the sampled endocardial contour points over the cardiac cycle.

3.2 Computational Considerations

Initial conditions: the use of our algorithm for the kinematics state estimation requires initial values for the state vectors and the matrices. The initialization of the error covariance matrix $P^-(0) = E[(x(0) - x^-(0))(x(0) - x^-(0))^T]$ is a critical factor as it drives the data association. In our current implementation, we use $P^-(0) = \lambda I, \lambda > 0$, which assumes no correlation between initial state estimates. In addition, we model the process noise Q_v and measurement noise R_e as diagonal matrix and use fixed values for both.

Further considerations for error covariance matrix: because the matrix $P(t)$ must be symmetric and non-negative definite, special attentions should be given in its recursive updating. If round-off errors should produce an

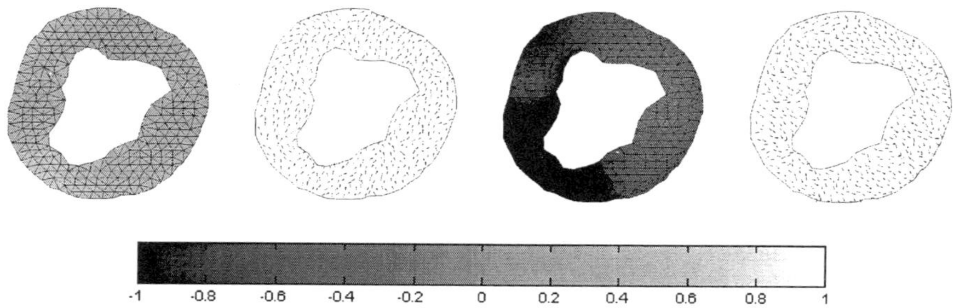

Figure 5. Estimated maximum principle strain/direction maps (left) and minimum principle strain/direction maps (right), ED to ES.

indefinite $P(t)$ matrix at given step, it is repaired with a *nearby* non-negative definite matrix or through $U - D$ factorization [14].

Boundary conditions: the dynamics equations are modified to account for the boundary conditions of the system. If the displacements of some nodal points are known to be $U_b = b$, say from shape-based boundary tracking or MR tagging images, the constraint $c(b)kU_b = c(b)kb$ is added to the governing equations [12], where $c(b)$ is related the confidence measurement on the displacement.

3.3 Experiment Results

The above described framework is used to estimated the kinematics parameters (displacement and strain) of the left ventricle from the MRI images of Figure 1, constrained by boundary displacement information of Figure 3. Figure 2 shows the estimated displacement fields between image frames #1 and #4 (middle), and frames #1 and #8 (right). Figure 4 and 5 show the strain and principle strain maps between image frames #1 and #8, respectively.

4 Conclusion

In this paper, we have described a biomechanically constrained framework for multiframe estimation of the cardiac kinematics from medical image sequence. The myocardial system dynamics equations are constructed from continuum biomechanics point of view, and Kalman filter is used to obtain optimal estimates of the kinematics state vectors.

This work is supported in part by the Hong Kong CERG Grant HKUST6057/00E, and by a HKUST Postdoctoral Fellowship Matching Fund.

References

1. A.F. Frangi, W.J. Niessen, and M.A. Viergever, *Three-dimensional modeling for functional analysis of cardiac images: a review*, IEEE Transactions on Medical Imaging **20**, 2–25(2001).
2. L. Glass and P. Hunter and A. McCulloch, *Theory of Heart* (Springer-Verlag, New York, 1991).
3. C.C. Moore, W.G. O'Dell, E.R. McVeigh, and E.A. Zerhouni, *Calculation of three-dimensional left ventricular strains from bi-planar tagged MR images*, Journal of Magnetic Resonance Imaging **2**, 165–175(1992).
4. P. Shi, A.J. Sinusas, R.T. Constable, and J.S. Duncan, *Point-tracked quantitative analysis of left ventricular motion from 3D image sequences*, IEEE Transactions on Medical Imaging **19**, 36–50(2000).
5. J. Park, D.N. Metaxas, and L. Axel, *Analysis of left ventricular wall motion based on volumetric deformable models and MRI-SPAMM*, Medical Image Analysis **1**, 53–71(1996).
6. P. Shi, A.J. Sinusas, R.T. Constable, and J.S. Duncan, *Volumetric deformation analysis using mechanics-based data fusion: applications in cardiac motion recovery*, International Journal of Computer Vision **35**, 87–107(1999).
7. J.C. McEachen, A. Nehorai, and J.S. Duncan, *Multiframe temporal estimation of cardiac nonrigid motion*, IEEE Transactions on Image Processing **9**, 651–665(2000).
8. F.G. Meyer, R.T. Constable, A.J. Sinusas, and J.S. Duncan, *Tracking myocardial deformation using phase contrast MR velocity fields: a stochastic approach*, IEEE Transactions on Medical Imaging **15**, 453–465(1996).
9. P. Clarysse, D. Friboulet, and I.E. Magnin, *Tracking geometrical descriptors on 3-D deformable surfaces - application to the left- ventricular surface of the heart*, IEEE Transactions on Medical Imaging **16**, 392–404(1997).
10. H. Yamada, *Strength of Biological Material* (Williams and Wilkins, Baltimore, 1970).
11. P. Shi and H.F. Liu, *Stochastic finite element framework for cardiac kinematics function and material property analysis*, Medical Image Computing and Computer Assisted Intervention , (in press) (2002).
12. K.-J. Bathe, *Finite Element Procedures* (Prentice Hall, Upper Saddle River, 1996).
13. E.W. Kamen and J.K. Su, *Introduction to Optimal Estimation* (Springer, London, 1999).
14. M.S. Grewal and A.P. Andrews, *Kalman Filtering* (Prentice Hall, 1993)

PARAMETERS IDENTIFICATION OF AN ELASTIC PLATE SUBJECTED TO DYNAMIC LOADING BY INVERSE ANALYSIS USING BEM AND KALMAN FILTER

MASA. TANAKA, T. MATSUMOTO AND H. YAMAMURA

Department of Mechanical Systems Engineering, Shinshu University, 4-17-1 Wakasato, Nagano, 380-8553 Japan

E-mail: dtanaka@gipwc.shinshu-u.ac.jp

There are many investigations in which computational software for direct analysis is successfully applied to the solution of inverse problems. In this study, a boundary element method (BEM) for analyzing the direct problems of elastic plates subjected to dynamic loadings is applied to the corresponding inverse problems of parameters identification under dynamic loadings. It is assumed that the lateral displacement of the plate is measured at several points in the plate domain. Using such measured data of deformation, inverse analysis is carried out to identify a series of unknown parameters for dynamic bending of plates. The extended Kalman filter is employed for iterative computation to modify the parameter values. A few examples are investigated by the proposed method of inverse analysis and the results obtained are discussed, whereby the potential usefulness of the proposed method is demonstrated.

1 Introduction

There are many investigations in which computational software so far developed for direct analysis is successfully applied to the solution of inverse problems [1,2,3,4,5]. In this study, a boundary element method (BEM) for analyzing the direct problems of elastic plates subjected to dynamic loadings is applied to the corresponding inverse problem of parameters identification. It is assumed that the lateral displacement of the plate is measured at several points in the spatial as well as temporal domains under consideration. The boundary element method combined with the Laplace transform, which was reported in authors' previous paper [6], is applied to compute the dynamic behavior of the elastic plate subjected to arbitrary dynamic loading under the known parameters. Using such measured data of deformation, inverse analysis is carried out to identify a series of unknown parameters for dynamic bending of plates. The extended Kalman filter [7,8,9] is employed for iterative computation to modify the parameter values toward the set of target values. A few examples of parameters identification are investigated by the proposed method of inverse analysis, and the results obtained are discussed, whereby the potential usefulness of the proposed method is demonstrated.

2 Boundary Element Analysis of Dynamic Bending of Elastic Plates

2.1 Integral Equation Formulation

The forced vibration of elastic plates subjected to dynamic loading is governed by the following differential equation:

$$D\nabla^4 w(x,t) + \rho h \frac{\partial^2 w(x,t)}{\partial t^2} + c_b \frac{\partial w(x,t)}{\partial t} = p(x,t) \tag{1}$$

where $w(x,t)$is the lateral displacement at point x and time t, ρ the density of mass, h the plate thickness, c_b the external damping coefficient, p the distributed exciting force per unit area of the plate middle plane, and ∇^4 the biharmonic differential operator. In addition, D is the flexural rigidity of the plate, which is related to E (Young's modulus), ν (Poisson's ratio) and h as follows:

$$D = \frac{Eh^3}{12(1-\nu^2)} \tag{2}$$

Using the Laplace transform as $\hat{F} = \int_0^\infty f(x,t)e^{-st}dt$, and assuming without loss of generality that the initial conditions are homogeneous, we can express Eq. (1) as follows:

$$D\nabla^4 \hat{W} + (\rho h s^2 + c_b s)\hat{W} = \hat{P} \tag{3}$$

In the boundary element method using the Laplace transform [6], Eq. (3) is solved for a series of the transform parameter s and then inverse transform is carried out to get the physical solution in space and time. In the present BEM we employ Durbin's method [10] for numerical Laplace inverse transform.

For the boundary integral equation formulation, we use the fundamental solution of static bending of elastic plate, i.e. the fundamental solution to the biharmonic operator:

$$W^*(x,y) = \frac{1}{8\pi D} r^2 \ln r \quad , \quad r = |x-y| \tag{4}$$

For the integral equation formulation, let us begin with the following identity obtained from Eq. (3) multiplied with the fundamental solution [5], that is,

$$\int_\Omega \{D\nabla^4 \hat{W} + (\rho h s^2 + c_b s)\hat{W} - \hat{P}\} W^*(x,y) d\Omega = 0 \tag{5}$$

Based on the procedure reported in authors' previous paper [12], we can eventually arrive at

$$\int_\Gamma \left[W^*\hat{V}_n - \Theta_n^*\hat{M}_n + M_n^*\hat{\Theta}_n - Q_n^*\{\hat{W} - \hat{W}(y)\} + M_{nt}^*\frac{\partial \hat{W}}{\partial t}\right] d\Gamma$$
$$+\left\{\int_\Gamma Q_n^* d\Gamma\right\}\hat{W}_{,k}(y)n_k(y) - \left(\rho h s^2 + c_b s\right)\int_\Omega \tilde{W}^*\hat{W}d\Omega$$
$$+\int_\Omega \tilde{W}^*\hat{P}d\Omega - \sum_{k=1}^{K_c}\left[W^*\hat{M}_{nt}\right]_k = 0 \tag{6}$$

In the above equation, $\sum[\ ]_k$ denotes the summation of jump of the variable in $[\ ]$ at a corner point k for all the corner points, and $\partial(\ )/\partial \mathbf{t}$ denotes the tangential derivative. Eq. (6) is the so-called regularized integral equation which holds equally when the source point y is located in the domain Ω as well as on the boundary Γ .

The counterpart of the integral equation for the normal derivative of deflection can be derived through differentiation of Eq. (6) with respect to the source point, and expressed as

$$\int_\Gamma \left[\tilde{W}^*\hat{V}_n - \tilde{\Theta}_n^*\hat{M}_n + \tilde{M}_n^*\{\hat{\Theta}_n - \hat{W}_{,k}(y)n_k\} - \tilde{Q}_n^*\{\hat{W} - \hat{W}(y) - r_k\hat{W}_{,k}(y)\}\right.$$
$$\left. +\tilde{M}_{nt}^*\left\{\frac{\partial \hat{W}}{\partial t} - \hat{W}_{,k}(y)t_k\right\}\right] d\Gamma - \left(\rho h s^2 + c_b s\right)\int_\Omega \tilde{W}^*\hat{W}d\Omega + \int_\Omega \tilde{W}^*\hat{P}d\Omega$$
$$-\sum_{k=1}^{K_c}\left[W^*\hat{M}_{nt}\right]_k = 0 \tag{7}$$

Since the domain integrals with the unknown deflection are included in the boundary integral equations Eq. (6) and Eq. (7), we have to supplement the integral equation which provides the relation between the deflection at an internal point and the quantities on the boundary. This integral equation can be obtained from Eq. (6), and expressed in a regularized form as follows:

$$\hat{W}(y) = \int_\Gamma \left[W^*\hat{V}_n - \Theta_n^*\hat{M}_n + M_n^*\hat{\Theta}_n - Q_n^*\{\hat{W} - \hat{W}(x_0)\} + M_{nt}^*\frac{\partial \hat{W}}{\partial t}\right] d\Gamma$$
$$+ \hat{W}(x_0) - \left(\rho h s^2 + c_b s\right)\int_\Omega W^*\hat{W}d\Omega + \int_\Omega W^*\hat{P}d\Omega - \sum_{k=1}^{K_c}\left[W^*\hat{M}_{nt}\right]_k \tag{8}$$

To make computation easier, we have introduced the nearest point x_0 to the source point y in the inner domain.

2.2 Boundary-Domain-Element Method

The consistent set of integral equations Eq. (6), Eq. (7) and Eq. (8) are discretized by the boundary-domain-element method [11]. The system of simultaneous equations thus obtained are solved under the given boundary conditions as well as the initial conditions.

After application of the boundary conditions, the discretized version of integral equations Eq. (6) and Eq. (7) can be expressed in the following matrix form:

$$[\mathbf{A}]\{\hat{\mathbf{X}}\} + [\mathbf{C}]\{\hat{\mathbf{W}}^i\} = [\mathbf{B}]\{\hat{\mathbf{Y}}\} + \{\hat{\mathbf{D}}\} \tag{9}$$

where $\{\hat{\mathbf{X}}\}$ is the column vector of unknown nodal values in the domain and on the boundary, $\{\hat{\mathbf{Y}}\}$ the column vector of known nodal values on the boundary, and $\{\hat{\mathbf{W}}^i\}$ the column vector of unknown displacement in the domain. In addition, $[\mathbf{A}]$, $[\mathbf{B}]$, and $[\mathbf{C}]$ are the coefficient matrices calculated from the fundamental solution, and $\{\hat{\mathbf{D}}\}$ is the column vector of known components.

If in Eq. (8) we locate the source point y at each nodal point in the domain, we can derive the following system of equations:

$$\{\hat{\mathbf{W}}^i\} + [\mathbf{a}]\{\hat{\mathbf{X}}\} + [\mathbf{c}]\{\hat{\mathbf{W}}^i\} = [\mathbf{b}]\{\hat{\mathbf{Y}}\} + \{\hat{\mathbf{d}}\} \tag{10}$$

From Eq. (9) and Eq. (10) we have

$$\begin{bmatrix} \mathbf{A} & \mathbf{C} \\ \mathbf{a} & \mathbf{I}+\mathbf{c} \end{bmatrix} \begin{Bmatrix} \hat{\mathbf{X}} \\ \hat{\mathbf{W}}^i \end{Bmatrix} = \begin{bmatrix} \mathbf{B} \\ \mathbf{b} \end{bmatrix} \{\hat{\mathbf{Y}}\} + \begin{Bmatrix} \hat{\mathbf{D}} \\ \hat{\mathbf{d}} \end{Bmatrix} \tag{11}$$

If Eq. (11) is solved for the unknown vectors $\{\hat{\mathbf{X}}\}$ and $\{\hat{\mathbf{W}}^i\}$, all the nodal unknowns on the boundary and in the boundary are calculated.

3 Inverse Analysis Using Kalman Filter

Identification problems of unknown parameters are in general nonlinear. In this study, a solution procedure based on BEM is applied to modify the parameters in an iterative manner using the extended Kalman filter [7,8,9]. It is assumed that there are no system errors, and that the necessary data are measured for the whole region in space and time. Identification of the parameters is carried out for the whole region, and hence the suffix k used in the following computational procedure can be interpreted as the index of iteration.

It can be assumed that the measured data, denoted by $\boldsymbol{y}_k$, on plate deflection at some selected points is a nonlinear function of the parameters $\boldsymbol{x}_k$ at the iteration k. The system under consideration can be expressed by the following state equation and observation equation, that is,

$$\boldsymbol{x}_{k+1} = \boldsymbol{F}_k(\boldsymbol{x}_k) + \boldsymbol{g}_k(\boldsymbol{x}_k)\boldsymbol{w}_k \tag{12}$$

$$\boldsymbol{y}_k = \boldsymbol{h}_k(\boldsymbol{x}_k) + \boldsymbol{v}_k \tag{13}$$

where $\boldsymbol{g}_k(\boldsymbol{x}_k)$ is called the system-noise coefficient, $\boldsymbol{w}_k$ the system noise, and $\boldsymbol{v}_k$ the measurement error. A linearized set of the above nonlinear equations provide the extended Kalman filter which is applicable to modification of the parameter values for the next iteration. The extended Kalman filter is composed of the following equations:

(i) Filter equations

$$\hat{\boldsymbol{x}}_{k+1} = \boldsymbol{f}_k(\hat{\boldsymbol{x}}_k) \tag{14}$$

$$\hat{\boldsymbol{x}}_{k/k} = \hat{\boldsymbol{x}}_{k/k-1} + \boldsymbol{K}_k[\boldsymbol{y}_k - \boldsymbol{h}_k(\hat{\boldsymbol{x}}_{k/k-1})] \tag{15}$$

(ii) Filter gain

$$\boldsymbol{K}_k = \boldsymbol{P}_{k/k-1}\boldsymbol{H}_k^T[\boldsymbol{H}_k\boldsymbol{P}_{k/k-1}\boldsymbol{H}_k^T + \boldsymbol{R}_k]^{-1} \tag{16}$$

(iii) Covariance matrices of estimation errors

$$\boldsymbol{P}_{k+1/k} = \boldsymbol{F}_k\boldsymbol{P}_{k/k}\boldsymbol{F}_k^T + \boldsymbol{G}_k\boldsymbol{Q}_k\boldsymbol{G}_k^T \tag{17}$$

$$\boldsymbol{P}_{k/k} = \boldsymbol{P}_{k/k-1} - \boldsymbol{K}_k\boldsymbol{H}_k\boldsymbol{P}_{k/k-1} \tag{18}$$

In the above extended Kalman filter, the new parameter values $\hat{\boldsymbol{x}}_{k/k}$ is estimated for the current parameter values $\boldsymbol{x}_k$ by the measured data $\boldsymbol{y}_k$. Starting from the iteration step $k = 0$, we may modify the parameter values iteratively by this solution procedure.

In the above expressions, $\mathbf{H}_k$ is called the sensitivity matrix, which is defined as

$$\boldsymbol{H}_k = \left.\frac{\partial \boldsymbol{h}_k(\boldsymbol{x}_k)}{\partial \boldsymbol{x}_k}\right|_{\boldsymbol{x}_k=\hat{\boldsymbol{x}}_{k/k-1}} = \begin{bmatrix} \frac{\partial h_1}{\partial x_1} & \frac{\partial h_1}{\partial x_2} & \cdots & \frac{\partial h_1}{\partial x_n} \\ \vdots & \vdots & \ddots & \vdots \\ \frac{\partial h_n}{\partial x_1} & \frac{\partial h_n}{\partial x_2} & \cdots & \frac{\partial h_n}{\partial x_n} \end{bmatrix}_{k/k-1} \tag{19}$$

The sensitivity matrix $\mathbf{H}_k$ depends on the estimated parameters $\hat{\boldsymbol{x}}_{k/k-1}$ at each iteration step, and hence this matrix should be calculated at each iteration step. The derivatives with respect to each parameter in the above expression is calculated by the finite difference scheme:

$$\left.\frac{\partial h_i}{\partial x_j}\right|_{k/k-1} \approx \left.\frac{h_i(x_1,\ldots,x_j+\Delta x_j,\ldots,x_n)-\mathbf{h}_i(x_1,\ldots,x_j,\ldots,x_n)}{\Delta x_j}\right|_{k/k-1} \quad (20)$$

The responses of the plate under the given parameters are computed by the Laplace-transform boundary element method [6].

4 Numerical Results and Discussion

To demonstrate usefulness of the proposed solution procedure for the inverse problems under consideration, let us apply it to identification of material constants and other parameters of the applied dynamic load. We shall consider the square plate of edge length $a = 1$ [m] and thickness $h = 0.01$ [m] with four edges simply supported, as shown in Figure 1. No external damping is assumed ($c_b = 0$). For numerical simulation's sake, we first compute via the BEM software the dynamic responses of the elastic plate under the given dynamic load for time $t = 0$ to 0.05 [s], assuming that Young's modulus $E = 2.0 \times 10^{11}$ [Pa], Poisson's ratio $\nu = 0.3$ and the density of mass $\rho = 7.8 \times 10^3$ [kg/m^3]. Figure 2 shows discretization of the plate via the boundary-domain-element method, in which quadratic elements are used both for the boundary and inner domain. For the numerical Laplace inverse transform, we place 20 sampling points in the time axis with an equal interval. The computational results thus obtained are used as the averaged values of measured data. Furthermore, it is assumed that the plate deflection is measured at the eight spatial points as shown in Figure 1 and also at several temporal points.

Now, we shall show the numerical results for the first example subjected to the dynamic concentrated load $P(t) = 50 \sin 40\pi t + 50$ [N]. It is assumed in this example that the deflection is measured at 20 points in time as in the Laplace inverse transform. In Figure 3 comparison is made between the exact and estimated results for the time variation of load. Computation is carried out by assuming that all the initial values of load at 20 points are equal to 50 [N] and that no measurement errors are included and the diagonal components of the estimation-error covariance matrix $\boldsymbol{P}_{k/k-1}$ is 1.0×10^4 at the step $k = 0$. Excellent agreement can be realized. It is noted that successful estimation can be made in this example even under a larger amount of measurement errors.

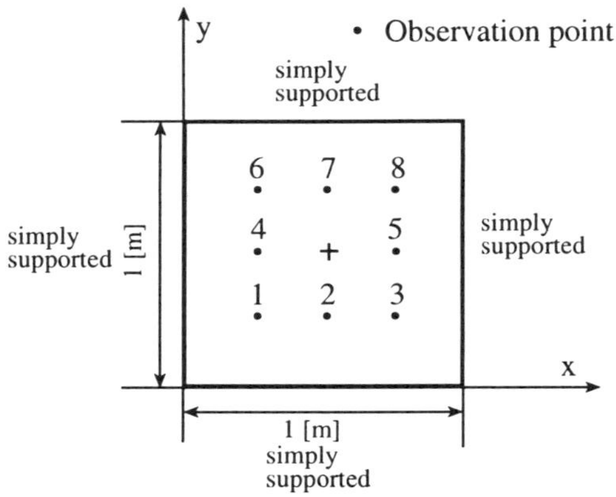

Figure 1. Analysis model.

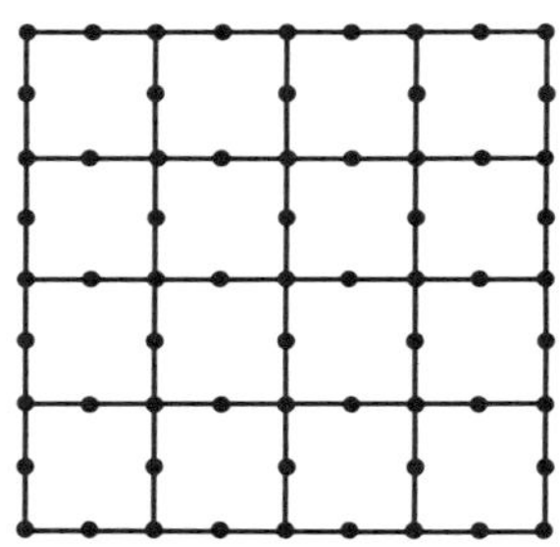

Figure 2. Analysis model.

Next, we shall show other numerical results under the dynamic load $P = P_0 H(t)$, where $P_0 = 100$ [N] and $H(t)$ is the Heaviside function. In Table 1 are shown the estimation results under the measurement errors $\mathbf{R} = 10^{-14}\mathbf{I}$, where $\mathbf{I}$ is the unit diagonal matrix. In this table, the numbers in parentheses in the estimation results indicate errors in percentage between the estimated and the target values of parameters. In Table 2 are shown similar estimation results for $\mathbf{R} = 10^{-13}\mathbf{I}$.

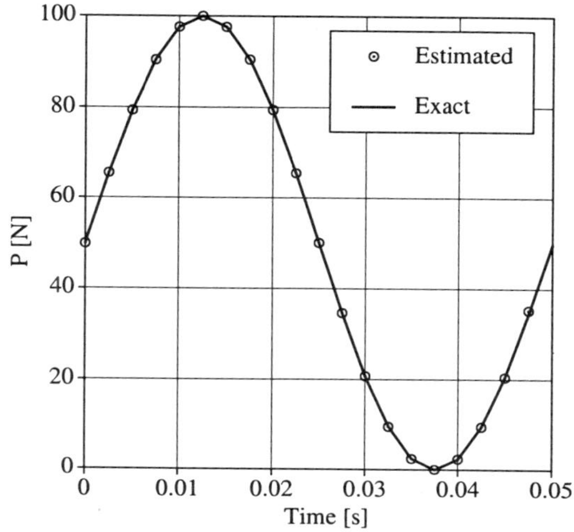

Figure 3. Estimation results on dynamic load.

Table 1. Estimation results on material constants.

Covariance of measurement errors $\mathbf{R} = 10^{-14}\mathbf{I}$;
Target values: $E = 2.0 \times 10^{11}$ [Pa], $\rho = 7.8 \times 10^3$ [kg/m^3]

Initial		Estimated	
E [Pa]	ρ [kg/m^3]	E [Pa]	ρ [kg/m^3]
1.8×10^{11}	7.02×10^3	1.996×10^{11} (−0.2)	7.838×10^3 (0.487)
2.2×10^{11}	8.58×10^3	1.996×10^{11} (−0.2)	7.839×10^3 (0.5)
1.4×10^{11}	5.46×10^3	1.990×10^{11} (−0.5)	7.821×10^3 (0.269)
2.6×10^{11}	10.14×10^3	1.995×10^{11} (−0.25)	7.837×10^3 (0.474)
1.0×10^{11}	3.9×10^3	1.856×10^{11} (−7.2)	7.896×10^3 (1.230)
3.0×10^{11}	11.7×10^3	1.993×10^{11} (−0.35)	7.824×10^3 (0.307)

Table 2. Estimation results on material constants.

Covariance of measurement errors $\mathbf{R} = 10^{-13}\mathbf{I}$;
Target values: $E = 2.0 \times 10^{11}$ [Pa], $\rho = 7.8 \times 10^{3}$ [kg/m^3]

Initial		Estimated	
E [Pa]	ρ [kg/m^3]	E [Pa]	ρ [kg/m^3]
1.8×10^{11}	7.02×10^{3}	1.975×10^{11} (−1.245)	7.910×10^{3} (1.417)
2.2×10^{11}	8.58×10^{3}	1.975×10^{11} (−1.240)	7.911×10^{3} (1.424)
1.4×10^{11}	5.46×10^{3}	1.966×10^{11} (−1.670)	7.900×10^{3} (1.284)
2.6×10^{11}	10.14×10^{3}	1.974×10^{11} (−1.275)	7.909×10^{3} (1.408)
1.0×10^{11}	3.9×10^{3}	1.798×10^{11} (−10.085)	8.063×10^{3} (3.371)
3.0×10^{11}	11.7×10^{3}	1.973×10^{11} (−1.320)	7.894×10^{3} (1.205)

5 Concluding Remarks

The method of inverse analysis based on the boundary element method and the extended Kalman filter has been applied to parameters identification of an elastic plate subjected to dynamic loading. The proposed method is rather robust even if some measurement errors are included in given additional information on measured data. It can be concluded that the proposed method could provide better estimation results more effectively, if the initial values of the parameters to be estimated are assumed close to the exact ones in an appropriate manner.

It is one of the most important subjects in the solution of inverse problems how to assume the initial values of parameters taking account of a priori information as much as possible. For this purpose, we may apply some knowledge-based methods to find an approximate solution of the inverse problem under consideration, and then apply the present method of inverse analysis based on sensitivity analysis. Such research can be recommended as future work along the present investigation.

References

1. M. Tanaka and H.D. Bui, *Inverse Problems in Engineering Mechanics* (Springer–Verlag, Berlin, 1992).
2. H.D. Bui and M. Tanaka, *Inverse Problems in Engineering Mechanics* (A.A. Balkema, Rotterdam/The Netherlands, 1994).
3. M. Tanaka and G.S. Dulikravich, *Inverse Problems in Engineering Mechanics* (Elsevier Science, Amsterdam-Oxford/UK, 1998).
4. K.A. Woodbury, *Inverse Problems in Engineering – Theory and Practice* (Engineering Foundation and ASME, New York, 1999).
5. M.Tanaka and G.S. Dulikravich, *Inverse Problems in Engineering Mechanics II* (Elsevier Science, Amsterdam–Oxford/UK, 2000).
6. M. Tanaka, T. Matsumoto and S. Judai, *Conf. on Computational Engineering, JSCES* **4**, 1015 (1999).
7. R.E. Kalman, *A New Approach to Linear Filtering and Prediction Problems*, Journal of Basic Engineering **82**, 35–45(1960).
8. A. Murakami and T. Hasegawa, *Proc. 6th Conf. on Numerical Methods in Geomechanics* **2**, 2051 (1988).
9. M. Tanaka, *Application of the boundary element method to some inverse problems in engineering mechanics. In Ref. [4]* , 9 (1999).
10. F. Durbin, *Numerical Inversion of Laplace Transforms: An Efficient Improvement to Dubner and Abate's Method*, The Computer Journal **17**, 371–376 (1974).
11. M. Tanaka, T. Matsumoto and A. Shiozaki, *Application of boundary-domain element method to the free vibration problem of plate structures*, Computers & Structures **66(6)**, 725–735 (1998).
12. T. Matsumoto, M. Tanaka and K. Hondo, *Some nonsingular direct formulations ofboundary integral equations for thin elastic plate bending analysis*, Applied Mathematical Modelling **17**, 586–594 (1993).

POSE TRACKING FOR VIRTUAL WALK-THROUGH ENVIRONMENT CONSTRUCTION

K.H. WONG AND S.H. OR

Dept. of Computer Science and Engineering, The Chinese University of Hong Kong, Shatin, Hong Kong

E-mail: khwong@cse.cuhk.edu.h, shor@cse.cuhk.edu.hk

M.M.Y. CHANG

Dept. of Information Engineering, The Chinese University of Hong Kong

E-mail: mchang@ie.cuhk.edu.hk

During the construction of a virtual walk-through environment, the most tedious work is to take pictures of the interior of an environment for forming the textures of the walls. The texture images must be taken with known intrinsic and extrinsic parameters of the camera used. Many existing methods can obtain the intrinsic parameters correctly. However, the inverse problem of tracking extrinsic parameters (or tracking the pose) of a camera from a long image sequence is a more difficult task because of the problem of point correspondence. Our work attempts to solve this problem by a probabilistic method that rejects poor correspondences to increase the accuracy for the tracking. Our approach is based on a recursive Bayesian filtering method called Condensation working in conjunction with a pose algorithm. Simulation results show that it is robust for tracking object poses in a long image sequence.

1 Introduction

Pose estimation is important in many applications such as robotics, model constructions and virtual reality system development. For most pose estimation problems we generally assume that we have the model of the object and an image sequence, and our target is to find the pose (rotation and translation) of the object with respect to the camera[1]. The problem is usually solved by an iterative algorithm based on the correspondences of the 3D model features and their 2D image points. And the pose of the object obtained is defined by 3 rotational angles and 3 translations in the 3D Cartesian space. The pose estimation problem has two variations: (a) pose estimation of an object based on one image. For example, if we want to recognize objects found in the Internet, we have to find the pose of the object before actual object recognition begins. (b) The second problem is the pose-tracking problem for an image sequence of an object in motion. Of course, one can solve it by finding the pose for each image separately, and then combine the result at the end. However, the dynamics of the object motion may give extra information for the tracking.

The Kalman filter method is an example for this kind of approaches[5].

We are interested in applying pose estimation in virtual environment system development. For example, in constructing a virtual environment for the interiors of a real building, developers have to capture large number of images of the environment and paste them onto a 3D graphics rendering system. This image taking process is a tedious one because the camera extrinsic and intrinsic parameters should be recorded for each picture taken. Our aim is to automate this picture taking process by a robot and a robust pose estimation algorithm. That is the reason why we studied the problem of pose tracking.

In fact, pose estimation algorithms for one single image already exist, the lowest number of point correspondence pairs needed is 3 and there are algorithms use 4, 5 or 8 or more[3,4]. By combining the pose estimation results of individual images within an image sequence we can have a pose tracking system. However, all these algorithms assume the point correspondences are available and accurate. Usually least square algorithms are employed for handling noise problems in these pose estimation algorithms. However, if a large percentage of mismatches occur these algorithms may fail. This mismatch problem becomes more serious as the video length becomes longer. Many researchers have experienced this lost track problem in dealing with long image sequences. That is, at the beginning of an image sequence the pose tracking is quite accurate, however, the error accumulates fast and the tracking will soon fail. The main problem is usually not the pose estimation itself, it is mainly caused by the fact that some 2D to 3D point correspondences are incorrect, and the least square pose estimation cannot recover this mistake. There are many causes to this mismatch problem in the correspondence process. Kalman filtering can provide a solution to reduce the effect of mismatch but it cannot eliminate the effect totally.

We use a Bayesian filtering method called Condensation[2] to solve this problem. First we know that the minimum number of point correspondences is 3, so if we have a large number of point correspondences available,we do not need to use all of them. The reason is, within the pool of correspondences some are erroneous. Our approach is to select randomly a subset of the n point correspondences (hoping to select only the good correspondences) say m (where n is greater than m), to form a pose. By repeating this selection process many times, we will have a collection of poses found. Then we can use a statistical method to determine the accurate result. However, if all possible choices (C^n_m) of selections are considered, it is too large. To solve this, a Monte Carlo type search method may be suitable. We use this method for tracking the pose of the object with mismatch noise and found that the algorithm can improve the accuracy of the tracking.

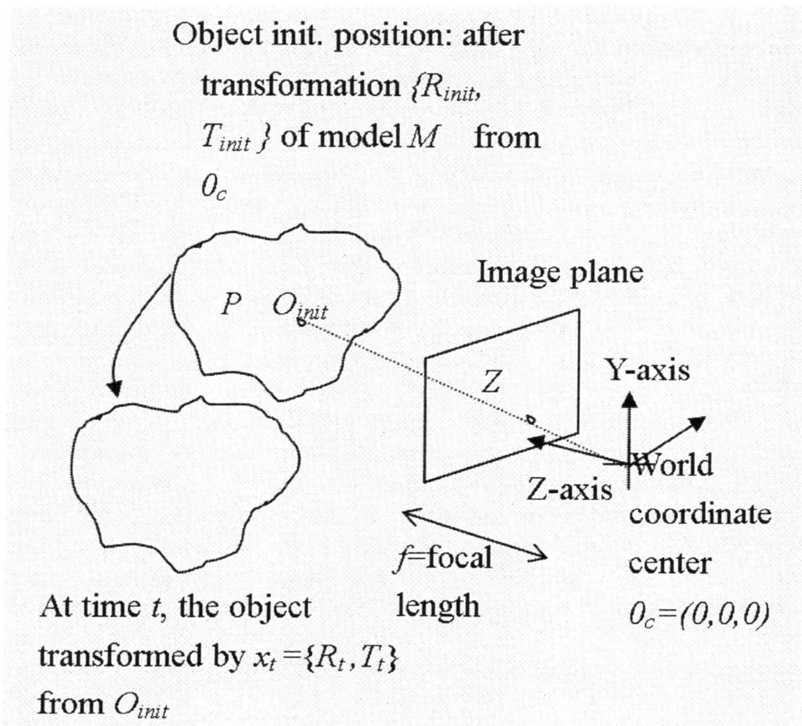

Figure 1. The object and the Camera.

In section 2, we will describe the theory used in our approach. In section 3, we will describe the implementation. And in section 4, we will show the experimental results. Conclusion can be found in section 5.

2 Theory

2.1 Problem Formulation

We have a camera viewing an object as in Figure 1, the camera is at the center of the 3D world coordinate center O_c. The movement of any point from $p = [p_x, p_y, p_z]^T$ to p' in the 3D space can be described by a transformation of $p' = Rp + T$ ($[\ ]^T$ is the matrix transpose operator). In here, $R(\theta_x, \theta_y, \theta_z)$ is a 3x3 rotation matrix and $T = [T_x, T_y, T_z]^T$ is a 3x1 translation vector, where $\theta_x, \theta_y, \theta_z$ are rotation angles about the X, Y, Z-axis, and T_x, T_y, T_z are translations along the X, Y, Z-axis, respectively. Also, the poses (or states) of the object up to time t is described by $\chi_t = \{x_1, x_2, ..., x_t\}$, where $x_t = (R_t, T_t)$.

The object centered at O_c has a model M containing a set of 3D feature points $p = \{p_1, p_2, ..., p_n\}$ on the surface. Before tracking, assume the object center is moved from O_c to O_{init} by a transformation governed by a rotation matrix R_{init} and a translation vector T_{init}. The 3D feature points are projected onto the image plane by the camera of focal length f to form a set of 2D features $Z_t = \{z_1, z_2, ..., z_n\}_t$, where $z = (u, v)$ and the perspective projection formulas $u = f\frac{p_x}{p_z}$ and $u = f\frac{p_y}{p_z}$. From the initial position the object moves to a new position at time t, therefore we will have a set of 2D-measurement (correspondences) history$\bar{Z}_t = \{Z_1, Z_2, ..., Z_t\}$. Our aim is to find χ_t from $\bar{Z}_t$ and M.

2.2 Noise Problems in Tracking Correspondences and Pose

Mismatch of correspondences is the major cause of tracking failure. We identify the major sources of mismatch as follows:

(1) 2D noise caused by lighting condition changes.

(2) Pose change may introduce mismatch error.

(3) 2D Clutter noise.

Our algorithm can reduce the effect of these noises.

2.3 The Probability Framework

A probabilistic approach is necessary to reduce the effect of the mismatch error. The following formulas describe the probability framework of the Condensation approach as in Isard and Blake[2].

$$p(x_t|\bar{Z}_t) = k_t p(Z_t|x_t) p(x_t|\bar{Z}_{t-1}) \tag{1}$$

$$p(x_t|\bar{Z}_{t-1}) = \int_{x_t-1} p(x_t|x_{t-1}) p(x_{t-1}|\bar{Z}_{t-1}) \tag{2}$$

Equation (1) calculates the probability of having the current pose as x_t if the measurement history $\bar{Z}_t$ is known. The Bayesian rule decomposes the conditional probability $p(x_t|\bar{Z}_t)$ into two components. (a) The density $p(x_t|\bar{Z}_{t-1})$

calculates the probability that the object is predicted to have the pose x_t if the measurement history up to the previous time frame is $\bar{Z}_{t-1}$; (b) $p(Z_t|x_t)$ is the likelihood function of finding the probability of producing the current measure $\bar{Z}_t$ if the current state is predicted to be x_t. We treat x_t as a temporal Markov chain, therefore Equation (2) decomposes into two terms: (a) $p(x_{t-1}|\bar{Z}_{t-1})$ is the probability of reaching state x_{t-1} based on the observations till the previous frame; (b) $p(x_t|x_{t-1})$ is the probability of all possible state transitions from the previous time to the current state x_t. Details of the formulation can be found in Isard and Blake[2].

Equation (1) and (2) serve as an iterative loop for finding x_t by maximizing $p(x_t|\bar{Z}_t)$ for the given measurement history if the initialized parameters are known. However, in actual calculation, the possible space for x_t over time is very large, and the complexity of finding $p(x_t|\bar{Z}_t)$ of all possible x_t is huge. So we use the Condensation algorithm to make the complexity limited to a manageable level.

2.4 The Condensation Algorithm for Pose Tracking

For the pose estimation of an object with n 3D features, we discussed that it is not practical to assume we can identify all 2D to 3D point correspondences without mistakes, because the matching may be corrupted by 2D and 3D noise. One approach is that we can select a subset m out of all n (where $m < n$) 3D-feature points on the object and use them to find their correspondences in the image. Then, we will use the correspondence set to find a pose by some known iterative methods such as the algorithm in Araujo *et al* [1]. We can repeat the above process many times, then we will have a set of poses. Then a voting algorithm can be used to give an overall result. It is essentially a statistical approach, however, this approach has two problems.

- **(a)** The number of combination of selecting m out of n features can be very large, say the combination of selection $m = 10$ out of $n = 20$ features is $C_{m=10}^{n=20} = 184756$, which is a huge number to be handled since each set has be processed by an iterative optimization method to find the pose.
- **(b)** Wrong matches using this method will still be bad, it will still corrupt the final result, and for the tracking of an object in an image sequence the error will accumulate and end up with intolerable errors.

In this work, we will solve it by the Condensation technique. It is a way to reduce the complexity of the algorithm and at the same time minimizes the error created by the mismatch correspondences. Assume that we have a set of 2D-measurement (correspondences) history $\bar{Z}_t = \{Z_1, Z_2, ..., Z_t\}$ and a Model M. At time t, we perform N_{cond} selections. For each selection, we select m out of the n features from Z_t. That is, for the j^{th}selection, it creates a set $I_t^{(j)} = \{i_1, i_2, .., i_m\}$ containing the indexes of the selected elements from the corresponding 2D measurements Z_t. Hence the selected set is $Z_t^{(j)} = \{z_{i_1}, z_{i_2}, .., z_{i_m}\}$, where $z = (u, v)$ is a 2D feature point and $j \in \{1, 2, ...N_{Cond}\}$. And from $z_t^{(j)}$ we can form a pose $x_t^{(j)} = (R_j, T_j)$ by the model M. So after each time frame, $Z_t = \{z_t^{(1)}, z_t^{(2)}, ...z_t^{(j)}, ..., z_t^{(N_{Cond})}\}$ and $X_t = \{x_t^{(1)}, x_t^{(2)}, ...x_t^{(j)}, ..., x_t^{(N_{Cond})}\}$ are formed.

By using the pose set X_t, M and its time derivative dX_t, we can find the prediction $X_{t+1}$and the possible measurement set Z_{t+1} for the next time frame. The probability $P(x_{t+1}^{(j)}|x_t^{(j)})$ tells how to predict new pose for thej^{th} selection. By matching the predicted measurement and real measurement, we can calculate $P(z_{t+1}^{(j)}|x_{t+1}^{(j)})$. If it is high, the jth selection is good enough and it can make prediction for the next selection of $z_{t+1}^{(j)}$, otherwise a new index set $I_{t+1}^{(k)}$ of randomly selected 2D image points $z_{t+1}^{(k)}$ will be selected for the replacement. The actual implementation can be found in the following section.

The advantage of this method is that we only handle N_{cond} processing steps for each time frame, therefore the complexity of the algorithm can be controlled to a tolerable level.

3 Implementation

3.1 Initialization and Sampling

During initialization, assume the model M with n 3D features points and (R_{init}, T_{init}) are known. Then we use a feature extraction method to locate the initial feature set Z_1 of n 2D feature points, and establish the 2D to 3D correspondences between Z_1 and M. From Z we perform N_{cond} selections and for each selection (e.g. the jth selection) we select m features out of the available n features to form the index set s_j and $Z_{t=1}^{(j)} = \{z_1, z_2, ..z_m\}$, and then by using M and an iterative algorithm[1] we can calculate the corresponding pose $x_{t=1}^{(j)}$. We also initialize all likelihood probabilities $P(z_{t=1}^{(j)}|x'^{(j)}_{t=1})$ and accumulated probabilities $C_{t=1}^{(j)}$ (for all j) as ones.

After initialization we have $S_{t=1} = \{s_1, s_2, ..s_j, .s_{N_{cond}}\}$, $X_{t=1} = \{x_1, x_2, ..x_j, x_{N_{cond}})$ and $P(X_2|X_1)$.

3.2 The Main Iteration Loop Modified from Isard and Blake[2]

From $(t = 2)$ to $(t = T^{(MAX)})$ frame, iterate the following steps:

1. Re-sampling for the jth selection

   (a) Generate a uniformly distributed random number $r \in [0, 1]$.

   (b) Find, by binary subdivision, the smallest q for which $c_t^{(q)} \geq r$.

   (c) Assign $s_t^{(j)} = s_{t-1}^{(q)}$

2. Prediction and measurement loop for constructing the normalized likelihood probability set $\{\pi_{t+1}^{(j)}\}$ for $j = \{1, 2, .., N_{Cond}\}$

   (a) Prediction for the jth selection { From $s_t^{(j)}$ and its related pose $x_t^{(j)}$ and $dx_t^{(j)}$ make predicted pose: ${x'_{t+1}}^{(j)} = x_t^{(j)} + dx_t^{(j)} + prediction\_noise$}

   (b) Transform the model points M according to ${x'_{t+1}}^{(j)}$ and project them to the image plane to form the prediction measurement ${z'_{t+1}}^{(j)}$.

   (c) Measurement for the jth selection

$$P(z_{t+1}^{(j)} | {x'_{t+1}}^{(j)}) = \frac{1}{\sqrt{2\pi\sigma}} \exp\left(\frac{-1}{2\sigma^2} E\right)$$

$$\text{where } E = \sqrt{\sum_{w=all\ m} \left({z'_{t+1}}^{(w)} - z_{t+1}^{(w)}\right)^2} \text{ and } \sigma = 0.398$$

   (d) Transforming the $P(z_{t+1}^{(j)} | x_{t+1}^{(j)})$ into a normalized probability function π_{t+1}, so that $\sum_m \pi_{t+1}^{(j)} = 1$.

3. Form a cumulative probability density c_{t+1}, where,

$$\begin{aligned} c_{t+1}^{(0)} &= 0, \\ c_{t+1}^{(j)} &= c_{t+1}^{(0)} + \pi_{t+1}^{(j)} \ (j = 1, ..., N_{Cond}) \end{aligned}$$

3.3 Result Generation

We can use the expected value of x_t as the output result.

4 Experiments

4.1 Synthetic Results

We generated an object of 20 features randomly at a range of 30cm^3 and 1 meter away from the camera. Then moved them in the 3D space, and projected them to form an image sequence. Our pose-condensation algorithm was used to track the object and we can see that the tracking was correct even mismatch noise was introduced. However when no Condensation was used, tracking failed. Parameters used for this test are as follows: total number of features is n=20, selected number of features for each Condensation sample is m=10, features corrupted by mismatch noise is 25 % of all features, the mismatch noise used is a unit random generator * 0.04 * image feature positions, N_{Cond}=200 .

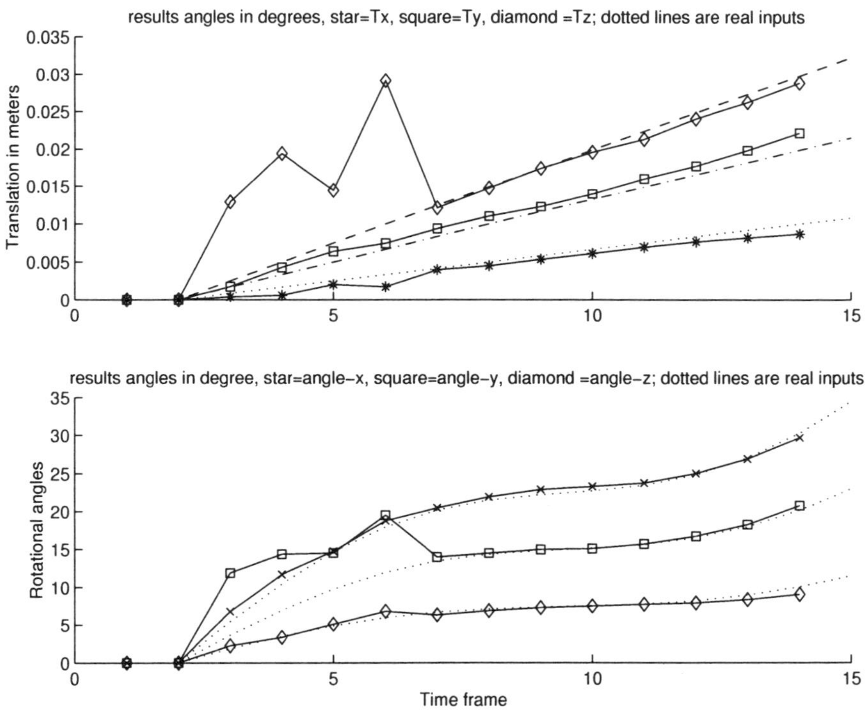

Figure 2. Tracking result of simulated data.

In the figure, the upper part are the translations and the lower part are the rotational angles. The dotted lines are real values and the solid lines are tracked result. The focal length used is 4 mm. The object is a cluster of points 1 meter away from the camera. All parameters were tracked except some initial error especially at the translation T_z, it can be explained that the iterative solution is not very sensitive to depth change. And it can be improved if good initial guess is given.

5 Conclusion

In this work we have successfully used the Condensation algorithm for pose tracking. The algorithm has been described and simulation results have been produced. The next step is to apply it to track poses of objects in real images.

Acknowledgements

The work described in this work was fully supported by a grant from the Research Grant Council of Hong Kong Special Administrative Region. (Project Number. CUHK4389/99E)

References

1. Helder Araujo , Rodrigo L. Carceroni and Christopher M. Brown, *A Fully Projective Formulation to Improve the Accuracy of Lowe's Pose-Estimation Algorithm*, Computer Vision and Image Understanding: CVIU **70(2)**, 227–238(1998).
2. Michael Isard and Andrew Blake ,*CONDENSATION – conditional density propagation for visual tracking*, Int. J. Computer Vision **29(1)**, 5–28 (1998).
3. M.L. Liu and K.H. Wong, *Pose Estimation Using Four Corresponding Points*, Pattern Recognition Letters **20(1)**, 69–74 (1999).
4. S.H. Or, W.S. Luk, K.H. Wong, I.King, *An efficient iterative pose estimation algorithm*, Image and Vision Computing Journal **16(5)**, 355–364(1998).
5. E. Koller-Meier and F. Ade, *Tracking Multiple Objects Using the Condensation Algorithm*, Journal of Robotics and Autonomous Systems **34(2-3)**, 93–105 (2001).
6. Camera Calibration Toolbox for Matlab: http://www.vision.caltech.edu/bouguetj/calib_doc/

A MATHEMATICAL METHOD TO SOLVE THE INVERSE PROBLEM OF A HEMODYNAMICS MODEL

WEI YAO AND GUANGHONG DING

Department of Mechanics and Engineering Science, Fudan University, 2500 Songhuajiang Road, Shanghai, PRC

E-mail: weiyao6@sohu.com

Based on a hemodynamic model of the Willis circulation, we developed a mathematical method to solve the inverse problem and got the values of parameters. Compared with the clinical data from hospital, most results agree with the clinical diagnoses.

1 Introduction

It is known that cerebrovascular disease (CVD) gives a severe threaten to human health and leads to huge burden to the society and economy. Many clinical experiments have indicated that, during the earlier stage of CVD, the dynamical index can change dramatically before the actual morphologic change (see Refs[1,2]). Some studies on the CVA hemodynamics model have been reported recently (see Refs[3,4]).

Based on the characters of Willis circulation: four entrances, three communicating arteries, in this paper we set up a hemodynamic model(see Refs[5]) with lumped parameters to study the cerebral circulation (Fig. 1).

In the Hemodynamic model of cerebral, the Willis circle is described as 18 arterial segments and 6 terminal resistance, each is regarded as a fluid resistor. In addition, according to the characters of pulsatile flow (see Refs[6,7]), we must consider elasticity and storage function of the blood vessel. the 8 fluid capacitors (Cc1 or Cc2, Cv1 or Cv2, Ca1 or Ca2, Cp1 or Cp2) present the compliance of left or right internal carotid, vertebral artery, cerebral anterior artery and posterior artery respectively. The fluid conductors (La1 or La2, Lp1 or Lp2) represent the inertia of left or right internal carotid and vertebral artery. The fluid conductor (Lpc1 or Lpc2) represents the inertia of left or right posterior communicating artery. The terminal resistance is added to its artery respectively. The numerical result of this model is identical with the experimental data either in time period or in frequent period (see Refs[5]).

We observe that the changes of Lpc1, Lpc2 play an insignificant role. Even when they are neglected, the theoretical result will not have any significant change. We also find that if Rb=0, Rp11=0, Rp21=0, the theoretical result has litter changes. In fact, their values can be reflected by other parameters.

We can obtain a simplified model if these parameters are neglected.

2 Mathematical Equation

According the hemodynamic model, the governing equations set up from its equivalent electrical circuit are

$$C_{c1}\frac{dP_{c1}}{dt} = -(\frac{R_{c1}+R_{m1}}{R_{c1}R_{m1}} + \frac{1}{R_{pc1}})P_{c1} - Q_{a1} + \frac{P_{p1}}{R_{pc1}} + \frac{P_1}{R_{c1}} \tag{1}$$

$$C_{c2}\frac{dP_{c2}}{dt} = -(\frac{R_{c2}+R_{m2}}{R_{c2}R_{m2}} + \frac{1}{R_{pc2}})P_{c2} - Q_{a2} + \frac{P_{p2}}{R_{pc2}} + \frac{P_2}{R_{c2}} \tag{2}$$

$$C_{a1}\frac{dP_{a1}}{dt} = -\frac{R_{a12}+R_{ac}}{R_{a12}R_{ac}}P_{a1} + \frac{1}{R_{ac}}P_{a2} + Q_{a1} \tag{3}$$

$$C_{a2}\frac{dP_{a2}}{dt} = -\frac{R_{a22}+R_{ac}}{R_{a22}R_{ac}}P_{a2} + \frac{1}{R_{ac}}P_{a1} + Q_{a2} \tag{4}$$

$$C_{v}\frac{dP_{v}}{dt} = -(\frac{R_{v1}+R_{v2}}{R_{v1}R_{v2}})P_{v} - Q_{lp1} - Q_{lp2} + \frac{P_3}{R_{v1}} + \frac{P_4}{R_{v2}} \tag{5}$$

$$C_{p1}\frac{dP_{p1}}{dt} = -(\frac{1}{R_{p12}} + \frac{1}{R_{pc1}})P_{p1} + \frac{P_{c1}}{R_{pc1}} + Q_{lp1} \tag{6}$$

$$C_{p2}\frac{dP_{p2}}{dt} = -(\frac{1}{R_{p22}} + \frac{1}{R_{pc2}})P_{p2} + \frac{P_{c2}}{R_{pc2}} + Q_{lp2} \tag{7}$$

$$L_{a1}\frac{dQ_{a1}}{dt} = -R_{11}Q_{a1} + P_{c1} - P_{a1} \tag{8}$$

$$L_{a2}\frac{dQ_{a2}}{dt} = -R_{21}Q_{a2} + P_{c2} - P_{a2} \tag{9}$$

$$L_{p1}\frac{dQ_{lp1}}{dt} = P_v - P_{p1} - R_b(Q_{lp1} + Q_{lp2}) \tag{10}$$

$$L_{p2}\frac{dQ_{lp2}}{dt} = P_v - P_{p2} - R_b(Q_{lp1} + Q_{lp2}) \tag{11}$$

Table 1 shows the meaning of parameters in these equations.

Table 1. Meaning of parameters in these equations

Parameter	meaning
R_{ci}	Internal carotid resistor
R_{vi}	Vertebral artery resistor
R_{a11}	Anterior cerebral artery I resistor
R_{a21}	Anterior cerebral artery II resistor
R_{mi}	Middle cerebral artery resistor
R_{pci}	Posterior communicating artery resistor
R_{pi}	Posterior cerebral artery resistor
R_{ac}	Anterior communicating artery resistor
Q_{ci}	Internal carotid blood flow
Q_{vi}	Vertebral artery blood flow
Q_{ai}	Anterior cerebral artery blood flow
Q_{mi}	Middle cerebral artery blood flow
C_{ci}	Internal carotid
C_{vi}	Vertebral artery compliance
C_{ai}	Anterior cerebral artery compliance
C_{pi}	Posterior cerebral artery compliance
L_{ai}	Anterior cerebral artery conductance
L_{pi}	Posterior cerebral artery conductance
P_{pi}	Posterior cerebral artery pressure
P_{ci}	Internal carotid pressure
Q_{pci}	Posterior communicating artery blood flow
Q_{lpi}	Posterior cerebral artery blood flow
P_{v}	Vertebral artery pressure
P_{ai}	Anterior artery pressure

3 Parameters Identification

In clinical application, we can obtain relatively accurate P_1, P_2, P_3, P_4, Q_{c1}, Q_{c2}, Q_{v1}, Q_{v2} ; obtain Q_{a1}, Q_{a2}, Q_{m1}, Q_{m2}, Q_{lp1}, Q_{lp2} through transcranial Doppler (TCD) and some morphologic data: obtain R_{a11}, R_{a12}, R_{pc1}, R_{pc2}, R_{ac}, L_{a1}, L_{a2}, L_{p1}, L_{p2} through calculating morphologic data directly or indirectly; obtain R_{c1}, R_{c2}, R_{v1}, R_{v2} through characteristic impedance formula

$$R_{ci} = \frac{1}{k-2}\sum_{j=3}^{k}|Z_{cij}|, \qquad R_{vi} = \frac{1}{k-2}\sum_{j=3}^{k}|Z_{vij}|, \qquad i = 1, 2.$$

Where

$$|Z_{cin}| = P_{cin}/Q_{cin}, \phi_{cin} = \phi_{pcin} - \phi_{qcin}$$

$$|Z_{vin}| = P_{vin}/Q_{vin}, \phi_{vin} = \phi_{pvin} - \phi_{qvin}$$

$$i = 1, 2, \cdots\cdots k$$

And we can obtain P_{c1}, P_{c2}, P_v through formula

$$P_{c1} = P_1 - Q_{c1}R_{c1}$$

$$P_{c2} = P_2 - Q_{c2}R_{c2}$$

$$P_v = (P_3 + P_4 - Q_{v1}R_{v1} - Q_{v2}R_{v2})/2$$

3.1 Calculate Resistance Parameters

R_{m1},R_{m2} can be obtained by steady fluid formula.

$$R_{mi} = \frac{P_i - Q_{ci}R_{ci}}{Q_{mi}}, \qquad i = 1, 2. \tag{12}$$

Integrating Eq. (3), (8), (9) or (4), (8), (9) from 0 to T, we get

$$R_{ai2} = \frac{R_{ac}(-R_{ai1}\int_0^T Q_{ai}dt + \int_0^T P_{ci}dt)}{-R_{a11}\int_0^T Q_{a1}dt + \int_0^T P_{c1}dt + R_{ac}\int_0^T Q_{ai}dt - R_{a12}\int_0^T Q_{a2}dt + \int_0^T P_{c2}dt}$$
$$i = 1, 2.$$

Integrating Eq. (1), (6), (10) or (2), (7), (11) from 0 to T, we get

$$R_{pi2} = \frac{\int_0^T P_v dt}{\int_0^T (\frac{R_{ci}+R_{mi}}{R_{ci}R_{mi}}P_{ci} - Q_{ai} + \frac{P_i}{R_{ci}})dt + \int_0^T Q_{lpi}dt}, \qquad i = 1, 2.$$

3.2 Compliance Identification

Multiplying Eq. (5) by $cos\frac{2\pi t}{T}$, integrating the result, then we obtain

$$C_v = \frac{\int_0^T (-\frac{R_{v1}+R_{v2}}{R_{v1}R_{v2}}P_v - Q_{lp1} - Q_{lp2} + \frac{P_3}{R_{v1}} + \frac{P_4}{R_{v2}})cos\frac{2\pi t}{T}dt}{\frac{2\pi}{T}\int_0^T P_v sin\frac{2\pi t}{T}dt}. \tag{13}$$

Multiplying Eq. (3) by $cos\frac{2\pi t}{T}$, integrating the result, then we obtain

$$C_{a1} = \frac{-\frac{R_{a12}+R_{ac}}{R_a 12R_{ac}}\int_0^T P_{a1}cos\frac{2\pi t}{T}dt + \frac{1}{R_{ac}}\int_0^T P_{a2}cos\frac{2\pi t}{T}dt + \int_0^T Q_{a1}cos\frac{2\pi t}{T}dt}{\frac{2\pi}{T}\int_0^T P_{a1}sin\frac{2\pi t}{T}dt}. \tag{14}$$

Where multiplying Eq. (8) by $cos\frac{2\pi t}{T}$ and integrating the result, we obtain

$$\int_0^T P_{a1}cos\frac{2\pi t}{T}dt = \int_0^T (-R_{a11}Q_{a1}+P_{c1})cos\frac{2\pi t}{T}dt - \frac{2\pi}{T}L_{a1}\int_0^T Q_{a1}sin\frac{2\pi t}{T}dt$$

$$\int_0^T P_{a1}sin\frac{2\pi t}{T}dt = \int_0^T (-R_{a11}Q_{a1}+P_{c1})sin\frac{2\pi t}{T}dt - \frac{2\pi}{T}L_{a1}\int_0^T Q_{a1}cos\frac{2\pi t}{T}dt$$

Multiplying Eq. (1) by $cos\frac{2\pi t}{T}$ and integrating the result, we obtain

$$\int_0^T P_{a2}cos\frac{2\pi t}{T}dt = \int_0^T (-R_{a21}Q_{a2}+P_{c2})cos\frac{2\pi t}{T}dt - \frac{2\pi}{T}L_{a2}\int_0^T Q_{a2}sin\frac{2\pi t}{T}dt$$

We can also obtain C_{a2} in this method.

Multiplying Eq. (1) by $cos\frac{2\pi t}{T}$, integrating the result, then we obtain

$$C_{c1} = \frac{\int_0^T (\frac{R_{c1}+R_{m1}}{R_{c1}R_{m1}}P_{c1} - Q_{a1} + \frac{P_1}{R_{c1}} + \frac{P_{p1}-P_{c1}}{R_{pc1}})cos\frac{2\pi t}{T}dt}{\frac{2\pi}{T}\int_0^T P_{c1}sin\frac{2\pi t}{T}dt} \quad (15)$$

Multiplying Eq. (6) by $cos\frac{2\pi t}{T}$, integrating the result, then we obtain

$$C_{p1} = \frac{\int_0^T (-R_{p12}P_{p1} + \frac{P_{c1}-P_{p1}}{R_{pc1}} + Q_{lp1})cos\frac{2\pi t}{T}dt}{\frac{2\pi}{T}\int_0^T P_{p1}sin\frac{2\pi t}{T}dt} \quad (16)$$

where

$$\int_0^T P_{p1}cos\frac{2\pi t}{T}dt = \int_0^T P_v cos\frac{2\pi t}{T}dt - \frac{2\pi}{T}L_{p1}\int_0^T Q_{lp1}sin\frac{2\pi t}{T}dt$$

$$\int_0^T P_{p1}sin\frac{2\pi t}{T}dt = \int_0^T P_v sin\frac{2\pi t}{T}dt - \frac{2\pi}{T}L_{p1}\int_0^T Q_{lp1}cos\frac{2\pi t}{T}dt$$

We can obtain C_{c2}, C_{p2} in this method.

4 Result and Discuss

Table 2 is the calculated result for a patient of SAH. Where Resistance unit is $10^2 dyn.s/cm^5$, compliance unit is $10^{-5} dyn.s^2/cm^5$.

There will exist error during non-invasive measurement. Method of calculating resistor based on integration, which can eliminate some man-made

Table 2. Calculating results for a SAH patient

Parameter	Result	Parameter	Result	Parameter	Result	Parameter	Result
R_{c1}	1090	R_{v1}	3650	R_{m1}	21700	R_{p12}	44500
R_{c2}	1400	R_{v2}	4780	R_{m2}	17000	R_{p22}	44000
C_{a1}	5.47	C_{p1}	0.58	C_{c1}	0.86	C_v	5.18
C_{a2}	6.81	C_{p2}	1.96	C_{c2}	4.32	R_{a22}	31700
R_{a12}	37900						

error in measurement and improve calculation stability, so the result is reliable. Method of calculating C_v, C_{a1}, C_{a2}, C_{c1}, C_{c2}, C_{p1} and C_{p2} based on the waveform of pressure and flow of the according vessel. Therefore, we multiply each input value with a random number between 0.95 1.05 when calculating. Calculating results show C_v, C_{a1}, C_{a2} will change less than 20%,but C_{c1}, C_{c2}, C_{p1} and C_{p2} in some examples will change more than 100%. That means we should continue the research and found an effective method to define the values of C_{c1}, C_{c2}, C_{p1} and C_{p2}.

We studied 30 cases from Hua Shang hospital. From statistical analysis of the theoretical results, we find that the RCI(right cerebral infarction)'s right cerebral middle artery resistance is 113% higher than left, CCI(cerebellum cerebral infarction)'s posterior cerebral artery resistance is 50% higher than RCI. These show the method of resistance calculation is identical to clinical diagnosis, it can be considered as a non-invasive method to clinical diagnosis.

References

1. F. C.Charles, S. R.Claudia and C. Jeffery, *Intracranial pressure waveform indices in transient and refractory intracranial hypertension*, J. of Neuroscience Methods **57**, 15–25 (1995).
2. A. M. Stepthan, E. T. Carole and E. D.Beverly, *Asymmetry of Intracranial Hemodynamics as an Indicator of Mass Effect in Acute Intracerebral Hemorrhage*, Stroke **27**, 1788–1792 (1996).
3. N.Westerhof , G. Elzinga and P. Sipkema, *An artificial arterial system for pumping hearts*, J. Appl. Physiol **31**, 776–781 (1971).
4. A.Noodergraf, *Circulatory system dynamics* (Academic Press, New York, 45–95, 1978).
5. G. H.Ding, K. R.Qin and J.Gao , *On hemodynamics of cerebral circulation, a mathematical model of the circle of Willis with steady flow*, Chinese Journal of Biomedical Ebgineering **17(1)**, 88–95, (1998).

6. G. H. Ding, C. Z. Lu and W. Yao, *A hemodynamic model and a mathematics method to calculate the dynamics index for cerebral circulation*, J. of Hydrodynamics **B**, 71–81(1997).
7. G. H.Ding and C. Z. Lu, *On hemodynamic model of cerebral circulation-a lumped parameters model of pulsatile flow*, Acta Mechanica Sinica **28(3)**, 336–346(1996).

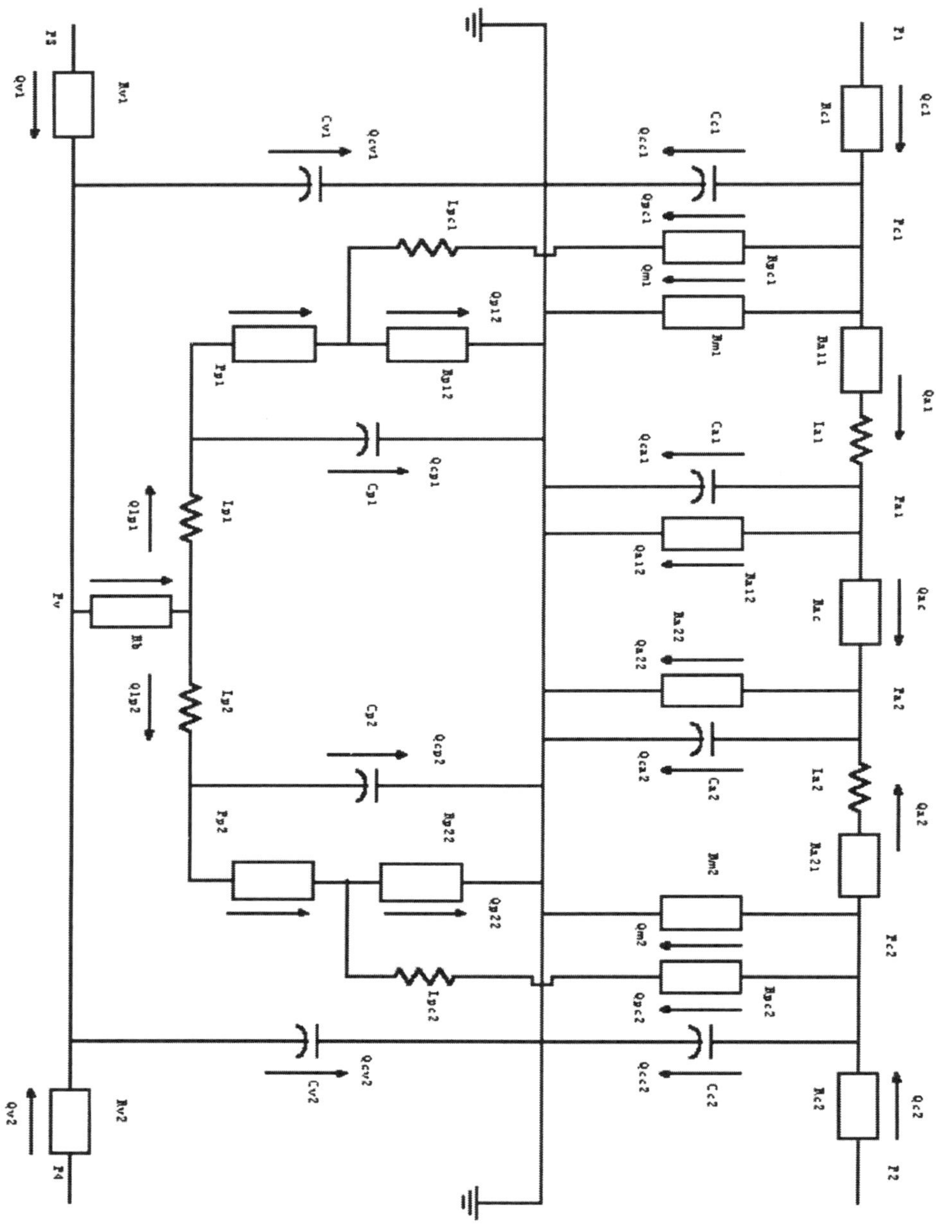

Figure 1. A Hemodynamic Model of Cerebral Circulation.

AN INVERSE PROBLEM OF DERIVATIVE SECURITY PRICING

GUANQUAN ZHANG

State Key Laboratory of Scientific and Engineering Computing, Institute of Computational Mathematics and Scientific/Engineering Computing, AMSS, CAS, Beijing, 100080, P.R.China
E-mail: zgq@lsec.cc.ac.cn

PEIJUN LI

Department of Mathematics, Michigan State University, East Lansing, MI, 48823, USA
E-mail: lipeijun@msu.edu

Suppose that interest rate is governed by a stochastic differential equation, a partial differential equation for the price of bond can be derived in a similar way to the derivation of the Black-Scholes equation. Valuation of bond with implied function in the equation, which is called the risk market price of interest rate, is known as the model of bond pricing. An inverse problem of bond pricing is to determined the risk market price of interest rate implied by current prices of bonds with different expirations. In this paper, numerical algorithm to solve this system is constructed and some numerical experiments are performed. The numerical results show that the algorithm is quite efficient and robust.

1 Introduction

Derivative securities are kinds of products whose values depend on other more underlying variables. Derivative security of interest rate is one whose pay off, to some extend, determined by interest rate. There is a large, and ever-going, number of different interest rate derivative products now. In view of our uncertainty about the future course of the interest rate, it is natural to model it as a random variable. This leads to a parabolic partial differential equation for price of bond. However, the function, $\lambda(t)$, is as yet unspecified in the equation, which is called the risk market of interest rate. To determine the function $\lambda(t)$, we must have additional market data. The problem will be formulated from the mathematical standpoint in the following.

Suppose that the short term interest rate, the spot rate, follows a random walk(ITO Process)

$$dr = \theta(r)dt + w(r)dz. \tag{1}$$

Where dz is a normally distributed random variable with zero mean and variance dt. In practice, the spot rate is never greater than a certain number, which is assumed R, and never less than or equal to zero. Therefore, we suppose that $r \in [0, R]$. $\theta(r)$ is a smooth bounded function, which satisfies

$$\theta(0) \geq 0, \qquad \theta(R) \leq 0. \tag{2}$$

So we can make the spot rate mean reverting, i.e., for large(small) r the interest rate will tend to decrease(increase) towards some mean value. $w(r)$ is a non-negative and smooth bounded function satisfies

$$w(0) = w(R) = 0. \tag{3}$$

Suppose that price of bond, $V(t, r; T)$, is a function of interest rate r and time t and maturity T. By pricing formula of general derivative security[1], we get the partial differential equation for zero-coupon bond in the form

$$\frac{\partial V}{\partial t} + \frac{w^2(r)}{2}\frac{\partial^2 V}{\partial r^2} + (\theta(r) + \lambda(t)w(r))\frac{\partial V}{\partial r} - rV = 0 \tag{4}$$

At $r = 0$ and $r = R$, the equation degenerates into a hyperbolic equation with positive and negative characteristics respectively

$$\frac{\partial V}{\partial t} + \theta(0)\frac{\partial V}{\partial r} = 0 \tag{5}$$

$$\frac{\partial V}{\partial t} + \theta(R)\frac{\partial V}{\partial r} = RV \tag{6}$$

The final condition is given by

$$V(T, r; T) = Z, \qquad 0 < T \leq T_{max} \tag{7}$$

Let L be a differential operator and suppose V be a differentiable function. We define

$$LV = \frac{w^2(r)}{2}\frac{\partial^2 V}{\partial r^2} + (\theta(r) + \lambda(t)w(r))\frac{\partial V}{\partial r} - rV \tag{8}$$

Then (4) can be rewritten in the form of operator

$$\frac{\partial V}{\partial r} + LV = 0 \tag{9}$$

Problem: *Derivative Security Pricing* — To determine the pair of functions $V(t, r; T)$ and $\lambda(t)$ that satisfy (4)–(7) from the current market prices

$$V(t = 0, r_0; T) = V(T), \qquad 0 < T \leq T_{max} \tag{10}$$

of zero-coupon bond with different expiration T.

2 Formulation of Integral Equation

Consider the adjoint equation of (4)

$$\begin{array}{r}\frac{\partial U}{\partial t}-\frac{1}{2}\frac{\partial^2}{\partial r^2}(w^2(r)U)+\frac{\partial}{\partial r}((\theta(r)+\lambda(t)w(r))U)+rU=0\\ (t,r)\in(0,\infty)\times[0,R]\end{array}\tag{11}$$

with given boundary conditions

$$U(t,0)=U(t,R)=0 \tag{12}$$

and initial condition

$$U(t=0,r)=\delta(r-r_0) \tag{13}$$

Where $\delta(r-r_0)$ is Dirac's Delta function concentrated at current interest rate r_0. Let L^* be a differential operator. If U is a differentiable function, we define

$$L^*U=\frac{1}{2}\frac{\partial^2}{\partial r^2}(w^2(r)U)-\frac{\partial}{\partial r}((\theta(r)+\lambda(t)w(r))U)-rU \tag{14}$$

So (11) can be rewritten as follows

$$\frac{\partial U}{\partial t}-L^*U=0 \tag{15}$$

By the definition of differential operators L, L^* (8), (14), boundary conditions (12), and using integration by parts, we get

$$\begin{aligned}\frac{d}{dt}\int_0^R V(t,r;T)U(t,r)dr &= \int_0^R\left(V\frac{\partial U}{\partial t}+U\frac{\partial V}{\partial t}\right)dr\\ &= \int_0^R(V\cdot L^*U-U\cdot LV)dr=0\end{aligned}\tag{16}$$

We integrate (16) with respect to t from $t=0$ to $t=T$, and using condition (10), (13) arrives at

$$\int_0^R U(T,r)dr=\frac{V(T)}{Z} \tag{17}$$

Differentiating (17) with respect to T, making use of the equation

$$\begin{array}{r}\frac{\partial U}{\partial T}=\frac{1}{2}\frac{\partial^2}{\partial r^2}(w^2(r)U)-\frac{\partial}{\partial r}((\theta(r)+\lambda(T)w(r))U)-rU\\ (T,r)\in(0,T_{max}]\times[0,R]\end{array}\tag{18}$$

and integration by parts, we have

$$\int_0^R rU(T,r)dr=-\frac{V'(T)}{Z} \tag{19}$$

Similarly, repeating the process for (19), we can obtain a non-linear integral equation in the form

$$\lambda(T)\int_0^R w(r)U(T,r)dr + \int_0^R (\theta(r) - r^2)U(T,r)dr = -\frac{V''(T)}{Z} \tag{20}$$

Lemma 1 *$\int_0^R w(r)U(T,r)dr > 0$ for $0 \le T \le T_{max} < \infty$, so integral equation (20) is well-defined.*

Proof: Define

$$U(t,r) = e^{\alpha t} \cdot W(t,r) \tag{21}$$

then W satisfies the following equation

$$\begin{aligned} \frac{\partial W}{\partial t} &= \frac{w^2(r)}{2}\frac{\partial^2 W}{\partial r^2} - [\theta(r) + \lambda(t)w(r) - (w^2)']\frac{\partial W}{\partial r} \\ &\quad -[\theta'(r) + \lambda(t)w'(r) + r - \frac{(w^2)''}{2} + \alpha]W \end{aligned} \tag{22}$$

$$W(t,0) = W(t,R) = 0 \tag{23}$$

$$W(t=0,r) = \delta(r - r_0) \tag{24}$$

According to the assumption of interest rate model, $\theta(r), w(r)$ are smooth functions. Therefore, there exists $\alpha > 0$, such that

$$\theta'(r) + \lambda(t)w'(r) + r - \frac{(w^2)''}{2} + \alpha > 0$$

So the extremum principle holds for equation (22), $W(t,r) \ge 0$. By (21), we get

$$U(t,r) \ge 0, \qquad 0 \le t \le T.$$

$w(r)$ is a non-negative function, so the following holds

$$\int_0^R w(r)U(T,r)dr > 0$$

This completes the lemma.

3 Numerical Implementation and Results

Let $\Omega = \{(T,r)|0 \leq T \leq T_{max}, 0 \leq r \leq R\}$, and cover Ω by the grid $\{(T_k, r_j)|T_k = k\Delta\tau, r_j = j\Delta r, \Delta\tau = \frac{T_{max}}{m}, \Delta r = \frac{R}{n}\}$. We denote $U_j^k = U(T_k, r_j), \lambda_k = \lambda(T_k)$. For U, λ, we use implicit and explicit difference scheme respectively. So the equation (18) can be discretized as follows

$$\begin{aligned}\frac{U_j^{k+1}-U_j^k}{\Delta\tau} &= \frac{1}{2}\frac{1}{\Delta r^2}(w_{j+1}^2 U_{j+1}^{k+1} - 2w_j^2 U_j^{k+1} + w_{j-1}^2 U_{j-1}^{k+1}) \\ &-\frac{1}{2}\frac{1}{\Delta r}((\theta_{j+1} + \lambda_k w_{j+1})U_{j+1}^{k+1} \\ &-(\theta_{j-1} + \lambda_k w_{j-1})U_{j-1}^{k+1}) - r_j U_j^{k+1}\end{aligned} \tag{25}$$

For each k, the equation (25) can be rewritten as

$$C_{j-1}U_{j-1}^{k+1} + A_j U_j^{k+1} + B_{j+1}U_{j+1}^{k+1} = U_j^k \tag{26}$$

Where

$$\begin{aligned}C_{j-1} &= -(\frac{w_{j-1}^2}{2\Delta r^2} + \frac{\theta_{j-1}+\lambda_k w_{j-1}}{2\Delta r}) \\ A_j &= 1 + (\frac{w_j^2}{\Delta r^2} + r_j)\Delta\tau \\ B_{j+1} &= -(\frac{w_{j+1}^2}{2\Delta r^2} + \frac{\theta_{j+1}+\lambda_k w_{j+1}}{2\Delta r}).\end{aligned} \tag{27}$$

Consider the boundary conditions, integral equation (20) can be discretized by numerical integral formula

$$\lambda_{k+1}\sum_{j=1}^{n-1} w_j U_j^{k+1} + \sum_{j=1}^{n-1}(\theta_j - r_j^2)U_j^{k+1} = -\frac{V''(T_{k+1})}{Z\Delta r} \tag{28}$$

The system of linear equation (26) is a tridiagonal system, which can be solved very efficiently. The numerical computation begins at the initial time $T = 0$ with initial values

$$U_j^0 = 0, \quad U_l^0 = \frac{1}{\Delta r}, \quad l = [\frac{r_0}{\Delta r}]$$

and advances forward. At each time T, $U(T,r)$ are obtained by solving linear equation (26). Using numerical integral equation (28), we can obtain $\lambda(T)$. In order to improve inversion accuracy, we can substitute $\lambda(T)$ into equation (26) and compute $U(T,r)$ again. Then using the new $U(T,r)$, we can get the improved $\lambda(T)$. The whole iterative algorithm is described in table 1.

Table 1. Iterative Algorithm of Inversion.

step 1. For $k = 0, i = 0, \lambda_k^i = 0$, where k is the index of time step and i is the index of iteration;
step 2. For known λ_k^i, solve direct problem (26) to obtain U;
step 3. For known U, solve the numerical integral formula (28) to get λ_k^{i+1};
step 4. If $\varepsilon = \| \lambda_k^{i+1} - \lambda_k^i \|$ is mall enough, stop the iteration, go to step 5; Otherwise let $i = i + 1$ go to step 2;
step 5. Let $k = k + 1$, if $T_k = T_{max}$ stop; Otherwise go to step 2.

We apply our method to CRSP issue. The compounded semiannually interest rates, from the quote date, Nov. 30, 1995 to maximum expiration, Aug. 15, 2025, are given. So the time stride is 30 years. The following are set of input parameters: $Z = 100, r \in [0, 0.2], T_{max} = 30, \Delta T = 0.05$, and $\theta(r) = 0.053 - r, w(r) = r(0.2 - r)$. Fig. 1 is the current prices of bonds with different expirations. Fig. 2 is the relative error of curve fitting the bond prices. To test the numerical stability of the inversion algorithm, random noise was added to the additional condition, i.e. $V(T)$.

$$V_\sigma = V \cdot (1 + \sigma \cdot rand())$$

Where $rand() \in [-1, 1]$ is a random function. σ is the level of noise. Numerical results are shown in Fig. 3–6.

In order to check the inversion algorithm, we compute the forward problem (4)–(7) using the numerically computed $\lambda(T)$. The original market data and computed numerical results $V(t = 0, r_0; T)$ are in figure 7 as follows.

Acknowledgments

We are grateful to Prof. Youlan Zhu at UNC for his providing the market data and helpful discussions.

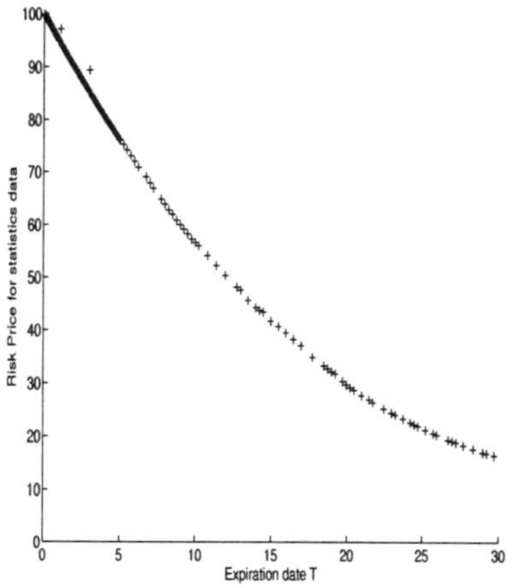

Figure 1. Current prices of bonds with different expirations.

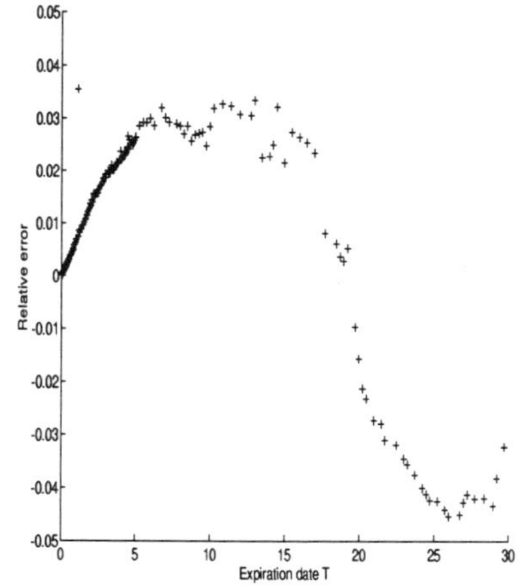

Figure 2. Relative error of curve fitting the bonds prices.

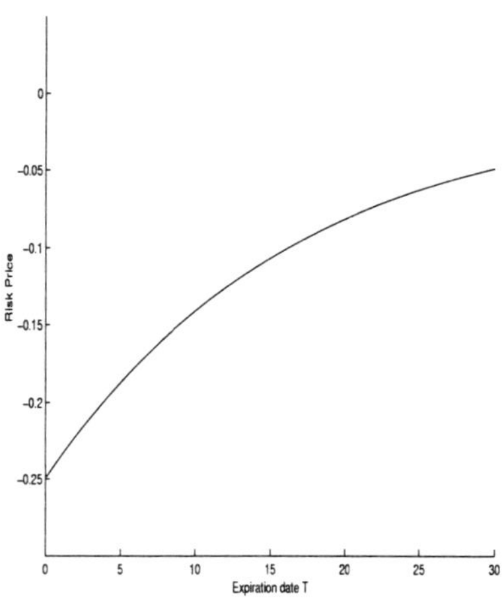

Figure 3. Result of inversion.

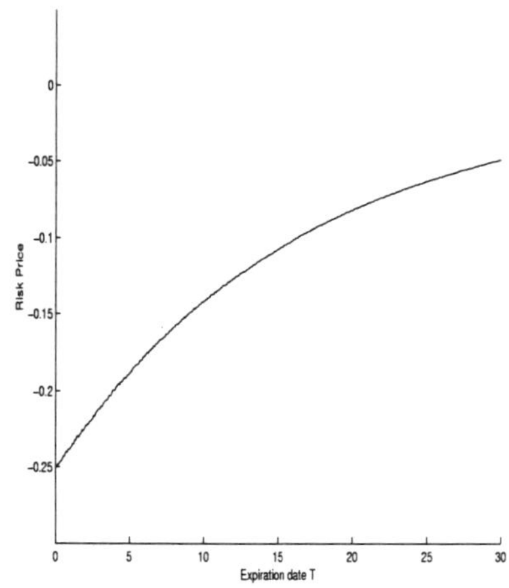

Figure 4. Result of inversion with random error 1%.

References

1. J. Hull, *Options, Futures, and Other Derivatives, 3rd ed.* (Prentice Hall,Upper Saddle River, N.J., 1997).

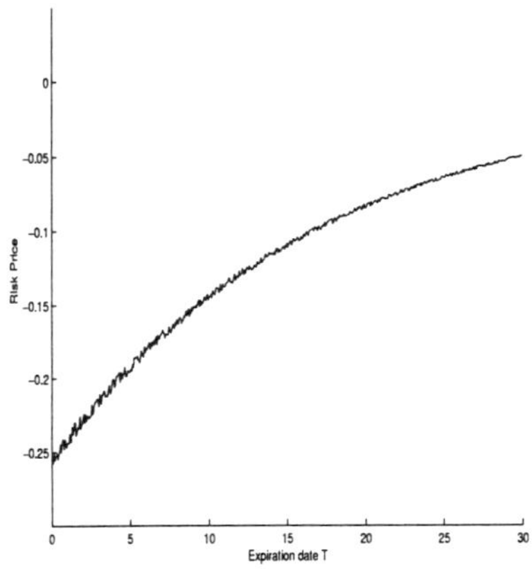

Figure 5. Result of inversion with random error 5%.

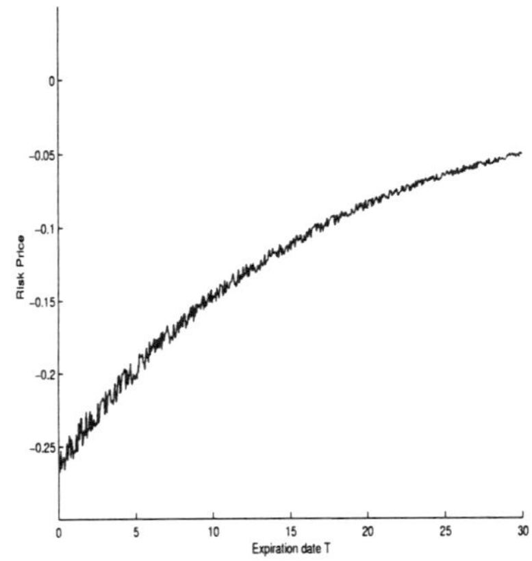

Figure 6. Result of inversion with random error 10%.

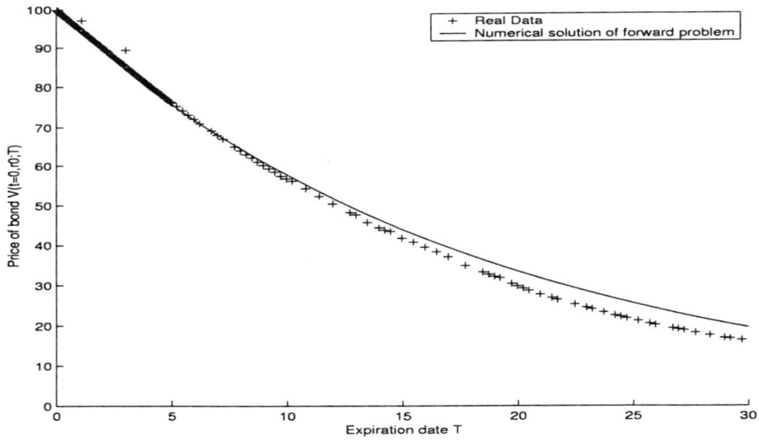

Figure 7. '+': Current market price of bond $V(t=0, r_0; T)$, '-': Numerical solution of forward problem.

2. I. Bouchouev & V. Isakov, *The Inverse Problem of Option Pricing*, Inverse Problems **13(5)**, L11–L17 (1997).
3. A. Friedman, *Partial Differential Equations of Parabolic Type* (Prentice-Hall, Englewood Cliffs, NJ, 1964).

4. R. Courant and D. Hilbert, *Methods of Mathematics Physics, Vol.2* (John Wiley, New York, 1962).
5. P. Wilmott, J. Dewynne and S. Howison, *Option Pricing: Mathematical Model and Computational*, (Oxford Financial Press, Oxford, 1994).
6. V. Isakov, *Inverse Problems for Partial Differential Equations* (springer-verlah, New York, 1998).
7. D. Richtmyer and K. W. Morton, *Difference Methods for Initial Value Problems, 2nd ed.* (Wiley-Interscience, 1967).
8. D.I. Richard, *Option Volatility & Pricing* (Higher Education Group, Inc., 1994).
9. G.Q. Zhang, *On an Inverse Problem for One-Dimensional*, Sci. China **Ser. A 32**, 257–274 (1989).

EFFICIENT INTERPRETATION OF LARGE-SCALE REAL DATA BY STATIC INVERSE OPTIMIZATION

HONG ZHANG AND MASUMI ISHIKAWA

Graduate School of Life Science & Systems Engineering
Kyushu Institute of Technology
Hibikino 2-4, Wakamatsu, Kitakyushu 808-0196, Japan
E-mail: {zhang, ishikawa}@brain.kyutech.ac.jp

We have proposed a method for static inverse optimization to interpret real data from a viewpoint of optimization. In this paper we propose an efficient method for generating constrains by divide-and-conquer to interpret large-scale real data by static inverse optimization. To evaluate its effectiveness, simulation experiments are carried out by using rented housing data (about 4,000 samples) with 4 attributes. Criterion functions for deciding housing of tenants living along Yamanote and Soubu-Chou lines in Tokyo are estimated.

1 Introduction

Behaviors of humans and animals seem to have rationality as a result of evolution[5,6]. We have proposed a new methodology based on a rationality hypothesis for interpreting real world data. The interpretation is carried out by *inverse optimization*. Inverse optimization is classified into static one and dynamic one[10,11,12,13]. In this paper we focus on the former, which estimates a criterion function under which given data become optimal subject to given constraints. A resulting criterion function provides interpretation of given data.

We have proposed a neural networks approach to static inverse optimization for estimating quadratic criterion functions corresponding to given data. A crucial idea here is neural network architecture representing the optimality conditions for both optimization and inverse optimization. Taking advantage of this duality, static inverse optimization problems can be solved by learning of neural networks.

This idea alone, however, is not sufficient for solving static inverse optimization. To overcome various difficulties, we have also proposed algorithms for generating constraints from given data, guaranteeing positive semi-definiteness of resulting criterion functions, estimating simple and understandable criterion functions, and interpreting non-Pareto optimal data. Although it can solve static inverse optimization problems and interpret real data, it still has a difficulty in interpreting large-scale real data due to computational complexity in generating constraints.

Generation of constraints requires computation of a convex-hull from given data. Although many algorithms for calculating a convex-hull from a set of points in 2-D and 3-D have been proposed, they are not applicable to real data in higher dimensions[1,7,2]. Existing brute force techniques for calculating a convex-hull from given data are not feasible due to excessive computation time even when the number of given data is fairly small[8].

To overcome this difficulty we propose an efficient method for generating constraints by divide-and-conquer. The main features of the proposed algorithm are the following. It randomly divides large-scale data into subsets, calculates Pareto optimal data for each subset, and calculates Pareto optimal data for the entire data by fusing them. It can be proved that resulting Pareto optimal data are the same as those obtained directly from the original data. By reducing non-Pareto optimal data as much as possible, computational cost for generating constraints becomes much smaller than that by an algorithm without divide-and-conquer.

To evaluate the effectiveness of the proposed method, simulation experiments are carried out by using rented housing data (about 4,000 samples) with 4 attributes. They are obtained from tenants living along Yamanote and Soubu-Chuo lines in Tokyo[9]. The proposed divide-and-conquer method requires less than 30 minutes for generating constraints from given data. In contrast a method without divide-and-conquer would require 3,330 years.

Section 2 presents formulation of optimality conditions. Section 3 shows a neural network architecture representing the optimality conditions. Section 4 describes a procedure for data interpretation. Section 5 illustrates an efficient method for generating constrants. Section 6 provides interpretation of large-scale real data. Section 7 concludes this paper.

2 Optimality conditions for static optimization

We consider the following static optimization with a quadratic criterion function,

$$\min_{x} \; f(\boldsymbol{x}) = \frac{1}{2}\boldsymbol{x}^T A \boldsymbol{x} + \boldsymbol{s}^T \boldsymbol{x} \tag{1}$$

$$s.t. \quad g_i(\boldsymbol{x}) = \boldsymbol{b}_i^T \boldsymbol{x} \le d_i, \quad i = 1, \cdots, m \tag{2}$$

where $\boldsymbol{x} \in \Re^n$ is a variable vector, $A \in \Re^{n \times n}$ is a symmetric positive semidefinite criterion matrix, $\boldsymbol{s} \in \Re^n$ is a criterion vector, $\boldsymbol{b}_i \in \Re^n$ is ith coefficient vector, and $d_i \in \Re^1$ is ith constant in the constraints.

A Lagrangian function, L, is,

$$L(\boldsymbol{x}, \boldsymbol{\lambda}) = f(\boldsymbol{x}) + \boldsymbol{\lambda}^T \boldsymbol{g}(\boldsymbol{x}). \tag{3}$$

where $\boldsymbol{\lambda}$ is a Lagrangian multiplier vector.

The following Kuhn-Tucker condition[3] is necessary and sufficient for static optimization.

$$\nabla_{\boldsymbol{x}} L(\boldsymbol{x}^o, \boldsymbol{\lambda}^o) = \mathbf{0} \tag{4}$$

$$\boldsymbol{g}(\boldsymbol{x}^o) \leq \mathbf{0}, \quad \boldsymbol{\lambda}^{oT} \boldsymbol{g}(\boldsymbol{x}^o) = 0 \tag{5}$$

$$\boldsymbol{\lambda}^o \geq \mathbf{0} \tag{6}$$

where $\boldsymbol{x}^o$ is the optimal solution and $\boldsymbol{\lambda}^o$ is the corresponding Lagrangian multiplier vector.

Since $\nabla_{\boldsymbol{x}} g_i(\boldsymbol{x}^o) = \boldsymbol{b}_i$ and $\nabla_{\boldsymbol{x}} f(\boldsymbol{x}^o) = A\boldsymbol{x}^o + \boldsymbol{s}$, Eq.(4) is rewritten as,

$$-(A\boldsymbol{x}^o + \boldsymbol{s}) = \sum_i \lambda_i^o \boldsymbol{b}_i. \tag{7}$$

Eq.(7) indicates that a gradient vector of a criterion function, $-(A\boldsymbol{x}^o + \boldsymbol{s})$, lies inside the polar cone formed by the coefficient vectors $\{\boldsymbol{b}_i\}(i = 1, \cdots, q; q \leq m)$ corresponding to active constraints. Here we assume, without loss of generality, that the first q constraints are active and the rest are inactive.

Based on the above formulation, Eqs.(5) (6) (7), and $A \geq 0$ are the necessary and sufficient conditions for optimality.

3 Neural network architecture

We propose the linear neural network architecture in Figure 1 representing the optimality conditions for static inverse optimization[10,11].

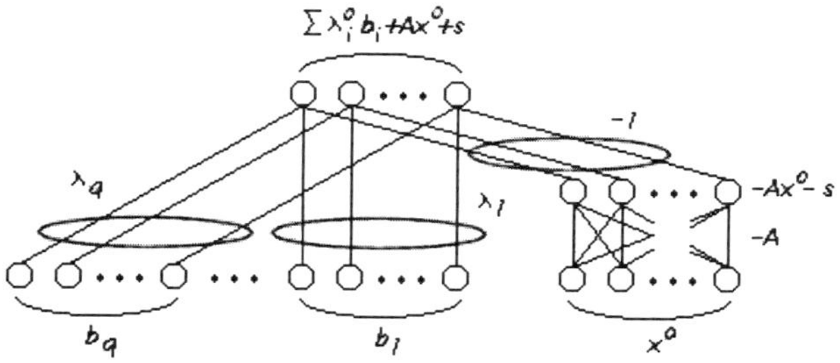

Figure 1. The structure of a neural network representing the optimality conditions

The solution, $\boldsymbol{x}^o$, is given to the rightmost block of the input layer in Figure 1. The vector $-(A\boldsymbol{x}^o + \boldsymbol{s})$ is produced at the next layer by propagating the activation through the connection weight matrix, A. $-\boldsymbol{s}$ corresponds to a bias. Therefore, the rightmost module in Figure 1 represents the left-hand side of Eq.(7). Similarly left modules with inputs, $\boldsymbol{b}_1, \cdots, \boldsymbol{b}_q$, corresponding to active constraints represent the right-hand side of Eq.(7). It is to be noted that both optimization and inverse optimization can be represented by the neural network in Figure 1. In optimization A, $\boldsymbol{s}$ and $\boldsymbol{b}_1, \cdots, \boldsymbol{b}_q$ are given, and $\boldsymbol{\lambda}_1, \cdots, \boldsymbol{\lambda}_q$ and $\boldsymbol{x}^o$ are determined by minimizing mean square output error. In inverse optimization $\boldsymbol{b}_1, \cdots, \boldsymbol{b}_q$ and $\boldsymbol{x}^o$ are given, and $\boldsymbol{\lambda}_1, \cdots, \boldsymbol{\lambda}_q$, A and $\boldsymbol{s}$ are obtained by learning of neural networks[4].

4 Procedure for data interpretation

A procedure of interpreting data by static inverse optimization is the following.

Step 1 It is assumed, without loss of generality, that the smaller the value of an attribute is, the the more preferable it is. Attribute values are transformed accordingly.

Step 2 Generate constraints from given data. During generation, an efficient method for generating constraints is used.

Step 3 Select a Pareto optimal sample, and obtain the corresponding active constraints.

Step 4 Estimate a criterion function matrix and a Lagrangian multiplier corresponding to the given sample. During learning, a criterion function matrix is modified to guarantee its positive-semidefiniteness.

Step 5 After learning by backpropagation, learning toward pseudo-diagonal is carried out for estimating a simple and understandable criterion function. Necessary modification of a criterion function matrix to guarantee positive-semidefiniteness is also done.

Step 6 A given sample is interpreted based on the resulting criterion function, lagrangian multiplier, marginal rates of substitution and so forth.

Steps 4 and **5** correspond to static inverse optimization, and **Step 2** corresponds to the proposed method described in Section 5.2.

5 Efficient method for generating constraints

5.1 Generation of constraints

In interpreting data, only data are given and constraints are not provided. It is, therefore, necessary to generate constraints from given data for their interpretation. A concept of Pareto optimality popular in welfare economics plays an important role.

We assume here, without loss of generality, that the smaller the value of a variable is, the more desirable it is. Under this assumption, $\boldsymbol{x}^*$ is Pareto optimal if $\boldsymbol{x}$ satisfying the following inequalities does not exist.

$$\boldsymbol{x}^* \geq \boldsymbol{x}, \quad \exists j \;\; x_j^* > x_j \tag{8}$$

Let the number of data be N and the number of attributes be M. A hyperplane in M-dimensional data space determined by the data $\{\boldsymbol{u}_{i_1}, \cdots, \boldsymbol{u}_{i_M}\}$ is given by,

$$g_r(\boldsymbol{x}) = \begin{vmatrix} \boldsymbol{x}^T & 1 \\ \boldsymbol{u}_{i_1}^T & 1 \\ \vdots & \vdots \\ \boldsymbol{u}_{i_M}^T & 1 \end{vmatrix} = 0, \quad \begin{cases} r = 1, 2, \cdots, {}_N C_M \\ 1 \leq i_1, \cdots, i_M \leq N \end{cases} . \tag{9}$$

Figure 2 illustrates the number of hyperplanes, r, as a function of the number of data, N, and the number of attributes, M. Those hyperplanes which satisfy the following two conditions constitute a set of Pareto optimal data. The first condition is that all data exist on one side of a hyperplane and the origin lies on the other side. The second condition is that the sign of all coefficients of the hyperplane are the same. Obtained hyperplanes correspond to partial surfaces of the convex-hull for given data.

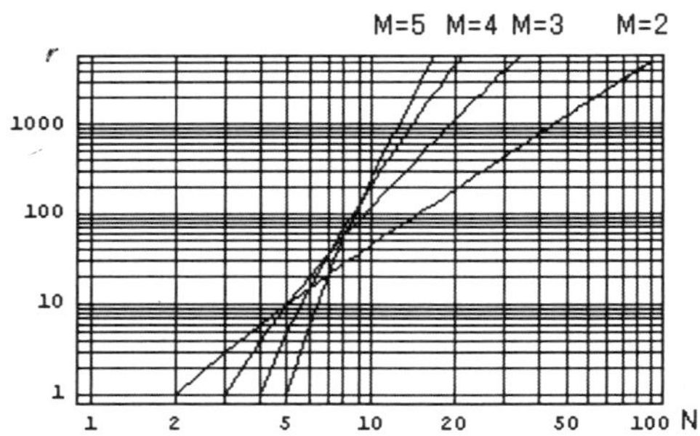

Figure 2. The number of hyperplanes, r, as a function of the number of data, N, and the number of attributes, M.

5.2 *Procedure for generating constraints*

We propose the following procedure for generating constraints by divide-and-conquer.

Step 2-1 Divide given data randomly into several subsets.

Step 2-2 Eliminate non-Pareto optimal data from each subset as much as possible by hyper-ellipsoid and hyperplane elimination algorithms.

Step 2-3 Obtain Pareto optimal data in each subset.
Step 2-4 Calculate Pareto optimal data for the entire data by fusing them.
Step 2-5 Generate constraints from Pareto optimal data.

It is proved that the resulting Pareto optimal data are the same as those obtained directly from the original data.

[Proof] Let D be a set of original data, D_i be ith subset of D, P be a set of Pareto optimal data of D, P_i be a set of Pareto optimal data of D_i. $D = D_1 \cup D_2 \cdots \cup D_k$, $P' = P_1 \cup P_2 \cup \cdots \cup P_k$.

Suppose $\boldsymbol{x} \in D_i \cap P$, i.e., $\boldsymbol{x}$ is Pareto optimal, this means that $\boldsymbol{y}$ satisfying $\boldsymbol{x} \geq \boldsymbol{y}$, $\exists j \ x_j > y_j$, does not exist in D. It is clear that $\boldsymbol{y} \in D_i$, satisfying $\boldsymbol{x} \geq \boldsymbol{y}$, $\exists j \ x_j > y_j$, does not exist. Accordingly $P_i \supseteq D_i \cap P$ is satisfied. Taking the union of both sides, we obtain

$$P' = P_1 \cup P_2 \cdots \cup P_k \supseteq P$$

Accordingly P is included in P', therefore we can obtain Pareto optimal data from P', and obtain P without loss. [End]

5.3 Two Elimination Algorithms

- Algorithm of hyper-ellipsoid elimination

  Calculate the average, $\hat{\boldsymbol{\mu}}$, and variance and covariance matrix, $\hat{\Sigma}$, from a set of data, U, with N samples and M attributes.

$$\hat{\mu}_i = \frac{1}{N}\sum_{j=1}^{N} u_{ji}, \quad i = 1, \cdots, M \tag{10}$$

$$\hat{\sigma}_{ik}^2 = \frac{1}{N-1}\sum_{j=1}^{N}(u_{ji} - \hat{\mu}_i)(u_{jk} - \hat{\mu}_k), \quad i, k = 1, \cdots, M \tag{11}$$

  Discard samples with Mahalanobis distance, $g_e(\boldsymbol{x}, \gamma)$, smaller than γ.

$$g_e(\boldsymbol{x}, \gamma) = (\boldsymbol{x} - \hat{\boldsymbol{\mu}})^T \hat{\Sigma}^{-1}(\boldsymbol{x} - \hat{\boldsymbol{\mu}}) \leq \gamma^2 \tag{12}$$

  U' is the remaining set of samples with N' samples.

- Algorithm of hyperplane elimination

  Find the minimum value of each attribute.

$$\underline{\boldsymbol{u}}_i = \boldsymbol{u}_{j_i}, \quad j_i = arg \min_j u_{ji}, \quad i = 1, \cdots, M \tag{13}$$

  Determine the hyperplane, $g_d(\boldsymbol{x}) = 0$, composed of M samples, $\underline{\boldsymbol{u}}_i (i = i_1, \cdots, i_M)$. Discard data satisfying $g_d(\boldsymbol{x}) > 0$. The remaining samples constitute the set U'' with N'' samples.

6 Application to rented housing data

It is assumed that a tenant of a rented house makes a decision by maximizing one's utility. Based on this assumption we interpret real data of rented houses in Tokyo[9].

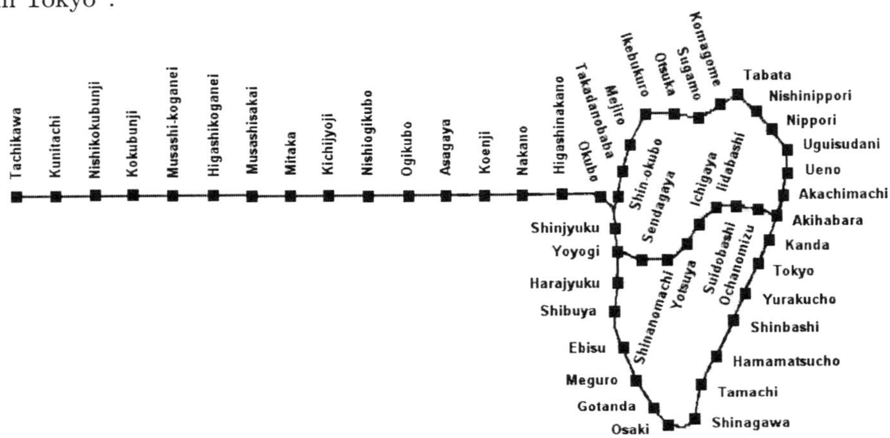

Figure 3. Yamanote and Soubu-Chuo lines in Tokyo

Figure 3 illustrates a map of Yamanote and Soubu-Chuo lines in Tokyo. The number of rented housing data, composed of separate house and apartment houses, along Yamanote and Soubu-Chuo lines is 3932. The attributes of the data are rent, commuting time to Shinjyuku station, area of housing and year of construction. Table 1 provides examples of data near Shinjyuku station.

Table 1. Examples of data near Shinjyuku station. y_1: rent(10^4 yen), y_2: commuting time(min.), y_3: area of housing (m^2), and y_4: year of construction (year).

data	attributes			
	y_1	y_2	y_3	y_4
1	5.8	16	14.13	1982.4
2	6.5	12	19.15	1976.4
⋮	⋮	⋮	⋮	⋮
86	40.0	13	107.6	1985.1
87	45.0	14	144.9	1989.3

Firstly, necessary modifications are made according to **Step 1** in Section 4. They are the area of housing and years after construction. New variables are: $x_1 = y_1$, $x_2 = y_2$, $x_3 = 287 - y_3$ and $x_4 = 2002 - x_4$. Paramenters in these transformations do not directly affect the interpretation, because only the marginal rates of substitution matter in interpretation as will be shown later.

Secondly, we generate constraints from modified data according to **Steps 2** and **3**. Because the number of the data is very large, the proposed method is iteratively carried out, i.e. 8 times, for generating constraints. 19 Pareto optimal data, and the following 21 constraints are obtained.

$$\begin{cases} g_1: & 305x_1 + 410.5x_2 + 65.0x_3 + 7.4x_4 = 21239 \\ g_2: & 1196x_1 + 621x_2 + 225x_3 + 635x_4 = 80537 \\ & \vdots \\ g_{21}: & 524x_1 + 588x_2 + 107x_3 + 2342x_4 = 50980 \end{cases}$$

Computation time is 30 minutes due to divide-and-conquer. It would require 3,330 years without divide-and-comquer [a].

Steps 4 and **5** are omitted here due to space limitation.

Table 2. Marginal rates of substitution between attribute x_1 (rent) and other attributes for Pareto optimal data. The 1st column is renumbered.

No.	$\rho^{(2)}_{(1)}$ ($10^4 yen/min.$)	$\rho^{(3)}_{(1)}$ ($10^4 yen/m^2$)	$\rho^{(4)}_{(1)}$ ($10^4 yen/year$)	region
1'	0.7∼1.9	4.7∼5.3	1.9∼41.	Shinjyuku
2'	0.7	4.7	41.	//
3'	0.7	4.7	41.	//
18'	0.1∼0.9	1.9∼32.	0.1∼0.3	//
4'	0.1∼0.7	1.9∼5.3	0.3∼41.	Shin-okubo
17'	0.1∼2.1	0.7∼5.3	0.1∼0.5	Takadanobaba
11'	0.5∼7.5	0.7∼6.5	0.3∼1.9	Mejiro
5'	1.5∼7.5	4.7∼6.5	0.4∼1.9	Ikebukuro
13'	0.1∼1.3	1.9∼32.	0.1∼0.3	Yoyogi
14'	0.9∼1.3	18.∼32.	0.1∼0.2	Shibuya
19'	0.5∼4.4	0.7∼1.0	0.1∼0.5	//
16'	0.9∼1.3	1.9∼18.	0.1∼0.3	Koenji
12'	0.9∼4.4	1.0∼18.	0.1∼0.3	Ogikubo
9'	3.1∼125	1.0∼5.9	0.1∼0.2	Kichijyoji
10'	1.5∼125	0.9∼5.9	0.1∼0.9	//
15'	0.9∼1.3	0.9∼32.	0.1∼0.3	//
7'	1.3∼7.5	3.3∼18.	0.2∼1.3	Musashi-koganei
6'	4.6∼125	2.1∼6.5	0.1∼1.3	Kunitachi
8'	125	2.1	0.1	//

Finally, we interpret the rented housing data according to **Step 6**. Table 2 presents the marginal rates of substitution between attribute x_1 (rent) and

[a]PC: CPU 1.4GHz, Memory 128MB with Mathematica ver.4.1

other attributes. Table 2 suggests that the decision maker $1'$ will pay $7,000 \sim 19,000$ yen to decrease commuting time by 1 munite. The decision maker $1'$ will also pay $47,000 \sim 53,000$ yen to increase the area of house by 1 m^2 and will pay $19,000 \sim 410,000$ yen to renovate a house by one year. Other Pareto optimal data can be interpreted in the same way.

The distribution of these Pareto optimal data has three characteristics. The first is that the Pareto optimal data alone Yamanote line are concentrated around Shinjyuku with commuting time of less than 11 minutes. The second is that the Pareto optimal data along Soubu-Chuo line are located west of Koenji. The third is that Pareto optimal data do not exist along Soubu-Chuo line between Sendagaya and Ochanomizu.

Table 3 presents the average values of marginal rates of substitution for Pareto optimal data along Yamanote and Soubu-Chuo lines.

Table 3. Average values of attributes and marginal rates of substitution for Pareto optimal data.

items	Yamanote	Soubu-Chuo
rent (10^4 yen)	32.7	23.9
commuting time (min.)	11.7	34.1
area of house (m^2)	86.8	106.
year after construction (year)	9.7	2.3
$\rho_{(1)}^{(2)}$ (10^4 yen/min.)	1.59	33.2
$\rho_{(1)}^{(3)}$ (10^4 yen/m^2)	8.19	7.50
$\rho_{(1)}^{(4)}$ (10^4 yen/year)	11.6	0.36

From Table 3 we can say the followings:

- The longer the commuting time is, the larger the monetary value of commuting time becomes. This is because those tenants who live far from Shinjyuku, have stronger desire to decrease the commuting time.
- The longer the commuting time is, the smaller the monetary value of years after construction becomes. This is because those tenants who live far from Shinjyuku have weaker desire to live in new houses.

7 Conclusions

We have proposed an efficient method for generating constraints by divide-and-conquer to interpret large-scale real data from a viewpoint of optimization. It is proved that the resulting Pareto optimal data are the same as those obtained directly from the original data.

We have applied the proposed method to large-scal real data, and have successfully estimated a criterion function governing decision making of the tenants living along Yamanote and Soubu-Chuo lines in Tokyo. These results well accord with data and our intuition.

References

1. D.R. Chand and S.S. Kapur, "An Algorithm for Convex Polytypes," Journal of the ACM **17-1**, 78-86 (1970).
2. A. Datta *et al.*, "A Connectionist Model for Convex-Hull of a Planar Set," Neural Networks **13**, 377-384 (2000).
3. H.W. Kuhn and A.W. Tucker, "Nonlinear Programming," Proceedings 2nd Berkeley Symposium on Mathematical Statistics and Probability J. Neyman (Ed.) , University of California Press (1951).
4. D.E. Rumelhart *et al.*, "Parallel Distributed Processing," The MIT Press (1986).
5. J.V. Neuman and O. Morgenstern, "Theory of Games and Economic Behavior," John Wiley & Sons, Inc. (1967).
6. H.A. Simon, "The Science of the Artificial," The MIT Press (1981).
7. E. Wennmyr, "A Convex Hull Algorithm for Neural Networks," IEEE Trans. on Circuits and Systems **36-11**, 64-68 (1989).
8. R. J.-B. Wets and C. Witzgall, "Towards an Algebraic Characterization of Convex Polyhedral Cones," Number. Math. **12**, 134-138 (1968).
9. Recruit Co., Ltd, "Rented Housing Information [Metropolican Area]," Ken Corp., Ltd , 2/2 (2000).
10. H. Zhang and M. Ishikawa, "A Neural Networks Approach to Inverse Optimization," The 2nd R.I.E.C. International Symposium on Design and Architecture of Information Processing Systems Based on the Brain Information Principles (DAIPS) , 197-200 (1998).
11. H. Zhang and M. Ishikawa, "A General Solution to Static Inverse Optimization Problems Using Neural Networks Learning," The Trans. of IEEJ **120-C(6)**, 857-864 (2000)(in Japanese).
12. H. Zhang and M. Ishikawa, "A Neural Networks Approach to Dynamic Inverse Optimization Problems," The Trans. of IEEJ **120-C(4)**, 481-488 (2000)(in Japanese).
13. H. Zhang and M. Ishikawa, "Structure Determination of a Criterion Function by Dynamic Inverse Optimization," Proceedings of 7th International Conference on Neural Information Processing (ICONIP-2000) , 662-666 (2000).

Section V

Related Topics

EGUCHI-OKI-MATSUMURA EQUATION FOR PHASE SEPARATION: NUMERICALLY GUIDED APPROACH

T. HANADA

Department of Mathematics, Chiba Institute of Technology, Narashino, Chiba 275-0023, Japan
E-mail: hanada@pf.it-chiba.ac.jp

H. IMAI

Department of Applied Physics and Mathematics, Faculty of Engineering, University of Tokushima, Tokushima 770-8506, Japan
E-mail: imai@pm.tokushima-u.ac.jp

N. ISHIMURA

Department of Mathematics, Faculty of Economics, Hitotsubashi University, Kunitachi, Tokyo 186-8601, Japan
E-mail: ishimura@math.hit-u.ac.jp

M.A. NAKAMURA

College of Science and Technology, Nihon University, Kanda-Surugadai, Tokyo 101-8308, Japan
E-mail: nakamura@media.cst.nihon-u.ac.jp

Eguchi-Oki-Matsumura (EOM) equations are introduced to describe the dynamics of pattern formation which arises from phase separation in some binary alloys. The model extends the well-known Cahn-Hilliard equation. We report our studies of the EOM equation, with an emphasis on numerical analysis.

1 Introduction

This is a report of our recent studies on the Eguchi-Oki-Matsumura (EOM) equation for phase separation [7,8], with an emphasis on numerical analysis from the inverse problem viewpoint.

The dynamics of pattern formation resulting from phase separation has been a fascinating topic for researches. Cahn and Hilliard [1], based on a continuum model in thermodynamics, made a phenomenological approach to explaining such kinetics and derive the fourth-order partial differential equations (PDEs), known as the Cahn-Hilliard equation. Many studies have been performed on this equation and much progress has been achieved so far from various points of view [2,3,5,9,10,11,12].

Eguchi, Oki, and Matsumura [4], on the other hand, introduced a system of equations, which extends the Cahn-Hilliard equation and consists of coupled two phase fields; one is the local concentration and the other is the local degree of order. After performing a suitable scaling of parameters presented shortly later, EOM equations in one-space dimension, with which we are mainly concerned, are expressed as follows.

$$\begin{cases} u_t = -\varepsilon^2 u_{xxxx} + ((a+v^2)u)_{xx} & \text{in } 0 < x < l, \quad t > 0 \\ v_t = v_{xx} + (b - u^2 - v^2)v & \text{in } 0 < x < l, \quad t > 0 \\ u_x = u_{xxx} = v_x = 0 & \text{at } x = 0 \text{ and } l, \quad t > 0 \\ u|_{t=0} = u_0, \quad v|_{t=0} = v_0 & \text{on } 0 \le x \le l, \end{cases} \tag{1}$$

where $u = u(x,t)$ and $v = v(x,t)$ denote unknown functions related to the local concentration and the local degree of order, respectively. The total concentration of u is conserved under the evolution of (1). Namely we have $(1/l)\int_0^l u(x,t)\,dx = m$, where m is a constant. u_0, v_0 are given initial data satisfying required compatibility conditions:

$$(u_0)_x = (u_0)_{xxx} = (v_0)_x = 0 \qquad \text{at } x = 0,\, l, \quad \text{and } \frac{1}{l}\int_0^l u_0(x)\,dx = m.$$

ε, a are positive constants depending on the temperature, and $b \in \boldsymbol{R}$ is the principal parameter which increases from negative to positive as the temperature crosses downward the critical one. Here we focus our attention on the case of positive b, since the negative b turns out to enjoy rather trivial behaviors; precisely, every solution converges to trivial solutions if b is nonpositive.

The free energy of the system serves as a Lyapunov functional, which is given by

$$F[u,v] := \int_0^l \left(\frac{\varepsilon^2}{2}u_x^2 + \frac{1}{2}v_x^2 + \frac{a}{2}u^2 + \frac{1}{4}v^4 - \frac{b}{2}v^2 + \frac{1}{2}u^2v^2\right) dx.$$

The direct calculation implies

$$\begin{aligned} \frac{d}{dt}F[u,v](t) &= -\int_0^l \{-\varepsilon^2 u_{xxx} + ((a+v^2)u)_x\}^2\,dx - \int_0^l v_t^2\,dx \\ &\le 0, \end{aligned}$$

for any solution (u,v) to (1), which leads us to expect that the solution $(u(x,t), v(x,t))$ for (1) converges as $t \to \infty$ to the solution $(u(x), v(x))$ for the steady state problem

$$\begin{cases} -\varepsilon^2 u_{xxxx} + ((a+v^2)u)_{xx} = 0 & \text{in } 0 < x < l \\ v_{xx} + (b - u^2 - v^2)v = 0 & \text{in } 0 < x < l \\ u_x = u_{xxx} = v_x = 0 & \text{at } x = 0 \text{ and } l \\ (1/l)\int_0^l u\,dx = m. \end{cases} \tag{2}$$

We remark that (2) always has a solution $u \equiv m$ and $v \equiv 0$. If $b \leq 0$, then it can be seen that this is the only solution to (2) by virtue of the maximum principle. If $b > m^2$, (2) has another solution $u \equiv m$ and $v \equiv \pm\sqrt{b - m^2}$. We call these solutions trivial. Solutions that are different from trivial ones will be called non-trivial solutions of the EOM equations; in other words, solution (u, v) to (2), both of which are not simultaneously constants, will be referred to as non-trivial solutions.

We conclude this introduction with our main analytical achievements.We refer to Hanada *et al* [8].

Theorem 1 *Suppose that* $u_0, v_0 \in H^2(0,l)$ *with* $(u_0)_x = (v_0)_x = 0$ *at* $x = 0, l$ *and* $(1/l)\int_0^l u_0\,dx = m$. *Then, for each* $T > 0$, *there exists a unique solution* (u, v) *to (1)such that*

$$u \in L^2((0,T); H^4(0,l)) \cap L^\infty([0,T); H^2(0,l)),$$
$$v \in L^2((0,T); H^2(0,l)) \cap L^\infty([0,T); H^1(0,l)).$$

For any initial data above, the solution (u, v) *converges as* $t \to \infty$ *to a solution of the steady state problem (2).*

(2) has at least one monotone non-trivial steady solution for all large $b \gg m^2$. *Moreover, for any integer* $k \geq 2$ *and for all large* $b \gg m^2$ *depending on* k, *(2) has at least one non-monotone non-trivial steady solution, each of whose derivatives changes sign exactly* $(k-1)$*-times.*

2 Existence of Solutions

To establish the local in time existence, a standard Galerkin approximation method is implemented.

Let $\mathcal{F}$ denote the complete orthonormal system in $L^2(0,l)$ with the even periodic boundary condition: $\mathcal{F} := \{1/\sqrt{l}, \sqrt{2/l}\cos(\pi x/l), \sqrt{2/l}\cos(2\pi x/l), \cdots, \sqrt{2/l}\cos(n\pi x/l), \cdots\}$. For every positive integer N, let W_N be the linear space spanned by $\{1/\sqrt{l}, \sqrt{2/l}\cos(\pi x/l), \cdots, \sqrt{2/l}\cos(N\pi x/l)\}$ and P_N denote the orthogonal projector in $L^2(0,l)$ onto W_N.

We are then looking for an approximate solution $(u^N(x,t), v^N(x,t))$ to (1) given by

$$u^N(x,t) = m + \sum_{n=1}^{N} u_n(t)\sqrt{\frac{2}{l}}\cos\frac{n\pi x}{l}$$

$$v^N(x,t) = \frac{1}{\sqrt{l}}v_0(t) + \sum_{n=1}^{N} v_n(t)\sqrt{\frac{2}{l}}\cos\frac{n\pi x}{l}.$$

The components $\{u_n(t), v_n(t)\}$ satisfy a system of ordinary differential equations, which has a unique solution on $[0, T_N)$ for some $T_N > 0$. Thanks to various a priori estimates, we are able to let $N \to \infty$; in particular, we have $\liminf_{N\to\infty} T_N \geq T > 0$ for some $T > 0$. Uniform bounds of $H^1(0,l)$-norms enable us to repeat the local solvability procedure and continue the solution. As a summary, our existence results are formulated as follows.We refer to Hanada *et al* [8] for the details.

Proposition 1 *Suppose that $u_0, v_0 \in H^1(0,l)$ with $(u_0)_x = (v_0)_x = 0$ at $x = 0, l$ and $(1/l)\int_0^l u_0\,dx = m$. Then, for each $T > 0$, there exists a unique solution (u,v) to (1) such that*

$$u \in L^2((0,T); H^3(0,l)) \cap L^\infty([0,T); H^1(0,l)),$$
$$v \in L^2((0,T); H^2(0,l)) \cap L^\infty([0,T); H^1(0,l)).$$

Concerning the long time behavior of the solution (u,v) to (1), rather routine inference involving the Lyapunov functional $F[u,v]$ works, and we conclude that (u,v) tends to an element of the ω-limit set of (u_0, v_0), on which $F[u,v]$ is constant; namely, (u,v) converges to an equilibrium solution of the steady state problem (2).

3 A Priori Estimates

We here collect some a priori estimates, which is needed to prove the existence and to determine the asymptotic profile of the solution (u,v) to (1). To start with, we introduce the following function spaces.

$$E_T := \{(u,v) \in L^2((0,T); H^4(0,l)) \times L^2((0,T); H^2(0,l)) \mid u_x = u_{xxx} = v_x = 0 \text{ at } x = 0, l\},$$

$$E_0 := \{(u_0,v_0) \in (H^2(0,l))^2 \mid (u_0)_x = (v_0)_x = 0 \text{ at } x = 0, l, \quad \frac{1}{l}\int_0^l u_0\,dx = m\},$$

where $T > 0$. The norm $\|\cdot\|$ denotes that of $L^2(0,l)$. Furthermore, C_0 stand for various constants depending only on the initial data and constants ε^2, a, b, which may differ from line to line. We understand that C_0 is independent of t.

Lemma 1 *There holds*

$$\|v(t)\|_\infty \leq \max\{\|v_0\|_\infty, \sqrt{b}\} \qquad \text{for } 0 < t < T.$$

In particular, $\|v(t)\|_\infty \leq \sqrt{b}$ *for all large* t.

Lemma 2 *For any initial data* $(u_0, v_0) \in E_0$, *the solution* $(u, v) \in E_T$ *to (1) verifies*

$$\|u(t)\|, \ \|u_x(t)\|, \ \|v_x(t)\| \leq C_0,$$

for $0 < t < T$, *and moreover*

$$\|u(t)\|_\infty \leq C_0.$$

Lemma 3 *It follows that for any* $0 \leq t \leq s \leq T$

$$\int_t^s (\|u_{xx}(\tau)\|^2 + \|u_x(\tau)\|^2 + \|v_x(\tau)\|^2)\, d\tau \leq C_0(1 + (s-t)).$$

Lemma 4 *There holds for any* $0 \leq t \leq s \leq T$

$$\int_t^s (\|u_{xxx}(\tau)\|^2 + \|v_{xx}(\tau)\|^2)\, d\tau \leq C_0(1 + (s-t)).$$

Lemma 5 *There hold for any* $0 \leq t \leq s \leq T$

$$\|u_{xx}(t)\|^2 \leq C_0(1+t), \qquad \int_t^s \|u_{xxxx}(\tau)\|^2\, d\tau \leq C_0(1 + (s-t)).$$

The proof of above lemmas are combinations of integration by parts and the application of Gronwall's inequality. The computations are tedious but straightforward; the detailed expositions are found in Hanada *et al* [8].

4 Structure of Steady Solutions

The structure of steady state solutions to EOM equations, that is, solutions $u = u(x)$ and $v = v(x)$ which verify (2) is investigated. Our results read as follows, which extends our previous establishments [7].

Proposition 2 *For all large $b \gg m^2$, there exists at least one monotone non-trivial steady solution for EOM equations. Furthermore, for any integer $k \geq 2$ and for all large $b \gg m^2$ depending on k, EOM equations have non-monotone non-trivial steady solutions, each of whose derivatives changes sign exactly $(k-1)$-times.*

We remark that the large values of b and m^2 stated in Proposition 2 can be computed explicitly.

5 Computational Study

Our numerical scheme is motivated in part by that for the Cahn-Hilliard equation [3]. Let $x_k = k\Delta x$ $(k = 0, 1, \cdots, n)$ with $\Delta x = l/n$. The discretized free energy $F^n[U, V]$ for the approximations (U_k, V_k) of $(u(x_k, t), v(x_k, t))$ is expressed as

$$F^n[U,V] = \Sigma(\frac{\varepsilon^2}{2}((\nabla_+ U_k)^2 + (\nabla_- U_k)^2) + \frac{1}{2}((\nabla_+ V_k)^2 + (\nabla_- V_k)^2) + \frac{a}{2}U_k^2 + \frac{1}{4}V_k^4 - \frac{b}{2}V_k^2 + \frac{1}{2}U_k^2 V_k^2)\Delta x.$$

Here ∇_+ and ∇_- denote the forward and backward difference in x, respectively:

$$\nabla_+ U_k = \frac{1}{\Delta x}(U_{k+1} - U_k), \quad \nabla_- U_k = \frac{1}{\Delta x}(U_k - U_{k-1}).$$

Σ represents the trapezoidal summation formula defined by

$$\Sigma U_k^2 = \frac{1}{2}U_0^2 + \sum_{k=1}^{N-1} U_k^2 + \frac{1}{2}U_N^2.$$

Now, for the approximations $(\bar{U_k}, \bar{V_k})$ of $(u(x_k, t+\Delta t), v(x_k, t+\Delta t))$, we mainly adopt the implicit scheme as follows:

$$
\begin{aligned}
\frac{\bar{U_k} - U_k}{\Delta t} &= \nabla^2 (-\varepsilon^2 \nabla^2 \frac{\bar{U_k} + U_k}{2} + a \frac{\bar{U_k} + U_k}{2} \\
&\quad + \frac{\bar{U_k}\bar{V_k} + U_k V_k}{2} \frac{\bar{V_k} + V_k}{2}) \\
\frac{\bar{V_k} - V_k}{\Delta t} &= \nabla^2 \frac{\bar{V_k} + V_k}{2} + b \frac{\bar{V_k} + V_k}{2} \\
&\quad - \frac{\bar{U_k}\bar{V_k} + U_k V_k}{2} \frac{\bar{U_k} + U_k}{2} - \frac{\bar{V_k}^2 + V_k^2}{2} \frac{\bar{V_k} + V_k}{2},
\end{aligned}
$$

where $\nabla^2 := \nabla_+ \nabla_-$ stands for the second order central difference in x. With this implicit scheme, we deduce that the discretized free energy is decreasing[6]:

$$
F^n[\bar{U}, \bar{V}] \leq F^n[U, V].
$$

This property is useful if the method is applied to the computation of inverse problems. Here we supplement our solver by the explicit scheme, since it is fast and a posteriori stable to implement.

$$
\begin{aligned}
\frac{\bar{U_k} - U_k}{\Delta t} &= -\varepsilon^2 \nabla^4 U_k + \nabla^2 ((a + V_k^2) U_k) \\
\frac{\bar{V_k} - V_k}{\Delta t} &= \nabla^2 V_k + (b - U_k^2 - V_k^2) V_k.
\end{aligned}
$$

In this case, the dissipation of the free energy holds only approximately. The discretized boundary conditions should be fixed as

$$
\begin{aligned}
U_{-1} &= U_1, \quad U_{n-1} = U_{n+1} \quad \text{in place of } u_x = 0 \text{ at } x = 0 \text{ and } l \\
V_{-1} &= V_1, \quad V_{n-1} = V_{n+1} \quad \text{in place of } v_x = 0 \text{ at } x = 0 \text{ and } l \\
U_{-2} &= U_2, \quad U_{n-2} = U_{n+2} \quad \text{in place of } u_{xxx} = 0 \text{ at } x = 0 \text{ and } l.
\end{aligned}
$$

We focus our interest on the question whether exists the variety of steady state solutions or not; taking constants

$$
l = 1, \quad \varepsilon = 1, \quad a = m = \frac{1}{4},
$$

several steady solutions are now illustrated in following Figures.

Figure 1 depicts the convergence of a solution (u, v) for (1) to a monotone steady solution. We set $b = 16/25$ and as initial function we employ

$$u_0(x) = m,$$
$$v_0(x) = \sqrt{b - m^2} - \frac{1}{1000}\cos(\pi x).$$

The computation is implemented under the mesh size $\Delta x = 1/64$ and $\Delta t = 1/256$ up to the time interval $0 \le t \le 4096$.

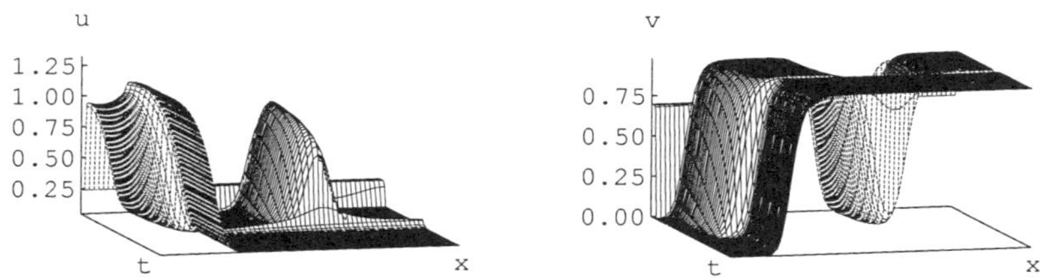

Figure 1. Convergence to a monotone steady solution.

The monotone steady solution corresponds to the case $k = 1$ in Theorem 1. It is numerically unstable with respect to the perturbation on initial data.

Figure 2, on the other hand, illustrates the convergence to a non-monotone steady solution. We set $b = 0.99$, and as initial function we take

$$u_0(x) = m,$$
$$v_0(x) = -\frac{1}{1000}\cos(\pi x).$$

The implementation data are the same as those of Figure 1, while we perform during the time interval $0 \le t \le 128$.

The limiting function is related to the case $k = 2$ in Theorem 1; however, the function v is monotone increasing in Figure 2. This apparent discrepancy is easily reconciled by virtue that the sign of v is irrespective to the problem. We hasten to remark that the non-monotone steady solution, which is constructed in Theorem 1 with $k = 2$, also is numerically realized.

Finally, we exhibit the energy diagram of various steady solutions in Figure 3. Here the notation $w-i$ $(i = 0, 1, \cdots, 8)$ means the steady solution to (1), which is akin to the one with $k = i + 1$ described in Theorem 1.

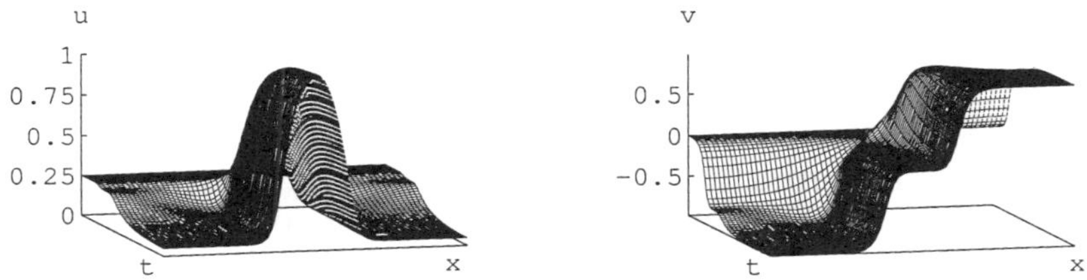

Figure 2. Convergence to a non-monotone steady solution.

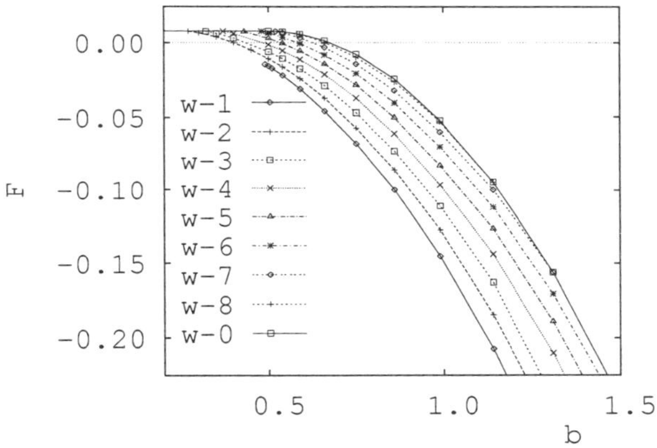

Figure 3. Energy diagram of steady solutions.

Acknowledgments

We are grateful to Professor Hiroshi Fujita for his interest in this research. Thanks are also due to the referee for various comments, which helps improving the manuscript. This work is partially supported by Grants-in-Aids for Scientific Research (Nos.10555023, 12640223, 13555021, 13640206), from the Japan Ministry of Education, Science, Sports and Culture.

References

1. J.W. Cahn and J.E. Hilliard, *Free energy of a nonuniform system, I., Interfacial free energy,* J. Chem. Phys. **28**, 258(1958).
2. J. Carr, M.E. Gurtin, and M. Slemrod, *Structured phase transitions on a finite interval,* Arch. Rational Mech. Anal. **86**, 317–351(1984).
3. C.M. Elliott and D.A. French, *Numerical studies of the Cahn-Hilliard equation for phase separation,* IMA J. Appl. Math. **38**, 97–128 (1987).
4. T. Eguchi, K.Oki, and S. Matsumura, *Kinetics of ordering with phase separation,* Mat. Res. Soc. Symp. Proc. **21**, Elsevier, 589–594(1984).
5. C.M. Elliott and Zheng S., *On the Cahn-Hilliard equation,* Arch. Rational Mech. Anal. **96**, 339–357 (1986).
6. D. Furihata and M. Mori, *A stable finite difference scheme for the Cahn-Hilliard equation based on a Lyapunov functional,* Z. angew. Math. Mech. **76**, S1, 405–406(1996).
7. T. Hanada, N. Ishimura, and M.A. Nakamura, *Note on steady solutions of the Eguchi-Oki-Matsumura equation,* Proc. Japan Acad. **Ser.A. 76**, 146(2000).
8. T. Hanada, N. Ishimura, and M.A. Nakamura, *On the Eguchi-Oki-Matsumura equation for phase separation in one-space dimension,* preprint, (2001), submited.
9. T. Hanada, M.A. Nakamura, and C. Shima, *On Eguchi-Oki-Matsumura equations,* GAKUTO Int. Ser. Math. Sci. Appl. **12**, 213(1999).
10. A. Novick-Cohen, *Energy methods for the Cahn-Hilliard equation,* Quart. Appl. Math. **46(4)**, 681–690(1988).
11. A. Novick-Cohen and L.A. Segel, *Nonlinear aspects of the Cahn-Hilliard equation,* Physica **D 10(3)**, 277–298(1984).
12. S. Zheng , *Asymptotic behavior of solution to the Cahn-Hilliard equation,* Applicable Anal. **23**, 165(1986).

SOME RESULTS ON THE EXACT BOUNDARY CONTROLLABILITY FOR QUASILINEAR HYPERBOLIC SYSTEMS

TA-TSIEN LI

Department of Mathematics, Fudan University Shanghai 200433, China
E-mail: dqli@fudan.edu.cn

BOPENG RAO

Institut de Recherche Mathématique Avancée Université Louis Pasteur de Strasbourg , Strasbourg, France

In this paper we present some results on the local exact boundary controllability for general one-dimensional first order quasilinear hyperbolic systems with general nonlinear boundary conditions and give corresponding applications to nonlinear vibrating string equations.

1 Introduction

First of all we recall the definition of exact boundary controllability for hyperbolic equations (systems).

For a given hyperbolic equation (system), for any given initial data φ and final data ψ, if we can find a time $T_0 > 0$ and suitable boundary input controls on the boundary $\partial\Omega$ of the domain Ω, such that the corresponding mixed initial-boundary value problem with the initial data φ admits a unique classical solution $u = u(t, x)$ on the whole domain $[0,\ T_0] \times \bar{\Omega}$, which verifies exactly the final condition

$$t = T_0 : \quad u = \psi(x), \qquad x \in \Omega, \tag{1}$$

namely, if by means of boundary input controls the system can drive any given initial state φ to any given final state ψ at $t = T_0$, then, we say that this system possesses the exact boundary controllability.

More precisely, if the exact boundary controllability can be realized only for initial and final states small enough in a certain sense, we say that the system possesses the local exact boundary controllability; Otherwise, we say the system possesses the global exact boundary controllability.

There are a number of publications concerning the exact controllability for linear hyperbolic equations (systems) (see J. L. Lions[1], D. L. Russell[2] etc.). For the nonlinear case, using the HUM method suggested by J. L. Lions and Schauder's fixed point theorem, E.Zuazua[3] proved the global (resp. local) exact boundary controllability for semilinear wave equations in the asymptotically linear case (resp. the super-linear case with suitable growth conditions). Furthermore, using a global inversion theorem, Lasiecka and Triggiani[4] established an abstract result on the exact controllability for semilinear equations. As applications, they gave the global exact boundary controllability for wave and plate equations in the asymptotically linear case. However, only a few results are known for quasilinear hyperbolic systems. In one-dimensional case, the exact boundary controllability for reducible quasilinear hyperbolic systems was proved in Li-Zhang[5] and Li-Rao-Jin [6,7] by a constructive method which does not work in the general case of quasilinear hyperbolic systems. In an earlier work, M. Cirinà[8,9] considered the zero exact boundary controllability for general quasilinear hyperbolic systems with linear boundary controls, but the author needed some very strong conditions on the coefficients of the system(globally bounded and globally Lipschitz continuous). Moreover, if one applies the result of M. Cirinà[8] twice to get the general exact boundary controllability, the corresponding controllability time should be doubled.

In this paper, we will present some results on the local exact boundary controllability for general one-dimensional quasilinear hyperbolic systems with general nonlinear boundary conditions and give corresponding applications to nonlinear vibrating string problems.

2 General Considerations

Since the hyperbolic wave has a finite speed of propagation, the exact boundary controllability of a hyperbolic equation (system) requires that the controllability time T_0 must be suitably large. In order to have a classical solution to the corresponding mixed initial-boundary value problem on the domain $[0,\ T_0] \times \bar{\Omega}$, we should first prove the existence and uniqueness of the semi-global classical solution, namely, the classical solution on the time interval $0 \leq t \leq T_0$, where $T_0 > 0$ is a preassigned and possibly quite large number. The exact boundary controllability will be based on the existence and uniqueness of semi-global classical solution to the mixed initial-boundary value problem of quasilinear hyperbolic equations (systems).

On the other hand, in order to realize the exact boundary controllability, it is only necessary to find a time $T_0 > 0$ such that the given hyperbolic equation (system) admits a classical solution $u = u(t, x)$ on the domain $[0,\ T_0] \times \bar{\Omega}$, which verifies simultaneously the initial condition

$$t = 0: \quad u = \varphi(x), \quad x \in \Omega \tag{2}$$

and the final condition (1). In fact, putting $u = u(t, x)$ into the boundary conditions, we get immediately the boundary controls. By uniqueness, the classical solution to the corresponding mixed initial-boundary value problem with the initial data φ must be $u = u(t,x)$, which automatically satisfies the given final data ψ.

Moreover, if the solution $u = u(t,x)$ constructed in the previous paragraph also satisfies a part of boundary conditions, then we need only to put $u = u(t,x)$ into the other part of boundary conditions to get the corresponding boundary controls, and, as a result, the number of boundary controls will be reduced and the boundary controls can be asked to act only on a part of boundaries, however, the controllability time will be enlarged.

Of course, for the purpose of application, the controllability time T_0 will be asked to be as small as possible.

3 Main Results

We now consider the following first order quasilinear hyperbolic system

$$\frac{\partial u}{\partial t} + A(u)\frac{\partial u}{\partial x} = F(u), \tag{3}$$

where $u = (u_1, \cdots, u_n)^T$ is a vector valued function of (t,x), $A(u) = (a_{ij}(u))$ is a $n \times n$ matrix with suitably smooth elements $a_{ij}(u)$ $(i,j = 1,\cdots,n)$, $F : \mathbf{R}^n \to \mathbf{R}^n$ is a vector valued function with suitably smooth components $f_i(u)(i = 1,\cdots,n)$ and

$$F(0) = 0. \tag{4}$$

By the definition of hyperbolicity, for any given u on the domain under consideration, the matrix $A(u)$ has n real eigenvalues $\lambda_i(u)(i = 1,\cdots,n)$ and a complete set of left eigenvectors $l_i(u) = (l_{i1}(u),\cdots,l_{in}(u))$ $(i = 1,\cdots,n)$:

$$l_i(u)A(u) = \lambda_i(u)l_i(u), \tag{5}$$

and, correspondingly, a complete set of right eigenvectors $r_i(u) = (r_{i1}(u),\cdots,r_{in}(u))^T$ $(i = 1,\cdots,n)$:

$$A(u)r_i(u) = \lambda_i(u)r_i(u). \tag{6}$$

We have

$$\det|l_{ij}(u)| \neq 0 \qquad (resp.\ \det|r_{ij}(u)| \neq 0). \tag{7}$$

Without loss of generality, we may assume that

$$l_i(u)r_j(u) \equiv \delta_{ij} \qquad (i,j = 1,\cdots,n) \tag{8}$$

and

$$r_i^T(u)r_i(u) \equiv 1 \qquad (i = 1, \cdots, n), \tag{9}$$

where δ_{ij} stands for the Kronecker symbol. Moreover, in this paper we assume that on the domain under consideration, the eigenvalues satisfy the following conditions:

$$\lambda_r(u) < 0 < \lambda_s(u) \qquad (r = 1, \cdots, m; \quad s = m+1, \cdots, n). \tag{10}$$

Let

$$v_i = l_i(u)u \qquad (i = 1, \cdots, n). \tag{11}$$

We consider the following mixed initial-boundary value problem for the quasilinear hyperbolic system (3) with the initial condition

$$t = 0: \quad u = \varphi(x), \qquad 0 \le x \le 1 \tag{12}$$

and the boundary conditions

$$x = 0: \quad v_s = G_s(t, v_1, \cdots, v_m) + H_s(t) \qquad (s = m+1, \cdots, n), \tag{13}$$

$$x = 1: \quad v_r = G_r(t, v_{m+1}, \cdots, v_n) + H_r(t) \qquad (r = 1, \cdots, m). \tag{14}$$

Without loss of generality, we assume that

$$G_i(t, 0, \cdots, 0) \equiv 0 \qquad (i = 1, \cdots, n). \tag{15}$$

For a preassigned and possibly quite large number $T_0 > 0$, we have the following existence and uniqueness of semi-global C^1 solution $u = u(t,x)$ to the mixed initial-boundary value problem (3) and (12)–(14) (See Li *et al* [10] and Li *et al* [11]).

Theorem 1 *Assume that $l_{ij}(u)$, $\lambda_i(u)$, $f_i(u)$, $G_i(t,\cdot)$, $H_i(t)$ $(i,j = 1, \cdots, n)$ and $\varphi(x)$ are all C^1 functions with respect to their arguments. Assume furthermore that (4), (7), (10) and (15) hold. Assume finally that the conditions of C^1 compatibility are satisfied at points (0,0) and (0,1) respectively. Then, for a given $T_0 > 0$, the mixed initial-boundary value problem (3) and(12)-(14) admits a unique C^1 solution $u = u(t,x)$ (called the semi-global C^1 solution) with sufficiently small C^1 norm on the domain*

$$R(T_0) = \{(t,x) | \quad 0 \le t \le T_0, \qquad 0 \le x \le 1\}, \tag{16}$$

provided that the C^1 norms $\|\psi\|_{C^1[0,1]}$ and $\|H\|_{C^1[0,T_0]}$ are small enough (depending on T_0).

Based on Theorem 1, we can get the following theorem on the local exact boundary controllability (See Li *et al* [11] and Li *et al* [12]).

Theorem 2 *Assume that* $l_{ij}(u)$, $\lambda_i(u)$, $f_i(u)$ *and* $G_i(t,\cdot)$ $(i,j=1,\cdots,n)$ *are all* C^1 *functions with respect to their arguments. Assume furthermore that (4), (7), (10) and (15) hold. Let*

$$T_0 > \max_{i=1,\cdots,n} \frac{1}{|\lambda_i(0)|}. \tag{17}$$

For any given initial data $\varphi \in C^1[0,1]$ *and finial data* $\psi \in C^1[0,1]$ *with small* C^1 *norm, the quasilinear hyperbolic system (3) admits a* C^1 *solution* $u = u(t,x)$ *with small* C^1 *norm on the domain* $R(T_0)$*, such that*

$$t=0: \quad u=\varphi(x), \quad 0 \le x \le 1$$

and

$$t=T_0: \quad u=\psi(x), \quad 0 \le x \le 1.$$

Therefore, we can find boundary input controls $H_i \in C^1[0,T_0]$ $(i=1,\cdots,n)$ *with small* C^1 *norm, such that the mixed initial-boundary value problem (3) and (12)–(14) admits a unique* C^1 *solution* $u=u(t,x)$ *on the domain* $R(T_0)$*, which verifies the final condition*

$$t=T_0: \quad u=\psi(x), \quad 0 \le x \le 1. \tag{18}$$

Remark 1 *The exact controllability time* T_0 *given in Theorem 2 is optimal.*

Remark 2 *In Theorem 2, the number of boundary controls is equal to* n*, the number of unknown functions.*

Remark 3 *In order to get the exact boundary controllability, the boundary controls are not unique.*

Remark 4 *In some special cases, for instance, for the reducible quasilinear hyperbolic system with linearly degenerate characteristics*

$$\{ \ \frac{\partial u_1}{\partial t} + \lambda_1(u_2)\frac{\partial u_1}{\partial x} = 0, \frac{\partial u_2}{\partial t} + \lambda_2(u_1)\frac{\partial u_2}{\partial x} = 0, \tag{19}$$

we can get the global exact boundary controllability (See Li-Zhang[5]*).*

We now consider the case

$$n = 2m. \tag{20}$$

We suppose that the boundary condition (13) can be equivalently rewritten as

$$x = 0: \quad v_r = \bar{G}_r(t, v_{m+1}, \cdots, v_n) + \bar{H}_r(t) \quad (r = 1, \cdots, m), \tag{21}$$

where

$$\bar{G}_r(t, 0, \cdots, 0) \equiv 0 \quad (r = 1, \cdots, m) \tag{22}$$

and

$$\textit{small } C^1 \textit{ norm of } H_s \Longleftrightarrow \textit{small } C^1 \textit{ norm of } \bar{H}_r, \tag{23}$$

where $r = 1, \cdots, m; \ s = m + 1, \cdots, n$.

Again based on Theorem 1, we can get the following theorem on the local exact boundary controllability (See Li *et al* [13]).

Theorem 3 *Under the assumptions of Theorem 2, suppose furthermore that conditions (20)–(23) hold and $\bar{G}_r(t, \ \cdot)$ $(r = 1, \cdots, m)$ are C^1 functions with respect to their arguments. Let*

$$T > 2 \max_{i=1,\cdots,n} \frac{1}{|\lambda_i(0)|}. \tag{24}$$

Suppose finally that $H_s(t)$ $(s = m+1, \cdots, n)$ are given $C^1[0, T]$ functions with small C^1 norm. Then, for any given initial data $\varphi \in C^1[0, 1]$ and final data $\psi \in C^1[0, 1]$ with small C^1 norm, such that the conditions of C^1 compatibility are satisfied at points (0,0) and (T,0) respectively, the quasilinear hyperbolic system (3) with the boundary condition (13) admits a C^1 solution $u = u(t, x)$ with small C^1 norm on the domain

$$R(T) = \{(t, x) \big| \ 0 \le t \le T, \ \ 0 \le x \le 1\}, \tag{25}$$

which verifies

$$t = 0: \quad u = \varphi(x), \quad 0 \le x \le 1$$

and

$$t = T: \quad u = \psi(x) \quad 0 \le x \le 1.$$

Therefore, there exist boundary controls $H_r(t) \in C^1[0, T]$ $(r = 1, \cdots, m)$ with small C^1 norm, such that the mixed initial-boundary value problem (3)and

(12)–(14) admits a unique C^1 solution $u = u(t,x)$ with small C^1 norm on the domain $R(T)$, which satisfies the final condition

$$t = T: \quad u = \psi(x), \quad 0 \le x \le 1. \tag{26}$$

Remark 5 *In Theorem 3, the number of boundary controls is equal to $m = \frac{n}{2}$ and the boundary controls act only on one end, however, the controllability time is doubled.*

Remark 6 *The exact controllability time T given in Theorem 3 is optimal.*

4 Application to A Class of Nonlinear Vibrating String Problem

Consider the following nonlinear vibrating string equation

$$\frac{\partial^2 u}{\partial t^2} - \frac{\partial}{\partial x}\Big(K(\frac{\partial u}{\partial x})\Big) = F(\frac{\partial u}{\partial x}, \frac{\partial u}{\partial t}), \tag{27}$$

where $K = K(v)$ is a given C^2 function of v, such that

$$K'(v) > 0, \tag{28}$$

and $F = F(v,w)$ is a C^1 function of v and w, satisfying

$$F(0,0) = 0. \tag{29}$$

Suppose that the boundary condition at the end $x = 0$ is of Dirichlet type:

$$u = h(t), \tag{30}$$

where $h(t)$ is a given C^2 function; while the boundary condition at the end $x = 1$ is one of the following types:

$$u = \bar{h}(t), \tag{31}$$

$$u_x = \bar{h}(t), \tag{32}$$

$$u_x + \alpha u = \bar{h}(t), \tag{33}$$

$$u_x + \alpha u_t = \bar{h}(t), \tag{34}$$

where α is a positive constant and $\bar{h}(t)$, as boundary control, is a C^2 function (in case (31)) or a C^1 function (in cases (32)–(34)).

Setting

$$v = \frac{\partial u}{\partial x}, \quad w = \frac{\partial u}{\partial t}, \tag{35}$$

equation (27) can be reduced to a first order quasilinear hyperbolic system, then, using Theorem 3 we can prove the following result on the local exact boundary controllability (See Li *et al* [13]).

Theorem 4 *Let*

$$T > \frac{2}{\sqrt{K'(0)}}. \tag{36}$$

Then, for any given initial data $\varphi \in C^2[0,1]$, $\psi \in C^1[0,1]$ *and final data* $\Phi \in C^2[0,1]$, $\Psi \in C^1[0,1]$ *with small* C^1 *norms* $\|(\varphi', \psi)\|_{C^1[0,1]}$ *and* $\|(\Phi', \Psi)\|_{C^1[0,1]}$ *and for any given function* $h(t) \in C^2[0,T]$ *with small* C^1 *norm* $\|h'\|_{C^1[0,T]}$, *satisfying the following conditions of* C^2 *compatibility at points (0,0) and* $(T,0)$ *respectively:*

$$\begin{aligned} &h(0) = \varphi(0), \quad h'(0) = \psi(0), \\ &h''(0) = K'(\varphi'(0))\varphi''(0) + F(\varphi'(0), \psi(0)) \end{aligned} \tag{37}$$

and

$$\begin{aligned} &h(T) = \Phi(0), \quad h'(T) = \Psi(0), \\ &h''(T) = K'(\Phi'(0))\Phi''(0) + F(\Phi'(0), \Psi(0)), \end{aligned} \tag{38}$$

there exists a boundary control $\bar{h}(t) \in C^2[0,T]$ *with small* C^1 *norm* $\|\bar{h}'\|_{C^1[0,T]}$ *in case (31) or* $\bar{h}(t) \in C^1[0,T]$ *with small* C^1 *norm* $\|\bar{h}\|_{C^1[0,T]}$ *in cases (32)–(34), such that the mixed initial-boundary value problem for equation (27) with the initial condition*

$$t = 0: \quad u = \varphi(x), \quad u_t = \psi(x), \tag{39}$$

the boundary condition (30) at the end $x = 0$ *and one of the boundary conditions (31)–(34) at the end* $x = 1$ *admits a unique* C^2 *solution* $u = u(t,x)$ *on the domain*

$$R(T) = \{(t,x) \big| \ 0 \le t \le T, \quad 0 \le x \le 1\},$$

which verifies the final condition

$$t = T: \quad u = \Phi(x), \quad u_t = \Psi(x). \tag{40}$$

Acknowledgments

The author Ta-tsien Li was supported by the Special Funds for Major State Basic Research Projects of China.

References

1. J. L. Lions, *Contrôlabilité Exacte, Perturbations et Stabilisation de Systèmes Distribués* (Vol. I, Masson, 1988).
2. D. L. Russell, *Controllability and stabilizability theory for linear partial differential equations, Recent progress and open questions*, SIAM Rev. **20**, 639–739(1978).
3. E. Zuazua, *Exact controllability for the semilinear wave equation*, J. Math. Pures et Appl. **69**, 1–32(1990).
4. I. Lasiecka and R. Triggiani, *Exact controllability of semilinear abstract systems with applications to waves and plates boundary control problems*, Appl. Math. Optim. **23**, 109–154 (1991).
5. Ta-tsien Li and Bing-yu Zhang, *Global exact boundary controllability of a class of quasilinear hyperbolic systems*, J. Math. Anal. Appl. **225**, 289–311(1998).
6. Ta-tsien Li, Bopeng Rao and Yi Jin, *Solution C^1 semi-globale et contrôlabilité exacte frontière de systèmes hyperboliques quasi linéaires réductibles*, C. R. Acad. Sci., Paris, t.330, Série I, 205–210(2000).
7. Ta-tsien Li, Bopeng Rao and Yi Jin, *Semi-global C^1 solution and exact boundary controllability for reducible quasilinear hyperbolic systems*, M2AN **34**, 399–408(2000).
8. M. Cirinà, *Boundary controllability of nonlinear hyperbolic systems*, SIAM J. Control Optim. **7**, 198–212(1969).
9. M. Cirinà, *Nonlinear hyperbolic problems with solutions on preassigned sets*, Michigan Math. J. **17**, 193–209(1970).
10. Ta-tsien Li, Yi Jin, *Semi-global C^1 solution to the mixed initial-boundary value problem for quasilinear hyperbolic systems*, Chin. Ann. of Math. **22B**, 325–336(2001).
11. and Yi Jin, *Solution C^1 semi-globale et contrôlabilité exacte frontière de systèmes hyperboliques quasi linéaires*, C. R. Acad. Sci. Paris, t.333, Série I , 219–224(2001).
12. Ta-tsien Li and Bopeng Rao, *Exact boundary controllability for quasilinear hyperbolic systems*, SIAM J. Control Optim, Submited.
13. Ta-tsien Li and Bopeng Rao, *Local exact boundary controllability for a class of quasilinear hyperbolic systems*, to appear in Chin. Ann. of Math., 23B(2002).

Author Index